VOLUME FIVE HUNDRED AND SIXTY TWO

METHODS IN ENZYMOLOGY

Analytical Ultracentrifugation

METHODS IN ENZYMOLOGY

VOLUME FIVE HUNDRED AND SIXTY TWO

METHODS IN ENZYMOLOGY

Analytical Ultracentrifugation

Edited by

JAMES L. COLE

Department of Molecular and Cell Biology and Department of Chemistry

University of Connecticut

Storrs, Connnecticut

ELSEVIER

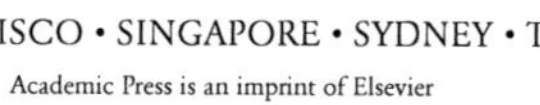

AMSTERDAM • BOSTON • HEIDELBERG • LONDON
NEW YORK • OXFORD • PARIS • SAN DIEGO
SAN FRANCISCO • SINGAPORE • SYDNEY • TOKYO
Academic Press is an imprint of Elsevier

Academic Press is an imprint of Elsevier
125 London Wall, London, EC2Y 5AS, UK
525 B Street, Suite 1800, San Diego, CA 92101–4495, USA
225 Wyman Street, Waltham, MA 02451, USA
The Boulevard, Langford Lane, Kidlington, Oxford OX5 1GB, UK

First edition 2015

ISBN: 978-0-12-802908-4
ISSN: 0076-6879

For information on all Academic Press publications
visit our website at http://store.elsevier.com/

CONTENTS

CONTRIBUTORS

Belinda M. Abbott
Department of Chemistry and Physics, La Trobe Institute for Molecular Science, La Trobe University, Melbourne Victoria, Australia

Gary G. Adams
National Centre for Macromolecular Hydrodynamics, and School of Health Sciences, University of Nottingham, Queen's Medical Centre, Nottingham, United Kingdom

Carlos Alfonso
Centro de Investigaciones Biológicas, Consejo Superior de Investigaciones Científicas (CSIC), Madrid, Spain

Fahad Almutairi
National Centre for Macromolecular Hydrodynamics, University of Nottingham, Nottingham, United Kingdom

Qushmua Alzahrani
National Centre for Macromolecular Hydrodynamics, University of Nottingham, Nottingham, United Kingdom

James Andya
Late Stage Pharamaceutical and Processing Development, Genentech

Kelly K. Arthur
Attribute Sciences, Amgen Inc., Longmont, Colorado, USA

David L. Bain
Department of Pharmaceutical Sciences, University of Colorado Anschutz Medical Campus, Aurora, Colorado, USA

Chad A. Brautigam
Department of Biophysics, The University of Texas Southwestern Medical Center, Dallas, Texas, USA

Cécile Breyton
Université Grenoble Alpes; CNRS, and CEA, IBS, Grenoble, France

Olwyn Byron
School of Life Sciences, College of Medical, Veterinary and Life Sciences, University of Glasgow, Glasgow, G12 8QQ, Scotland, United Kingdom

Carlos Enrique Catalano
Skaggs School of Pharmacy and Pharmaceutical Sciences, University of Colorado, Anschutz Medical Campus, Aurora, Colorado, USA

Jonathan B. Chaires
James Graham Brown Cancer Center, University of Louisville, Louisville, Kentucky, USA

Catherine T. Chaton
Graduate Program in Molecular Genetics, Biochemistry & Microbiology, University of Cincinnati College of Medicine, Cincinnati, Ohio, USA

Janni B. Christensen
Department of Biochemistry and Genetics, and La Trobe Institute for Molecular Science, La Trobe University, Melbourne, Victoria, Australia

Helmut Cölfen
Physical Chemistry, Department of Chemistry, University of Konstanz, Konstanz, Germany

Keith D. Connaghan
Department of Pharmaceutical Sciences, University of Colorado Anschutz Medical Campus, Aurora, Colorado, USA

John J. Correia
Department of Biochemistry, University of Mississippi Medical Center, Jackson, Mississippi, USA

Rolando W. De Angelis
Department of Pharmaceutical Sciences, University of Colorado Anschutz Medical Campus, Aurora, Colorado, USA

William L. Dean
James Graham Brown Cancer Center, University of Louisville, Louisville, Kentucky, USA

Gregory D. Degala
Department of Pathology, University of Colorado Anschutz Medical Campus, Aurora, Colorado, USA

Urko del Castillo
Unidad de Biofisica (CSIC/UPV-EHU), Departamento de Bioquímica y Biología Molecular, Universidad de País Vasco-Euskal Herriko Unibertsitatea (UPV-EHU), Bilbao, Biscay, Spain

Aysha K. Demeler
Department of Biochemistry, The University of Texas Health Science Center at San Antonio, San Antonio, Texas, USA

Borries Demeler
Department of Biochemistry, The University of Texas Health Science Center at San Antonio, San Antonio, Texas, USA

Barthélemy Demeule
Late Stage Pharamaceutical and Processing Development, Genentech

Sebastien Desbois
Department of Biochemistry and Genetics, and La Trobe Institute for Molecular Science, La Trobe University, Melbourne, Victoria, Australia

Matthew T. Downton
IBM Research Collaboratory for Life Sciences-Melbourne, Victorian Life Sciences Computation Initiative, Carlton, Victoria, Australia

Christine Ebel
Université Grenoble Alpes; CNRS, and CEA, IBS, Grenoble, France

Tayyibe Erten
National Centre for Macromolecular Hydrodynamics, University of Nottingham, Nottingham, United Kingdom

Michael G. Fried
Center for Structural Biology, Department of Molecular and Cellular Biochemistry, University of Kentucky, Lexington, Kentucky, USA

John P. Gabrielson
Attribute Sciences, Amgen Inc., Longmont, Colorado, USA

Chamodi K. Gardhi
Department of Chemistry and Physics, La Trobe Institute for Molecular Science, La Trobe University, Melbourne Victoria, Australia

Richard B. Gillis
National Centre for Macromolecular Hydrodynamics, and School of Health Sciences, University of Nottingham, Queen's Medical Centre, Nottingham, United Kingdom

Gary E. Gorbet
Department of Biochemistry, The University of Texas Health Science Center at San Antonio, San Antonio, Texas, USA

Shane E. Gordon
Department of Biochemistry and Genetics, and La Trobe Institute for Molecular Science, La Trobe University, Melbourne, Victoria, Australia

Michael D.W. Griffin
Department of Biochemistry and Molecular Biology, Bio21 Molecular Science and Biotechnology Institute, University of Melbourne, Parkville, Victoria, Australia

Ruchi Gupta
Department of Biochemistry and Genetics, and La Trobe Institute for Molecular Science, La Trobe University, Melbourne, Victoria, Australia

Dirk Haffke
Physical Chemistry, Department of Chemistry, University of Konstanz, Konstanz, Germany

Jeffrey C. Hansen
Department of Biochemistry and Molecular Biology, Colorado State University, Fort Collins, Colorado, USA

Stephen E. Harding
National Centre for Macromolecular Hydrodynamics, University of Nottingham, Nottingham, United Kingdom

Andrew B. Herr
Division of Immunobiology and Center for Systems Immunology, and Division of Infectious Diseases, Cincinnati Children's Hospital Medical Center, Cincinnati, Ohio, USA

John J. Hill*
Amgen, Seattle, Washington, USA

Campbell J. Hogan
Department of Biochemistry and Genetics, and La Trobe Institute for Molecular Science, La Trobe University, Melbourne, Victoria, Australia

Geoffrey J. Howlett
Department of Biochemistry and Molecular Biology, Bio21 Molecular Science and Biotechnology Institute, University of Melbourne, Parkville, Victoria, Australia

Brent S. Kendrick
Attribute Sciences, Amgen Inc., Longmont, Colorado, USA

Frank Krause
Nanolytics GmbH, Potsdam, Germany

James R. Lambert
Department of Pathology, University of Colorado Anschutz Medical Campus, Aurora, Colorado, USA

Thomas M. Laue
Center to Advance Molecular Interaction Science, University of New Hampshire, Durham, New Hampshire, USA

Aline Le Roy
Université Grenoble Alpes; CNRS, and CEA, IBS, Grenoble, France

Huy T. Le
James Graham Brown Cancer Center, University of Louisville, Louisville, Kentucky, USA

JiaBei Lin
Department of Chemistry, The University of Alabama at Birmingham, Birmingham, Alabama, USA

Jun Liu
Late Stage Pharamaceutical and Processing Development, Genentech

Aaron L. Lucius
Department of Chemistry, The University of Alabama at Birmingham, Birmingham, Alabama, USA

Nasib Karl Maluf
Alliance Protein Laboratories, San Diego, California, USA

Ianire Martín
Unidad de Biofísica (CSIC/UPV-EHU), Departamento de Bioquímica y Biología Molecular, Universidad de País Vasco-Euskal Herriko Unibertsitatea (UPV-EHU), Bilbao, Biscay, Spain

*Current address: Department of Bioengineering, University of Washington, Seattle, WA 98195.

Yee-Foong Mok
Department of Biochemistry and Molecular Biology, Bio21 Molecular Science and Biotechnology Institute, University of Melbourne, Parkville, Victoria, Australia

Arturo Muga
Unidad de Biofísica (CSIC/UPV-EHU), Departamento de Bioquímica y Biología Molecular, Universidad de País Vasco-Euskal Herriko Unibertsitatea (UPV-EHU), Bilbao, Biscay, Spain

Tao G. Nelson
Department of Biochemistry and Genetics, and La Trobe Institute for Molecular Science, La Trobe University, Melbourne, Victoria, Australia

Santosh Panjikar
Australian Synchrotron, and Department of Biochemistry and Molecular Biology, Monash University, Clayton, Victoria, Australia

Joseph Z. Pearson
Physical Chemistry, Department of Chemistry, University of Konstanz, Konstanz, Germany

Matthew A. Perugini
Department of Biochemistry and Genetics, and La Trobe Institute for Molecular Science, La Trobe University, Melbourne, Victoria, Australia

Germán Rivas
Centro de Investigaciones Biológicas, Consejo Superior de Investigaciones Científicas (CSIC), Madrid, Spain

Mattia Rocco
Biopolimeri e Proteomica, IRCCS AOU San Martino-IST, Istituto Nazionale per la Ricerca sul Cancro, Genova, Italy

Ryan A. Rogge
Department of Biochemistry and Molecular Biology, Colorado State University, Fort Collins, Colorado, USA

M. Samil Kök
National Centre for Macromolecular Hydrodynamics, University of Nottingham, Nottingham, United Kingdom, and Department of Food Engineering, Abant Izzet Baysal University, Bolu, Turkey

Béatrice Schaack
Université Grenoble Alpes; CNRS, and CEA, IBS, Grenoble, France

Kristian Schilling
Nanolytics GmbH, Potsdam, Germany

Peter Schuck
Dynamics of Macromolecular Assembly Section, Laboratory of Cellular Imaging and Macromolecular Biophysics, National Institute of Biomedical Imaging and Bioengineering, National Institutes of Health, Bethesda, Maryland, USA

David J. Scott
National Center for Macromolecular Hydrodynamics, School of Biosciences, University of Nottingham, Nottingham, and ISIS Spallation Neutron and Muon Source and Research Complex at Harwell, Rutherford-Appleton Laboratory, Oxford, United Kingdom

Steven J. Shire
Late Stage Pharamaceutical and Processing Development, Genentech

Tatiana P. Soares da Costa
Department of Biochemistry and Genetics, and La Trobe Institute for Molecular Science, La Trobe University, Melbourne, Victoria, Australia

Walter F. Stafford
Department of Systems Biology, Boston Biomedical Research Institute and Harvard Medical School, Boston, Massachusetts, USA

Ingrid Tessmer
Rudolf Virchow Center for Experimental Biomedicine, University of Würzburg, Würzburg, Germany

John O. Trent
James Graham Brown Cancer Center, University of Louisville, Louisville, Kentucky, USA

John Wagner
IBM Research Collaboratory for Life Sciences-Melbourne, Victorian Life Sciences Computation Initiative, Carlton, Victoria, Australia

Kai Wang
Université Grenoble Alpes; CNRS, and CEA, IBS, Grenoble, France

Donald J. Winzor
School of Chemistry and Molecular Biosciences, University of Queensland, Brisbane, Queensland, Australia

Sandeep Yadav
Late Stage Pharamaceutical and Processing Development, Genentech

Qin Yang
Department of Pharmaceutical Sciences, University of Colorado Anschutz Medical Campus, Aurora, Colorado, USA

Teng-Chieh Yang
Pfizer, Pharmaceutical Research and Development, Chesterfield, Missouri, USA

PREFACE

Analytical ultracentrifugation (AUC) is a powerful method to characterize the size, shape, interactions, and homogeneity of macromolecules in free solution. AUC is widely used to characterize biomacromolecules, including proteins, nucleic acids, and carbohydrates. It can be applied over a broad range of molecular sizes, from short peptides and oligonucleotides to large assemblies such as viruses, chromatin assemblies, and nanoparticles.

It is worth noting that AUC has a long history, beginning with Svedberg's investigations of colloids and hemoglobin in the early 1920s. With the development of commercial instrumentation after World War II, the use of AUC exploded in subsequent decades. However, by 1980 interest in AUC waned as new methods to determine molecular weights became available. During the ensuing "dark ages," only a few laboratories maintained active research programs in AUC using aging instruments. The development of a new generation of analytical ultracentrifuges in the 1990s combined with powerful data analysis algorithms running on fast computers has catalyzed a renaissance in AUC research that continues to this day. The goal of this volume of *Methods in Enzymology* is to provide a broad overview of this rapidly evolving field, with chapters devoted to developments in instrumentation, theory, data analysis, and applications in basic biology and biopharmaceutical research and development.

Instrumentation and Analysis. The development of a multiwavelength absorption detector (Chapter 1) and associated data analysis methods (Chapter 2) adds an additional dimension of spectral characterization to complement hydrodynamic information. Chapter 3 provides the theoretical basis for analysis of sedimentation velocity. Sedimentation velocity experiments provide shape information that can be precisely interpreted in the context of high-resolution structures with hydrodynamic modeling (Chapter 4). Organization of large AUC data sets has become cumbersome and Chapter 5 describes software that simplifies the processing and presentation of AUC data.

Proteins. A major application of AUC is the analysis of proteins and protein association reactions. The self-association behavior of proteins is affected by crowding, which can be studied by sedimentation equilibrium and complementary light-scattering measurements (Chapter 6). Self-association is often thermodynamically linked to small-molecule binding

(Chapter 7). The resolution of complex heterointeractions by sedimentation velocity experiments is improved by generating two-dimensional size-and-shape distributions from sedimentation velocity data (Chapter 8). AUC, complemented by structural studies and molecular dynamics simulations, can provide insight into the regulation of enzymes by changes in quaternary structure (Chapter 9). AUC has important applications in the analysis of proteins containing intrinsically disordered regions (Chapter 10), amyloid oligomers and fibrils (Chapter 11), and membrane proteins (Chapter 12).

Nucleic Acids. DNAs, RNAs, and their interactions with proteins are also amenable to investigation by AUC. The hydrodynamic properties of G-quadruplex DNAs determined from sedimentation velocity provide valuable structural information (Chapter 13). AUC has been applied to the characterization of nonspecific protein–nucleic acid interactions (Chapter 14), the cooperative assembly of protein clusters on DNAs (Chapter 15), and large protein–DNA complexes formed during chromatin compaction (Chapter 16). In the case of steroid receptors, protein self-association is linked to sequence-specific, cooperative assembly at DNA promoters (Chapter 17).

Other Biomacromolecules. Chapter 18 provides an example of the strength of AUC in the analysis of complex biomacromolecules. Unlike proteins, carbohydrates are highly polydisperse and exhibit significant nonideality.

Biopharmaceuticals. AUC is widely used in the biotechnology industry. Chapter 19 provides a broad overview of the application of AUC in formulation development and research on protein therapeutics. A key application of AUC is detection of potentially harmful protein aggregates (Chapter 20). *In vivo,* biopharmaceuticals function in complex, crowded solutions. Using fluorescence detection, AUC is one of the few biophysical techniques capable of probing the association state and interactions of protein therapeutics in complex media such as serum (Chapter 21).

Finally, I would like to thank the authors for their excellent chapters and the staff at Elsevier for their assistance in assembling this volume. I hope that it will serve as a useful resource to both new and experienced researchers.

James L. Cole

CHAPTER ONE

Next-Generation AUC Adds a Spectral Dimension: Development of Multiwavelength Detectors for the Analytical Ultracentrifuge

Joseph Z. Pearson*, Frank Krause†, Dirk Haffke*, Borries Demeler‡, Kristian Schilling†, Helmut Cölfen*,[1]

*Physical Chemistry, Department of Chemistry, University of Konstanz, Konstanz, Germany
†Nanolytics GmbH, Potsdam, Germany
‡Department of Biochemistry, The University of Texas Health Science Center at San Antonio, San Antonio, Texas, USA
[1]Corresponding authors: e-mail address: Helmut.Coelfen@uni-konstanz.de

Contents

Methods in Enzymology, Volume 562
ISSN 0076-6879
http://dx.doi.org/10.1016/bs.mie.2015.06.033

Abstract

We describe important advances in analytical ultracentrifugation (AUC) hardware, which add new information to the hydrodynamic information observed in traditional AUC instruments. In contrast to the Beckman-Coulter XLA UV/visible detector, multi-wavelength (MWL) detection is able to collect sedimentation data not just for one wavelength, but for a large wavelength range in a single experiment. The additional dimension increases the data density by orders of magnitude, significantly improving the statistics of the measurement and adding important information to the experiment since an additional dimension of spectral characterization is now available to complement the hydrodynamic information. The new detector avoids tedious repeats of experiments at different wavelengths and opens up new avenues for the solution-based investigation of complex mixtures. In this chapter, we describe the capabilities, characteristics, and applications of the new detector design with biopolymers as the focus of study. We show data from two different MWL detectors and discuss strengths and weaknesses of differences in the hardware and different data acquisition modes. Also, difficulties with fiber optic applications in the UV are discussed. Data quality is compared across platforms.

1. INTRODUCTION

Analytical ultracentrifugation (AUC) is a powerful tool for the analysis of (bio)polymers and nanoparticles since the days of its invention in the 1920s by The Svedberg. A fundamental advantage is the fractionation of complex mixtures into their components. For nanoparticles, it has a size resolution in the Angström range (Cölfen & Pauck, 1997). Current detectors are based on the Beckman-Coulter Optima XL-A/I, which presently is the only commercially available AUC (UV/visible single wavelength and Rayleigh interference). A fluorescence detector is also available as a retrofit (MacGregor, Anderson, & Laue, 2004), and for nanoparticles, special turbidity detectors were developed on the basis of preparative ultracentrifuges, which allow for the detection of very broad particle size distributions by the application of gravitational sweep techniques (Mächtle, 1999; Müller, 1989). These detectors extend the range of applications for AUC and make it possible to exploit the presence of different chromophores, measure refractive index or turbidity for nonabsorbing molecules, or detect fluorescently tagged molecules with exquisite selectivity and high sensitivity. With the exception of the gravitational sweep technique (Mächtle, 1999), the observed signal is recorded as a function of radius and time and forms the basis for the extraction of hydrodynamic parameters. For the Beckman-Coulter XLA instrument, the wavelength of the light used to measure absorbance in the ultracentrifuge cell can be adjusted to match the chromophore of the analyte. However, this method of data acquisition has several important shortcomings: (1) In the XLA, a single scan requires about 2–5 min, depending on rotor speed and radial resolution, a time far too long to avoid temporal distortion for fast sedimenting analytes; (2) the total number of scans that can be collected, especially when multiple samples are measured, is limited by the slow scanning speed, reducing the number of scans available for analysis; and (3) the inability to collect data at multiple wavelengths during a single run in the XLA prevents the acquisition of spectral information. Many mixed systems display strong spectral diversity due to the extinction properties of their individual components. These properties could be exploited, but the XLA's design makes a multiwavelength (MWL) analysis prohibitive in terms of instrument time and sample requirements since each wavelength measurement requires an individual run. Though the Beckman-Coulter data acquisition software permits collection of three different wavelengths during a single experiment, the use of this feature is not practical because (a) the number of scans for each individual wavelength is reduced by two-thirds and (b) the monochromator is not guaranteed to reset to the same wavelength while cycling through the different wavelengths, causing changes in recorded absorbance due to changes in the extinction coefficients. Together, these limitations hinder scientific progress in the high-resolution analysis of UV/visible absorbance data from the analytical ultracentrifuge. To overcome these limitations, an MWL detector was developed for the AUC (Bhattacharyya, 2006; Bhattacharyya et al., 2006; Karabudak, 2009; Strauss et al., 2008) within the framework of the open AUC project (Colfen et al., 2010). The central part of this detector is a CCD array-based spectrometer, which is able to acquire a full UV/visible spectrum in 1 ms. Each of the 2048 pixels of the linear CCD array corresponds to a fixed wavelength reading. The basic design of the current detector is described below and by (Strauss et al., 2008). The data acquisition and control software of this detector was recently improved (Walter et al., 2014). This detector adds a spectral dimension to the hydrodynamic characterization by AUC and has enabled AUC experiments with so far unsurpassed information content (Backes et al., 2010; Karabudak, Backes, et al., 2010; Karabudak, Wohlleben, & Colfen, 2010). Due to the attenuation of UV light intensity by fiber optic cables, the detector has so far primarily been used for wavelengths >300 nm, which is sufficient for many colloidal samples, which absorb light in the visible range. However, limited measurements of important biopolymers like proteins have been made (Bhattacharyya, 2006; Walter et al., 2014, 2015), and no measurements of DNA have yet been published (Bhattacharyya, 2006; Walter et al., 2014, 2015). So far, the study of biopolymers with MWL AUC has not been the target of research. Here, we show that a fiber-based MWL detector system offers sufficient data quality to permit the investigation of biopolymer samples absorbing in the UV, despite the relatively low light intensity available in the UV. We predict this technology will prove particularly useful for the study of mixtures and heterointeracting biopolymer systems with components having distinct spectra, and build off of the work previously explored by AUC methods using several wavelengths for analysis (Cole, 2004; Lewis, Shrager, & Kim, 1994).

2. DEVELOPMENT OF MWL ABSORBANCE DETECTORS

Note: In the literature, the terms MWL and MWA are used interchangeably for the same type of MWL detector. We reserve the term "MWL" for the early detector generations and refer to MWA as the third generation.

2.1 First-Generation MWL

In order to address the shortcomings of the current XLA design, the user community has focused on new detector development and new data management paradigms. The design of the MWL system for the Beckman-Coulter Optima L, XL, and XL-A platform has been iterated through several generations. The first-generation design, described in Bhattacharyya et al. (2006), featured the flash lamp and spectrometer outside of the vacuum chamber, each connected to the optical system by fiber optic cables passing through the vacuum chamber wall. This resulted in low light intensities.

2.2 Second-Generation MWL Developments

A second-generation design, described in Strauss et al. (2008), removed the fiber optic component on the detector side and installed the spectrometer directly above the objective lens of the optical path. The detection systems profiled in this chapter have evolved from the second-generation design, highlighting advancements of two independent implementations. One, a continuation of the open source device developed in the Cölfen lab (Bhattacharyya, 2006; Karabudak, 2009), and a second redesign developed at Nanolytics' in 2013. Both implementations are based on the Beckman-Coulter ultracentrifuge platforms (Nanolytics: Optima L and Open AUC: Optima XL/XL-A), and share similar hardware architecture. They differ mainly in the data acquisition software. In these designs, at incremental radial positions, a complete UV/visible spectrum is captured on a CCD chip, resulting in radial scans of a sample sector. The Open AUC MWL developments have proceeded with incremental advances to the second-generation design of (Strauss et al., 2008), outlined below, and have been demonstrated with effective results in several published applications (Backes et al., 2010; Karabudak, Backes, et al., 2010; Karabudak, Wohlleben, et al., 2010; Voelkle, Gebauer, & Coelfen, 2015; Walter et al., 2014, 2015).

In parallel to the open AUC developments, a second-generation redesign from Nanolytics' features several distinct changes based on the following objectives: (1) Allow for the accommodation of eight-hole-rotors by using a newly calculated, shorter lens system; (2) optimize detector arm geometry, minimizing required space; (3) optimize optical resolution and light intensity; (4) replace LabView control of multiplexing by a standalone, primitive, non-Windows computer; (5) create new user interface software, offering more flexibility than XL-A software; (6) optimize time performance for scanning and calibration routines, exploiting full spectrometer and lamp performance. An important constraint for all changes was to modify the hosting Beckman-Coulter preparative centrifuge as little as possible. In fact, the only invasive measures were the following: (a) drilling two feedthroughs into the heat sink (one electric, one optical); (b) providing a cutout in the can and safety plate to allow the detector to move below the rotor; and (c) one single electronic connection, where the pickup's Hall signal is collected from the safety plate's circuit board. Though these changes will still void any warranty on the centrifuge, they still leave the centrifuge technically unchanged, permitting service by trained professionals.

2.3 Third-Generation MWA

A third-generation design is poised for introduction with the newest AUC platform in the final stages of development by Spin Analytical, branded as the Centrifugal Fluid Analyzer (CFA) (Laue & Austin, n.d.). The MWA detection system on the CFA does not need fiber optics or lens focusing, but instead makes use of parabolic mirrors for light collimation and imaging, avoiding chromatic abberration, and uses a linear diode array to capture the entire radial domain without the need for radial scanning. A single wavelength is selected by a precision adjustment monochromatic light source. As in the XL-A, this design requires that each wavelength and each channel is individually measured, but because the entire radial domain is captured with a brief burst of light flashes at a single wavelength, lasting between one to a few seconds depending on rotor speed, the data are essentially devoid of temporal distortion. Since multiple flashes can be integrated on the linear diode array, each wavelength captured can be normalized to an optimum intensity. At the time of this writing, the CFA design is still in prototype stage and was not available for further testing, so all data shown here refer to the two fiber-based designs, which are described in further detail below.

Importantly, all MWL designs provide rapid scanning and allow capturing of a fourth dimension, the spectral information of the sample. As a result, instead of a single wavelength, an entire spectrum can be acquired for each radial position, increasing the data volume by two to three orders of magnitude. With this enormous increase in data density, not only is the statistical confidence in the results greatly enhanced, but also the additional information contained in the spectral dimension opens up possibilities for exciting new investigations.

3. HARDWARE AND CONTROL SYSTEM OF SECOND-GENERATION MWL DETECTORS

3.1 Second-Generation MWL Open AUC Design

The essential elements of the second-generation MWL optical detector (Strauss et al., 2008; Walter et al., 2014) are as follows, beginning with the light source: A 5-W Xenon flash lamp module (Hamamatsu 9455) is coupled to an OZ optics 200 μm UV/visible fiber with a vacuum feed-through adapter. The fiber end is positioned before a 10-mm biconvex quartz lens and the resulting collimated light is passed to a 45-degree flat mirror, directing it up through the optical path of the sample cell housing. The collimated light passes through the sample channel of the cell housing. A 40-mm objective quartz lens positioned above the rotor refocuses the light from the image plane within the cell of the spinning rotor onto the entrance of the spectrometer (Ocean Optics USB2000+). A 25-μm slit at the spectrometer entrance provides the aperture for calibrated spectral resolution in addition to sampling a portion of the light source beam representative of a radial increment. The entire optical assembly (fiber end to spectrometer) is fixed to one piece of an aluminum arm. The arm is attached to an aluminum chassis by a sliding rail. A Zaber T-LA-28-SV stepper motor positions the optical assembly on the sliding rail over the radial dimension of the sample channel.

This open AUC device has been described previously (Strauss et al., 2008), and all of the relevant design documentation is freely available as open source materials. The electronic control board (NIPCI-6602) and connector block (NIBNC2121) are now integrated into a 64-bit PC running LabView 2013. A new version of the data acquisition and control software written by Johannes Walter is currently used (Walter et al., 2014). The spectrometers are Ocean Optics 16 bit USB2000+, upgraded from the 12-bit USB2000 spectrometers previously described (Strauss et al., 2008), and can be configured with gratings to optimize to various spectral regions. A higher powered Xenon flash lamp (Hamamatsu 9455-11) improves the UV signal over a lower powered lamp (Hamamatsu 9455-13) previously employed. For measurements in the visible, the higher powered lamp saturates the detector and use of the lower powered lamp is necessary. The experimental data provided below demonstrates the capability of the device to now make high-quality measurements of samples absorbing in the UV.

3.2 Second-Generation MWL from Nanolytics

The redesigned MWL second-generation device available commercially from Nanolytics' labs (Nanolytics), is based on the open AUC design, but is implemented on a Beckman-Coulter Optima L preparative centrifuge and includes some distinct changes. Similar to the open AUC design, but unlike the Beckman-Coulter monochromator/laser in the Optima XL-A, the MWL detector assembly is not meant to be removed from the vacuum chamber; thus, it is desirable to gain as much space as possible for inserting and removing the rotor. These concerns are supported by the intention to introduce yet more stationary detectors (interference or fluorescence) into the same instrument.

3.3 Nanolytics Hardware

The following hardware changes were made to the open AUC design by Nanolytics: (1) the chassis was redesigned to be a single part, including the characteristic "duck feet," giving more stability and allowing the feet to be much smaller; (2) using a vacuum rated Zaber T-NA08A25-SV1 step motor, thus making its large heat sink obsolete; (3) recalculating the optical system, making the spectrometer optics shorter and giving more room above the rotor; (4) choosing a smaller adjustable stage carrying the spectrometer; (5) adding three parking positions for safe fastening of spare fibers. The detector arm is fastened to the heat sink with four screws that usually fasten the can to the heat sink. Thus, no additional holes in the heat sink were required (Fig. 1).

3.4 Nanolytics Data Acquisition

The optical system was also redesigned based on observations at an early stage that a single lamp flash provides enough intensity to saturate the spectrometer, allowing for a different concept of hardware triggering. Rather than bringing the step motor to a given radial position, flashing until enough intensity has been collected, stopping data acquisition, and moving to the next radial coordinate, it is possible to set the motor to a continuous movement, flashing at every rotor revolution, and stopping the procedure when the motor has reached its final position. This reduces the time required for scan acquisition drastically, to only several seconds at moderate angular velocities. Scanning speed is limited by the minimum spectrometer integration time, the maximum flash lamp frequency, and the rotor speed. Therefore, at higher angular velocities not every rotor revolution can be used for a flash. The system can still operate at 250 Hz (maximum lamp frequency in practice) under many circumstances, where the acquisition of 1300 data

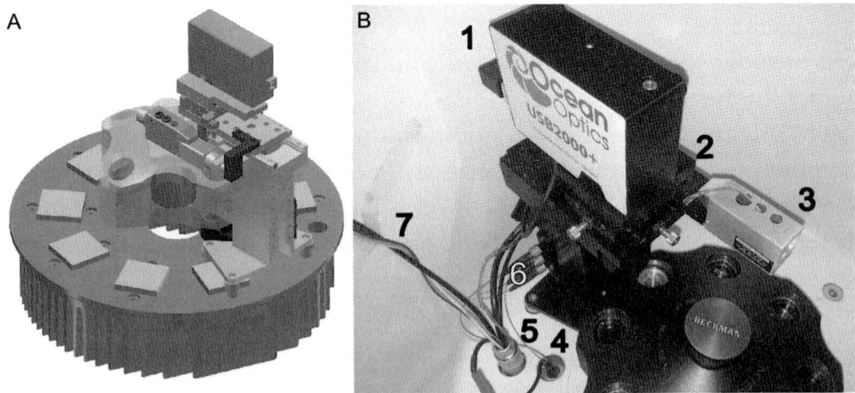

Figure 1 (A) Schematic and (B) picture of Nanolytics detector hardware. (1) Ocean Optics USB2000+ spectrometer, (2) X–Y stage, (3) step motor, (4) optical feedthrough, (5) electrical feedthrough, (6) spare fibers, and (7) connections for optional interference optics. (See the color plate.)

points (typical scan at maximum resolution) will require only approximately 5 s. Additional time is required for bringing the stepper motor into position, initializing the flash series, and transferring the acquired data. Table 1 summarizes the time required for scanning a standard cell for different angular velocities. Delay calibration, consuming several minutes on the XLA, takes only 30 s, due to a sophisticated procedure that fully exploits maximum flash lamp frequency.

The signal-to-noise ratio (SNR) of the open AUC system is superior because it averages multiple lamp flashes (typically eight for higher angular velocities or larger radial increments, but subject to users' requirements), while the Nanolytics system collects only one flash per point (further discussed below). However, the shorter scan duration in the Nanolytics system allows for averaging scan replicates to compensate for lower SNR. Users can choose an optimal balance between duration and noise according to the application's needs. Besides a considerably faster operation, continuous scanning has the advantage that the step motor is not started and stopped continuously with short intervals during each scan, thus extending lifetime.

3.5 Nanolytics Omega Device

The timing problems observed with the first-generation MWL, using a 32-bit LabView interface, has inspired two lines of further development. In the open AUC design, the problems have been resolved by advancing to 64 bit LabView programming. The Nanolytics solution is an independent self-constructed hardware solution, named Omega Device for the sake of its target parameter. It is virtually, an independent, primitive computer, which is controlled by the data acquisition program on the PC, but will carry out fast operations independently until further notice. Thus, the tasks of (a) measuring the time of rotor revolution and (b) multiplexing light source and spectrometer are performed by this device in a robust and failsafe

Table 1 Scan Durations of 1 Cell (Two Channels) for Both MWL Systems and XLA at Different Rotational Speeds and a Typical Radial Increment of 30 μm

	Open AUC			Nanolytics			XLA		
Rotor speed (rpm)	3000	25,000	60,000	3000	25,000	60,000	3000	25,000	60,000
Scan time (s)	216	110	90	48	6	5	100	80	80

In this example, the open AUC system takes eight flash averages and Nanolytics and XLA only one.

manner. The device also contains the Hall-to-TTL converter and provides dials for adjustment of threshold voltage and pulse width. Other dials allow one to adjust the light source's intensity (or to mute it), likewise for the optional interference system. The device also contains power supplies for all hardware components (in the case of MWL, the stepper motor and the Xe flash lamp). Taking all computer connections on the right, directing all connectors to the detections system's hardware components on the left, and taking in one 220 VAC power supply on the rear, the housing accommodates all components in a single housing. Schematics and appearance of this device are shown in Fig. 2. The device had been designed to control a previous AUC system equipped with the Advanced Interference Detection Array. It was then upgraded to control an MWL detection system as well and is capable of controlling both detection systems simultaneously.

3.6 Nanolytics Acquisition and Control Software

The operating software, C++ based running on Windows 7, is a complex package, consisting of modules for each hardware component, and a controlling module with a graphical user interface. The control GUI provides somewhat higher flexibility than the XL-A control software, allowing for individual scan numbers and scan delay times for each cell. This allows for more frequent scanning of more rapidly sedimenting systems along with slower sedimenting systems in other cells by choosing a shorter delay and, possibly, a smaller number of scans to be collected. Furthermore, any setting can be changed while scanning is in progress, and will take effect on the next possible occasion. Thus, new data directories are not created upon any

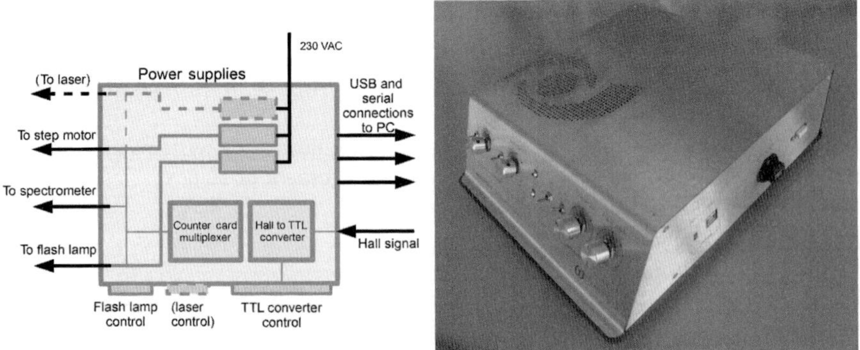

Figure 2 The Omega device. Left: schematics of operation (illustrating the optional interference detector control parts). Right: photograph of the device.

change in settings, and it is not necessary to renumber scans in each new directory.

4. MWL DATA FROM AUC

4.1 Open AUC Data Format and Management

The MWL data are written by the acquisition software in a binary file format shown in SI 1 (http://dx.doi.org/10.1016/bs.mie.2015.06.033). Any version changes will be made available on the web (wiki.bcf2.uthscsa, n.d.-b). This step serves multiple purposes. First, it averages scans that were collected with adjacent wavelengths. The averaging of adjacent scans is reasonable since the resolution of the diffraction grating used in the optics is less than the pixel density of the CCD array, which generates fractional wavelengths. By averaging fractional wavelengths, the data density is reduced to a level actually supported by the underlying optics, which is around 1 nm for the standard UV/visible diffraction grating, and lends itself to the "triple" concept used in UltraScan-III that expects an integral wavelength, see Colfen et al. (2010). This step also gives the investigator an opportunity to discard data, which has too little signal or data that exceeds the dynamic range of the detector. The intermediate file is accompanied by a time-state object, see SI 1 (http://dx.doi.org/10.1016/bs.mie.2015.06.033), where the data acquisition software can record instrument statistics with intervals as short as 1 s. This is used to record metrics such as rotor speed, temperature, vacuum, and acceleration settings. Any version changes will be made available on the UltraScan wiki Web site (wiki.bcf2.uthscsa, n.d.-c). This format is flexible and allows other metrics to be included.

4.2 Nanolytics Data Format and Management

Generally, established conventions were maintained as far as possible. Data are saved in a Beckman-Coulter-like directory structure, numbering starts from 1 and Beckman-Coulter-like filenames were used: the extension (typically "RAn") was changed to "MWn," the scans are simply numbered with five digits as usual, only preceded by a channel specification: a preceding "A" for the sample and "B" for the reference sector. The concept of separate storage for either sector was taken from the Open AUC project data format; it is useful for researchers who implement the pseudo-absorbance technique (Kar, Kingsbury, Lewis, Laue, & Schuck, 2000). The data can simply be combined to classical absorbance data if required.

The data format was kept as close to the format suggested by the open AUC project group (Colfen et al., 2010), but it allows for the storage of negative intensities (that might result from dark current subtraction) and saves the entire data scope as recorded by the spectrometer, using fractional wavelengths. It also contains all information for the scan, such as sample description and runtime, as in the header from Beckman-Coulter data files. The complete format definition is provided in SI 1 (http://dx.doi.org/10.1016/bs.mie.2015.06.033). The software can also export data in the open AUC data format described in section. Also, Beckman-Coulter single wavelength radial scans can be exported in legacy format; up to four wavelengths per cell can be selected for display and immediate export during scan acquisition.

5. DATA VISUALIZATION AND EXPERIMENTAL RESULTS

The data collected from a MWL experiment are first examined as the raw intensity counts from the spectrometer versus the radial dimension of a cell channel for one wavelength, compiling all of the scans from a run on to a single plot. An example is shown in Fig. 3, for an experiment with a mixture of nucleic acid and protein biopolymers, here DNA fragments and bovine serum albumin (BSA).

The right and left vertical indicators of Fig. 3 designate the region above the air–liquid meniscus where the reference intensity is recorded. A notable intensity decline is observed over the course of an experiment, due to attenuation phenomena in the optical fiber system; see Section 6.

By selecting the reference intensity of the air region above the meniscus, the data can be converted to pseudo-absorbance units (Kar et al., 2000). This procedure removes the effects of the intensity attenuation in an experimental run, revealing the well-known AUC sedimentation profile as illustrated in Fig. 4.

The sedimentation profile of Fig. 4 is from absorbance data collected at 270 nm. At this wavelength, both DNA and BSA absorb and contribute to the moving boundary traces. However, the MWL detection system collects a continuous spectrum from the sample space at each radial position of a scan. In order to gain a full view of the experimental MWL data, all four dimensions are needed. The experimental data signal, here light intensity as proportional to analyte concentration, extends over time, radius, and wavelength. With wavelength added as an additional dimension,

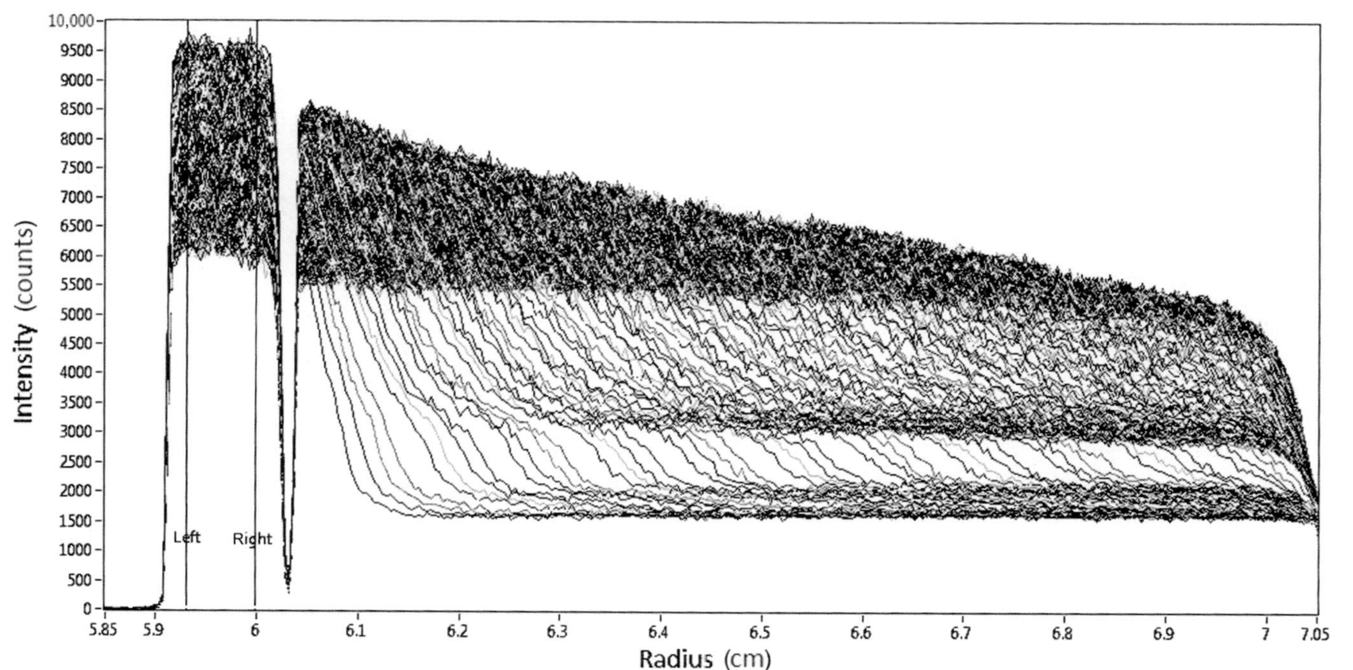

Figure 3 Raw intensity data from 200 MWL scans for a mixture of two DNA fragments and BSA at 270 nm collected over 15 h. Data from Cölfen lab second-generation MWL open AUC design, using 16 bit USB2000+ Ocean Optics spectrometer with #1 grating and 30 μm radial step size. *Image generated by LabView-based MWL data viewer (http://wiki.bcf2.uthscsa.edu/openAUC/wiki/WikiStart) written by Dirk Haffke.*

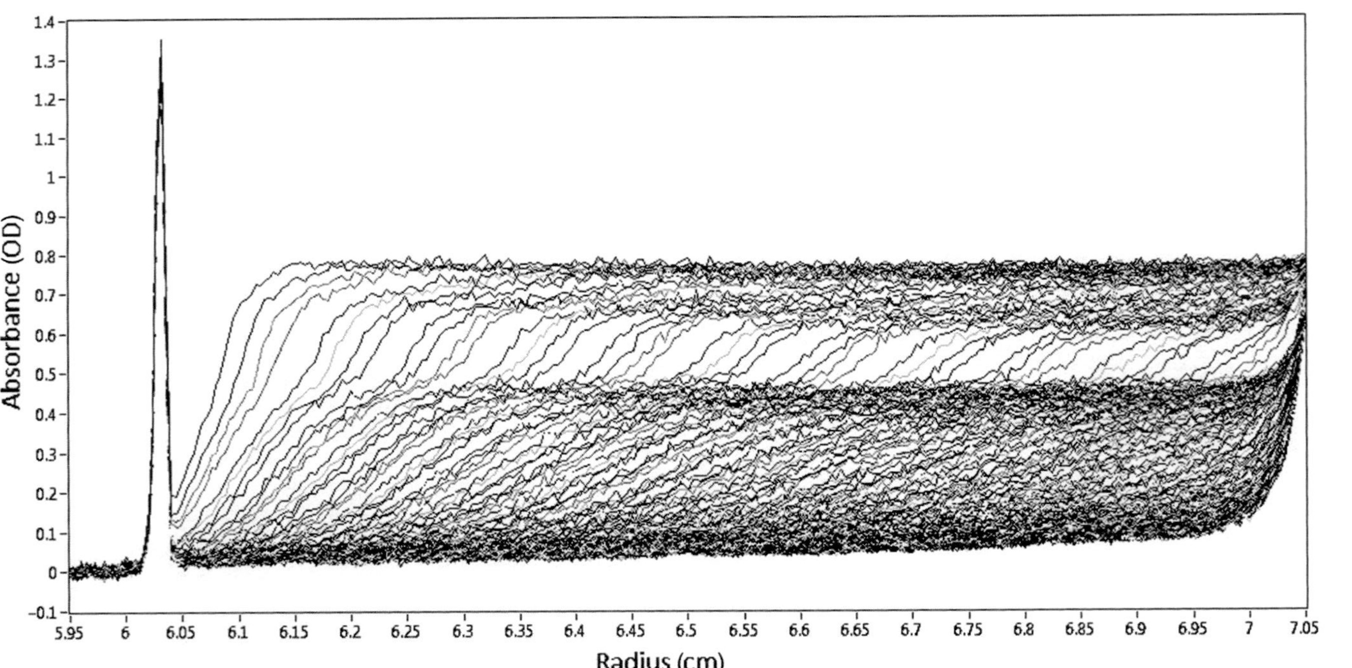

Figure 4 Sedimentation profile of BSA and DNA absorbance data at 270 nm. *Image generated by LabView-based MWL data viewer (http://wiki.bcf2.uthscsa.edu/openAUC/wiki/WikiStart) written by Dirk Haffke.*

three-dimensional representations are no longer sufficient to visualize the entire MWL experiment. UltraScan now includes 3D graphics and 3D movies to provide a comprehensive four-dimensional view of experimental data and analysis results, allowing a three-dimensional view of experimental data: absorbance versus radius and wavelength, with each scan time rendered as a frame in the movie, for details see the chapter in this volume (Gorbet, Pearson, Demeler, Cölfen, & Demeler, 2015). An MWL data viewer has also been developed in the Cölfen lab by Dirk Haffke, written in LabView (wiki.bcf2.uthscsa, n.d.-a). The concentration is represented by a color gradient. An example of several frames (DNA and BSA mixture) from the 3D MWL data movie is shown in Fig. 5.

The complete movie of a three-dimensional view of all scans as a function of wavelength is shown in SI 2 (http://dx.doi.org/10.1016/bs.mie.2015.06.033). The example DNA and BSA data highlighted in the above figures are from a mixture of two DNA fragments and BSA, provided by Aysha Demeler and Blanca Hernandez-Uribe at UTHSCSA. The two DNA fragments are from a restriction digest of a plasmid and are highly purified by HPLC. The DNA sample has approximately equal ODs of a 208-bp fragment and a 2811-bp fragment. A 1 OD sample of DNA is mixed with a 1 OD sample of BSA (Sigma) by a ratio of 80/20 total DNA optical density to protein. The samples are dissolved in 20 m*M* $NaPO_4$ buffer pH 7.8 and 150 m*M* NaCl. The sample was run at 28,000 rpm, 20 °C in a 12-mm titanium centerpiece.

To accompany the DNA and BSA measurements from the open source MWL AUC, examples of data from further biopolymer systems were acquired with the Nanolytics MWL AUC. Two noninteracting binary systems out of three model proteins each exhibiting oligomers were selected: BSA, hemoglobin (Hb), and myoglobin (Mb). A sample of human Hb and horse Mb is shown in Fig. 6 and the corresponding movie in supporting information SI 3 (http://dx.doi.org/10.1016/bs.mie.2015.06.033). The absorbance spectra of these two heme proteins are very similar, whereas their major size-variants have a clearly different molar mass at the measured loading concentrations, Hb 65 kDa and Mb 17 kDa.

A sample with human Hb and BSA is shown in Fig. 7 and the corresponding movie in supporting information SI 4 (http://dx.doi.org/10.1016/bs.mie.2015.06.033). The major species have a similar molar mass (66 vs. 65 kDa) and sedimentation coefficients at the measured loading concentrations, whereas the absorbance spectra are distinct from each other. BSA features a typical protein spectrum confined to the UV region, while

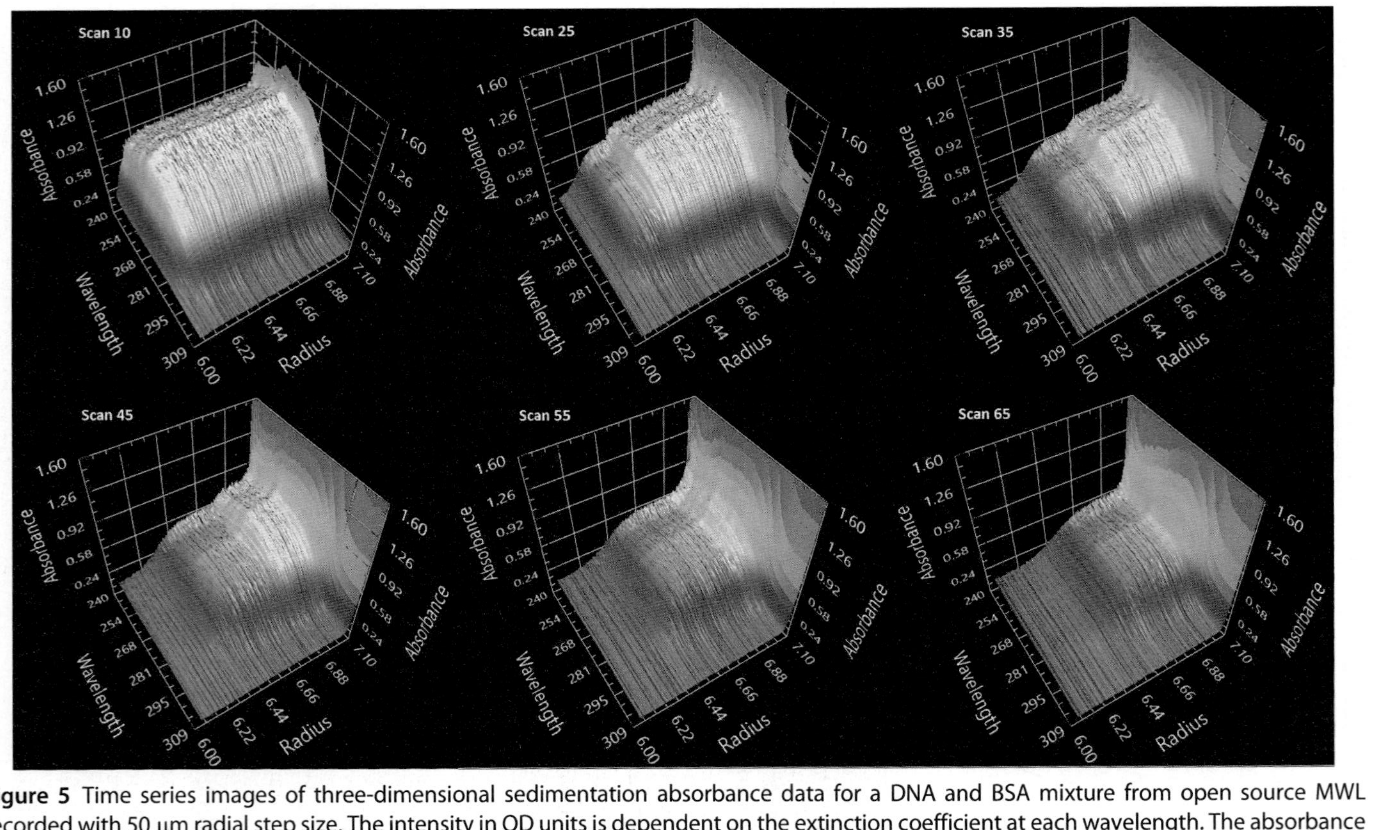

Figure 5 Time series images of three-dimensional sedimentation absorbance data for a DNA and BSA mixture from open source MWL recorded with 50 μm radial step size. The intensity in OD units is dependent on the extinction coefficient at each wavelength. The absorbance spectra of the species in solution are evident across the wavelength range. The yellow coded species at ca. 270 nm across the spectra is indicative of a more rapidly sedimenting species, in this case a larger DNA fragment. *Images generated by LabView-based MWL data viewer written by Dirk Haffke.* (See the color plate.)

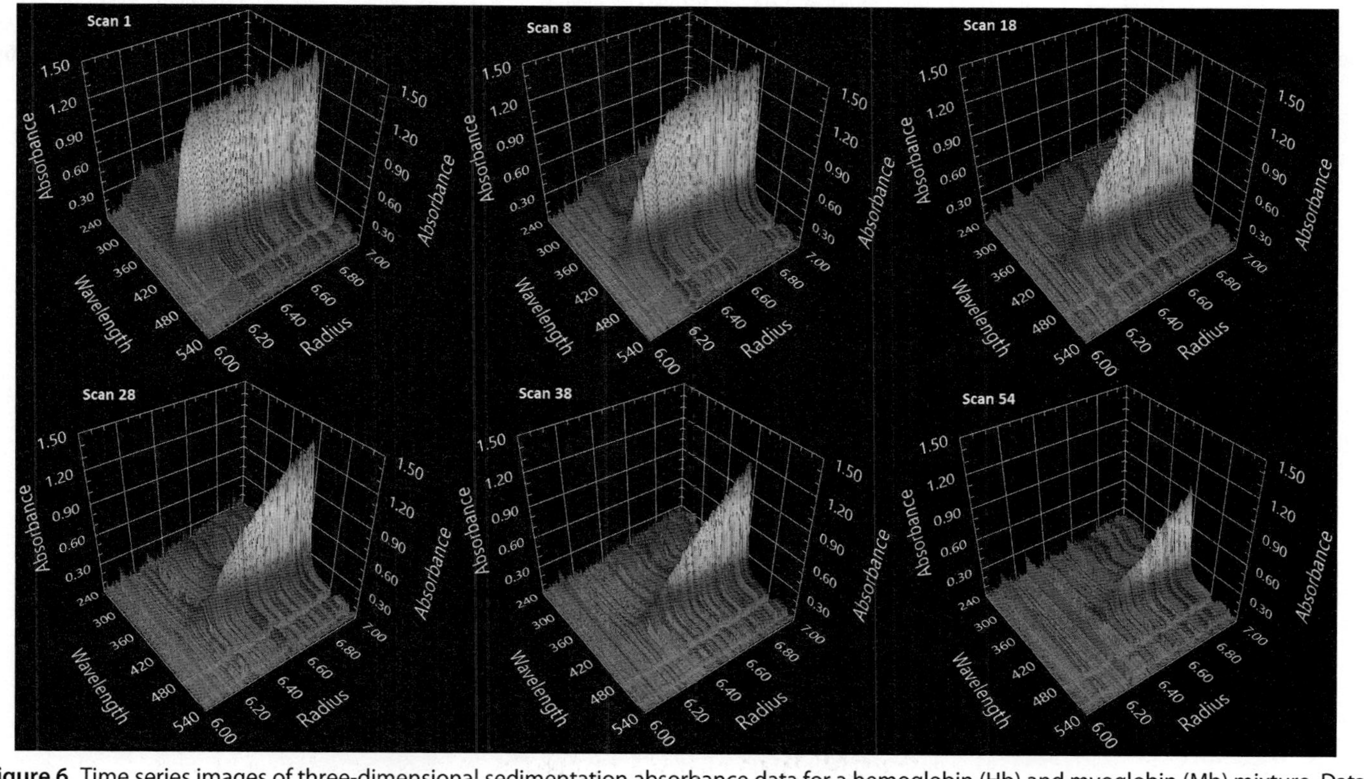

Figure 6 Time series images of three-dimensional sedimentation absorbance data for a hemoglobin (Hb) and myoglobin (Mb) mixture. Data from Nanolytics second-generation MWL, using 16 bit USB2000+ Ocean Optics spectrometer with #1 grating and 10 µm radial step size. *Images generated by LabView-based MWL data viewer written by Dirk Haffke.* (See the color plate.)

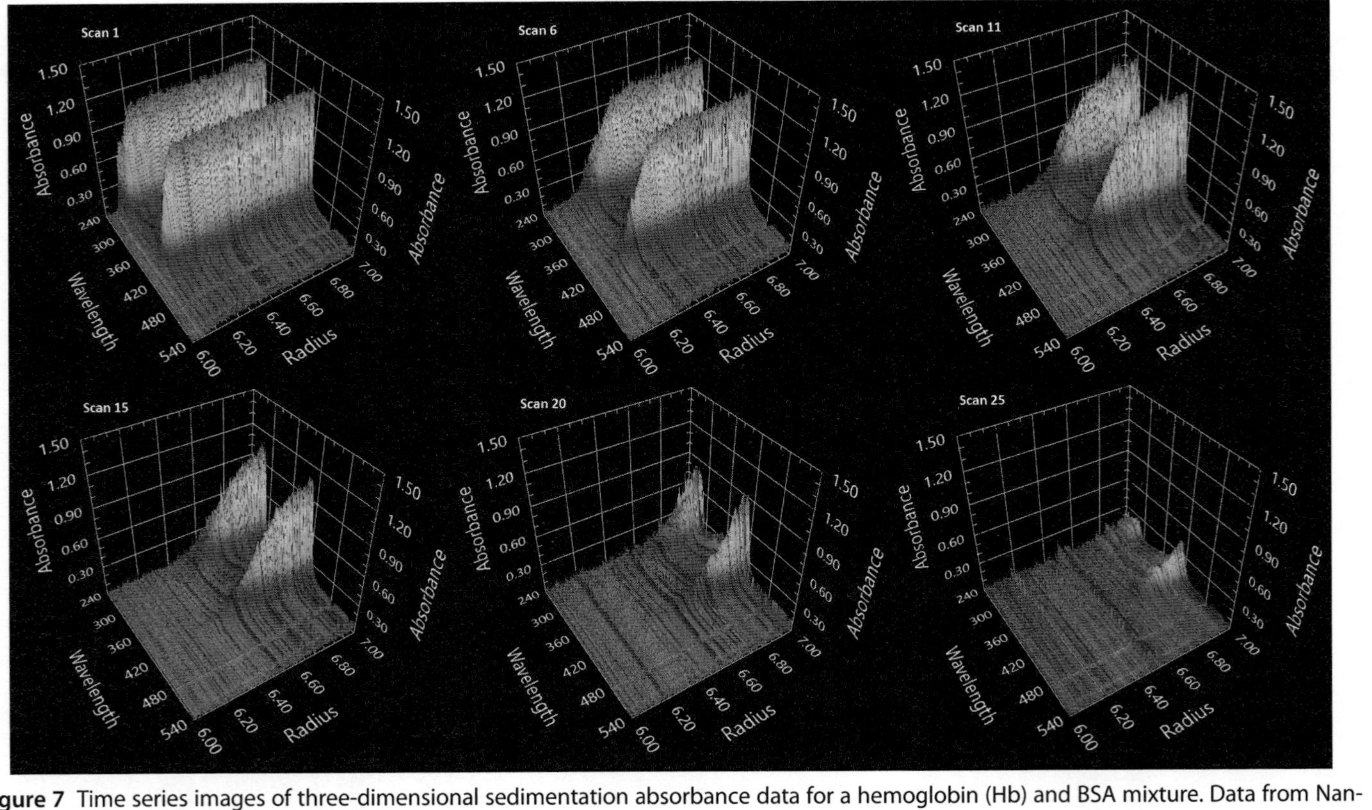

Figure 7 Time series images of three-dimensional sedimentation absorbance data for a hemoglobin (Hb) and BSA mixture. Data from Nanolytics second-generation MWL, using 16 bit USB2000+ Ocean Optics spectrometer with #1 grating and 10 μm radial step size. *Images generated by LabView-based MWL data viewer written by Dirk Haffke.* (See the color plate.)

Hb combines the chromophoric properties of its protein and heme components. Thus, Hb strongly absorbs in the visible region ($\lambda_{max} = 409$ nm, oxygenated Hb), due to the heme group, but also has an absorbance band in the UV. However, the absorption of the Hb heme group is much higher than the protein absorption part in the UV.

The Nanolytics samples were from stock solutions of BSA and Heme proteins, prepared in PBS buffer. Loading concentrations for experiments were 5 mg/mL BSA:0.6 mg/mL Hb and 0.3 mg/mL Mb:0.3 mg/mL Hb. Samples were run at 50,000 rpm, 20 °C in custom-produced titanium centerpieces with sapphire windows and path lengths of 3 mm.

5.1 SNR and Stability

An indicator of the intrinsic noise contributions from a flash lamp and detector system is provided by a comparison of the reference intensity and associated SNR (Skoog, Holler, & Nieman, 1998) across platforms at a number of wavelengths (Table 2). An average intensity is recorded for a non-absorbing reference scan across the radial dimension of a cell, at the two wavelengths most important for biopolymers as well as one in the visible. These scans are shown in the supporting information, SI 5 (http://dx.doi.org/10.1016/bs.mie.2015.06.033). The standard deviation and subsequent SNR is included.

The noise of the baseline absorbance signal has been reported previously for MWL machines in a comparison of data quality with the XLA machine (Strauss et al., 2008; Walter et al., 2014). Here, we report the intensity for the researcher also interested in signal availability at different wavelengths. Intensity is easily converted to absorbance, and the present devices give values close to those reported previously. However, this metric is insufficient as a measure of overall data quality in comparisons with the XLA. Appropriate

Table 2 Intensity and SNR Comparison of the Two MWL Platforms at the Three Wavelengths

	260 nm			280 nm			430 nm		
Wavelength	**Ave. Intensity**	**Std. Dev.**	**SNR**	**Ave. Intensity**	**Std. Dev.**	**SNR**	**Ave. Intensity**	**Std. Dev.**	**SNR**
Open AUC	9254	157	59	10,169	186	55	21,080	284	74
Nanolytics	5067	213	24	5336	210	25	35,334	1034	34

In this example of one scan, the open AUC system uses eight replicate averaging at each radial position, and the Nanolytics system uses just one (see Section 3).

and effective tools are still needed to fully evaluate data quality for absorbance-based AUC detection instruments. One approach to evaluate overall data quality comes out of the recent analytical work for MWL data (Gorbet et al., 2015).

An intensity instability phenomena associated with optical fibers has been observed (Fig. 3). For this example, in the reference intensity region of the figure, the first scans of a run begin with an intensity of ~9500 counts. The intensity is then observed to be dropping over 200 scans of a 15 h run to ~6000 counts, an approximate 37% decrease from the originally observed intensity. The decreasing intensity is illustrated in the figure included in SI 6 (http://dx.doi.org/10.1016/bs.mie.2015.06.033), where it is shown to be a linear phenomenon. A chemical instability in the optical fibers is thought to be responsible for attenuating the light through the optics over the course of a run, see Section 6.

5.2 Radial Resolution

The radial resolution of an AUC optical configuration can be estimated from the sharpness of the meniscus, but is not simple to quantify. A test of the radial resolution of the device is made by examining the inner edge of a cell. Considering the example of Fig. 3, the steep increase in the signal intensity at the far left of the scan is marking the edge of the sample channel in a cell. By zooming in on a single scan of this region, the steepness of the edge as recorded by the MWL optics can be evaluated. We define the edge resolution as the radial range over which the signal increases from 10% to 90% of the plateau intensity. The example of a single scan from Fig. 3 is illustrated in supporting information SI 7 (http://dx.doi.org/10.1016/bs.mie.2015.06.033), zoomed to the left edge. Either of the available MWL data viewers provide this function. At 280 nm, this resolution is approximately 70 μm for the open AUC, 70 μm for the Nanolytics MWL, and about 30 μm for the XL-A.

6. DISCUSSION

6.1 Qualitative Visualization of Data without Analysis

The absorbance data constituting a sedimentation profile provides the visual result familiar to the AUC researcher, as in Fig. 4. The 2D sedimentation profile is compiled to include wavelength spectra as a third visual dimension in a MWL data viewer such as UltraScan or the LabView-based viewer profiled above. Playing the scans through as the time domain of a 3D movie provides an instant and intuitive understanding of the time lapse sedimentation process in an experiment. The absorbance spectra characteristic of the analytes under investigation are observed moving toward the cell bottom as sedimentation proceeds. In addition, the presence of multiple species can be revealed as inflections in the sedimentation boundaries with characteristic absorbance spectra emerging across the radial dimension. This simple evaluation immediately discloses the spectral range of interest to the investigator. Furthermore, this preliminary analysis may uncover a species that would have gone unnoticed if only a single wavelength experiment had been recorded.

The increased capabilities of the MWL ultracentrifuge are illustrated by the examples of several mixed biopolymer samples. The first data ever from a sample mixture of DNA and protein as analyzed by MWL optics are illustrated in Figs. 3–5. The qualitative data visualization of this sample demonstrates the potential for the combination of spectral information of MWL data and fractionation of AUC as applied to mixed biopolymers. One can clearly see three steps in the sedimentation profiles, which immediately tells that this mixture contains (at least) three species without any evaluation. In this example, the smaller DNA fragment has a sedimentation coefficient very similar to that of BSA; therefore, this preliminary analysis alone cannot distinguish the equally sedimenting species although the relatively constant absorption plateau between 250 and 280 nm in Fig. 5 indicates that this species can neither be a pure protein nor a pure DNA but is likely a mixture of them sedimenting equally. In addition, the observable species suggest the fastest sedimenting species recognizable around 270 nm being DNA. However, the analytical power of hydrodynamic AUC theory must be applied to this system to realize the full possibilities of uncovering the molecular information concealed within the sedimentation boundaries. The wavelength deconvolution power available to MWL data can be used to identify unique species even with overlapping spectral information (Gorbet et al., 2015; Lewis et al., 1994; Walter et al., 2015).

Examples are also provided of two noninteracting binary systems out of three model proteins each exhibiting monomers and oligomers: BSA, Hb, and Mb. The absorbance spectra of Hb and Mb are characteristic for the oxygenated form. The sedimentation profile of the Hb/Mb mixture represents a superposition of the results from the two independently analyzed proteins. It is observed by the inflection steps appearing in the sedimentation profiles around the 409 nm absorption peak as the scans progress. This becomes more obvious when taking 2D radial slices at 409 nm SI 8 (http://dx.doi.org/10.1016/bs.mie.2015.06.033). Differences in comparison with the single component experiments observed in the mixtures can be mainly attributed to hydrodynamic interactions resulting in nonideal sedimentation behavior.

Observing the BSA–Hb mixture in Fig. 7, two almost equally sedimenting boundaries can be observed with maxima at 278 and 409 nm. Taking the absorption ratio of BSA and Hb into account, we can state that the species detectable at 409 nm is exclusively Hb, while that at 278 nm is almost exclusively BSA due to the very low Hb absorption at 278 nm in this mixture. We can see from the last scan in Fig. 7 that the BSA sediments slightly faster than the Hb, although their sedimentation coefficients are very similar ($BSA_{s20,w}=4.3$ S, $Hb_{s20,w}=4.2$ S). In addition, both proteins appear to be pure and form no noticeable complex.

6.2 MWL Data Quality

Raw intensity data provide a first qualitative assessment of the data. Here, we can note the signal intensity in the reference region. In Fig. 3, a reference intensity of ~9500 counts is recorded at 270 nm wavelength. With 9500 counts, we are using only ~15% of the dynamic range of the 16-bit spectrometer chip. The light intensity recorded as counts from the spectrometer is a function of the emission spectrum from the Xenon flash lamp and losses through the optical system, in addition to the efficiency of the grating, CCD chip, and A/D within the spectrometer. The limited intensity apparent below 300 nm is primarily an issue of the spectrometer response. The effects can be mitigated by selection of a spectrometer grating optimized for a wavelength region of interest.

The raw intensity data also reveal a visual assessment of scan quality. The open AUC MWL is shown to have a higher SNR at the wavelengths of importance for the proteins studied here. It is about twice as high as that of the Nanolytics MWL (see Table 2) and is largely due to the averaging routine within the open AUC control software. However, these results must be seen in the light of the scanning speed given in Table 1, which is much higher for the Nanolytics MWL. The resulting much higher number of scans from the Nanolytics MWL will compensate for the lower SNR as compared to the open AUC MWL. We therefore consider the quality of both MWL designs as comparable. Since the fundamental architecture of the two MWL optics systems is the same, discrepancies in the data quality between the two systems are expected to vanish as the implementations are further refined.

A chemical instability in the optical fibers is thought to be responsible for attenuating the light through the optics over the course of a run. The known problem of UV light attenuation in fiber optics is referred to as solarization. However, the decline in recorded intensity detailed in Section 5 is dependent not only on the power and duration of the flash but also on the presence of vacuum. A recording of the light intensity at atmospheric pressure does not exhibit the decline as illustrated in SI 6 (http://dx.doi.org/10.1016/bs.mie.2015.06.033), but in fact a small and steady increase in intensity is recorded, which can in part reverse the negative effects of intensity loss through the fibers. The chemistry of this phenomenon is still unknown. A fiber option that does not exhibit this effect is being investigated. Nevertheless, by correcting the data with a conversion to pseudo-absorbance Fig. 4, the quality was sufficient for a thorough evaluation of the species present in the sample. Solarization and the intensity instability phenomena are observed to be more pronounced with higher lamp power, but are almost completely eliminated when light below 300 nm is blocked by a high-pass filter. Therefore to maximize fiber life, the flash lamp power is set as low as possible, and UV light is filtered out when only measuring in the visible. This improves both the intensity instability phenomena, as well as long-term fiber solarization effects.

A first approximation of radial resolution is provided by an image of the upper cell edge. The ideal evaluation of radial resolution must take into account the light passing through the entire event space of a sample channel, which is typically 12 mm. The effects of alignment and focusing will be further complicated by the speed dependence of a sample meniscus formed at the window boundary, making the upper cell edge a simpler measure of radial resolution. The resolution determined at 280 nm for the open AUC design with 50 μm step size, as shown in SI 7 (http://dx.doi.org/10.1016/bs.mie.2015.06.033), is ~70 μm which is comparable with the Nanolytics design, but considerably worse than the 30 μm of the XLA. Reasons for this could be chromatic abberation or focusing issues in the MWL designs. However, considering that the radial step size was 50 μm, one can say that the edge resolution of all designs is in the order of magnitude of the typical radial step size chosen for sedimentation velocity experiments.

7. SUMMARY AND GENERAL DISCUSSION OF THE IMPACTS OF MWL AUC DATA IN THE FIELD OF BIOPOLYMERS

Two independent implementations of an MWL optics are detailed above. Both systems have an acceptable data quality at the wavelengths relevant for biopolymer analysis, despite the obstacle of fiber solarization, which significantly decreases the dynamic range for measurements at these wavelengths. Most importantly, the scope for multiplex analyses, by exploiting distinct absorbance spectra of various components in mixtures, is significantly increased and the number of addressable chromophores is theoretically not limited. It could already be shown that the analysis of three distinct chromophoric analytes using the XL-A ultracentrifuge with three different wavelengths significantly increased the quality of the analysis and information content obtained (Balbo et al., 2005).

The raw data from MWL experiments contain far more information than the data from conventional experiments at a single wavelength. The data viewers allow a convenient access to the data in the form of a movie, to first observe the approximate number of components, heterogeneities, impurities, etc., directly in the raw data and without any prior evaluation. In addition, the relevant wavelength range for a further evaluation can be selected, to constrain the large dataset to those data, which contain relevant information. With this tool, even difficult mixtures like BSA and Hb, with almost equal sedimentation coefficients, become accessible. Our evaluations of the raw data provide a glimpse of what information will become available if such data are evaluated by powerful evaluation software such as UltraScan (Demeler, Gorbet, Zollars, & Dubbs, 2015) or Sedanal (Walter et al., 2015).

Similarly to the already demonstrated huge impact in the field of nanoparticle analysis (de Roo et al., 2014; Karabudak, Backes, et al., 2010; Karabudak, Wohlleben, et al., 2010; Walter et al., 2014), we anticipate that MWL analysis will have a major impact on biopolymer analysis with AUC. Due to the added spectral dimension and the huge data space, which allows averaging, unsurpassed data quality can be expected and therefore, the accessible information from AUC experiments is expected to increase significantly. Indeed, the thorough analysis of the BSA–DNA system presented in another study using UltraScan shows, how much more information can be obtained from this type of experiment (Gorbet et al., 2015). We are therefore confident that new possibilities of biopolymer analysis will emerge in the future.

ACKNOWLEDGMENTS

J.P. and H.C. thank the Center for Applied Photonics (CAP) at the University of Konstanz for financial support of this work. F.K. and K.S. gratefully acknowledge Tina Franke at Nanolytics for excellent technical assistance. We thank Aysha Demeler and Blanca Hernandez-Uribe for providing the DNA sample used in this study. B.D. wishes to acknowledge support by NSF Grant ACI-1339649.

SUPPLEMENTAL INFORMATION

SI 1: MWL Raw Data file details (http://dx.doi.org/10.1016/bs.mie.2015.06.033)

SI 2: Movie of BSA–DNA mixture detected with the open AUC MWL (http://dx.doi.org/10.1016/bs.mie.2015.06.033).

SI 3: Movie of Myoglobin–Hemoglobin detected with the Nanolytics MWL (http://dx.doi.org/10.1016/bs.mie.2015.06.033).

SI 4: Movie of BSA–Hemoglobin detected with the Nanolytics MWL (http://dx.doi.org/10.1016/bs.mie.2015.06.033).

SI 5: Data of an empty rotor hole for calculation of signal-to-noise ratio (http://dx.doi.org/10.1016/bs.mie.2015.06.033).

SI 6: Reference Intensity vs. Scan Number; Intensity of scans at 270 nm for a single radial position in the reference region of the cell channel over 200 scans within 15 h (http://dx.doi.org/10.1016/bs.mie.2015.06.033).

SI 7: Left edge single scan from a 12-mm sample channel at 280 nm and 30 μm radial step (http://dx.doi.org/10.1016/bs.mie.2015.06.033).

SI 8: 2D radial scans at 409 nm (http://dx.doi.org/10.1016/bs.mie.2015.06.033)

REFERENCES

Backes, C., Karabudak, E., Schmidt, C. D., Hauke, F., Hirsch, A., & Wohlleben, W. (2010). Determination of the surfactant density on SWCNTs by analytical ultracentrifugation. *Chemistry—A European Journal, 16*(44), 13176–13184.

Balbo, A., Minor, K. H., Velikovsky, C. A., Mariuzza, R. A., Peterson, C. B., & Schuck, P. (2005). Studying multiprotein complexes by multisignal sedimentation velocity analytical ultracentrifugation. *Proceedings of the National Academy of Sciences of the United States of America, 102*(1), 81–86.

Bhattacharyya, S. K. (2006). *Development of detector for analytical ultracentrifuge.* PhD, Universität Potsdam.

Bhattacharyya, S. K., Maciejewska, P., Börger, L., Stadler, M., Gülsün, A. M., Cicek, H. B., et al. (2006). Development of a fast fiber based UV–vis multiwavelength detector for an ultracentrifuge. *Vol. VIII. Analytical ultracentrifugation* (pp. 9–22). Springer.

Cole, J. L. (2004). Analysis of heterogeneous interactions. *Methods in Enzymology, 384*, 212–232.

Colfen, H., Laue, T. M., Wohlleben, W., Schilling, K., Karabudak, E., Langhorst, B. W., et al. (2010). The open AUC project. *European Biophysics Journal with Biophysics Letters, 39*(3), 347–359.

Cölfen, H., & Pauck, T. (1997). Determination of particle size distributions with Angström resolution. *Colloid and Polymer Science, 275*(2), 175–180.

de Roo, T., Haase, J., Keller, J., Hinz, C., Schmid, M., Seletskiy, D. V., et al. (2014). A direct approach to organic/inorganic semiconductor hybrid particles via functionalized polyfluorene ligands. *Advanced Functional Materials, 24*(18), 2714–2719.

Demeler, B., Gorbet, G., Zollars, D., & Dubbs, B. (2015). *UltraScan-III version 3.3: A comprehensive data analysis software package for analytical ultracentrifugation experiments.* from, http://www.utrascan3.uthscsa.edu/.

Gorbet, G. E., Pearson, J. Z., Demeler, A. K., Cölfen, H., & Demeler, B. (2015). Next-generation AUC: Analysis of multiwavelength analytical ultracentrifugation data. *Methods in Enzymology, 562,* 27–48.

Kar, S. R., Kingsbury, J. S., Lewis, M. S., Laue, T. M., & Schuck, P. (2000). Analysis of transport experiments using pseudo-absorbance data. *Analytical Biochemistry, 285*(1), 135–142.

Karabudak, E. (2009). *Development of MWL CCDC SLS detectors for the analytical ultracentrifuge.* PhD, Universität Potsdam.

Karabudak, E., Backes, C., Hauke, F., Schmidt, C. D., Colfen, H., Hirsch, A., et al. (2010). A universal ultracentrifuge spectrometer visualizes CNT-intercalant-surfactant complexes. *Chemphyschem, 11*(15), 3224–3227.

Karabudak, E., Wohlleben, W., & Colfen, H. (2010). Investigation of beta-carotene-gelatin composite particles with a multiwavelength UV/vis detector for the analytical ultracentrifuge. *European Biophysics Journal with Biophysics Letters, 39*(3), 397–403.

Laue, T., & Austin, B. (In Press). The CFA analytical ultracentrifuge architecture. Analytical ultracentrifugation: Instrumentation software and application.

Lewis, M., Shrager, R., & Kim, S. (1994). Analysis of protein-nucleic acid and protein-protein interactions using multi-wavelength scans from the XL-A analytical ultracentrifuge. In T. M. Schuster & T. M. Laue (Eds.), *Modern analytical ultracentrifugation: acquisition and interpretation of data for biological and synthetic polymer systems emerging biochemical and biophysical techniques: Part I* (pp. 94–118). Boston: Birkhäuser.

MacGregor, I. K., Anderson, A. L., & Laue, T. M. (2004). Fluorescence detection for the XLI analytical ultracentrifuge. *Biophysical Chemistry, 108*(1–3), 165–185.

Mächtle, W. (1999). High-resolution, submicron particle size distribution analysis using gravitational-sweep sedimentation. *Biophysical Journal, 76*(2), 1080–1091.

Müller, H. (1989). Automated determination of particle-size distributions of dispersions by analytical ultracentrifugation. *Colloid & Polymer Science, 267*(12), 1113–1116.

Nanolytics. from http://www.nanolytics.de/.

Skoog, D. A., Holler, F. J., & Nieman, T. A. (1998). *Principles of instrumental analysis.* Orlando, Fl: Saunders College Pub.; Harcourt Brace College Publishers.

Strauss, H. M., Karabudak, E., Bhattacharyya, S., Kretzschmar, A., Wohlleben, W., & Colfen, H. (2008). Performance of a fast fiber based UV/Vis multiwavelength detector for the analytical ultracentrifuge. *Colloid and Polymer Science, 286*(2), 121–128.

Voelkle, C., Gebauer, D., & Coelfen, H. (2015). FD Nucleation: High-resolution insights into the early stages of silver nucleation and growth. *Faraday Discussions.*

Walter, J., Lohr, K., Karabudak, E., Reis, W., Mikhael, J., Peukert, W., et al. (2014). Multidimensional analysis of nanoparticles with highly disperse properties using multiwavelength analytical ultracentrifugation. *ACS Nano, 8*(9), 8871–8886.

Walter, J., Sherwood, P., Lin, W., Segets, D., Stafford, W., & Peukert, W. (2015). Simultaneous analysis of hydrodynamic and optical properties using analytical ultracentrifugation equipped with multiwavelength detection. *Analytical Chemistry, 87*(6), 3396–3403.

http://wiki.bcf2.uthscsa.edu/openAUC/wiki/WikiStart. Open AUC project. From http://wiki.bcf2.uthscsa.edu/openAUC/wiki/WikiStart.

http://wiki.bcf2.uthscsa.edu/ultrascan3/wiki/MwrsFormat. OpenAUC standard for multiwavelength analytical ultracentrifugation data storage used in UltraScan-III.

http://wiki.bcf2.uthscsa.edu/ultrascan3/wiki/TimeState. Time-state object for multiwavelength recorded files.

CHAPTER TWO

Next-Generation AUC: Analysis of Multiwavelength Analytical Ultracentrifugation Data

Gary E. Gorbet*, Joseph Z. Pearson†, Aysha K. Demeler*, Helmut Cölfen†, Borries Demeler*,[1]

*Department of Biochemistry, The University of Texas Health Science Center at San Antonio, San Antonio, Texas, USA

†Physical Chemistry, Department of Chemistry, University of Konstanz, Konstanz, Germany

[1]Corresponding author: e-mail address: demeler@biochem.uthscsa.edu

Contents

Abstract

We describe important advances in methodologies for the analysis of multiwavelength data. In contrast to the Beckman-Coulter XL-A/I ultraviolet–visible light detector, multiwavelength detection is able to simultaneously collect sedimentation data for a large wavelength range in a single experiment. The additional dimension increases the data density by orders of magnitude, posing new challenges for data analysis and management. The additional data not only improve the statistics of the measurement but also provide new information for spectral characterization, which complements the hydrodynamic information. New data analysis and management approaches were integrated into the UltraScan software to address these challenges. In this chapter, we describe the enhancements and benefits realized by multiwavelength analysis and compare the results to those obtained from the traditional single-wavelength detector. We illustrate the advances offered by the new instruments by comparing results from mixtures

Methods in Enzymology, Volume 562
ISSN 0076-6879
http://dx.doi.org/10.1016/bs.mie.2015.04.013

that contain different ratios of protein and DNA samples, representing analytes with distinct spectral and hydrodynamic properties. For the first time, we demonstrate that the spectral dimension not only adds valuable detail, but when spectral properties are known, individual components with distinct spectral properties measured in a mixture by the multiwavelength system can be clearly separated and decomposed into traditional datasets for each of the spectrally distinct components, even when their sedimentation coefficients are virtually identical.

1. INTRODUCTION

Analytical ultracentrifugation (AUC) has long been an indispensable tool for the characterization of macromolecules in the solution phase. The study of molecules in the solution phase permits the investigator to closely replicate physiological conditions; modulate solution conditions such as pH, ionic strength, and redox potential; or investigate the effect of added ligands or drugs. Biopolymers such as nucleic acids and proteins exhibit distinct spectra that offer potential for high resolution by multiwavelength detection in the AUC instrument. A plethora of options exist to impart additional unique chromophores to systems under investigation: Heme proteins offer unique spectral information in the visible range, and many nanoparticles exhibit size-dependent spectral changes. It is common practice to modify nucleic acids and proteins by attaching fluorescent tags to add unique chromophores. Fusion proteins with green, yellow, or red fluorescent protein allow the investigator to express color-coded proteins *in vivo*. Traditional AUC detectors reveal information about the hydrodynamic parameters of the analytes in solution. Subject to limitations in the resolution of the analysis method and the quality of the data, it is possible to extract several parameters for each analyte present in a mixture when performing sedimentation velocity (SV) experiments: the sedimentation and diffusion coefficients, and the partial concentration (Demeler et al., 2014; Correia & Stafford, 2015). A further interpretation of these parameters to obtain molecular weight and anisotropy is possible if the partial specific volume of the analyte as well as the buffer density and viscosity is known. If two molecules interact, their interactions give rise to mixtures of free and complexed species, each with unique hydrodynamic properties. These properties can also be extracted, yielding information such as stoichiometry of association and binding strengths of such interactions. For large molecules, and slow interactions, even rate constants can be obtained (Demeler,

Brookes, Wang, Schirf, & Kim, 2010; Correia & Stafford, 2015). With increases in computer power, new high-resolution analysis methods were developed in our laboratory (Brookes, Boppana, & Demeler, 2006; Brookes, Cao, & Demeler, 2010; Brookes & Demeler, 2006, 2007, 2008; Demeler & Brookes, 2008; Demeler, Brookes, & Nagel-Steger, 2009; Demeler et al., 2010, 2014; Gorbet et al., 2014) that allow investigators to characterize increasingly complex mixtures and interactions between molecules. Here, we discuss the application of those analysis methods to data obtained from the open AUC multiwavelength detector described earlier (Bhattacharyya et al., 2006; Pearson et al., 2015; Strauss et al., 2008; Walter et al., 2014) and show how these advances will further broaden the appeal of AUC and open up new avenues for the analysis of ever more complex systems.

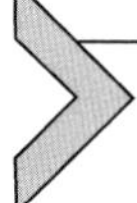

2. COMPUTATIONAL TREATMENT OF MULTIWAVELENGTH DATA

2.1 Data Representation

The increase in data density from the new multiwavelength detector poses major challenges to the management and analysis of multiwavelength data (MWLD). Instead of a single-wavelength, three-dimensional (3D) dataset of time, radius, and absorbance, the additional spectral dimension multiplies the traditional dataset size by hundreds of wavelengths. The original ASCII-based format from the XL-A's Beckman-Coulter data acquisition software for storing sedimentation data is no longer suitable for the large amount of data stored in multiwavelength experiments due to inherent inefficiencies present in ASCII formatted storage. While alternatives have been proposed that retain the legacy ASCII format (Walter et al., 2015), we feel that a binary format is much more suitable, providing rapid loading, efficient storage, and reduced network transfer speeds. The open AUC standard binary format (Cölfen et al., 2010) used in the UltraScan software (Demeler & Gorbet, 2015; Demeler, Gorbet, Zollars, & Dubbs, 2015) offers an efficient data representation that is readily adapted for radial scans collected by the multiwavelength detector. Compared to the Beckman-Coulter ASCII format, and depending on data type, our binary representation achieves between 12:1 and 14:1 lossless compression by taking the following steps: (a) replacing the radial vector with a radial start and end point, and defining the radial increment; (b) missing data points in the radial grid are interpolated and marked by a boolean flag as interpolated; and (c) using a properly scaled

4-byte integer representation for absorbance, intensity, or fringe values. All metadata are carried in the file header and a separate binary file is created for each triple, where a triple is defined as the collection of data originating from a unique channel, rotor hole, and wavelength. A converter exists in UltraScan-III to export binary open AUC data to legacy Beckman-Coulter ASCII format. A separate intensity profile XML file carries the reference intensities for each wavelength as a function of time.

2.2 Data Preprocessing

We found that accounting for time-dependent intensity changes requires individual corrections when converting raw intensity data to absorbance data (Pearson et al., 2015). These corrections are applied on a scan-by-scan basis to all triples using a reference intensity obtained by averaging a few adjacent radial points in the air region above the meniscus from the corresponding scan time. This assures that the intensity level available at the scan's time is used for the conversion to absorbance data. A detailed description of the open AUC storage format is shown in SI-1 (http://dx.doi.org/10.1016/bs.mie.2015.04.013) (any version changes will be made available on the web in the UltraScan-III wiki: http://wiki.bcf2.uthscsa.edu/ultrascan3/wiki/Us3Formats). In UltraScan-III, the open AUC formatted data are ordinarily stored in a relational database which creates links to experimental details such as analytes, buffer composition, centerpiece type, rotor type, instrument, data owner, and other related information. Before importing MWLD into UltraScan-III, the data are written by the multiwavelength data acquisition software to an intermediate binary file format described earlier (Pearson et al., 2015). The intermediate file is accompanied by a time state object (see Pearson et al., 2015) which is used to record metrics such as rotor speed, temperature, vacuum, and acceleration settings with time increments as short as 1 s. For analysis purposes, the need for this time state object is its utility for the fitting of multispeed experiments. When modeling speed transitions using finite element solutions, it is critical to know the precise point of acceleration or deceleration, and the acceleration rate used. During acceleration the optical detectors typically do not record data, and if one would rely only on the time points recorded in the scan header files, the lag time before and after the last and first recorded scan on either side of the acceleration period is too large for an accurate modeling of the speed transition.

2.3 Data Analysis

The analysis of MWLD can be tailored to the specifics of the experiment. It is possible to distinguish two experimental conditions that can be approached with separate analysis strategies. In the general case, the extinction profiles for one or more sedimenting components are unknown. In this case, data from each wavelength should be analyzed independently. In our analysis approach, hydrodynamic separation of unlike solutes will provide pure spectra for each solute. Complexes formed by different solutes will present composite spectra with proportional contributions from each solute involved in the complex. Data quality permitting this will, in principle, facilitate extraction of complex stoichiometries. In the second case, extinction profiles for all wavelengths covered by the collected data are known *a priori* for all components in the mixture. If the extinction profiles for all components are sufficiently different, the vectors describing these profiles can be considered orthogonal, and they can be used for a linear decomposition of the measured absorbance profile. Here, the 4D sedimentation profile of the MWLD (wavelength, radius, time, and absorbance) can be reduced to regular 3D sedimentation profiles (radius, time, and absorbance) for each component in the mixture, which can then be analyzed further by conventional single-wavelength methods. This approach is discussed in further detail below. In both cases, the hydrodynamic analysis is based on whole boundary fitting methods using high-resolution finite element modeling (Cao & Demeler, 2005, 2008), according to the workflow sequence described in Demeler (2010). Alternatively, MWLD can be analyzed by methods such as the van Holde–Weischet (Demeler & van Holde, 2004) or DCDT (Walter et al., 2015) analysis. Our workflow implements algebraic elimination of time- and radially invariant noise contributions (Schuck & Demeler, 1999) using the 2D spectrum analysis (2DSA), a meniscus position fit, and an iterative refinement by 2DSA. Methods proposed in Walter et al. (2015) for MWLD analysis rely on the subtraction of pairwise scans in order to remove time-invariant noise. We believe that this approach is unsatisfactory since it increases stochastic noise by ~1.4-fold. Furthermore, the presence of radially invariant noise contributions cannot be corrected by this approach. This turns out to be a serious drawback for the fiber optics multiwavelength detection in the UV where time-dependent changes in the light intensity due to fiber instability have been identified (Pearson et al., 2015). This process causes a reduction of the recorded intensity over the course of a run and requires fine-grained adjustments of the radially invariant baseline offsets for each scan. Our approach correctly handles this type of noise and corrects any time-dependent baseline fluctuations due to intensity changes (see Section 2.2). Since the data from each wavelength can be treated as a separate dataset for analysis with the 2DSA method, this is a demanding, but tractable computational challenge. In the UltraScan workflow, the datasets for each individual wavelength are analyzed in parallel, except the meniscus fit, where the meniscus determination for a single triple (generally selected from the center of the wavelength range) is sufficient to extract the meniscus position for all other wavelengths of a unique channel. We determined that the meniscus position varies slightly with wavelength due to chromatic aberration in the lens-based MWL detectors (<0.003 cm, see SI-2 (http://dx.doi.org/10.1016/bs.mie.2015.04.013)), but this variation is so small that it can be safely ignored. Meniscus corrected data are then refined using the iterative 2DSA approach with up to 10 iterations. In the last step, a 2DSA-Monte Carlo analysis with 100 iterations is used to obtain a concentration distribution for the sedimentation and diffusion profile of any solutes absorbing at a given wavelength. The Monte Carlo analysis will provide statistics that are needed to calculate confidence regions for the extracted parameters (Demeler & Brookes, 2008). If prior knowledge about the experimental system is available and can be used as a global constraint, the *Custom Grid* approach (Demeler et al., 2014) can be applied to map the 2DSA results to other hydrodynamic parameters, such as partial specific volume, frictional ratio, and molar mass. Once the analysis is complete, finite element models are posted back to the database using the Airavata grid middleware (Pierce, Marru, Demeler, Singh, & Gorbet, 2014). Models can then be visualized on the desktop and meta-analysis can be performed to extract amplitudes of each sedimenting species at any wavelength in order to compile an absorbance spectrum for each species present in the mixture and to obtain a partial concentration for each species.

2.4 Analysis of MWLD Containing Species with Known Extinction Profiles

In the case where extinction profiles for all components are known, MWLD can be decomposed into separate sedimentation profiles for each individual component. This requires that the intrinsic extinction profile of each pure component can be measured separately, and that any shifts in the absorbance spectrum occurring as a result of complex formation are negligible. This would assure that the spectral properties of each component stay constant, even when complexes are formed. In our approach, a dilution series of each spectrally unique component is wavelength scanned in a benchtop spectrophotometer over the entire range measured in the MWLD dataset:

$$S_j = c_j \sum_{i=1}^{n} a_i G(\lambda_i, \sigma_i). \tag{1}$$

The absorption signal S from each dilution j is globally fitted to an intrinsic extinction profile composed of a sum of n Gaussian terms with amplitudes a_i, wavelength center λ_i, peak width σ_i, and concentration scaling factor c_j (see Eq. 1) using nonlinear least squares optimization implemented in UltraScan (Demeler et al., 2015). Using 50 Gaussian terms, this approach faithfully represents the entire spectrum observed in the MWL instruments (see SI-3 (http://dx.doi.org/10.1016/bs.mie.2015.04.013)). Inclusion of greatly varying concentrations assures that the dynamic range of the spectrophotometer can be optimally used for even greatly varying extinction ranges at different wavelengths. Our implementation in UltraScan allows the user to exclude data above a threshold optical density to assure that only linear data are included in the fit. A benefit of the global fitting of multiple concentrations at multiple wavelengths is the ability to span greater concentration ranges, because extinction coefficients vary considerably over a large wavelength range for most biological macromolecules, especially in the low UV range below 230 nm. To stay within the dynamic range of the detector, combined multiple concentrations which optimally exploit the dynamic range of the detector, lead to a more reliable extinction profile for the entire wavelength range. Examples for global fits of a dilution series of mixtures containing different ratios of bovine serum albumin and a DNA plasmid restriction digest are shown in SI-3 (http://dx.doi.org/10.1016/bs.mie.2015.04.013).

$$W_{r,t} = \left[\sum_{l=1}^{k} x_l S_l \right]_{r,t}. \tag{2}$$

In the next step, intrinsic extinction profiles for each species n are now fitted to a linear decomposition of the wavelength spectrum W at a radial point r and scan time t (see Eq. 2) in order to obtain the amplitudes x_l for each of the k components. The problem is described by the linear system $\boldsymbol{Ax} = \boldsymbol{W}$, where $\boldsymbol{A}$ is the matrix composed of k columns containing the values of the extinction profiles S for each of the k components, and $\boldsymbol{W}$ is the wavelength spectrum to be decomposed, and $\boldsymbol{x}$ is a vector with the unknown amplitudes to be fitted. We point out that the approach for fitting this linear system suggested by Walter et al. (2015) does not impose the needed constraint that would guarantee that contributions for all solutes will always be ≥ 0. A nonnegatively constrained least squares (NNLS, as proposed by Lawson & Hanson, 1974) algorithm should be used, which guarantees $\boldsymbol{x} \geq 0$ for all x. Since wavelength scans from the MWLD experiment are not available as a dilution series, each wavelength scan W from the MWL instrument is used directly in the NNLS fit. Once the amplitudes $\boldsymbol{x}_l$ have been determined for each radial position and each time point, a new transformed dataset B_k can be constructed for each component k, yielding the concentration of component k as a function of radius and time. Conventional analysis of the resulting 3D datasets now provides the sedimentation profiles for spectrally distinct components, as well as any formed complexes containing the component. Once relative concentrations are determined, equilibrium constants for the complex formation can be readily derived from the mass action. Interpretation of these reduced datasets will be greatly facilitated since only one free component and its complexes are immediately identified by the column vector in $\boldsymbol{A}$. Identical hydrodynamic components found in multiple datasets correspond to complexes to which the species of interest is bound; the relative concentrations determined in Eq. (2), together with the hydrodynamic fitting methods will directly provide the stoichiometry of the association.

2.5 Parallelization

Due to the large data volume encountered in MWLD, we employ special submission mechanisms in a high-performance computing (HPC) infrastructure called the UltraScan Laboratory Information Management System (USLIMS) (Brookes & Demeler, 2008; Demeler et al., 2009). The USLIMS allows the user to manage all of their experimental data and submit experimental data for high-resolution analysis to a remote HPC resource. It allows the user to process computationally demanding data analysis algorithms in parallel using the UltraScan Science Gateway (Memon et al., 2014). Integration of HPC solutions benefits the investigator on three levels: (1) increased resolution, (2) faster calculations, and (3) analysis throughput, which is significantly higher since many data sets can be analyzed simultaneously. Conventional single-wavelength data consist of multiple time scans (typically 50–500 scans) of the radial domain (~300–800 datapoints/scan). MWLD contain hundreds of different wavelengths, but because of UltraScan's efficient open AUC data storage algorithms, the data can still be transferred over networks without noticeable delay. At the lowest level, the optimization algorithm involves fitting finite element solutions to the experimental data (see Brookes & Demeler, 2006, 2007, 2008; Brookes et al., 2006, 2010; Demeler & Brookes, 2008; Demeler et al., 2009, 2010, 2014; Gorbet et al., 2014). On the next level, jobs containing groups of up to 50 triples are submitted to the HPC resource queue. Depending on the load of the queue, all or some of the jobs may execute in parallel. On the third level (depending on the availability of resources), each job is further subdivided into parallel master groups (PMGs), resulting in a scalable HPC parallelization. On large HPC resources many submissions can be processed in parallel. On the final level, Monte Carlo calculations can be performed in parallel, adding a fourth level of parallelization. Proper selection of the PMG level allows the queuing system to optimally load-balance the submissions for different architectures. The user can even distribute portions of each analysis over multiple HPC resources. For a run with 100 scans performed at 350 individual wavelengths (250–600 nm) at 50 μm resolution, a total of 9.45 million datapoints (assuming a 13.5-mm column) need to be analyzed. This amounts to 12,600 NNLS computations when a standard 60×60 resolution 2DSA grid with 100 grid points is used. By splitting the computational load along individual wavelengths, nearly linear speedup is achieved, and 12,600 NNLS computations are performed in the same time as just 36. Parallelization is achieved using the Message Passing Interface standard (The Open MPI Project, 2004-2015) on distributed HPC architectures (available to UltraScan users from NSF-XSEDE resources and from UTHSCSA). For the analysis of MWLD containing species with known extinction profiles, each wavelength scan needs to be decomposed into individual contributions from known extinction profiles (Eq. 1). Since each wavelength scan can be decomposed independently, this process is also performed in parallel, using multithreaded fitting implemented in the desktop version of UltraScan-III which takes advantage of modern multicore architectures.

2.6 Visualization

2.6.1 Experimental Data

In order to gain a full view of either the experimental MWLD or the analysis results, four dimensions are needed (Pearson et al., 2015). We introduced 3D movies to provide a comprehensive 4D view of experimental data

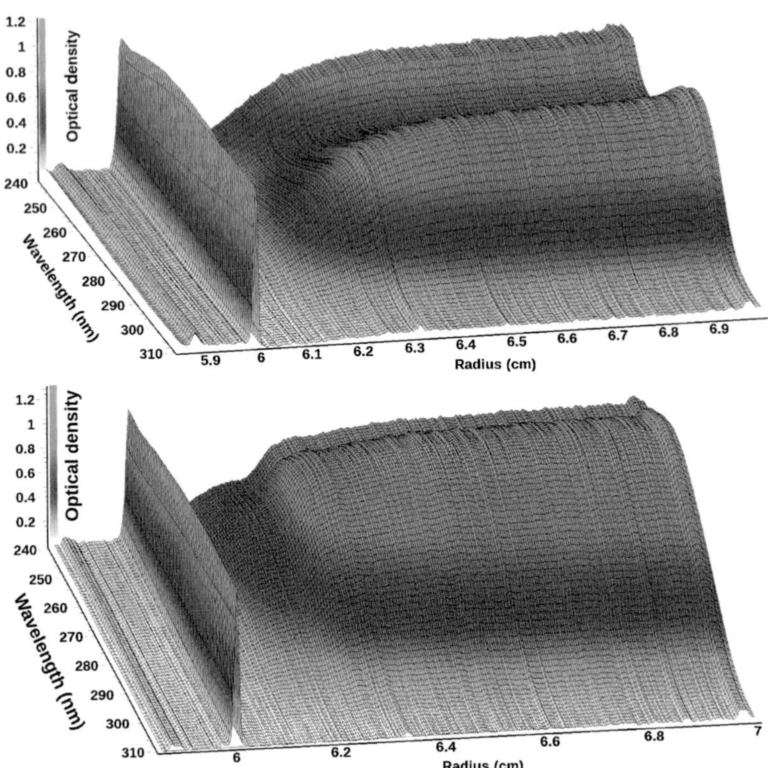

Figure 1 Three-dimensional view of scan 35 from a multiwavelength sedimentation velocity experiment of BSA (top) and scan 16 from a mixture of DNA fragments (bottom) measured between 240 and 310 nm. The meniscus is clearly visible at ~6.0 cm. For BSA, an absorbance maximum is visible at 278 nm from tyrosine and tryptophan residues, and below 250 nm absorbance from the peptide backbone increases for BSA, while a 258 nm peak typical for DNA results in a pronounced difference between the two biological macromolecules.

and analysis results. UltraScan offers the following movies in its data viewer: A 3D view of experimental data: Absorbance vs. radius and wavelength, with each scan time rendered as a frame in the movie. The concentration is represented by a color gradient. An example of a single frame is shown in Fig. 1. Movies over all scans of each DNA:BSA ratio are shown in SI-4 (http://dx.doi.org/10.1016/bs.mie.2015.04.013). Movies of the 2D view of all scans as a function of wavelength are shown in SI-5 (http://dx.doi.org/10.1016/bs.mie.2015.04.013), and movies of the absorbance at all wavelengths as a function of radius are shown in SI-6 (http://dx.doi.org/10.1016/bs.mie.2015.04.013).

2.6.2 Analysis Results

The results from a 2DSA generate a surface of amplitudes for the two fitted dimensions for each wavelength. These data can be shown either as a 3D plot (Fig. 2) or as a pseudo-3D plot where a color gradient is used to project the amplitudes into a 2D plane (Fig. 3). In UltraScan, model files for individual fits can be combined to generate a global view of each analyzed wavelength. Solutes appearing in the global model can be integrated visually in UltraScan, and the global model can be displayed either in a 3D viewer (see Fig. 4) or as a pseudo-3D plot (see SI-7 (http://dx.doi.org/10.1016/bs.mie.2015.04.013)). Alternatively, the pseudo-3D plot from each wavelength can be shown as an individual frame in a movie, and an example of this is shown in SI-8 (http://dx.doi.org/10.1016/bs.mie.2015.04.013). The same approaches work for other methods based on Lamm equation modeling, such as the PCSA (Gorbet et al., 2014), the custom grid analysis (Demeler et al., 2014), or genetic algorithms (GAs) (Brookes & Demeler, 2006, 2007). In addition, multiwavelength data can be quite informative when used with the model independent van Holde–Weischet method. Figure 5 shows an integral distribution plot for the 50:50 mixture of DNA and BSA. A differential distribution of the same analysis is shown in Fig. 6.

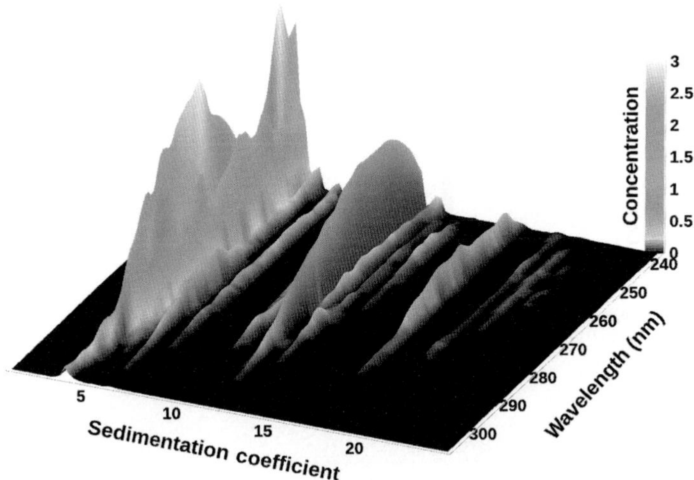

Figure 2 3D view of the sedimentation profile as a function of wavelength for the 50:50 DNA–BSA mixture after 2DSA-Monte Carlo analysis. The protein absorbance spectrum at 4.3 S (two yellow (white in the print version) peaks) can be clearly distinguished from the DNA peak with absorbance maximum around 258 nm. Minor species can be identified based on their spectrum.

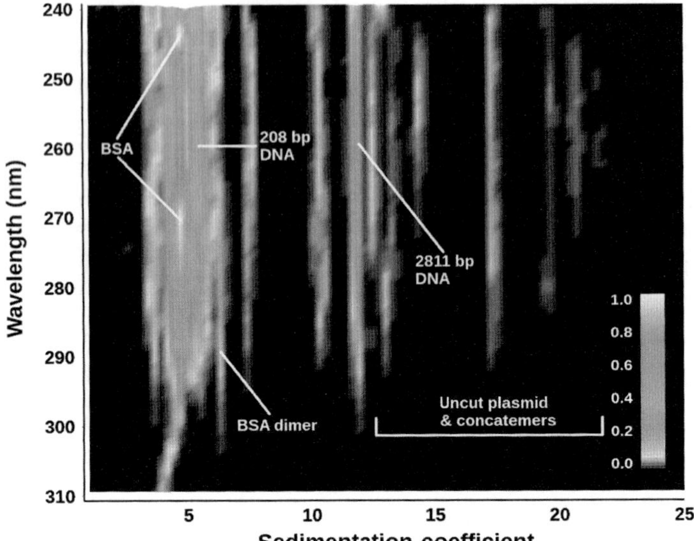

Figure 3 Projection view of the 2DSA-Monte Carlo sedimentation profile as a function of wavelength for the 50:50 DNA–BSA mixture. Remarkably, the protein absorbance spectrum at 4.3 S (two yellow peaks) can be clearly distinguished from the adjacent DNA peak with absorbance maximum around 258 nm, despite the proximity of the peaks (4.5 S vs. 5.2 S). Minor species can be identified based on their spectra. The straight lines attest to the high resolution and robustness of this approach to fit multiwavelength data (each wavelength is separately analyzed). (See the color plate.)

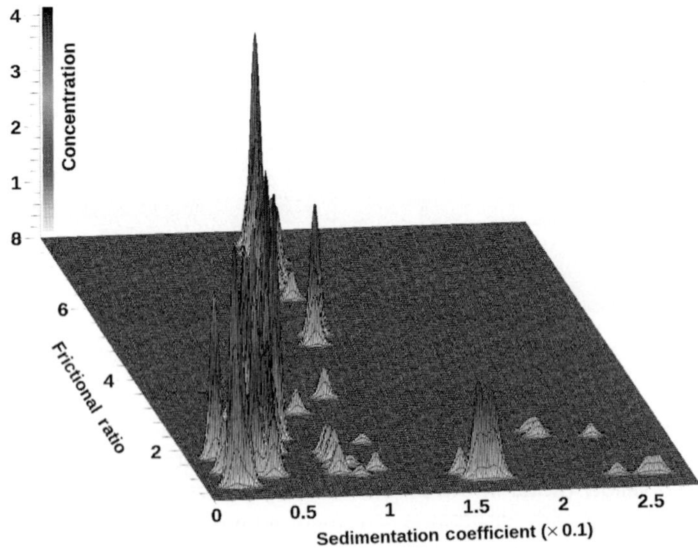

Figure 4 3D plot of a globally combined s vs. f/f_0 distribution for a 50:50 mixture of BSA and DNA fragments. Data collected on a Beckman-Coulter XL-A, at 258 and 278 nm. Data from three repeat experiments and from each wavelength were combined to generate this global view.

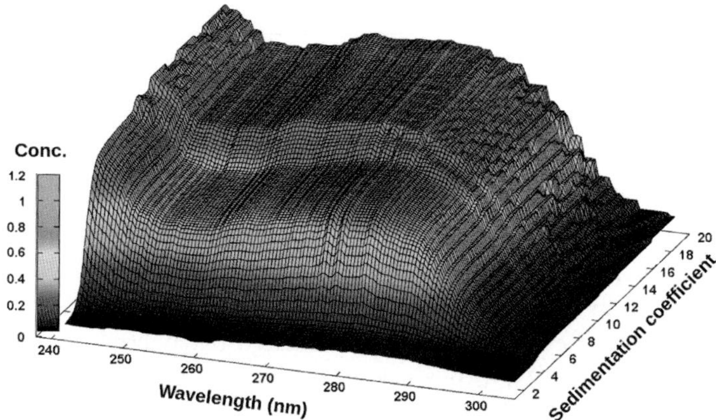

Figure 5 van Holde–Weischet integral distribution plots with boundary fractions scaled to the total concentration measured at each wavelength from the BSA–DNA 50:50 mixture. The shape of the BSA extinction profile is clearly visible at 4 S, and the DNA peak for the larger fragment shows maximum contribution around 258 nm at 11 S.

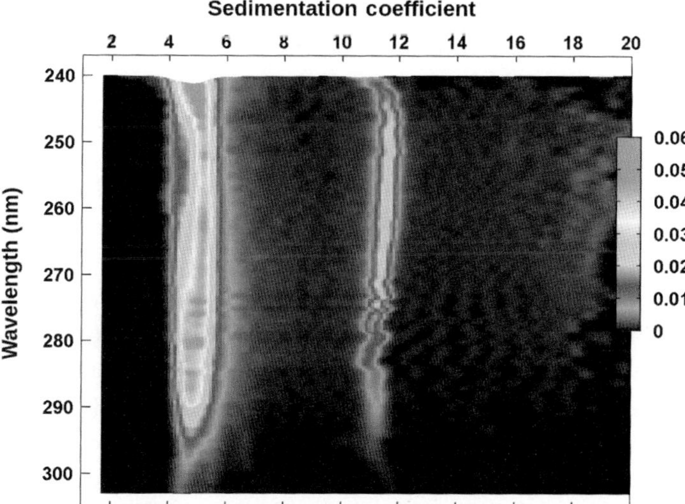

Figure 6 van Holde–Weischet differential distribution projection plot as a function of wavelength for the BSA–DNA 50:50 mixture. The asymmetric shape of the peak between 4 and 6 S clearly indicates a heterogeneity in spectral properties originating from the closely sedimenting BSA and 208 bp DNA peaks. On the left side of the peak, the BSA extinction dominates; on the right side of the peak, the DNA extinction profile dominates. The peak at 11 S displays a DNA extinction profile only. The color (different gray shades for the print version) gradient indicates relative concentration.

3. APPLICATIONS

To illustrate the capabilities of the MWL instrument and to compare MWL performance to traditional single-wavelength AUC, we characterized mixtures containing different ratios of DNA and protein (20:80, 35:65, 50:50, 65:35, and 80:20 on a per volume basis) and determined both the hydrodynamic properties of all components and identified the ratios of DNA and protein in each mixture from the AUC analysis. As a first step, the composition of each mixture was validated by spectral decomposition described in Section 2.4 using a benchtop spectrophotometer (see SI-3 (http://dx.doi.org/10.1016/bs.mie.2015.04.013)).

1. *Hydrodynamic characterization by traditional AUC:* After validating the composition via spectral decomposition, the same samples were measured by AUC in the Beckman-Coulter XL-A at 258 and 278 nm, at 20 °C, 28,000 rpm using epon-charcoal 2-channel centerpieces (Beckman-Coulter) to verify the protein:DNA ratios with an independent method. The data were analyzed with UltraScan according to the workflow described in Demeler (2010), and the relative amounts of protein and DNA were quantified by integration of global models built from the GA results (Brookes & Demeler, 2007). The resulting globally fitted data are shown in Fig. 4, and results from individual mixtures are presented as pseudo-3D plots in SI-7 (http://dx.doi.org/10.1016/bs.mie.2015.04.013) (right panels).
2. *Hydrodynamic and spectral characterization by MWL:* Next, the same mixtures were measured on the open AUC MWL instrument at the University of Konstanz, using identical conditions that were used in the Beckman-Coulter XL-A. The resulting data were analyzed both by parallel finite element analysis of each wavelength's velocity dataset by 2DSA-Monte Carlo and by spectral decomposition using the spectral components for BSA and DNA determined previously (see SI-3 (http://dx.doi.org/10.1016/bs.mie.2015.04.013)), with subsequent standard analysis according to the workflow described in Demeler (2010). The parallel analysis of all wavelengths resulted in well-separated hydrodynamic species, showing characteristic wavelength spectra for BSA and DNA (Figs. 2–4). A van Holde–Weischet analysis of the noise-corrected MWLD is shown in Figs. 5 and 6. The decomposition of MWLD from the six samples measured in the open AUC MWL instrument resulted in 12 datasets, 6 corresponding to the DNA and 6 corresponding to BSA contributions for each dataset. The decomposed data are shown on SI-9

(http://dx.doi.org/10.1016/bs.mie.2015.04.013). The 12 datasets obtained by the decomposition were analyzed as described in Demeler (2010) with refinement steps terminated by a global GA–Monte Carlo analysis for all 6 samples from each species. This resulted in two pseudo-3D plots which are shown in Fig. 7.

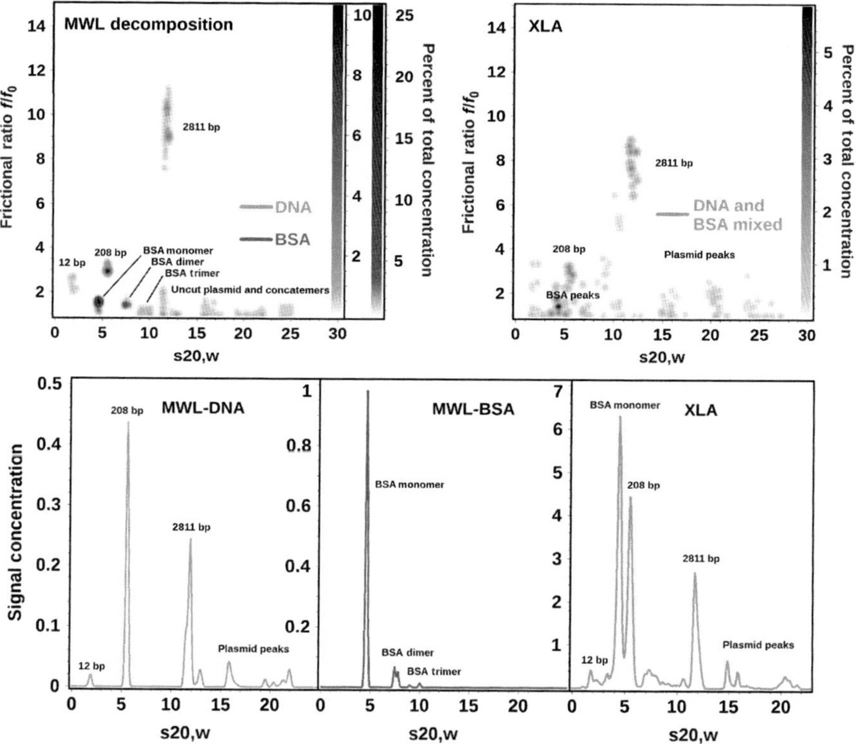

Figure 7 Global genetic algorithm Monte Carlo analysis of decomposition results obtained from six different DNA and BSA mixtures analyzed on the open AUC MWL instrument (top left) and the dual-wavelength results obtained from the Beckman-Coulter XL-A (right panel). The separate decomposition results for DNA (red) and BSA (blue) are combined in the left panel pseudo-3D plot to illustrate the exceptional separation achieved by spectral decomposition which even separates species with nearly identical sedimentation coefficients (the two major species sedimenting near 5 S). This approach demonstrates the superior resolution obtained from MWL analysis compared to the global two-wavelength analysis performed on the Beckman-Coulter XL-A (right panel, green). Lower panel: differential distributions from the same data shown above (red): decomposition for DNA, blue: decomposition for BSA, green: nonseparated XL-A data for DNA–BSA mixtures globally fitted to genetic algorithm–Monte Carlo analysis. A color representation of this figure is provided in the color plate section to allow the reader to distinguish the DNA and BSA contributions based on color. (See the color plate.)

3. *Statistical analysis:* To gain a better appreciation of the different noise levels present in these disparate datasets, we performed a statistical analysis of the fitting results as a function of total concentration. It is reasonable to expect different noise levels at different concentrations on different instruments; hence, we compared the root mean square deviations (RMSDs) obtained from the degenerate 2DSA analysis of the iterative refinement step, which produced random residuals in all cases. We plotted the RMSDs obtained for different concentrations from each of the six conditions: (a) Beckman-Coulter XL-A measurements at 258 and 278 nm, (b) open AUC MWL measurements at 258 and 278 nm, and (c) fits from the decomposition results for DNA and BSA as obtained from the open AUC MWL instrument. Here, the total concentration provides some measure of how much light passes through to the detector, and the RMSD level from the 2DSA fit provides a measure of residual noise. A plot of these data for each of the six conditions can be fitted to straight lines and is shown in SI-10 (http://dx.doi.org/10.1016/bs.mie.2015.04.013).
4. *Quantification of BSA and DNA:* Each experimental method was then used to estimate the known ratios of DNA to BSA from the experimental data. A summary plot of percentages observed in the dual-wavelength GA, the MWLD decomposition, and the spectrophotometer decomposition is shown in Fig. 8.

4. DISCUSSION

Our first result shows that the spectral profile from a dilution series of any chromophore or mixture of chromophores can be accurately parameterized by a linear combination of Gaussian terms that are globally fitted to the wavelength scans from all dilutions. The resulting intrinsic extinction profiles can be scaled to any concentration scale desired (e.g., optical density, mg/ml, molar concentration). Second, the intrinsic extinction profiles of known species in a mixture can then be used as orthogonal basis vectors in a nonnegatively constrained least squares fit of any unknown wavelength scan to identify the precise quantity of each contributing species. A decomposition of spectra measured from carefully prepared mixtures of DNA and protein resulted in an average error of only 0.7% in the expected concentrations of the deconvoluted component spectra, proving that the NNLS decomposition is both precise and robust. We suspect the residual error is likely due to pipetting error, and not instrument or algorithm related. These results demonstrate that different ratios of a mixture of protein and nucleic acid can be accurately resolved by spectral decomposition, even

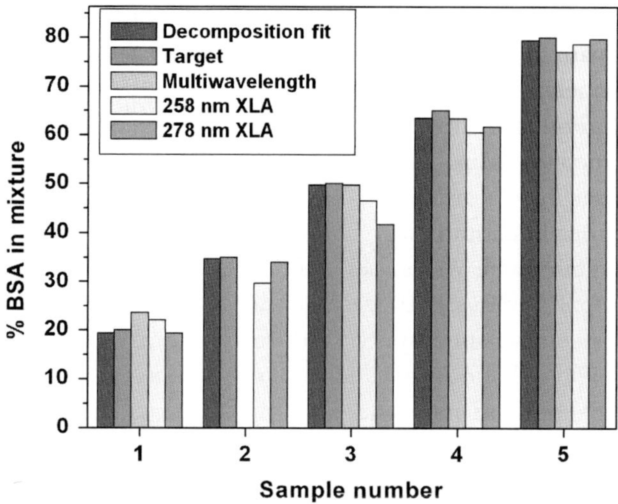

Figure 8 Comparison of results from the dual-wavelength AUC/GA analysis with the spectral decomposition method to ascertain the relative composition of spectrally different components in mixtures. The results clearly show that the spectral decomposition achieves even higher fidelity in reproducing the known target values than single-wavelength AUC/GA. Samples 1–5 refer to ratios of 20%, 35%, 50%, 65%, and 80% BSA in a BSA:DNA mixture with average errors of around 0.7%. Errors observed from the decomposition MWLD analysis results are on par with the spectral decompositions of the Genesys spectrophotometer data and are at least as good as the data obtained from the Beckman-Coulter XL-A.

though their spectra show significant overlap (SI-3 (http://dx.doi.org/10.1016/bs.mie.2015.04.013); Fig. 8).

AUC analysis of SV experiments provides the concentration distribution of any two hydrodynamic parameters (Demeler et al., 2014), which may be molar mass, anisotropy (frictional ratio), frictional coefficient, hydrodynamic radius, partial specific volume, sedimentation, and diffusion coefficient. When spectral information is added, the information from multiple wavelengths (up to 800 wavelengths per sample, depending on diffraction grating used) significantly increases the amount of data to be analyzed. We presented different approaches for the analysis of MWLD, depending on the availability of known spectral properties for all constituents. Even when spectral properties of all constituents are not available, impressive new detail can be obtained compared to the single-wavelength Beckman-Coulter XL-A. When analyzed by the parallel submission mechanisms implemented in UltraScan, even large numbers of wavelengths can be processed in a matter of a few minutes in batch mode on large parallel supercomputers available through XSEDE. High-resolution 2DSA modeling of

MWLD provides exceptional detail (for example, Fig. 3) by resolving both the hydrodynamic and the spectral domain, even for complex mixtures with many components. But also lower resolution methods like the van Holde–Weischet analysis can already provide valuable detail for hydrodynamic properties of components and resolve these components based on spectral differences (see Figs. 5 and 6).

Even higher resolution can be obtained when spectral properties of all components can be decomposed into separate, traditional 3D datasets (see SI-9 (http://dx.doi.org/10.1016/bs.mie.2015.04.013)). In this case, the NNLS fit of the wavelength spectra reduces the number of triples to be analyzed from the number of wavelengths to the number of constituents. These datasets can be analyzed independently (Fig. 7), providing unrivaled resolution, even of components that have nearly the same sedimentation coefficient as long as they are spectrally diverse. As can be seen in Fig. 7, essentially 100% separation between BSA and DNA is achieved by our method, imparting exceptional resolving power to optically diverse mixtures. This resolution is achieved despite significant overlaps in the absorbance spectra from BSA and DNA. It should be emphasized that theoretically many more than two spectrally distinct components could be separated, limited only by the extinction difference between individual components. It should be noted that the difference in extinction properties can be experimentally enhanced by labeling molecules with similar absorbance spectra with different chromophores or fluorophores. We predict that the impact of this new capability will be important for the analysis of complex mixtures such as nucleic acid–protein assemblies, e.g., chromatin, ribosomes, DNA- or RNA-binding proteins, heme proteins, and any number of assemblies that can be labeled with unique chromophores. The improvement in data quality is evident when decomposed data are fitted. As is shown in SI-10 (http://dx.doi.org/10.1016/bs.mie.2015.04.013), the RMSD levels from decomposed DNA data are significantly lower than the RMSD levels from XL-A data, but also from individual wavelengths from the open AUC MWL detector. The improvement in data quality is due to the averaging that occurs when datasets acquired at multiple wavelengths are decomposed into a single dataset. This reduced noise level is critically important to improve hydrodynamic resolution. As Fig. 7 shows, the data obtained from the Beckman-Coulter XL-A and from the MWLD show similar species, except in the MWLD decomposition process, these species are much better resolved hydrodynamically.

5. CONCLUSION

In this study, we have shown that analysis of multiwavelength data is superior to that of traditional single-wavelength data from AUC. Even if no spectral information is known about the components prior to the analysis, a standard whole boundary modeling approach like 2DSA coupled with Monte Carlo analysis or the enhanced van Holde–Weischet analysis reveals important information about the system under investigation which is complemented by the spectral information. By taking advantage of the earlier determined extinction spectra of DNA and BSA, the BSA–DNA mixtures investigated here could be decomposed, and resolved into all major components, even if their sedimentation rates were very similar, achieving 100% separation between DNA and BSA. By separating signals from distinct spectral species into individual datasets by spectral decomposition, the hydrodynamic resolution is also enhanced, and the characterization is greatly improved. The information content of such evaluation is superior to existing methods, even with the most sophisticated evaluation methods. We have demonstrated this case for molecules with two different base spectra, but this capability can be readily extended to more components and will be the subject of future work.

While we have seen the dramatic improvements in data quality in the current second-generation MWL systems, which now provide sufficient signal in the UV down to 240 nm, there is still room for improvement, and fiber instability issues need to be overcome. Yet, these results demonstrate clearly the important potential of multiwavelength detection. Further improvements in data quality are expected in the near future when third-generation MWA systems like the CFA (SpinAnalytical) become available, which no longer rely on fibers.

We predict that MWLD of AUC data will prove to be a very useful technique in a wide range of applications where complex mixtures with spectral diversity need to be examined. The multiwavelength approach promises much improved resolution for the study of multicomponent assemblies and hetero-associating systems, and importantly, can resolve molecular weight ambiguities. Specific chromophore labeling will pave the way to the analysis of complex mixtures, where multiwavelength detection will be able to resolve a larger number of compounds, by far exceeding the capabilities of traditional AUC.

ACKNOWLEDGMENTS

We acknowledge the invaluable contributions from our colleagues at the Texas Advanced Computing Center, in particular Chris Hempel, who have provided important assistance with the use of XSEDE supercomputing resources. We also would like to thank the Science Gateway group at Indiana University (Marlon Pierce, Raminder Singh, and Suresh Marru) for their help with implementing the parallel submissions. This research was in part supported by NSF Grant ACI-1339649 to B.D., and computer time on XSEDE resources was funded by NSF allocation Grant TG-MCB070039N to B.D. J.P. and H.C. acknowledge financial support by the Center for Applied Photonics (CAP) at the University of Konstanz.

REFERENCES

Bhattacharyya, S. K., Maciejewska, P., Börger, L., Stadler, M., Gülsün, A. M., Cicek, H. B., et al. (2006). Development of fast fiber based UV-Vis multiwavelength detector for an ultracentrifuge. *Progress in Colloid and Polymer Science*, *131*, 9–22.

Brookes, E., Boppana, R. V., & Demeler, B. (2006). Computing large sparse multivariate optimization problems with an application in biophysics. In *Supercomputing '06 ACM 0-7695-2700-0/06*.

Brookes, E., Cao, W., & Demeler, B. (2010). A two-dimensional spectrum analysis for sedimentation velocity experiments of mixtures with heterogeneity in molecular weight and shape. *European Biophysics Journal*, *39*(3), 405–414.

Brookes, E., & Demeler, B. (2006). Genetic algorithm optimization for obtaining accurate molecular weight distributions from sedimentation velocity experiments. In C. Wandrey & H. Cölfen (Eds.), *Progress in Colloid and Polymer Science: 131. Analytical ultracentrifugation VIII* (pp. 33–40): Berlin Heidelberg: Springer-Verlag. http://dx.doi.org/10.1007/2882_004.

Brookes, E., & Demeler, B. (2007). Parsimonious regularization using genetic algorithms applied to the analysis of analytical ultracentrifugation experiments. In *GECCO proceedings ACM 978-1-59593-697-4/07/0007*.

Brookes, E., & Demeler, B. (2008). Parallel computational techniques for the analysis of sedimentation velocity experiments in UltraScan. *Colloid & Polymer Science*, *286*(2), 138–148.

Cao, W., & Demeler, B. (2005). Modeling analytical ultracentrifugation experiments with an adaptive space-time finite element solution of the Lamm equation. *Biophysical Journal*, *89*(3), 1589–1602.

Cao, W., & Demeler, B. (2008). Modeling analytical ultracentrifugation experiments with an adaptive space-time finite element solution for multi-component reacting systems. *Biophysical Journal*, *95*(1), 54–65.

Cölfen, H., Laue, T. M., Wohlleben, W., Schilling, K., Karabudak, E., Langhorst, B. W., et al. (2010). The open AUC project. *European Biophysics Journal*, *39*(3), 347–359.

Correia, J. J., & Stafford, W. F. (2015). Sedimentation velocity: A classical perspective. In J. L. Cole (Ed.), *Analytical ultracentrifugation*. (pp. 49–80). Amsterdam: Elsevier.

Demeler, B. (2010). Methods for the design and analysis of sedimentation velocity and sedimentation equilibrium experiments with proteins. *Current Protocols in Protein Science*, 60:7.13.1–7.13.24.

Demeler, B., & Brookes, E. (2008). Monte Carlo analysis of sedimentation experiments. *Colloid & Polymer Science*, *286*(2), 129–137.

Demeler, B., Brookes, E., & Nagel-Steger, L. (2009). Analysis of heterogeneity in molecular weight and shape by analytical ultracentrifugation using parallel distributed computing. *Methods in Enzymology*, *454*, 87–113.

Demeler, B., Brookes, E., Wang, R., Schirf, V., & Kim, C. A. (2010). Characterization of reversible associations by sedimentation velocity with UltraScan. *Macromolecular Bioscience, 10*(7), 775–782, PMID: 20486142.

Demeler, B., & Gorbet, G. (2015). Analytical ultracentrifugation data analysis with UltraScan-III. In S. Uchiyama, W. F. Stafford, & T. Laue (Eds.), *Analytical ultracentrifugation: Instrumentation, software, and applications*: Springer, (to appear in Dec. 2015).

Demeler, B., Gorbet, G., Zollars, D., & Dubbs, B. (2015). *UltraScan-III version 3.3: A comprehensive data analysis software package for analytical ultracentrifugation experiments.* http://www.utrascan3.uthscsa.edu/.

Demeler, B., Nguyen, T. L., Gorbet, G. E., Schirf, V., Brookes, E. H., Mulvaney, P., et al. (2014). Characterization of size, anisotropy, and density heterogeneity of nanoparticles by sedimentation velocity. *Analytical Chemistry, 86*(15), 7688–7695.

Demeler, B., & van Holde, K. E. (2004). Sedimentation velocity analysis of highly heterogeneous systems. *Analytical Biochemistry, 335*(2), 279–288.

Gorbet, G., Devlin, T., Hernandez Uribe, B., Demeler, A. K., Lindsey, Z., Ganji, S., et al. (2014). A parametrically constrained optimization method for fitting sedimentation velocity experiments. *Biophysical Journal, 106*(8), 1741–1750. http://dx.doi.org/10.1016/j.bpj.2014.02.022.

Lawson, C. L., & Hanson, R. J. (1974). *Solving least squares problems.* Englewood Cliffs, NJ: Prentice-Hall, Inc.

Memon, S., Riedel, M., Janetzko, F., Demeler, B., Gorbet, G., Marru, S., et al. (2014). Advancements of the UltraScan scientific gateway for open standards-based cyberinfrastructures. *Concurrency and Computation: Practice & Experience, 26*(13), 2280–2291. http://dx.doi.org/10.1002/cpe.3251, Wiley.

Open AUC standard for analytical ultracentrifugation data storage used in UltraScan-III: http://wiki.bcf2.uthscsa.edu/ultrascan3/wiki/Us3Formats. 2015.

Pearson, J., Krause, J., Haffke, D., Demeler, B., Schilling, K., & Cölfen, H. Next generation AUC adds a spectral dimension: Development of multiwavelength detectors for the analytical ultracentrifuge.

Pierce, M., Marru, S., Demeler, B., Singh, R., & Gorbet, G. (2014). The Apache Airavata application programming interface: Overview and evaluation with the UltraScan science gateway. In *Proceedings of the 9th gateway computing environments workshop (GCE '14)* (pp. 25–29). Piscataway, NJ: IEEE Press. http://dx.doi.org/10.1109/GCE.2014.15.

Schuck, P., & Demeler, B. (1999). Direct sedimentation boundary analysis of interference optical data in analytical ultracentrifugation. *Biophysical Journal, 76*, 2288–2296.

Strauss, H. M., Karabudak, E., Bhattacharyya, S., Kretzschmar, A., Wohlleben, W., & Cölfen, H. (2008). Performance of a fast fiber based UV/Vis multiwavelength detector for the analytical ultracentrifuge. *Colloid and Polymer Science, 286*, 121–128.

The Open MPI Project (2004-2015). A high performance message passing library. http://www.open-mpi.org/community/license.php.

Walter, J., Löhr, K., Karabudak, E., Reis, W., Mikhael, J., Peukert, W., et al. (2014). Multidimensional analysis of nanoparticles with highly disperse properties using multiwavelength analytical ultracentrifugation. *ACS Nano, 8*, 8871–8886.

Walter, J., Sherwood, P. J., Lin, W., Segets, D., Stafford, W. F., & Peukert, W. (2015). Simultaneous analysis of hydrodynamic and optical properties using analytical ultracentrifugation equipped with multi-wavelength detection. *Analytical Chemistry, 87*(6), 3396–3403. http://dx.doi.org/10.1021/ac504649c.

CHAPTER THREE

Sedimentation Velocity: A Classical Perspective

John J. Correia*, Walter F. Stafford[†,1]
*Department of Biochemistry, University of Mississippi Medical Center, Jackson, Mississippi, USA
[†]Department of Systems Biology, Boston Biomedical Research Institute and Harvard Medical School, Boston, Massachusetts, USA
[1]Corresponding author: e-mail address: wstafford3@walterstafford.com

Contents

Abstract

Here we give an overview of the history of sedimentation velocity analysis focusing on seminal and fundamental contributions that derived from early ultracentrifugation studies. We introduce the concepts of nonequilibrium thermodynamics and outline the derivation of the Svedberg and the Lamm equations and the requirements for including both hydrodynamic and thermodynamic nonideality. We introduce the phenomenological equations for coupled flows as developed from the principles of nonequilibrium or irreversible thermodynamics and derive a form of the Lamm equation that incorporates cross-diffusion coefficients and coupled gradient terms. We give an historical overview of solutions to the Lamm equation including Fujita–MacCosham solutions and Claverie finite-element numerical solutions and discuss the software that have implemented these solutions. We discuss the three major optical systems (absorbance, interference, and fluorescence) and recently developed multiwavelength systems. We also suggest a number of experimental practices and guidelines for optimizing the determination of *s* and *D* and discuss the appropriate centerpiece components and

Methods in Enzymology, Volume 562
ISSN 0076-6879
http://dx.doi.org/10.1016/bs.mie.2015.06.042

their utility. This chapter complements other recent reviews submitted by the authors (Correia, Lyons, Sherwood, & Stafford, 2015; Stafford, 2015) and should be considered an effort to revive the importance of irreversible thermodynamics in the understanding and analysis of sedimentation velocity ultracentrifugation data.

1. EARLY HISTORY

The fields of centrifugation and protein science had their birth in 1924 in Uppsala, Sweden, where Theodore (The) Svedberg built the first analytical ultracentrifuge, as described in "The ultracentrifuge," by Svedberg and Pederson (1940) also reviewed in Van Holde and Hansen (1998). This first instrument utilized a hydrogen atmosphere to reduce heating, and sector shaped cells to avoid convection. Svedberg had directed his interests from gold sols to proteins and performed sedimentation equilibrium experiments that led to the concepts of tertiary and quaternary structure. The observation was that proteins are not suspensions of colloidal material but rather homogeneous materials with well determined and large molecular weights. For example, Svedberg and Nichols (1926) showed that purified egg albumin was homogeneous with a molecular weight of 34,000 g/mol establishing the concept of a tertiary structure. The first experiment on hemoglobin (Svedberg & Fahraeus, 1926) and hemocyanins Svedberg & Heyroth, (1929a, 1929b) was consistent with multisubunit complexes and suggested the concept of quaternary structure. To verify the sedimentation equilibrium results on hemoglobin and egg albumin, an oil turbine ultracentrifuge was developed to achieve speeds sufficient to perform the first sedimentation velocity runs (Svedberg & Nichols, 1927). These initial sedimentation-diffusion experiments verified a hemoglobin mass identical to the 68,000 g/mol measured by equilibrium and confirmed the presence of a single diffusing component. Significant theoretical contributions to our understanding of the sedimentation velocity technique were also made around the same time by Lamm (1929; a student of Svedberg's) and Faxen (1929; see Faxén solution to the Lamm equation below). Svedberg and his collaborators laid the foundation for analytical ultracentrifugation and in general, macromolecular transport techniques that allowed the determination of size, shape, density, and the effect of charge (Svedberg & Pederson, 1940; Tiselius, 1937), thus establishing a practical and philosophical basis for protein science and molecular biology. Svedberg was awarded the Noble prize in 1926 for his work on disperse or colloidal systems, although his Noble lecture extensively discusses the details of these early ultracentrifuge experiments (Svedberg, 1926).

During the 1930s and 1940s, ultracentrifuge experiments were primarily performed in instruments based on the Svedberg oil turbine design. Many of these experiments used sedimentation velocity methods to investigate the behavior of readily available proteins from eggs, serum, and plasma and led to the discovery of a boundary anomaly known as the Johnston–Ogston effect (Harrington & Schachman, 1953; Johnston & Ogston, 1946; McFarlane, 1935; Ogston, 1937; Trautman, Schumaker, Harrington, & Schachman, 1954). This era coincided with the development of new methods focused on amino acid, peptide, and protein research (Cohn & Edsall, 1943). During World War II, the national plasma fractionation program was a central area of research that drove the large-scale fractionation of protein products including human albumin as a "plasma extender" for transfusions (Simoni, Hill, & Vaughan, 2002). This applied research also produced valuable fundamental knowledge of protein solubility properties and techniques for protein fractionation and understanding macromolecular hydrodynamics. The schlieren optical system was also developed during this time by Philpot (1938), Svensson (1939, 1940), and Longsworth (1939); for a full history, see Schachman (1959).

The Spinco Model E analytical ultracentrifuge, introduced in 1950, descended from an air turbine ultracentrifuge design by Beams and Pickels (1935). This allowed the development of the Rayleigh interferometric optical system (Richards & Schachman, 1959) and the photoelectric scanning absorbance optical system, also in Schachman's laboratory (Hanlon, Lamers, Lauterbach, Johnson, & Schachman, 1962). This paralleled the development of new methodology like short column techniques for sedimentation equilibrium (Van Holde & Baldwin, 1958; Yphantis, 1960) and density gradient equilibrium (Meselson, Stahl, & Vinograd, 1957). The latter led directly to the discovery of fundamental concepts like semiconservative replication of DNA (Meselson & Stahl, 1958). Velocity methods were also essential in the discovery of 80S yeast ribosomes (Chao & Schachman, 1956) and the relationship between 30S and 50S subunits and the 70S ribosome (Tissieres & Watson, 1958). The necessity of using a second moment position for sedimentation velocity measurements was first rigorously derived during this time (Goldberg, 1953). Baldwin also developed a data analysis approach for generating a distribution of sedimentation coefficients and extrapolating them to infinite time and zero concentration (Baldwin, 1954), thus preceding the Stafford DCDT method by four decades (Stafford, 1992). This era also utilized sedimentation velocity analytical banding techniques to discover DNA supercoiling and the impact of ethidium bromide titrations on plasmid structure (Bauer & Vinograd, 1968; Vinograd, Bruner, Kent, & Weigle, 1963; Vinograd, Lebowitz, Radloff, Watson, & Laipis, 1965). It is worth noting that these velocity experiments look at changes in shape due to super coiling and uncoiling in the absence of significant changes in mass. Likewise difference sedimentation velocity was also useful in identifying allosteric conformational changes in ATCase by substrate binding (Kirschner & Schachman, 1971a, 1971b, 1973a, 1973b; Richards & Schachman, 1957). It was also during this era when Schumaker and Lindgren developed flotation sedimentation techniques to look at the mass and polydispersity of lipoproteins from human subjects and as a function of diet. This is an excellent early example of the application of biophysical hydrodynamic methods to translational research (Adams & Schumaker, 1969a, 1969b; Anderson, Nichols, Forte, & Lindgren, 1977; Gofman, Lindgren, & Elliott, 1949; Talwinder, Kahlon, Glines, & Lindgren, 1986). Sedimentation velocity was also extremely useful for understanding the mechanism of virus assembly (Phillips, 1969; Phillips, Summers, & Maizel, 1968; Shire, Steckert, Adams, & Schuster, 1979; Shire, Steckert, & Schuster, 1979).

This brief and incomplete early history of the seminal importance of analytical ultracentrifugation, and especially sedimentation velocity methods in the discovery of fundamental principles of biochemistry and molecular biology, coincided with the development of a rigorous theory of biophysical transport processes, which is the focus of this review.

2. IRREVERSIBLE THERMODYNAMICS

Sedimentation velocity theory is firmly based on the principles of irreversible thermodynamics, the validity of which was established in the mid-nineteenth century in terms of conjugate flows and forces (De Groot & Masur, 1962). This chapter is concerned with the development and theory of sedimentation velocity in an analytical ultracentrifuge. We will outline the derivation of the Svedberg and Lamm equations for ideal solutions beginning with force balance, equations for flows, and inserting the flow equation into the continuity equation. This will be expanded to include binary solutions, ternary solutions, and the complexity of coupled flows required to understand and interpret the sedimentation and diffusion of mixtures of nonideal and interacting macromolecules. The theory of the irreversible or nonequilibrium thermodynamics of diffusion and sedimentation is rigorously described in Gosting (1956), Williams, vanHolde, Baldwin, and Fujita (1958), Fujita (1962), De Groot and Masur (1962), and Katchalsky and Curran (1965).

3. SEDIMENTATION VELOCITY

Sedimentation velocity analytical ultracentrifugation is a fundamental hydrodynamic method that measures the rate of sedimentation of macromolecular species under the influence of a high centrifugal field. It has the advantage of being a measurement made in free solution (no matrices or surface effects) under native and biochemically relevant solution conditions. As we will outline below, sedimentation velocity is based upon a rigorous hydrodynamic and thermodynamic theory. Sedimentation velocity allows the determination of purity or homogeneity of a macromolecular species (see the chapter by Arthur, Kendrick, & Gabrielson, 2015 in this volume). If there are contaminants or aggregates, these can often be quantified by velocity methods down to 0.1–1%. For example, Fig. 1 shows a simulation of a sedimentation velocity experiments for a 6S monomer with 50% contaminant of a 9S irreversible dimer. From the sedimentation and diffusion coefficient, the so-called s/D method, the mass or buoyant molecular weight

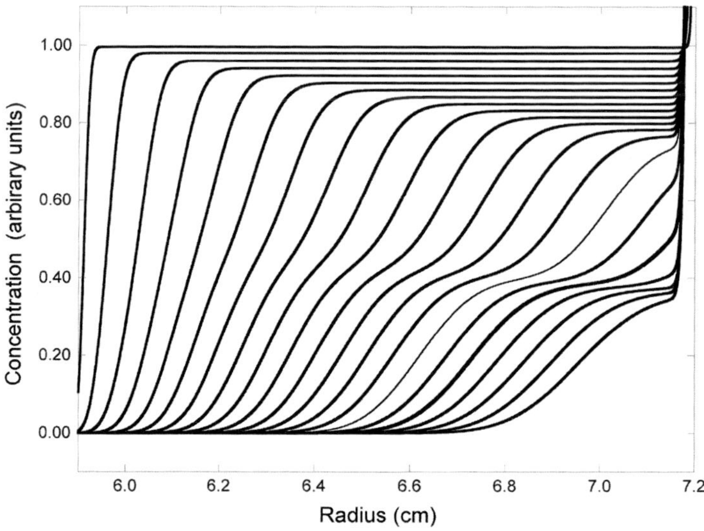

Figure 1 Sedimentation velocity simulation of a noise-free boundary for a 100,000 6S species and 50% contaminant of a 200,000 9S dimeric species.

can be determined. Given the molecular weight, measurements of s and D each allow the frictional coefficient or shape information to be determined (see the chapter by Chaton & Herr, 2015 in this volume) including detailed comparisons to PDB files and crystal structures (see the chapter by Rocco & Byron, 2015 in this volume). If the sedimentation coefficient is concentration dependent, then the association mechanism, affinity, and thermodynamic properties of those reactions can all be determined by sedimentation velocity methods (see the discussion below and the chapter by Lin & Lucius, 2015 in this volume). There are decades of published applications in basic biology, biochemistry, molecular biology, and polymer sciences that demonstrate the utility of sedimentation velocity. In the last 40 years, analytical ultracentrifugation has also become a central method in the biotech industry with applications to polymer development and therapeutic antibodies (see chapters by Hill & Laue, 2015; Liu, Yadav, Andya, Demeule, & Shire, 2015 in this volume). Here, we present a review of the early history of sedimentation velocity analytical ultracentrifugation with derivations of all the relevant equations including the Svedberg and the Lamm equation and the inclusion of hydrodynamic and thermodynamic nonideality. The inquiring reader should also refer to other recent reviews of the field (Berkowitz, 2006; Brautigam, 2011; Cole, Correia, & Stafford, 2011; Cole, Lary, Moody, & Laue, 2008; Howlett, Minton, & Rivas, 2006; Laue & Stafford, 1999; Lebowitz, Lewis, & Schuck, 2002; Stafford, 2015).

4. BALANCED FORCES IN DIFFUSION

Diffusion is a transport process that can be treated as a mechanical balanced force process or as a classical nonequilibrium thermodynamic process. A mechanical treatment describes a steady-state velocity, v_i, where the total force on the molecule is zero

$$f_i v_i + F_i = 0 \tag{1}$$

and the frictional coefficient f_i can be related to size and shape, for example, by Stokes law where η is the viscosity and R_s is the so-called Stokes radius which is the radius of the corresponding hydrodynamically equivalent sphere,

$$f_i = 6\pi\eta R_s \tag{2}$$

The theory of irreversible processes expresses the flow, J_i, at any point in the system as being proportional to the gradient of an appropriate potential

$$J_i = -L_i(\partial U_i/\partial r) \tag{3}$$

where U_i is a generalized potential (temperature for heat conduction, chemical potential for diffusion, or centrifugal potential for sedimentation) and L_i is a generalized conductivity (or diffusional mobility or sedimentation velocity). Assuming conservation of mass and treating flow of a concentration through a volume element per unit time as net mass in minus net mass out across a surface, one derives the differential continuity equation

$$\frac{\partial c_i}{\partial t} = -\frac{\partial J_i}{\partial x} \tag{4}$$

which states the general relationship between a change in mass concentration, c_i, per unit time and the flow into or out of a region or volume element.

To apply this approach to diffusion, one must experimentally set up a sharp boundary and then observe the concentration attempt to achieve equilibrium by establishing a constant value across the system (Gosting, 1956). Thermodynamically, this means that the gradient or change in chemical potential with distance in the boundary is the driving force for the diffusional flow of solute, giving rise to

$$J_2 = -L_2\frac{\partial \mu_2}{\partial x} \tag{5}$$

where chemical potential μ_2 has the usual meaning

$$\mu_2 = \mu_2^{\circ} + RT\ln(\gamma_2 c_2) \tag{6}$$

Taking the derivative gives

$$(\partial \mu_2/\partial r) = (\partial \mu_2/\partial c_2)(\partial c_2/\partial r) \tag{7}$$

and thus

$$J_2 = -\frac{L_2 RT}{c_2}\left[1 + c_2\left(\frac{\partial \ln(\gamma_2)}{\partial c_2}\right)\right]\left(\frac{\partial c_2}{\partial r}\right) \tag{8}$$

This gives rise to Fick's first law

$$J_2 = -D_2\left(\frac{\partial c_2}{\partial r}\right) \tag{9}$$

where the diffusion coefficient is defined as

$$D_2 = \frac{RT}{Nf_2}\left[1 + c_2\left(\frac{\partial \ln(\gamma_2)}{\partial c_2}\right)\right] \tag{10}$$

We replaced L_2/c_2 with $1/Nf_2$, where N is Avogadro's number, f_2 is the frictional coefficient, and D_2 is the diffusion coefficient of component 2 (where component 1 is solvent). The mechanical connection is that f_2 and thus D_2 are functions of size and shape. For ideal solutions for which $\gamma_2 = 1$, D_2 is RT/Nf_2 consistent with the Stokes–Einstein equation as derived from either a kinetic or a Brownian motion approach (Berg, 1993; Einstein, 1905; Gosting, 1956). Here we use two-component numbering where 1 is solvent and 2 is solute. Below we alter this nomenclature (based upon Fujita, 1962; Williams et al., 1958) to correspond to numbers of solutes or macromolecular species where solvent is 0 and macromolecules are 1, 2, and 3 for a ternary system.

5. BALANCED FORCES IN SEDIMENTATION

Consider a particle suspended in solvent in the ultracentrifuge at rest and then acted upon by centrifugal force. By Newton's first law, it will initially be at rest and remain that way until acted upon by an external force. When the centrifugal force is applied, it will start to accelerate according to Newton's second law, $F = m(\mathrm{d}v/\mathrm{d}t)$. As it accelerates the acceleration is counteracted by the frictional force of the solvent which is proportional to its velocity, $f = f_o(\mathrm{d}r/\mathrm{d}t)$. As the particle continues to accelerate, its velocity will increase until the frictional force, f, is just exactly balanced by the centrifugal force, F. At this point, after a few nanoseconds of acceleration, with the forces in balance, the net force on the particle will become zero and the particle will sediment according to Newton's first law (Fig. 2), moving at essentially constant velocity over short time periods. In the cylindrical coordinate system of the centrifuge, the centrifugal acceleration is given by

$$a = \omega^2 r \tag{11}$$

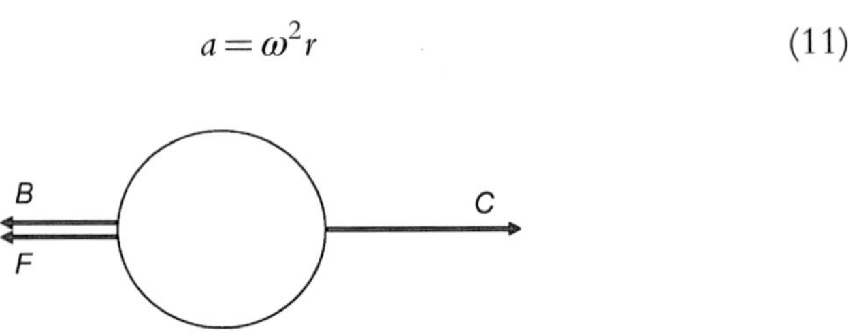

Figure 2 Depiction of the forces on a molecule under a centrifugal field, where C is centrifugal force, B is buoyant force, and F is frictional drag as described in Section 4.

By Newton's second law, the centrifugal force is given by

$$F_i = M_p\omega^2 r - M_s\omega^2 r = M_p\omega^2 r\left(1 - v_p\rho_s\right) \tag{12}$$

where M_p is the molar mass of the particle, M_s is the molar mass of the solvent, $\bar{v}$ is the partial specific volume of the particle, and ρ is the density of the solvent.

The total centrifugal force is given by the contribution from the mass of the particle minus the mass of water displaced by the particle according to Archimedes' buoyancy principle. The terminal velocity of the particle is achieved when the centrifugal force is balanced by the viscous drag on the particle, i.e., when $F_i = f\,\mathrm{d}r/\mathrm{d}t$, and under that condition, we can write

$$M(1 - \bar{v}\rho)\omega^2 r = f\left(\frac{\mathrm{d}r}{\mathrm{d}t}\right) \tag{13}$$

From the Stokes–Einstein equation, we can write that $f = RT/D$ and substituting gives

$$M(1 - \bar{v}\rho)\omega^2 r = \frac{RT}{D}\left(\frac{\mathrm{d}r}{\mathrm{d}t}\right) \tag{14}$$

The operational definition of the sedimentation coefficient is the velocity divided by the centrifugal field strength:

$$s = \frac{1}{\omega^2 r}\frac{\mathrm{d}r}{\mathrm{d}t} \tag{15}$$

Equation (15) implies that the slope of a plot of the logarithm of the maximum in a schlieren peak or a plot of the logarithm of the second moment of the boundary position versus time will give $\omega^2 s$ (Goldberg, 1953):

$$\ln\left(r_b(t)/r_b(t_o)\right) = s\omega^2(t - t_o) \tag{16}$$

Substituting $\omega^2 sr$ above for $(\mathrm{d}r/\mathrm{d}t)$ allows us to write the Svedberg equation (for a two-component solvent and solute system), which relates the sedimentation coefficient, the diffusion coefficient, and the buoyant mass of the particle:

$$\frac{s}{D} = \frac{M(1 - \bar{v}\rho)}{RT} \tag{17}$$

6. DERIVATION OF THE LAMM EQUATION

Starting with the continuity equation, which is essentially a statement of the conservation of mass,

$$\left(\frac{\partial c}{\partial t}\right)_r = -\mathrm{Div}\left[J_{\mathrm{total}}\right] \tag{18}$$

we can also write that

$$J_{\mathrm{total}} = J_{\mathrm{sed}} + J_{\mathrm{dif}} \tag{19}$$

where

$$J_{\mathrm{sed}} = c\frac{\partial r}{\partial t} = c\omega^2 sr \tag{20}$$

and

$$J_{\mathrm{dif}} = -D\left(\frac{\mathrm{d}c}{\mathrm{d}r}\right) \tag{21}$$

and substituting gives

$$\left(\frac{\partial c}{\partial t}\right)_r = -\nabla\left[c\omega^2 sr^2 - Dr\left(\frac{\mathrm{d}c}{\mathrm{d}r}\right)_t\right] \tag{22}$$

In cylindrical coordinates (sedimentation takes place in a sector shaped cell), the divergence operator in the radial direction is

$$\nabla = \frac{\partial}{r\partial r} \tag{23}$$

and the above equation becomes the Lamm equation (Lamm, 1929)

$$\left(\frac{\partial c}{\partial t}\right)_r = -\frac{\partial}{r\partial r}\left[c\omega^2 sr^2 - \mathrm{D}r\left(\frac{\mathrm{d}c}{\mathrm{d}r}\right)_t\right] \tag{24}$$

7. HETEROGENEOUS PAUCIDISPERSE SYSTEMS

For multiple noninteracting species, we can write simultaneous Lamm equations

$$\sum\left(\frac{\partial c_i}{\partial t}\right)_r = -\sum\frac{\partial}{r\partial r}\left[c_i\omega^2 s_i r^2 - D_i r\left(\frac{\mathrm{d}c_i}{\mathrm{d}r}\right)_t\right] \tag{25}$$

As described in Section 10, many software packages treat multiple noninteracting systems in this manner (e.g., SVEDBERG, SEDANAL, and SEDFIT). However, this is a two-component approximation that treats the diffusion coefficient as either a self-diffusion or an apparent average diffusion coefficient. Using the nomenclature of flows developed above for a two macromolecular component system (where solvent is component zero), we get the phenomenological equations

$$J_1 = -L_{11}\frac{\partial\mu_1}{\partial x} - L_{12}\frac{\partial\mu_2}{\partial x} \tag{26}$$

$$J_2 = -L_{21}\frac{\partial\mu_1}{\partial x} - L_{22}\frac{\partial\mu_2}{\partial x} \tag{27}$$

It is these L_{ij} terms that obey Onsager's reciprocal relations, where by microscopic reversibility Onsager (1931a, 1931b) showed that $L_{ij} = L_{ji}$ (see also Miller, 1959 for experimental evidence). Substituting for chemical potentials and differentiating as above we get

$$J_1 = -D_{11}\frac{c_1}{RT}\frac{\partial\mu_1}{\partial r} - D_{12}\frac{c_2}{RT}\frac{\partial\mu_2}{\partial r} \tag{28}$$

$$J_2 = -D_{21}\frac{c_1}{RT}\frac{\partial\mu_1}{\partial r} - D_{22}\frac{c_2}{RT}\frac{\partial\mu_2}{\partial r} \tag{29}$$

This is Fick's equation for two-independent diffusional flows. Irreversible thermodynamics as expressed in these equations says that flow in a multicomponent system is coupled to the total gradient of the chemical potentials of all the components. The D_{ii} terms are self-diffusion coefficients and the D_{ij} terms are the cross-diffusion coefficients. Combining these two equations shows that $D_{12} \neq D_{21}$

$$D_{12} = L_{11}\frac{\partial\mu_1}{\partial c_2} + L_{12}\frac{\partial\mu_2}{\partial c_2} \tag{30}$$

$$D_{21} = L_{21}\frac{\partial\mu_1}{\partial c_1} + L_{22}\frac{\partial\mu_2}{\partial c_1} \tag{31}$$

This is in the format of Katchalsky and Curran (equation 9-26, 1975; where we neglect a similar term for coupled flow of sedimentation, equation 9-50). Coupled flows apply even in the absence of nonideality although the magnitude of the chemical potential gradient and thus the concentration gradient will dictate the importance of the cross terms on the total flows. It is also assumed that these cross terms are small (Gosting, 1956; Miller, 1959) although there is mostly experimental diffusion data on salts and very little data on macromolecules. A complete flow expression can be derived and inserted into the Lamm equation where for each species or component multiple diffusion terms are included

$$\sum\left(\frac{\partial c_i}{\partial t}\right)_r = -\sum\frac{\partial}{r\partial r}\left[c_i\omega^2 s_i r^2 - \sum_{j=1}^{n} D_{i,j} r\left(\frac{\mathrm{d}c_j}{\mathrm{d}r}\right)_t\right] \tag{32}$$

This is essentially equation (1.77) in Fujita (1962). This format has not been implemented in any modern software package, although Stafford (2015) recently derived these expressions for a ternary or three macromolecular component system. In general, AUC work is performed under dilute solution conditions and even for mixtures of components the ideal simultaneous solution of the Lamm equation works well. Efforts to investigate hydrodynamics of therapeutic and diagnostic proteins in serum are underway in many labs (Correia, Lyons, Sherwood, & Stafford, 2015; Kingsbury & Laue, 2011; Kingsbury, Laue, Chase, & Connors, 2011; Kroe & Laue, 2009) and biotech companies (see Hill & Laue, 2015; Liu et al., 2015 in this volume). Therefore, incorporation of these flow equations may be required for a complete understanding and interpretation of sedimentation velocity data in serum, cell, and tissue extracts and other high concentration systems.

8. NONIDEAL SYSTEMS

Nonideality can arise from two sources, hydrodynamic and thermodynamic. The molecular origin of nonideality is generally thought to be excluded volume and charge effects, although thermodynamic interpretation often requires the application of the theory of preferential interactions. In sedimentation velocity studies, charge effects are often described in terms of the primary charge effect (Pederson, 1958; Svedberg & Pederson, 1940) which emphasizes the need for sufficient salt concentrations to suppress primary charge effects. Primary charge effects can be attributed to the backflow caused by the larger ionic atmosphere of the protein at low ionic strength relative to that at higher ionic strengths. Fuoss and Onsager showed that the Stokes radius is effectively equal to the ionic radius at low ionic strength (Fuoss, 1959; Onsager & Fuoss, 1932). Although the primary charge effect has been attributed to the differences in sedimentation and diffusion of the protein and its counterions, they will migrate as a single electroneutral particle whose Stokes' radius is effectively equal to the radius of the ionic atmosphere. Hydrodynamic nonideality is manifest through the concentration dependence of the frictional coefficient, which is due to the backflow of solvent because of the closed cell (where c_i is mg/ml and K_s is in ml/mg units):

$$f_i = f_i^{\mathrm{o}}\left(1 + K_{s,i}c_i\right) \tag{33}$$

Thermodynamic nonideality arises through concentration dependence of the activity coefficient, y_i

$$\left(1 + \frac{\partial \ln(y_i)}{\partial \ln(c_i)}\right) \tag{34}$$

Consequently, nonideality from both sources is reflected in the concentration dependence of both the sedimentation and the diffusion coefficient

$$s_1 = \frac{s_1^{\mathrm{o}}}{(1 + K_s c_1)} \tag{35}$$

$$D_i = D_i^0 \frac{\left(1 + \frac{\partial \ln(y_i)}{\partial \ln(c_i)}\right)}{(1 + K_s c_i)} \tag{36}$$

Note the hydrodynamic terms are assumed here to be the same for both s and D (Goldberg, 1953). Previously, LaBar and Baldwin (1963) experimentally demonstrated that the friction coefficients (the $1 + K_s c$ term) determined for s and for D from sedimentation studies of sucrose are the same. Scott, Harding, and Winzor (2015) have recently suggested $K_s \neq K_D$ (when K_D is estimated from dynamic light scattering). However, it has been shown that it must be the case that $K_s = K_D$ by a proof recently published by Stafford (2015). The thermodynamic concentration dependence of the activity coefficient is often expressed as a power series called a virial expansion; for many of the systems we study, the expansion can be truncated after the linear term:

$$\left(1 + \frac{\partial \ln(y_i)}{\partial \ln(c_i)}\right) = (1 + 2B_i M_i c_i) \tag{37}$$

so that

$$D_i = D_i^0 \frac{(1 + 2B_i M_i c_i)}{(1 + K_s c)} \tag{38}$$

These expressions can be incorporated directly into the Lamm equation and numerically solved as a function of concentration along the centrifuge cell (see below).

For heterogeneous systems, the hydrodynamic and thermodynamic terms for species i are also dependent upon the concentrations of all the other species. This is a direct outcome of irreversible thermodynamics and the coupling between the various potential gradients including the chemical potential gradient for each species, $\partial\mu_i/\partial c_j$. This is reflected in the sedimentation coefficient as a denominator with a series of concentration terms

$$s_1 = \frac{s_1^{\mathrm{o}}}{(1 + K_{11}c_1 + K_{12}c_2 + K_{13}c_3 + \ldots)} \tag{39}$$

where K_{ii} is the self-nonideality term and K_{ij} terms are referred to as cross-term nonideality (Correia et al., 2015; Stafford, 2015). A similar equation can be written for the diffusion coefficient that involves both the hydrodynamic nonideality expansion in the denominator and the thermodynamic nonideality expansion in the numerator

$$D_1 = \frac{D_1^{\mathrm{o}}(1 + 2B_{11}M_1c_1 + 2B_{12}M_2c_2 + 2B_{13}M_3c_3 + \ldots)}{(1 + K_{11}c_1 + K_{12}c_2 + K_{13}c_3 + \ldots)} \tag{40}$$

As stated above, these expressions for $s(c)$ and $D(c)$, Eqs. (39) and (40), can be incorporated into the Lamm equation and numerically solved as a function of local concentration along the cell (see below). The empirical determination of K_{ij} and B_{ij} terms by pairwise interactions has recently been discussed by Correia et al. (2015). An attempt to fit directly for these terms in a ternary or more complex system quickly becomes intractable. Note however, as discussed above, simply adding these concentration-dependent terms for $s(c)$ and $D(c)$ may not be sufficient and multiple diffusion coefficients with concentration dependence may be required for some systems. This would give rise to flow equations of the form (Stafford, 2015)

$$\begin{aligned} J_1 = &-D_{11}\frac{c_1}{RT}\frac{\partial\mu_1}{\partial r}(1 + 2B_{11}M_1c_1 + 2B_{12}M_2c_2) \\ &-D_{12}\frac{c_2}{RT}\frac{\partial\mu_2}{\partial r}(1 + 2B_{21}M_1c_1 + 2B_{22}M_2c_2) \end{aligned} \tag{41}$$

where only thermodynamic nonideality is incorporated. If we included both hydrodynamic and thermodynamic nonideality and coupled flows, this would give rise to equations of the following form:

$$\Sigma\left(\frac{\partial c_i}{\partial t}\right)_r = -\Sigma\frac{\partial}{r\partial r}\left[c_i\omega^2 s_i r^2 - \sum_{j=1}^{n} D_{i,j} r\left(\frac{dc_j}{dr}\right)_t\right] \tag{42}$$

where explicit $s(c)$ and $D(c)$ terms would become

$$\sum_i^n\left(\frac{\partial c_i}{\partial t}\right)_r = -\sum_{i=1}^{n}\frac{\partial}{r\partial r}\left[c_i\omega^2 r^2 \frac{s_1^o}{(1+K_{11}c_1+K_{12}c_2+K_{13}c_3+\ldots)}\right.$$
$$\left.-\sum_{j=1}^{n}\frac{D_{i,j}^{o}(1+2B_{11}M_1c_1+2B_{12}M_2c_2+2B_{13}M_3c_3+\ldots)}{(1+K_{11}c_1+K_{12}c_2+K_{13}c_3+\ldots)}r\left(\frac{dc_j}{dr}\right)_t\right] \tag{43}$$

Devising experimental strategies to measure and test the need for all these parameters might include sedimentation velocity experiments as a function of pairwise interactions, sedimentation equilibrium experiments that measure B_{ij} terms, and measuring accurate heterogeneous concentration profiles by new multiwavelength absorbance systems (see Section 11). This last method has been implemented in Sedanal (Walter et al., 2015).

9. INTERACTING SYSTEMS

Interacting systems comprise either self-interactions between like particles to form oligomers or heterologous interactions between different particles to form complexes. Self-interactions are of the following type:

$$\begin{aligned}
&2A_1 = A_2K_{1,2}\\
&A_2 + A_1 = A_3K_{2,3}\\
&A_3 + A_1 = A_4K_{3,4}\\
&\vdots\\
&A_{n-1} + A_1 = A_nK_{n-1,n}
\end{aligned}$$

Heterologous interactions are of the following general type:

$${}_nA + {}_mB = A_nB_mK = [A_nB_m]/[A]^n[B]^m$$

Since the most probable interactions are binary, this type of interaction is treated as a stepwise series of binary interactions so that, in practice, we have relations of the following form, with their associated mass action expressions:

$$
\begin{aligned}
&A + B = AB \\
&AB + A = A_2B \\
&AB + B = AB_2 \\
&AB_2 + A = A_2B_2 \\
&AB_2 + B = A_2B_3 \\
&A_2B + A = A_3B \\
&A_2B + B = A_2B_2 \\
&\vdots \\
&A_{n-1}B_m + A = A_nB_m \\
&A_nB_{m-1} + B = A_nB_m
\end{aligned}
$$

Attempts to fit sedimentation velocity data involving reversible interactions must solve these equilibrium expressions locally as part of the numerical solution to the Lamm equation (Stafford & Sherwood, 2004). The ModelEditor in SEDANAL that facilitates the building of complex self- and hetero-association fitting functions as described above has recently been described by Sherwood and Stafford (2015). It is also possible to incorporate kinetics in the analysis over a limited range of relaxation times from 0.005 to 10^{-5} s^{-1} (Correia & Stafford, 2009; Stafford & Sherwood, 2004). When reactions are kinetically limiting during a velocity run, the boundaries begin to spread due to the lack of local equilibrium and to partially resolve into individual peaks corresponding to the reaction species. This can be seen in Fig. 3 for a monomer–dimer system

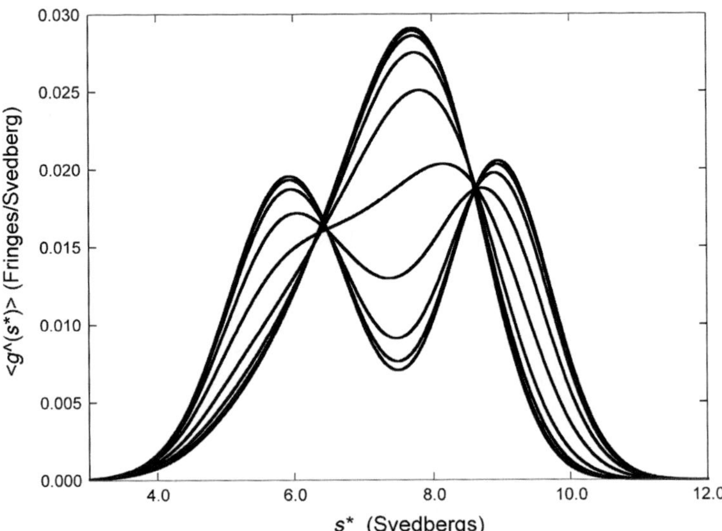

Figure 3 Simulation of a monomer–dimer equilibrium where $K = 1/c_o$ and k_{off} is varied from 1 to 10^{-5} s^{-1}. The data are plotted as a family of $g(s)$ curves as determined with the DCDT function in SEDANAL.

where k_{off} is varied from 1 to $3\times10^{-5}\,\text{s}^{-1}$. Prior to the computational ability to solve the Lamm equation for interacting systems, the analysis of weight average sedimentation (s_{w} vs. c(mg/ml)) was the preferred method for analyzing velocity data (Correia, 2000) and it still has great utility. We use software developed in FITALL or in SCIENTIST, but weight average applications are available in SEDPHAT (see the chapter by Brautigam, 2015 in this volume).

10. SOLUTIONS OF THE LAMM EQUATION

Faxen (1929) was the first to suggest an approximate solution to the Lamm equation. His approach applies to small D or large macromolecules, and times early in the run. In addition, the approximation assumes an "infinite cell" condition to avoid base effects or effects of any back diffusion from the buildup of solute. One final assumption is that s and D are independent of concentration (see Williams et al., 1958 for more details). Given these assumptions Faxén's approximate solution is

$$\frac{c}{c_{\text{o}}}=\frac{\text{e}^{-\tau}}{2}[1-\phi(\xi)]+\frac{\text{e}^{-\xi^2}}{\sqrt{\gamma}}\left[\frac{\epsilon\text{e}^{-\tau}}{16\pi}(1-\text{e}^{-\tau})\right]^{1/2}+\text{higher terms} \qquad (44)$$

where $\tau=2\omega^2st$, $\gamma=(r/r_{\text{o}})^2$, $\xi=\left[1-(\gamma\text{e}^{-\tau})^{1/2}\right]/[\epsilon(1-\text{e}^{-\tau})]^{1/2}$, $\epsilon=2D/s\omega^2r_{\text{o}}^2$

with the assumptions that $\epsilon\ll1$ and $\tau\ll1$ giving rise to the small D and early times in the run as stated above. It can be shown that this solution gives rise to a Gaussian concentration gradient, $\partial c/\partial r$, which allows determination of s, from a plot of the peak position $r*$ versus time as $\ln r*=\ln r_o+s\omega^2t$, and D by plotting $(A/H)^2=4\pi Dt$, where A is the area under the Gaussian and H is the height of the Gaussian. These were the standard methods applied for extracting s and D from sedimentation velocity runs with schlieren optics. The Fujita and MacCosham (1959) solution provides a zero time correction to what is referred to as "semi-infinite cell" conditions that adjusts for the time of acceleration during a run. Fujita (1956) further developed a solution that allows s to vary linearly with c ($s=s_{\text{o}}(1-kc)$) and where D is independent of c that results in the experimentally observed "sharpening effect" on a concentration-dependent boundary.

In parallel to these Faxen solutions, Archibald (1942, 1947a, 1947b) developed a solution to the Lamm equation that pertained to small solutes. In the Faxen nomenclature, ϵ is small and corresponds to solute that develops broad boundaries that are difficult to analyze by traditional methods to determine s and D. This approach-to-equilibrium method was developed to allow the determination of D and molecular weight M (Archibald, 1947a, 1947b; Klainer & Kegeles, 1955; Yphantis, 1959). Waugh and Yphantis (1953) applied the Archibald method to low molecular weight solutes (1000–3000 Da) by implementing a differential analyzer and a separation cell that allowed measurement of centrifugal and centripetal solute concentration ratios. Mason and Weaver (1924) developed a theory for a semi-infinite, rectangular cell in a uniform field to investigate the settling of spherical particles under gravity in a finite bounded cell. This approximation was adjusted by Yphantis and Waugh (1956a, 1956b) and Yphantis (1959) for first-order effects of the sectorial shaped cell and applied to the Archibald approach-to-equilibrium method. This approximation predicted $c(r, t)/c_0$ at the cell meniscus and cell base for solutes of various s and D or D and M values and became a routine method for molecular weight average calculations for schlieren and interference optical data. Fujita and MacCosham (1959), as mentioned above, developed a similar approximation that was also applicable to initial stages of ultracentrifugation.

The implementation of approximate solutions to the Lamm equation has continued even as more exact numerical approaches, discussed below, paralleled the development of fast computers. Philo (1997) developed the software program SVEDBERG that improves on the Fujita–MacCosham solution to the Lamm equation and works well for low MW solutes both early and late into the run. The new function was developed with simulated, noise-free data obtained from the Claverie finite-element method (Claverie, 1976; see below). This modified Fujita–MacCosham function introduced two new terms of the order $(1+\alpha\tau)$ and $(1+\beta\epsilon\tau)$, where τ and ϵ are as defined above, and α and β were determined by a numerical best-fit approach. Svedberg is highly appropriate for fitting paucidisperse, two, three, or more noninteracting species, or kinetically slow interacting systems. It may not be appropriate for systems that exhibit $s(c)$ or nonideal concentration dependence. Behlke and Ristau (1997, 1998) developed a similar software package LAMM utilizing five different model functions described by Fujita (1962) that best apply to low molecular weight proteins (10–20 kDa). Two of these equations were appropriate to synthetic boundary or overlay experiments. Behlke and Ristau (2002) updated this approach to provide an improved whole boundary solution to the Lamm equation that works over a wide range of molecular weights.

These approximate solutions to the Lamm equation have now been surpassed by numerical solutions that began to appear in the 1960s as computational power increased, initially in the form of IBM 360/65. Weiss and

Yphantis published a series of papers on finite difference numerical solutions to the Lamm equation (Dishon, Weiss, & Yphantis, 1966, 1967; Weiss & Yphantis, 1965, 1972) that included investigations of pressure effects (Dishon, Stroot, Weiss, & Yphantis, 1971), pressure effects in monomer–polymer systems (Johnson, Yphantis, & Weiss, 1973), and studies of the Johnston–Ogston effect in a two-component system (Correia, Johnson, Weiss, & Yphantis, 1976). The FORTRAN code was implemented on an IBM 360/65. While the solutions were reasonably accurate, they were probably slower than subsequent methods, although no one has attempted to implement finite difference solutions on a modern platform.

Claverie, Dreux, & Cohen, (1975) and Claverie (1976) developed a very fast matrix-based numerical method for solving the Lamm equation. Rapid numerical solutions to the Lamm equation allowed the development of nonlinear least squares curve-fitting methods. Todd and Haschemeyer (1981, 1983) were the first to develop curve-fitting methods using Claverie's rapid numerical solutions to the Lamm equation. Minimization of the sum of the squares of the residuals was computed using a modified Gauss–Newton algorithm. Global fitting to multiple data sets was first introduced by Stafford (1998) and Rivas, Stafford, and Minton (1999) who also used an adaptive time interval to speed up the calculations; Schuck (1998) developed a method by least squares fitting to single data sets employing moving hat and adaptive time interval and then in 2004 developed SEDPHAT to allow global fitting to multiple data sets.

All methods of analysis that use numerical solutions of the Lamm equation, for each species present in the sample, are fitting for its sedimentation coefficient and diffusion coefficient internally, in addition to its concentration at each radial position at each time point during the run. For interacting systems, the concentration of each species is recomputed according to the Law of Mass Action using either the equilibrium or kinetic relations for each time increment used in the simulation. Molar masses obtained by fitting with the Lamm equation are derived from the corresponding s and D values through the Svedberg equation.

11. EQUIPMENT AND OPTICAL SYSTEMS

There are two refractometric optical systems, the schlieren system, which displays the refractive index gradient as a function of radial position (proportional to the concentration gradient), and the interference system, which displays the refractive index difference between sample and reference solutions as the vertical displacement of horizontal fringes as a function of radial position. Use of interference optics almost always requires dialysis as well as blank subtraction. Blank subtraction of the time-independent background signal, caused by inhomogeneities in the optical components, is mandatory and is accomplished either by performing a blank run with water versus water and subtracting that curve from each scan in the run or by doing time derivative analysis (DCDT+ or SEDANAL) or by least squares fitting for the background (SEDFIT). Although it is possible in principle to fit for the buffer redistribution if the molecular parameters (s, D, and activity coefficients) of all the buffer components are known for a particular buffer composition, exhaustive dialysis and meniscus matching will eliminate the need to include the buffer in the model, unless there is a significant dynamic density gradient setup by the buffer or compressibility of the solvent (Schuck, 2004a, 2004b).

The standard absorbance optical system on the Beckman–Coulter ProteomeLab ultracentrifuge displays absorbance at selected wavelengths as a function of position. Absorbance optics allows selection of up to three wavelengths affording higher selectivity than either a single wavelength or interference optics. Use of absorbance optics usually does not require dialysis for optical compensation for an absorbing buffer; however, global analysis of data does require uniform buffer composition in all samples, and correction of sedimentation-diffusion data to $s_{20,w}$ does require knowledge of density and viscosity and thus buffer equilibration is required for accurate correction to standard conditions. Moreover, dialysis is required for the correct definition of components in a binary or ternary system (Casassa & Eisenberg, 1964) especially if high concentrations of buffer components are being used. Dialysis to equilibrium allows one to treat the buffer as a single component (Casassa & Eisenberg, 1964) as assumed in the derivations above (Eqs. 42 and 43).

There are also two additional absorbance systems that have multi-wavelength capability (from 500 up to 2048 wavelengths), one type in Helmut Colfen's and in Johannes Walter's laboratories (Karabudak & Cölfen, 2014; Pearson et al., 2015; Walter et al., 2014), respectively, and the other type in Kristian Schilling's laboratory at Nanolytics (Schilling & Krause, 2015). In addition, an absorbance optical system using a linear photodiode array capable of scanning a cell in less than 10 s is being developed by Tom Laue at Spin Analytical, Inc. The monochromator has highly accurate and repeatable wavelength selection capability and without limit to the number of wavelengths at which scans are to be taken.

Fluorescence optics, also developed by Tom Laue at the University of New Hampshire (Kingsbury & Laue, 2011; Kroe & Laue, 2009; MacGregor, Anderson, & Laue, 2004), is available from Aviv Instruments, NJ (Aviv-FDS) and provides a third method to access the hydrodynamic behavior of fluorescently labeled macromolecules. Fluorescence optics requires time derivative methods or a fit for time-independent noise to eliminate the time-independent background components of the signal. Moreover, if there are fluorescent buffer components that redistribute significantly during the run, they would have to be included in the model being fitted. Recent advances attempt to correct for inner filter effects and optical artifacts at the cell base (Zhao, Casillas, Shroff, Patterson, & Schuck, 2013). FDS data do not produce an optical artifact at the meniscus. To establish the meniscus position, some users fit for the meniscus position during analysis (SEDFIT or SEDANAL); some scan the cell with intensity mode after the run (Correia et al., 2015); while some add a fluorescent dye to the solution that floats and generates a peak at the meniscus position in the scan (Bailey, Angley, & Perugini, 2009). New FDS systems have a 50.2 mW laser installed that produces a change in Raman scattering at the meniscus position (Zhao et al., 2014).

The ability to process and analyze data from these new systems has been implemented in various modern software packages. The most common modern, comprehensive software packages for analyzing and fitting AUC data are SEDANAL (Stafford,1998, 2015; Stafford & Sherwood, 2004), SEDFIT/SEDPHAT (Schuck, 1998; Schuck & Demeler, 1999; Schuck & Rossmanith, 2000), and ULTRASCAN (Demeler, 2005a, 2005b; Demeler & Saber, 1998). These packages fit various models to data using the numerical solutions of the Lamm equation developed by Claverie et al. (1975) and Claverie (1976) and the curve-fitting method pioneered by Todd and Haschemeyer (1981, 1983) who were the first to use numerical solutions of the Lamm equation for curve fitting to sedimentation velocity data. DCDT-PLUS (Philo, 2006) is an implementation of the model-independent time derivative method of Stafford (1992) that is particularly convenient to use.

12. SEDIMENTATION VELOCITY EXPERIMENTAL DESIGN

One should choose a speed and running time that will allow the boundary of interest to migrate at least 2/3 of the way down the cell. For many analyses, the run should be allowed to proceed long enough to completely clear the solution column in order to obtain the most reliable information. This is because many methods of analysis will be influenced by the existence of very slowly sedimenting material that has not been allowed to sediment a sufficient distance down the cell and therefore its presence will distort the analysis unless it is accounted for correctly. Slowly sedimenting material can be confounded with systematic baseline errors by some least squares procedures. For very high molecular weight, rapidly sedimenting components, a speed should be chosen sufficiently low enough to allow time for significant diffusion to take place during the run. Depending on what information is being measured from a particular sample, the choice of speed will be a compromise between relatively high speed for maximum resolution and relatively lower speed for accurate determination of the diffusion coefficients.

We propose a rule of thumb for choosing the optimal speed to determine molar mass from both s and D. Starting with the definition of sigma, Eq. (45),

$$\sigma \equiv \frac{\omega^2 s}{D} = \frac{M(1-\bar{v}\rho)\omega^2}{RT} \tag{45}$$

σ can be rewritten as, after multiplying numerator and denominator by t,

$$\sigma = \frac{\omega^2 st}{Dt} \tag{46}$$

We also have from the relations above, Eq. (16), that $\omega^2 st = \ln(r/r_m)$ and that the mean squared distance moved for diffusing particles is given by $\overline{x^2} = 2Dt$. Assuming the boundary standard deviation is about 0.15 cm, the value of σ for a velocity run that will give optimal determination for both s and D for a single sedimenting species would be given by the following equation:

$$\sigma = \frac{\omega^2 st}{Dt} = \frac{\ln(6.5/5.9)}{(0.15)^2/2} \cong \frac{0.1}{0.01} = 10.0\,\text{cm}^{-2} \tag{47}$$

This condition would allow the boundary to migrate a little more than half way to the bottom of the cell and still have zero gradient at the meniscus and the plateau region.

Sedimentation velocity should be conducted in double sector centerpieces, and if interference optics is being used, a meniscus-matching centerpiece gives a properly matched baseline and the best results. Window holders with narrow slits in the upper window (Spin Analytical) are also recommended for interference optics. It is also possible to perform velocity runs in double sector centerpieces with intensity data, referred to as pseudo-absorbance (Kar, Kingsbury, Lewis, Laue, & Schuck, 2000). This allows 2 samples to be run in each cell, 6 samples in a four-hole rotor, and 14 samples in an eight-hole rotor. The precaution is that the OD of the sample on the reference side should not be more than 0.5, which can cause the machine to readjust the PM tube gain. Using intensity data increases the signal to noise by a factor of sqrt(2) and allows the running of twice as many samples as with absorbance optics. The Aviv Au-FDS fluorescence optical system does not collect or require a reference cell and thus twice as many samples can be run with the AU-FDS as well.

Velocity centerpieces from Beckman are rated at 44,000 RPM, but in reality, the community ignores the rating and runs at 50,000 or 60,000 RPM, routinely, when necessary. If the menisci are matched in height, there will be no net force on the septum between the sectors and, therefore, negligible deformation of the septum. Spin Analytical sells centerpieces rated for 50,000 (Spin50) and 60,000 RPM (Spin60) centerpieces, although 60K centerpieces can be problematic with absorbance optics due to the beam clipping the thicker sector walls in the Spin60 centerpieces (see their Web site).

Centerpieces are available in 12 and 3 mm thickness from Beckman and 12, 3, and 1 mm thickness from Spin Analytical. This difference in optical path provides an opportunity to vary the macromolecular concentration by two orders of magnitude. The 3 and 1 mm cells must be torqued tighter, ~150 in. lbs or more, to prevent leaking and warping of the centerpiece. The 1 mm cells are especially prone to warp, and good centerpieces must be selected and tested from a batch. The 3 and 1 mm cells require that spacers be placed in the housing to properly position the filling holes. These spacers also provide appropriate weight to match a 1.2 cm assembly. For absorbance and interference, two equal height spacers are placed above and below each window assembly to match centerpiece positions. For fluorescence, only a single longer spacer is used to position the upper face of the 3 or 1 mm centerpieces at the top of the assembly. The cone of light from the FDS optics is focused down into the solution and this arrangement preserves the depth of that focus. While it is possible to run FDS with a 3 mm centerpiece in the Abs position, the depth of focus increases from 2000 to 5000 μm. It is also possible to run absorbance optics with a 3 or 1 mm centerpiece positioned at the FDS upper position. In general, we try to adhere to the manufacturers' recommendations.

All optical systems and cells require that the samples be properly equilibrated with the buffer by exhaustive dialysis, gel filtration, or spun columns. This is a rigorous and well-founded requirement for proper biophysical interpretation of hydrodynamic and thermodynamic data (Casassa & Eisenberg, 1964). Interference optics is especially sensitive to buffer mismatch since refractive index gradients of high concentrations of buffer and salt components will produce background fringe displacements. Thus, it is especially important to use meniscus-matching centerpieces for interference optics as we explain in detail here.

There are two special requirements for interference optics: The first is that the macromolecular solution under investigation must be equilibrated with its reference buffer; the second is that the menisci must be matched as closely as possible. In the first instance, if the solution and the buffer are not equilibrated, there will be a signal from buffer redistribution superimposed on the signal from macromolecule sedimentation. With interference optics, one is subtracting two very large numbers (i.e., the refractive index of the solvent from the refractive index of the solution) to give the contribution of the macromolecule to the observed refractive index. The importance of careful meniscus matching can be seen in a synthetic boundary experiment to measure protein concentration. Not only is exhaustive dialysis required but careful handling of the solution is also required to avoid evaporation. For example, if one is handling a solution of protein in 0.5 *M* NaCl, a 1% (i.e. 5 m*M*) change in NaCl concentration due to either evaporation or mismatch from incomplete dialysis will result in about four fringes of error, while a 1 mg/mL solution of protein will itself produce only about 3.3 fringes of signal (W. F. Stafford, personal observation, 1975). Equilibration is usually accomplished by exhaustive dialysis; however, other methods like gel filtration or spun columns can also be used successfully. Second, when the menisci are exactly matched, the sedimentation of buffer will start at exactly the same radius in both the reference and solution sectors allowing exact optical cancellation of the buffer's contribution to the refractive index as it redistributes. However, if the menisci are not matched, exact cancellation will not occur and that difference signal will be superimposed on the relatively smaller signal from the protein.

Meniscus matching can be accomplished with centerpieces that allow flow between the sectors through a capillary channel that otherwise prevents free communication between the sectors. The procedure is to fill the reference sector with a slightly greater volume than the sample sector. When the centrifuge is started spinning at, say 5000 RPM, buffer will flow through the capillary from the reference sector to the solution sector until the menisci on each side are matched hydrostatically at the same radial position. For example, in standard 12 mm centerpieces, the reference side is filled to 0.460 mL and the sample side is filled to 0.440 mL and 0.010 mL is transferred from the reference side to the sample side resulting in a slight dilution of the sample (<2.3%) but accomplishing the desired effect of matching the menisci. At this point, it is important to stop the centrifuge and shake up the sample by removing the rotor and shaking the rotor gently so as not to dislodge the cell. If the cell is loose in its hole and moves even slightly, the cell should be resecured in its hole and the meniscus-matching procedure repeated. After the menisci are matched and the rotor returned to the centrifuge, the rotor should be allowed to temperature-equilibrate for about 1 h before starting the run. It is possible, in principle, to fit for the buffer redistribution, however, that would require adding several parameters to the fit to account for the properties of the buffer components complicating the fitting process. Meniscus-matching centerpieces rated for both 50,000 RPM (Spin50-MM) and 60,000 RPM (Spin60-MM) can be acquired from Spin Analytical, Inc.

Fluorescence data, if collected in the linear range (Lyons, Lary, Husain, Correia, & Cole, 2013), will in general be least sensitive optically to mismatched buffer components unless the buffer contains fluorescent compounds. An example is bilirubin bound to serum albumin in serum samples, although it is often invisible at low gain settings. In addition, centerpieces often do exhibit autofluorescence that raises the baseline. For absorbance optics, as long as the buffer does not absorb at the wavelength being used, neither dialysis nor meniscus matching is required, with the caveat outlined above about proper interpretation of data. Absorbance mismatch can occur due to optical changes in reducing agents, or colored contaminants in commercial compounds or samples. In general, these compounds produce raised baselines that can be corrected for by software adjustments, but their presence can and does increase uncertainty in the analysis. If the buffer does absorb at the wavelength being used, then the same considerations about meniscus matching apply to absorbance optics. If one is using intensity data, it may be necessary to fit for buffer redistribution if it absorbs at the wavelength being used.

It is recommended that the cell be filled as full as possible to provide the longest possible solution column. Filling to put the meniscus at 5.9 cm will give a sedimentation path of 1.3 cm. Filling cells to only 6.1 cm or higher radii will reduce significantly the resolution obtainable. The effects of boundary broadening due to diffusion go as the square root of time (and therefore the square root of the distance traveled), while the effects of boundary spreading due to heterogeneity go as the first power of time. This means that the resolution between noninteracting species will go as the square root of time. In other words, the effects of diffusion become less and less important as the boundary proceeds down the cell, while the

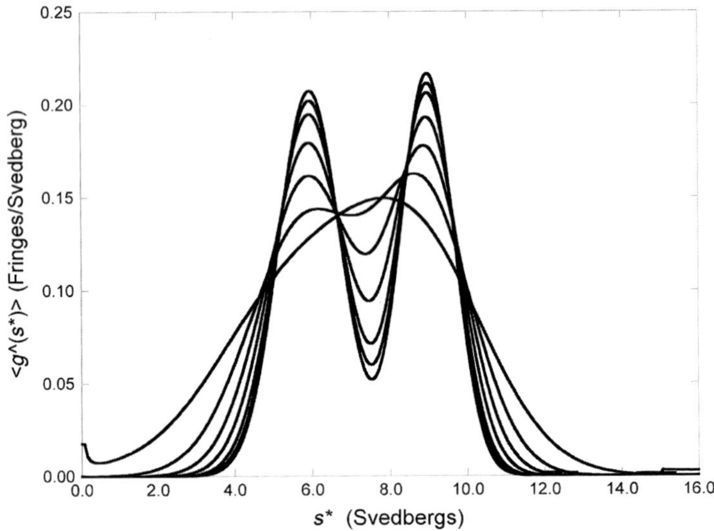

Figure 4 DCDT+ analysis of the data in Fig. 1 over different scan ranges to demonstrate the need to use a long column and wait until late in the run to maximize the estimates of *S* and improve the resolution power of sedimentation velocity.

resolution increases (Fig. 4). This is seen in Fig. 4 where DCDT analysis of the simulated data from Fig. 1 as a function of later and later scans during the run shows better peak resolution and thus better determination of *s* values.

ACKNOWLEDGMENTS

We thank Tom Laue, Jim Cole, and Sharon Lobert for comments and suggestions that improved this chapter. We apologize for the omission of a large number of excellent investigations that helped formulate the field.

REFERENCES

Adams, G. H., & Schumaker, V. N. (1969a). Rapid molecular weight estimates for low-density lipoproteins. *Analytical Biochemistry*, *29*, 117–129.

Adams, G. H., & Schumaker, V. N. (1969b). Polydispersity of human low-density lipoproteins. *Annals of the New York Academy of Sciences*, *164*, 130–146.

Anderson, D. W., Nichols, A. V., Forte, T. M., & Lindgren, F. T. (1977). Particle distribution of human serum high density lipoproteins. *Biochimica et Biophysica Acta*, *493*(1), 55–68.

Archibald, W. J. (1942). The integration of the differential equation of the ultracentrifuge. *Annals of the New York Academy of Sciences*, *43*, 211–227.

Archibald, W. (1947a). An approximate solution of the differential equation of the ultracentrifuge. *Journal of Applied Physics*, *18*(4), 362–367.

Archibald, W. (1947b). A demonstration of some new methods of determining molecular weights from the data of the ultracentrifuge. *Journal of Physical and Colloid Chemistry*, *51*(5), 1204–1214.

Arthur, K. K., Kendrick, B. S., & Gabrielson, J. P. (2015). Guidance to achieve accurate aggregate quantitation in biopharmaceuticals by SV-AUC. *Methods in Enzymology, 562*, 477–500.

Bailey, M. F., Angley, L. M., & Perugini, M. A. (2009). Methods for sample labeling and meniscus determination in the fluorescence-detection analytical ultracentrifuge. *Analytical Biochemistry, 390*, 218–220.

Baldwin, R. L. (1954). Boundary spreading in sedimentation velocity experiments. II. The correction of sedimentation coefficient distributions for the dependence of sedimentation coefficient on concentration. *Journal of the American Chemical Society, 76*, 402–407.

Bauer, W., & Vinograd, J. (1968). The interaction of closed circular DNA with intercalative dyes: I. The superhelix density of SV40 DNA in the presence and absence of dye. *Journal of Molecular Biology, 33*, 141–171.

Beams, J. W., & Pickels, E. G. (1935). The production of high rotational speeds. *Review of Scientific Instruments, 6*, 299–308.

Behlke, J., & Ristau, O. (1997). Molecular mass determination by sedimentation velocity experiments and direct boundary fitting of the concentration profiles. *Biophysical Journal, 72*, 428–434.

Behlke, J., & Ristau, O. (1998). An improved approximate solution of the Lamm equation for the simultaneous estimation of sedimentation and diffusion coefficients from sedimentation velocity experiments. *Biophysical Chemistry, 70*, 133–146.

Behlke, J., & Ristau, O. (2002). A new approximate whole boundary solution of the Lamm differential equation for the analysis of sedimentation velocity experiments. *Biophysical Chemistry, 95*, 59–68.

Berg, H. C. (1993). *Random walks in biology*. Princeton, NJ: Princeton University Press.

Berkowitz, S. A. (2006). Role of analytical ultracentrifugation in assessing the aggregation of protein biopharmaceuticals. *The AAPS Journal, 8*, E590–E605.

Brautigam, C. A. (2011). Using Lamm-equation modeling of sedimentation velocity data to determine the kinetic and thermodynamic properties of macromolecular interactions. *Methods, 54*, 4–15.

Brautigam, C. A. (2015). Calculations and publication-quality illustrations for analytical ultracentrifugation data. *Methods in Enzymology, 562*, 109–134.

Casassa, E. F., & Eisenberg, H. (1964). Thermodynamic analysis of multicomponent systems. *Advances in Protein Chemistry, 19*, 287–395.

Chao, F.-C., & Schachman, H. K. (1956). The isolation and characterization of a macromolecular ribonucleoprotein from yeast. *Archives of Biochemistry and Biophysics, 61*, 220–230.

Chaton, C. T., & Herr, A. B. (2015). Elucidating complicated assembling systems in biology using size-and-shape analysis of sedimentation velocity data. *Methods in Enzymology, 562*, 187–204.

Claverie, J. M. (1976). Sedimentation of generalized systems of interacting particles. III. Concentration dependent sedimentation and extension to other transport methods. *Biopolymers, 15*, 843–857.

Claverie, J. M., Dreux, H., & Cohen, R. (1975). Sedimentation of generalized systems of interacting particles. I. solution of systems of complete Lamm equations. *Biopolymers, 14*(8), 1685–1700.

Cohn, E. J., & Edsall, J. T. (1943). *Proteins, amino acids and peptides as ions and dipolar ions*. New York: Reinhold.

Cole, J. L., Correia, J. J., & Stafford, W. F. (2011). The use of analytical sedimentation velocity to extract thermodynamic linkage. Special issue on 25th Gibbs conference on biothermodynamics. *Biophysical Chemistry, 159*, 120–128.

Cole, J. L., Lary, J. W., Moody, T. P., & Laue, T. M. (2008). Analytical ultracentrifugation: Sedimentation velocity and sedimentation equilibrium. *Methods in Cell Biology, 84*, 143–1179.

Correia, J. J. (2000). Analysis of weight average sedimentation velocity data. *Methods in Enzymology, 321*, 81–100.

Correia, J. J., Johnson, M. L., Weiss, G. H., & Yphantis, D. A. (1976). Numerical study of the Johnston-Ogston effect in two-component systems. *Biophysical Chemistry, 5*, 255–264.

Correia, J. J., Lyons, D. F., Sherwood, P., & Stafford, W. F. (2015). Techniques for dissecting the Johnston–Ogston effect. In Susumu Uchiyama, Fumio Arisaka, Tom Laue, & Walter Stafford (Eds.), *Analytical ultracentrifugation—Instrumentation, analysis and applications*. Japan, Tokyo: Springer.

Correia, J. J., & Stafford, W. F. (2009). Extracting equilibrium constants from kinetically limited reacting systems. *Methods in Enzymology: 455*, pp. 419–446, Biothermodynamics, Part A.

De Groot, S. R., & Masur, P. (1962). *Non-equilibrium thermodynamics*. Amsterdam: North-Holland Publishing Company.

Demeler, B. (2005a). UltraScan a comprehensive data analysis software package for analytical ultracentrifugation experiments. In D. J. Scott, S. E. Harding, & A. J. Rowe (Eds.), *Modern analytical ultracentrifugation: Techniques and methods* (pp. 210–229). UK: Royal Society of Chemistry.

Demeler, B. (2005b). Hydrodynamic methods. In H. Rashidi & L. Buehler (Eds.), *Bioinformatics basics: Applications in biological science and medicine* (2nd ed., pp. 226–255). Boca Raton, Florida: CRC Press LLC.

Demeler, B., & Saber, H. (1998). Determination of molecular parameters by fitting sedimentation data to finite-element solutions of the Lamm equation. *Biophysical Journal, 74*(1), 444–454.

Dishon, M., Stroot, M. T., Weiss, G. H., & Yphantis, D. A. (1971). New approach to effects of pressure dependence on sedimentation velocity experiments. *Journal of Polymer Science Part A2—Polymer Physics, 9*, 939–945.

Dishon, M., Weiss, G. H., & Yphantis, D. A. (1966). Numerical solutions of the Lamm equation. I. Numerical procedure. *Biopolymers, 4*, 449–455.

Dishon, M., Weiss, G. H., & Yphantis, D. A. (1967). Numerical solutions of Lamm equation 3. Velocity centrifugation. *Biopolymers, 5*, 697–705.

Einstein, A. (1905). On the motion of small particles suspended in a stationary liquid, as required by the molecular kinetic theory of heat. *Annalen der Physik, 17*(8), 549–560.

Faxen, O. H. (1929). Uber eine differentialgleichung aus der physikalischen. *Arkiv foer Matematik, Astronomi, och Fysik, 21B*, 1–6.

Fujita, H. (1956). Effects of a concentration dependence of the sedimentation co-efficient in velocity ultracentrifugation. *The Journal of Chemical Physics, 24*(5), 1084–1090.

Fujita, H. (1962). *Mathematical theory of sedimentation analysis*. New York: Academic Press.

Fujita, H., & MacCosham, V. (1959). Extension of sedimentation velocity theory to molecules of intermediate sizes. *The Journal of Chemical Physics, 30*(1), 291–298.

Fuoss, R. W. (1959). The velocity field in electrolytic solutions. *Journal of Physical Chemistry, 63*(4), 633–636.

Gofman, J. W., Lindgren, F. T., & Elliott, H. (1949). Ultracentrifugal studies of lipoproteins of human serum. *The Journal of Biological Chemistry, 179*(2), 973–979.

Goldberg, R. J. (1953). Sedimentation in the ultracentrifuge. *The Journal of Physical Chemistry, 57*, 194–202.

Gosting, L. J. (1956). Measurement and interpretation of diffusion coefficients of proteins. *Advances in Protein Chemistry, 11*, 429–554.

Hanlon, S., Lamers, K., Lauterbach, G., Johnson, R., & Schachman, H. K. (1962). Ultracentrifuge studies with absorption optics: I. An automatic photoelectric scanning absorption system. *Archives of Biochemistry and Biophysics, 99*, 157–174.

Harrington, W. F., & Schachman, H. K. (1953). Analysis of a concentration anomaly in the ultracentrifugation of mixtures. *Journal of the American Chemical Society, 75*, 3533–3539.

Hill, J. J., & Laue, T. M. (2015). Protein assembly in serum and the differences from assembly in buffer. *Methods in Enzymology, 562*, 501–528.

Howlett, G. J., Minton, A. P., & Rivas, G. (2006). Analytical ultracentrifugation for the study of protein association and assembly. *Current Opinion in Chemical Biology, 10*, 430–436.

Johnson, M., Yphantis, D. A., & Weiss, G. H. (1973). Instability in pressure-dependent sedimentation of monomer-polymer systems. *Biopolymers, 12*, 2477–2490.

Johnston, J. P., & Ogston, A. G. (1946). A boundary anomaly found in the ultracentrifugal sedimentation of mixtures. *Transactions of the Faraday Society, 42*, 789–799.

Kar, S. R., Kingsbury, J. S., Lewis, M. S., Laue, T. M., & Schuck, P. (2000). Analysis of transport experiments using pseudo-absorbance data. *Analytical Biochemistry, 285*(1), 135–142.

Karabudak, E., & Cölfen, H. (2014). The multiwavelength UV/Vis detector: New possibilities with an added spectral dimension. In Susumu Uchiyama, Fumio Arisaka, Tom Laue, & Walter Stafford (Eds.), *Analytical ultracentrifugation—Instrumentation, analysis and applications*. Japan, Tokyo: Springer.

Katchalsky, A., & Curran, P. F. (1975). *Nonequilibrium thermodynamics in biophysics*. Cambridge: Harvard University Press.

Kingsbury, J. S., & Laue, T. M. (2011). Fluorescence-detected sedimentation in dilute and highly concentrated solutions. *Methods in Enzymology, 492*, 283–304.

Kingsbury, J. S., Laue, T. M., Chase, S. B., & Connors, L. H. (2011). Detection of high-molecular-weight amyloid serum protein complexes using biological on-line tracer sedimentation. *Analytical Biochemistry, 425*, 151–156.

Kirschner, M. W., & Schachman, H. K. (1971a). Conformational changes in proteins as measured by difference sedimentation studies. I. A technique for measuring small changes in sedimentation coefficient. *Biochemistry, 10*, 1900–1919.

Kirschner, M. W., & Schachman, H. K. (1971b). Conformational changes in proteins as measured by difference sedimentation studies. II. Effect of stereospecific ligands on the catalytic subunit of aspartate transcarbamylase. *Biochemistry, 10*, 1919–1926.

Kirschner, M. W., & Schachman, H. K. (1973a). Local and gross conformational changes in aspartate transcarbamylase. *Biochemistry, 12*, 2997–3004.

Kirschner, M. W., & Schachman, H. K. (1973b). Conformational studies on the nitrated catalytic subunit of aspartate transcarbamylase. *Biochemistry, 12*(16), 2987–2997.

Klainer, S. M., & Kegeles, G. (1955). Simultaneous determination of molecular weights and sedimentation constants. *Journal of Physical Chemistry, 59*, 952–955.

Kroe, R. R., & Laue, T. M. (2009). NUTS and BOLTS: Applications of fluorescence-detected sedimentation. *Analytical Biochemistry, 390*, 1–13.

LaBar, F. E., & Baldwin, R. L. (1963). The sedimentation coefficient of sucrose. *Journal of the American Chemical Society, 85*, 3106–3108.

Lamm, O. (1929). Die differentialgleichung der ultrazentrifugierung. *Arkiv för Matematik, Astronomi och Fysik, 21B*(2), 1–4.

Laue, T. M., & Stafford, W. F. (1999). Modern applications of analytical ultracentrifugation. *Annual Review of Biophysics and Biomolecular Structure, 28*, 75–100.

Lebowitz, J., Lewis, M. S., & Schuck, P. (2002). Modern analytical untracentrifugation in protein science: A tutorial review. *Protein Science, 11*, 2067–2079.

Lin, J. B., & Lucius, A. L. (2015). Analysis of linked equilibria. *Methods in Enzymology, 562*, 161–186.

Liu, J., Yadav, S., Andya, J., Demeule, B., & Shire, S. J. (2015). Analytical ultracentrifugation and its role in development and research of therapeutical proteins. Methods in Enzymology, 562, 441–476.

Longsworth, L. J. (1939). A modification of the schlieren method for use in electrophoretic analysis. *Journal of the American Chemical Society, 61*, 529–530.

Lyons, Daniel F., Lary, Jeffrey W., Husain, Bushra, Correia, John J., & Cole, James L. (2013). Are fluorescence-detected sedimentation velocity data reliable? *Analytical Biochemistry, 437*(2), 133–137.

MacGregor, I. K., Anderson, A. L., & Laue, T. M. (2004). Fluorescence detection for the XLI analytical ultracentrifuge. *Biophysical Chemistry, 108*, 165–185.
Mason, M., & Weaver, W. (1924). The settling of small particles in a fluid. *Physics Reviews, 23*, 412–426.
McFarlane, A. S. (1935). The ultracentrifugal analysis of normal and pathological serum fractions. *The Biochemical Journal, 29*, 1209–1226.
Meselson, M., & Stahl, F. W. (1958). The replication of DNA in *Escherichia coli*. *PNAS, 44*, 671–682.
Meselson, M., Stahl, F. W., & Vinograd, J. (1957). Equilibrium sedimentation of macromolecules in density gradients. *Proceedings of the National Academy of Sciences of the United States of America, 43*, 581–588.
Miller, D. G. (1959). Thermodynamics of irreversible processes. The experimental verification of the Onsager reciprocal relations. *Chemical Reviews, 60*, 15–37.
Ogston, A. G. (1937). Some observations on mixtures of serum albumin and globulin. *The Biochemical Journal, 31*, 1952–1957.
Onsager, L. (1931a). Reciprocal relations in irreversible processes. I. *Physics Review, 37*, 405–426.
Onsager, L. (1931b). Reciprocal relations in irreversible processes. II. *Physics Review, 38*, 2265–2279.
Onsager, L., & Fuoss, R. W. (1932). Irreversible processes in electrolytes diffusion, conductance, and viscous flow in arbitrary mixtures of strong electrolytes. *Journal of Physical Chemistry, 36*, 2689–2778.
Pearson, J., Krause, F., Haffke, D., Demeler, B., Schilling, K., & Cölfen, H. (2015). Next-generation AUC adds a spectral dimension: Development of multiwavelength detectors for the analytical ultracentrifuge. *Methods in Enzymology, 562*, 1–26.
Pederson, K. O. (1958). On charge and specific ion effects on sedimentation in the ultracentrifuge. *The Journal of Physical Chemistry, 62*, 1282–1290.
Phillips, B. A. (1969). In vitro assembly of polioviruses: I. Kinetics of the assembly of empty capsids and the role of extracts from infected cells. *Virology, 39*, 811–821.
Phillips, B. A., Summers, D. F., & Maizel, J. V., Jr. (1968). In vitro assembly of poliovirus-related particles. *Virology, 35*, 216–226.
Philo, J. S. (1997). An improved function for fitting sedimentation velocity data for low-molecular-weight solutes. *Biophysical Journal, 72*, 435–444.
Philo, J. S. (2006). Improved methods for fitting sedimentation coefficient distributions derived by time-derivative techniques. *Analytical Biochemistry, 354*, 238–246.
Philpot, J. St. L. (1938). Direct photography of ultracentrifuge sedimentation curves. *Nature, 141*, 283–284.
Richards, E., & Schachman, H. (1957). A differential ultracentrifuge technique for measuring small changes in sedimentation coefficients. *Journal of the American Chemical Society, 79*, 79–80.
Richards, E. G., & Schachman, H. K. (1959). Ultracentrifuge Studies with Rayleigh Interference Optics. I. General Application. *Journal of Physical Chemistry, 63*, 1578–1591.
Rivas, G., Stafford, W. F., & Minton, A. P. (1999). Characterization of heterologous protein-protein interaction via analytical ultracentrifugation. *Methods: A Companion to Methods in Enzymology, 19*, 194–212.
Rocco, M., & Byron, O. (2015). Hydrodynamic modeling and its application in AUC. *Methods in Enzymology, 562*, 81–108.
Schachman, H. K. (1959). *Ultracentrifugation in biochemistry*. New York: Academic Press.
Schilling, K., & Krause, F. (2015). Analysis of antibody aggregate content at extremely high concentrations using sedimentation velocity with a novel interference optics. *PLoS One, 10*(3), e0120820. http://dx.doi.org/10.1371/journal.pone.0120820.
Schuck, P. (1998). Sedimentation analysis of noninteracting and self-associating solutes using numerical solutions to the Lamm equation. *Biophysical Journal, 75*(3), 1503–1512.
Schuck, P. (2004a). A model for sedimentation in inhomogeneous media. I. Dynamic density gradients from sedimenting co-solutes. *Biophysical Chemistry, 108*(1–3), 187–200.

Schuck, P. (2004b). A model for sedimentation in inhomogeneous media. II. Compressibility of aqueous and organic solvents. *Biophysical Chemistry, 108*(1–3), 201–214.

Schuck, P., & Demeler, B. (1999). Direct sedimentation analysis of interference optical data in analytical ultracentrifugation. *Biophysical Journal, 76*(4), 2288–2296.

Schuck, P., & Rossmanith, P. (2000). Determination of the sedimentation coefficient distribution by least-squares boundary modeling. *Biopolymers, 54*(5), 328–341.

Scott, D. J., Harding, S. E., & Winzor, D. J. (2015). Concentration dependence of translational diffusion coefficients for globular proteins. *The Analyst, 139*, 6242–6248.

Sherwood, P. J., & Stafford, W. F. (2015). SEDANAL: Model-dependent and model-independent analysis of sedimentation data, chapter 6. In S. Uchiyama, F. Arisaka, T. Laue, & W. Stafford (Eds.), *Analytical ultracentrifugation—Instrumentation, analysis and applications*. Japan, Tokyo: Springer.

Shire, S. J., Steckert, J. J., Adams, M. L., & Schuster, T. M. (1979). Kinetics and mechanism of tobacco mosaic virus assembly: Direct measurement of relative rates of incorporation of 4S and 20S protein. *PNAS, 76*, 2745–2749.

Shire, S. J., Steckert, J. J., & Schuster, T. M. (1979). Mechanism of self-assembly of tobacco mosaic virus protein. II. Characterization of the metastable polymerization nucleus and the initial stages of helix formation. *Journal of Molecular Biology, 127*, 487–506.

Simoni, R. D., Hill, R. L., & Vaughan, M. (2002). John T. Edsall: Biochemist, Teacher, Journal of Biological Chemistry Editor, and Responsible Scientist. *Journal of Biological Chemistry, 277*, 29351–29354.

Stafford, W. F. (1992). Boundary analysis in sedimentation transport experiments: A procedure for obtaining sedimentation coefficient distributions using the time derivative of the concentration profile. *Analytical Biochemistry, 203*, 295–301.

Stafford, W. F. (1998). Time difference sedimentation velocity analysis of rapidly reversible interacting systems: Determination of equilibrium constants by non-linear curve fitting procedures. *Biophysical Journal, 74*, A301.

Stafford, W. F. (2015). Analysis of non-ideal, interacting and non-interacting systems by sedimentation velocity analytical ultracentrifugation, chapter 24. In Susumu Uchiyama, Fumio Arisaka, Tom Laue, & Walter Stafford (Eds.), *Analytical ultracentrifugation-instrumentation, analysis and applications*. Japan, Tokyo: Springer.

Stafford, W. F., & Sherwood, P. J. (2004). Analysis of heterologous interacting systems by sedimentation velocity: Curve fitting algorithms for estimation of sedimentation coefficients, equilibrium and kinetic constants. *Biophysical Chemistry, 108*, 231–243.

Svedberg, T. (1926). http://www.nobelprize.org/nobel_prizes/chemistry/laureates/1926/svedberg-lecture.pdf.

Svedberg, T., & Fahraeus, R. (1926). A new method for the determination of the molecular weight of the proteins. *Journal of the American Chemical Society, 48*(1), 430–438.

Svedberg, T., & Heyroth, F. F. (1929a). The molecular weight of the hemocyanin of limulus polyphemus. *Journal of the American Chemical Society, 51*, 539–550.

Svedberg, T., & Heyroth, F. F. (1929b). The influence of the hydrogen-ion activity upon the stability of the hemocyanin of helix pomatia. *Journal of the American Chemical Society, 51*, 550–561.

Svedberg, T., & Nichols, J. (1926). The molecular weight of egg albumin in electrolyte-free condition. *Journal of the American Chemical Society, 48*, 3081–3092.

Svedberg, T., & Nichols, J. (1927). The application of the oil turbine type of ultracentrifuge to the study of the stability region of carbon monoxide-hemoglobin. *Journal of the American Chemical Society, 49*(2), 2920–2934.

Svedberg, T., & Pederson, K. O. (1940). *The ultracentrifuge*. Oxford: Oxford University Press.

Svensson, H. (1939). Direct photographic recording of electrophoresis diagrams. *Kolloid-Zeitschrift, 87*(2), 181–186.

Svensson, H. (1940). The theory of the observation method of crossed dissociation. *Kolloid-Zeitschrift, 90*(2), 145–160.

Talwinder, S., Kahlon, S., Glines, L. A., & Lindgren, F. T. (1986). Analytical ultracentrifugation of plasma lipoproteins. *Methods in Enzymology*, *129*, 26–45.

Tiselius, A. (1937). A new apparatus for the electrophoretic analysis of colloidal mixtures. *Transactions of the Faraday Society*, *33*, 524–531.

Tissieres, A., & Watson, J. D. (1958). Ribonucleoprotein particles from *Escherichia coli*. *Nature*, *182*, 778–780.

Todd, G. P., & Haschemeyer, R. H. (1981). General solution to the inverse problem of the differential equation of the ultracentrifuge. *PNAS*, *78*, 6739–6743.

Todd, G. P., & Haschemeyer, R. H. (1983). Generalized finite element solution to One-Dimensional Flux Problems. *Biophysical Chemistry*, *17*, 321–336.

Trautman, R., Schumaker, V. N., Harrington, W. F., & Schachman, H. K. (1954). The determination of concentrations in the ultracentrifugation of two-component systems. *The Journal of Chemical Physics*, *22*, 555–559.

Van Holde, K. E., & Hansen, J. C. (1998). Analytical ultracentrifugation from 1924 to the present: A remarkable history. *Chemtracts—Biochemistry Molecular Biology*, *11*, 933–943.

Van Holde, K. E., & Baldwin, R. L. (1958). Rapid attainment of sedimentation equilibrium. *Journal of Physical Chemistry*, *62*(6), 734–743.

Vinograd, J., Bruner, R., Kent, R., & Weigle, J. (1963). Band-centrifugation of macromolecules and viruses in self-generating density gradients. *PNAS*, *49*, 902–910.

Vinograd, J., Lebowitz, J., Radloff, R., Watson, R., & Laipis, P. (1965). The twisted circular form of polyoma viral DNA. *Proceedings of the National Academy of Sciences of the United States of America*, *53*, 1104–1111.

Walter, J., Lohr, K., Karabudak, E., Reis, W., Mikhael, J., Peukert, W., et al. (2014). Multidimensional analysis of nanoparticles with highly disperse properties using multiwavelength analytical ultracentrifugation. *ACS Nano*, *8*, 8871–8886.

Walter, J., Sherwood, P. J., Lin, W., Segets, D., Stafford, W. F., & Peukert, W. (2015). Simultaneous analysis of hydrodynamic and optical properties using analytical ultracentrifugation equipped with multiwavelength detection. *Analytical Chemistry*, *87*(6), 3396–3403.

Waugh, D. F., & Yphantis, D. A. (1953). Transient solute distributions from the basic equation of the ultracentrifuge. *The Journal of Physical Chemistry*, *57*, 312–318.

Weiss, G. H., & Yphantis, D. A. (1965). Rectangular approximation for concentration-dependent sedimentation in the ultracentrifuge. *The Journal of Chemical Physics*, *42*, 2117–2123.

Weiss, G. H., & Yphantis, D. A. (1972). Pressure-dependent diffusion in velocity sedimentation. *Journal of Polymer Science Part A2—Polymer Physics*, *10*, 339–344.

Williams, J. W., vanHolde, K. E., Baldwin, R. L., & Fujita, H. (1958). The theory of sedimentation analysis. *Chemical Reviews*, *58*, 715–806.

Yphantis, D. A. (1959). Ultracentrifuge molecular weight averages during the approach to equilibrium. *The Journal of Physical Chemistry*, *63*, 1742–1747.

Yphantis, D. A. (1960). Rapid determination of molecular weights of peptides and proteins. *Annals of the New York Academy of Sciences*, *88*, 586–601.

Yphantis, D. A., & Waugh, D. F. (1956a). Ultracentrifugal characterization by direct measurement of activity. 1. Theoretical. *The Journal of Physical Chemistry*, *60*, 623–629.

Yphantis, D. A., & Waugh, D. F. (1956b). Ultracentrifugal characterization by direct measurement of activity. 2. Experimental. *The Journal of Physical Chemistry*, *60*, 630–635.

Zhao, H., Casillas, E., Shroff, H., Patterson, G. H., & Schuck, P. (2013). Tools for the quantitative analysis of sedimentation boundaries detected by fluorescence optical analytical ultracentrifugation. *PLoS One*, *8*, e77245.

Zhao, H., Ma, J., Ingaramo, M., Andrade, E., MacDonald, J., Ramsay, G., et al. (2014). Accounting for photophysical processes and specific signal intensity changes in fluorescence-detected sedimentation velocity. *Analytical Chemistry*, *86*, 9286–9292.

CHAPTER FOUR

Hydrodynamic Modeling and Its Application in AUC

Mattia Rocco*^,[1], Olwyn Byron†,[1]

*Biopolimeri e Proteomica, IRCCS AOU San Martino-IST, Istituto Nazionale per la Ricerca sul Cancro, Genova, Italy

†School of Life Sciences, College of Medical, Veterinary and Life Sciences, University of Glasgow, Glasgow, G12 8QQ, Scotland, United Kingdom

[1]Corresponding authors: e-mail address: mattia.rocco@hsanmartino.it; olwyn.byron@glasgow.ac.uk

Contents

Abstract

The hydrodynamic parameters measured in an AUC experiment, $s_{(20,\mathrm{w})}$ and $D^0_{\mathrm{t}(20,\mathrm{w})}$, can be used to gain information on the solution structure of (bio)macromolecules and their assemblies. This entails comparing the measured parameters with those that can be computed from usually "dry" structures by "hydrodynamic modeling." In this chapter, we will first briefly put hydrodynamic modeling in perspective and present the basic physics behind it as implemented in the most commonly used methods. The important "hydration" issue is also touched upon, and the distinction between rigid bodies versus those for which flexibility must be considered in the modeling process is then made. The available hydrodynamic modeling/computation programs, HYDROPRO, BEST, SoMo, AtoB, and *Zeno*, the latter four all implemented within the US-SOMO suite, are described and their performance evaluated. Finally, some literature examples are presented to illustrate the potential applications of hydrodynamics in the expanding field of multiresolution modeling.

Methods in Enzymology, Volume 562
ISSN 0076-6879
http://dx.doi.org/10.1016/bs.mie.2015.04.010

1. INTRODUCTION

Our understanding of the structural basis for biomacromolecular function relies heavily on high-resolution atomic coordinates (derived either from X-ray crystallography or NMR spectroscopy studies), yet the molecules of life generally function in complex, crowded solutions. Analytical ultracentrifugation (AUC) is one of several low- or meso-resolution methods that can generate absolute parameters representative of molecular conformation and oligomeric state in conditions that more closely resemble, e.g., the cell cytoplasm or the formulation of a biopharmaceutical. The role for the AUC and the interpretation of hydrodynamic data described in this chapter are unlikely to be diminished even with the success of free electron lasers that offer the possibility of determining atomic structures (Boutet et al., 2012) without the need for the comparatively large crystals required by even microfocus macromolecular crystallography. Indeed, despite decades of progress in structural biology, AUC continues to offer complementary structural insights even when atomic resolution data are available, since the crystal structure is not always representative of the solution conformation (Nakasako, Fujisawa, Adachi, Kudo, & Higuchi, 2001; Smolle et al., 2006; Trewhella, Carlson, Curtis, & Heidorn, 1988; Vestergaard et al., 2005) nor of the interactions key to biological function. The AUC is not alone in providing experimentally determined hydrodynamic parameters. It is the primary source of the sedimentation coefficient s, and sometimes of the translational diffusion coefficient D_t, dynamic light scattering (DLS) being a more widely used alternative for the latter. Other, more shape-sensitive hydrodynamic parameters are the rotational correlation (or relaxation) time τ and the intrinsic viscosity $[\eta]$, but the former can be determined from NMR spectroscopy only for a limited size range (fluorescence spectroscopy being an alternative, requiring labeling with its associated issues), while the latter comes from precise viscosity measurements.

The challenge then is how to relate such parameters with some structural descriptors. The field of "hydrodynamic modeling" (HM) started by using "single" simple geometrical bodies, such as spheres, ellipsoids of revolution, and cylinders, for which exact or very accurate expressions to calculate their frictional properties are available, to give a low-resolution representation of an otherwise unknown structure (Broersma, 1960; Burgers, 1938; Perrin, 1936). The development of theory to permit the use of arrays (Kirkwood, 1949, 1954) of the simplest geometrical body available, a sphere, to produce more detailed, although still very low-resolution, representations of biomacromolecules and their complexes paved the way for modern HM (Rotne & Prager, 1969; Yamakawa, 1970). This effort was led by Bloomfield and García de la Torre (reviewed by them in García de la Torre & Bloomfield, 1981) with many other contributors playing major roles (e.g., Brookes, Demeler, Rosano, & Rocco, 2010; Byron, 1997; Rai et al., 2005; Spotorno et al., 1997; Swanson, Teller, & de Haën, 1978; Wegener, 1982; Wegener, Dowben, & Koester, 1980; Zimm, 1982). Hydrodynamic bead modeling (HBM), as this technique is usually referred to, is still the major workhorse in the field, and it has undergone some fundamental changes since its elevation from the rough modeling of an unknown structure (e.g., Gregory et al., 1987) to the accurate calculation of the hydrodynamic parameters starting from atomic-resolution structures/models (e.g., Moeller et al., 2012).

It is very important to emphasize that comparing the kinds of monodimensional experimental parameters listed earlier with those calculated from a model cannot prove that the model is unique. Rather, such comparisons can more profitably be used to disprove alternative models, or to reinforce models derived from other techniques. Prime examples are the verification of a crystallographic structure with respect to its behavior in solution, the module/domain arrangement in modular/multidomain structures, the assessment of segmental flexibility, or the definition of conformational ensembles for intrinsically disordered structures. It should be also mentioned that over the years some alternative (to HMB) methods for the computation of hydrodynamic parameters from 3D structures/models have been developed; as this chapter unfolds, they will be referred to properly.

2. AUC AND HYDRODYNAMICS

Among the experimental techniques used to determine hydrodynamic parameters, AUC is prime: it is able to provide s and D_t, or, better, their infinite-dilution values corrected to standard conditions, $s^0_{(20,w)}$ and $D^0_{t(20,w)}$, both of which are directly related to the translational friction coefficient $f^0_{t(20,w)}$ of a macromolecule (the 20,w subscript indicates that the parameter has been standardized to solvent conditions of water at 20 °C). However, while $s^0_{(20,w)}$ can be determined with remarkable accuracy, that achievable for $D^0_{t(20,w)}$ is usually less so. This in unfortunate, because while $D^0_{t(20,w)}$ is directly related to $f^0_{t(20,w)}$ through the Stokes–Einstein relation

$$D^0_{t(20,w)} = \frac{k_b T}{f^0_{t(20,w)}} \tag{1}$$

where k_b is Boltzmann's constant and T is the absolute temperature (293.15 K), obtaining $s^0_{(20,w)}$ from $f^0_{t(20,w)}$ requires the precise knowledge of both the molecular weight M and the partial specific volume $\bar{v}_{(20,w)}$ of the macromolecule,

$$s^0_{(20,w)} = \frac{M\left(1 - \bar{v}_{(20,w)}\rho_{(20,w)}\right)}{f^0_{t(20,w)} N_A} \tag{2}$$

where $\rho_{(20,w)}$ is the density of water at 20 °C and N_A is Avogadro's number. It is the precise knowledge of $\bar{v}_{(20,w)}$, which is difficult to measure (Durchschlag & Jaenicke, 1982; Kratky, Leopold, & Stabinger, 1973; Perkins, 1986) and whose computation is notoriously complicated and often not very reliable (Durchschlag & Zipper, 1997; Laue, Shah, Ridgeway, & Pelletier, 1992), that limits the usefulness of measured $s^0_{(20,w)}$ values when comparing experimental and model-derived hydrodynamic data, since $f^0_{t(20,w)}$ is the parameter that can be computed from a structure. Without going into details, which can be found in the literature (e.g., García de la Torre & Bloomfield, 1981), the core relations of the hydrodynamic parameter computations for a rigid body of arbitrary shape is given by:

$$\begin{pmatrix} \mathbf{F} \\ \mathbf{T_O} \end{pmatrix} = \begin{pmatrix} \Xi_t & \Xi^T_{c,O} \\ \Xi_{c,O} & \Xi_{r,O} \end{pmatrix} \begin{pmatrix} \mathbf{u_O} \\ \omega \end{pmatrix} \tag{3}$$

which relates the force $\mathbf{F}$ and the torque $\mathbf{T_O}$ vectors to the linear and angular velocities $\mathbf{u_O}$ and ω, respectively, through the translational, rotational, and roto-translational coupling frictional tensors, Ξ_t, $\Xi_{r,O}$, $\Xi_{c,O}$, respectively, and the transpose of the latter, $\Xi^T_{c,O}$ (the "O" subscripts indicate their origin-dependence) (García de la Torre & Bloomfield, 1981). For instance, $f^0_{t(20,w)}$ is determined from the trace of $\Xi_{t(20,w)}$ as:

$$f^0_{t(20,w)} = (1/3)tr\left(\Xi_{t(20,w)}\right) \tag{4}$$

2.1 Hydrodynamic Bead Modeling

In the HBM method, the frictional tensors are obtained by solving a system of n linear equations with $3n$ unknowns (where n is the number of beads in the model), which stem from the fact that the contribution of each bead to the velocity field is described by a hydrodynamic interaction tensor (see García de la Torre & Bloomfield, 1981 and references therein). Several forms of this tensor have been developed over the years, the two most important being that developed by Rotne and Prager (1969) and Yamakawa (1970) for beads of equal radius σ (valid for non-overlapping and overlapping beads, alike)

$$\mathbf{T}_{ij} = \frac{1}{6\pi\eta_0\sigma}\left(\left(1 - \frac{9R_{ij}}{32\sigma}\right)\mathbf{I} + \frac{3\mathbf{R}_{ij}\mathbf{R}_{ij}}{32\sigma R_{ij}}\right) \tag{5}$$

where $\mathbf{T}_{ij}$ is the hydrodynamic interaction tensor, η_0 is the solvent viscosity, $\mathbf{R}_{ij}$ is the distance vector, and R_{ij} is the distance between the centers of beads i and j and $\mathbf{I}$ is the unit tensor, and another, the most widely used (developed by García de la Torre and Bloomfield (1977)), that can account for non-overlapping beads of different sizes

$$\mathbf{T}_{ij} = \frac{1}{8\pi\eta_0 R_{ij}}\left(\mathbf{I} + \frac{\mathbf{R}_{ij}\mathbf{R}_{ij}}{R_{ij}^2} + \frac{\sigma_i^2 + \sigma_j^2}{R_{ij}^2}\left(\frac{\mathbf{I}}{3} - \frac{\mathbf{R}_{ij}\mathbf{R}_{ij}}{R_{ij}^2}\right)\right). \tag{6}$$

Usually, the system is solved by matrix inversion techniques (García de la Torre & Bloomfield, 1981), but a more precise variational treatment is also available, which, however, does not include the calculation of [η] (Goldstein, 1985) and thus is seldom employed (see Spotorno et al., 1997). When matrix inversion procedures are utilized in HBM, the hydrodynamic interaction effect also leads to other issues, such as the need to introduce a "volume correction" when computing the rotational frictional coefficients and [η] (García de la Torre & Rodes, 1983).

2.2 Boundary Element Modeling

Two other major alternative HM methods have been developed. The first, known as the "boundary element" (BE) method, is based on solving directly the resistance problem (instead of the mobility problem, as for the HBM approach) (Youngren & Acrivos, 1975). It starts by writing the velocity field of the solvent flow as an integral over the particle surface, from which the surface stress force can then be obtained. Since the integral equation contains the original Oseen (1927) and Burgers (1938) hydrodynamic interaction tensor

$$\mathbf{T}_{ij} = \frac{1}{8\pi\eta_0 R_{ij}}\left(\mathbf{I} + \frac{\mathbf{R}_{ij}\mathbf{R}_{ij}}{R_{ij}^2}\right) \tag{7}$$

which is exact in the limit of infinitesimal surface elements, the BE method (implemented in the program BEST; Aragon, 2004, 2011, 2015) can yield more accurate results than HBM and can be applied to bodies of arbitrary shapes. However, it requires the discretization of the body surface into very small elements (hence its name), and, since for computational reasons this cannot be pushed over a certain limit, the generation of several discretized surfaces (with BEs of decreasing size) followed by extrapolation of the hydrodynamic parameters computed from these surfaces to the value thus predicted for a surface composed of infinitely small BEs (Aragon, 2004, 2011; Aragon & Hahn, 2006).

2.3 Electrostatic-Hydrodynamic Analogy

The second non-bead HM method is based on the electrostatic-hydrodynamic analogy (EHA), which stems from the fact that while electrostatic and hydrodynamic properties are governed by the Laplacian and the Navier–Stokes equations, respectively, a specific orientational averaging of the Navier–Stokes equations brings them into the form of the Laplace equation (Mansfield & Douglas, 2008; Mansfield, Douglas, & Garboczi, 2001). Thus, an approximate analogy exists between certain hydrodynamic and electrostatic properties. In particular, the hydrodynamic radius R_h and $[\eta]$ of a macromolecule are proportional, respectively, to the capacitance and polarizability of a perfect conductor whose shape is the same as the macromolecule. Moreover, there is a fundamental relationship between the Laplacian operator and random paths whose step size has a finite variance. This correspondence allows for the solution of the Laplacian equations to be formally expressed as averages over random walk trajectories. The advantage of the EHA approach is that it allows for the calculation of some transport properties ($D^0_{t(20,w)}$ and $[\eta]$, but not yet τ) for objects having essentially arbitrary shape. However, it requires the generation of properly "hydrated" models (see later).

3. THE HYDRATION ISSUE

There is at present no method to experimentally directly quantify hydrodynamic hydration and yet $D^0_{t(20,w)}$ and $s^0_{(20,w)}$ embody a modulation of $f^0_{t(20,w)}$ as the result of hydration effects: the higher the hydration for a molecule of given shape and mass, the greater the increase in its frictional surface and the greater the reduction in $D^0_{t(20,w)}$ and $s^0_{(20,w)}$. For the accurate and reliable computation of hydrodynamic parameters from an HBM, BE, or EHA model, some level of hydration has to be objectively assigned and modeled.

A static picture of protein hydration was proposed in the 1970s by Kuntz and Kauzmann (1974), based on data from NMR freezing studies, and further developed with particular reference to hydrodynamics by Squire and Himmel (1979). Typical proteins were thought to include about 0.3-0.4 g water/g protein, with glycosylation elevating this value. However, this model, in which the water of hydration was viewed as being relatively tightly bound to the protein surface, is inconsistent with subsequent magnetic relaxation dispersion measurements (Denisov & Halle, 1995) and molecular dynamics simulations (Henchman & McCammon, 2002; Makarov, Andrews, Smith, & Pettitt, 2000), both of which portrayed a highly dynamic interface between protein and water at which the rate of rotation and exchange of most water molecules are several orders of magnitude higher than biomolecular diffusion. This suggested that it would not be possible to easily model hydrodynamic hydration for proteins (or, indeed, other biomacromolecules). However, Halle and Davidovic (2003) developed a dynamic hydration model linking protein hydrodynamics with hydration dynamics. On the basis of their work, it appears that the increase in density of water molecules approaching charged/polar groups on the protein surface results in a local viscosity change. While directly taking into account such an effect would be quite complicated, it turns out that placing strongly bound, denser water molecules at well-defined positions on the protein surface constitutes a fortuitous but tractable method for objective protein hydration, resulting in the computation of accurate hydrodynamic parameters. This approach can be, and has been, extended to other types of biomacromolecules (e.g., nucleic acids, carbohydrates, etc.). For instance, the volume of theoretically bound water molecules can be added to each corresponding bead in the HBM prior to hydrodynamic computations, thereby accounting for the local variation in hydration, and hydration can be conceived as a non-uniform shell of water that moves with the macromolecule. This method of accounting for hydration was proposed by Rai et al. (2005) and is implemented in US-SOMO (Brookes, Demeler, & Rocco, 2010; Brookes, Demeler, Rosano, et al., 2010). But this approach to objective hydration has not been universally adopted and indeed cannot be applied when an HBM is not based on atomic coordinates. When the HMB is ultra-low-resolution and is based on, e.g., EM images or a molecular envelope restored from small-angle scattering data (see later), the distribution of hydration and hence the true hydrodynamic surface is unlikely to be known. Such an HBM could be hydrated prior to hydrodynamic computation by a process of uniform expansion (Byron, 1997) such that the volume of its constituent beads equals that estimated for the hydrated particle and its coordinates modified to represent the necessarily larger particle. This works tolerably well for representing the hydration of ultra-low-resolution globular particles but fails in the case of elongated molecules because it disproportionately lengthens the model and results in too great a reduction in $D^0_{t(20,w)}$ and $s^0_{(20,w)}$. Similarly, computed anhydrous values of s (s_0) can be very approximately "hydrated" by a nominal level δ (g water/g macromolecule) as follows (Byron, 1997):

$$s_\delta = s_0\left(\frac{\bar{\nu}}{\bar{\nu} + \delta\nu^0}\right)^{1/3} \quad (8)$$

where $\bar{\nu}$ is the partial specific volume of the protein and ν^0 is the specific volume of the solvent (both at the same temperature). The same applies to D_t.

In finite element (either bead-shell HMB or BE modeling), hydrodynamic hydration is represented by a uniform surface layer by virtue of the use of the atomic element radius (AER) of radius ≈2.9 Å in bead-shell HBM (García de la Torre, Huertas, & Carrasco, 2000; Ortega, Amorós, & García de la Torre, 2011) or by the addition of 1.1 Å to the atomic radii (in the case of proteins, for nucleic acids a slightly different hydration strategy is used) in BE modeling (Aragon & Hahn, 2006). Extrapolation to infinitely small shell beads or BE yields the final values for the hydrodynamic parameters that should match their experimentally determined counterparts.

4. SEGMENTAL FLEXIBILITY AND INTRINSICALLY DISORDERED STRUCTURES

Most of this chapter concerns the HM of rigid bodies, i.e., compact biomacromolecules for which flexibility should not be an issue. But for more complex structures that include domains connected by flexible linkers, significant mobile surface loops, or regions of intrinsic disorder, flexibility should be evaluated and taken into account when interpreting experimentally determined values for $D^0_{t(20,w)}$ and $s^0_{(20,w)}$. There are several approaches that can be taken in this instance (see e.g., Byron, 2015 for details). In Monte Carlo (MC) rigid body modeling (as in the program MONTEHYDRO; García de la Torre, Ortega, Pérez Sánchez, & Hernández Cifre, 2005) different conformations of models comprising beads joined by conceptually flexible connectors are generated using MC methods. Solution parameters (including $D^0_{t(20,w)}$ and $s^0_{(20,w)}$) are computed (using rigid body HBM) and reported as a conformational average that can be compared with the experimentally determined counterparts. An analogous strategy is adopted when using the discrete molecular dynamics (DMD) tool in US-SOMO (Brookes, Demeler, & Rocco, 2010; Brookes, Demeler, Rosano, et al., 2010), reviewed by Byron (2015). Numerous conformers are generated not by MC methods but through perturbations to an initial starting structure driven by thermal processes. Very precise solution parameters have been computed, using BEST (Aragon, 2004, 2011, 2015), for models generated using atomistic (as opposed to discrete) molecular dynamics. This is a computationally intensive process but will increasingly become tractable for most users. The final mainstream approach to modeling the hydrodynamics of flexible systems is to simulate the internal dynamics of the molecular system by connecting the rigid parts of the bead model with conceptual flexible linkers. Brownian dynamics simulations are then used (by the program SIMUFLEX; García de la Torre et al., 2009) to compute the molecular trajectory from which solution parameters are then generated.

5. AVAILABLE PROGRAMS FOR RIGID BODY HM

5.1 Modeling in the Absence of Atomic Coordinates

It is not always possible to utilize atomic coordinates (e.g., from the Protein Data Bank (PDB; http://www.rcsb.org) (Berman et al., 2000)) as a starting point for HM. For instance, there may not be a high-resolution structure for the molecule of interest or for a sufficiently similar homolog. Instead, modeling can be done based on much lower resolution information such as an electron density map (using HYDROMIC; García de la Torre, Llorca, Carrascosa, & Valpuesta, 2001) or a geometric-shape populated with beads by HYDROPIX (García de la Torre, 2001) or small-angle X-ray or neutron scattering (SAXS, SANS) data from which an *ab initio* multisphere model can be generated (using, e.g., DAMMIF; Franke & Svergun, 2009) or a completely *de novo* model created bead-by-bead with AtoB (Byron, 1997; Kreiner, Byron, Domingues, & van der Walle, 2009). For both of these latter two routes, $D^0_{t(20,w)}$ and $s^0_{(20,w)}$ can then be computed using US-SOMO (Brookes, Demeler, & Rocco, 2010; Brookes, Demeler, Rosano, & Rocco, 2010) or HYDRO++ (García de la Torre, del Rio, & Ortega, 2007; see, e.g., Ackerman, Harnett, Harnett, Svergun, & Byron, 2003).

5.2 Modeling Based on Atomic Coordinates

More often, the user will have an atomic resolution structure or model to use as the basis for HM in which case there are five main methods that can be used to compute $D^0_{t(20,w)}$, $s^0_{(20,w)}$, and other solution parameters for the model. These are summarized in Table 1 and the implementation of four of the methods (in US-SOMO) is described more fully by Brookes and Rocco (2015). All methods require the input of solvent properties (e.g., density, viscosity, etc.) which should, ideally be determined experimentally but can, alternatively, be quite reliably computed using the program SEDNTERP (Hurton et al., 2012; Laue et al., 1992).

The following sections summarize the inputs required for four of the five main rigid body HM methods, the key steps in the construction of the models and computation of solution parameters and options available to the user for controlling these processes. *Zeno* is excluded since it is, in this context, simply an alternative to the supermatrix inversion method for computation of $D^0_{t(20,w)}$ and $s^0_{(20,w)}$ (and other solution parameters). AtoB models may also be evaluated using HYDRO++ (García de la Torre et al., 2007).

5.2.1 HYDROPRO

Inputs	PDB file (no checks for completeness are performed by the program and so should ideally be performed by the user), molecular weight (M), and partial specific volume ($\bar{v}_{(20,\,w)}$) of the macromolecule.
Process	**1.** Replaces all atoms in the input PDB file with equally sized beads (primary model). The default radius (2.84 Å) of these "Atomic Elements" (AER) provided in the current HYDROPRO was arrived at by comparing experimental and calculated values for some hydrodynamic properties for a series of test proteins (Ortega et al., 2011). **2.** Generates a series of shell models by covering the primary model with touching shell beads of equal and diminishing size, whose radius is either automatically calculated or based on the user input (a typical model is shown in Fig. 1A). **3.** Computes the hydrodynamic properties of each shell model. **4.** Automatically performs the extrapolation of the computed parameters to zero shell bead size to arrive at final outputs. No visualization of the extrapolation is provided.
Options	AER value; manual minimum and maximum shell bead radius; intervals between them (default is automatic, with a maximum of 2000 shell beads and 7 shell bead sizes).

Table 1 Summary of the Five Main Methods for the Computation of Solution Parameters for Atomic Coordinate Structures or Models

Method	Reference(s) and url	Overview of Capabilities
HYDROPRO	Ortega et al. (2011) http://leonardo.inf.um.es/macromol/programs/hydropro/hydropro.htm	Constructs shell models from PDB files (or bead-per-atom/residue models) and computes $D^0_{t(20,w)}$ and $s^0_{(20,w)}$ (and other solution parameters) for shell models of decreasing sphere size, extrapolating values to those that would be obtained for infinitely small spheres to arrive at the final outputs.
BEST	Aragon (2004) and Aragon (2011) http://esmeralda.sfsu.edu/	Constructs surface models composed of tessellated triangles from PDB files and computes $D^0_{t(20,w)}$ and $s^0_{(20,w)}$ (and other solution parameters) for models of decreasing triangle size, extrapolating values to those that would be obtained for infinitely small triangles to arrive at the final outputs. BEST is available in US-SOMO.
AtoB	Byron (1997) http://www.somo.uthscsa.edu/	Constructs bead models from PDB files using the modified AtoB algorithm wherein atoms are averaged into beads of dimension governed by the user choice of cubic grid size. $D^0_{t(20,w)}$ and $s^0_{(20,w)}$ (and other solution parameters) are computed for these models using either the supermatrix inversion method to solve the system of linear equations taking into account the modified García de la Torre & Bloomfield hydrodynamic interaction tensor or the *Zeno* method (below). AtoB is available in US-SOMO.
SoMo	Rai et al. (2005), Brookes, Demeler, and Rocco (2010), and Brookes, Demeler, Rosano, and Rocco (2010) http://www.somo.uthscsa.edu/	Constructs bead models from PDB files using the *SoMo* algorithm representing, e.g., amino acid residues with one bead for the main-chain segment and one bead for the side-chain. $D^0_{t(20,w)}$ and $s^0_{(20,w)}$ (and other solution parameters) are computed for these models using the supermatrix inversion method or the *Zeno* method (below). *SoMo* is available in US-SOMO.
Zeno	Kang, Mansfield, and Douglas (2004) http://web.stevens.edu/zeno/	Computes $D^0_{t(20,w)}$ and $[\eta]$ (and other parameters) for any model by enclosing it in a sphere from whose internal surface a series of random walks is launched. The fraction of these walks that reaches the model determines the value of the solution parameter. *Zeno* is available in US-SOMO (providing also $s^0_{(20,w)}$).

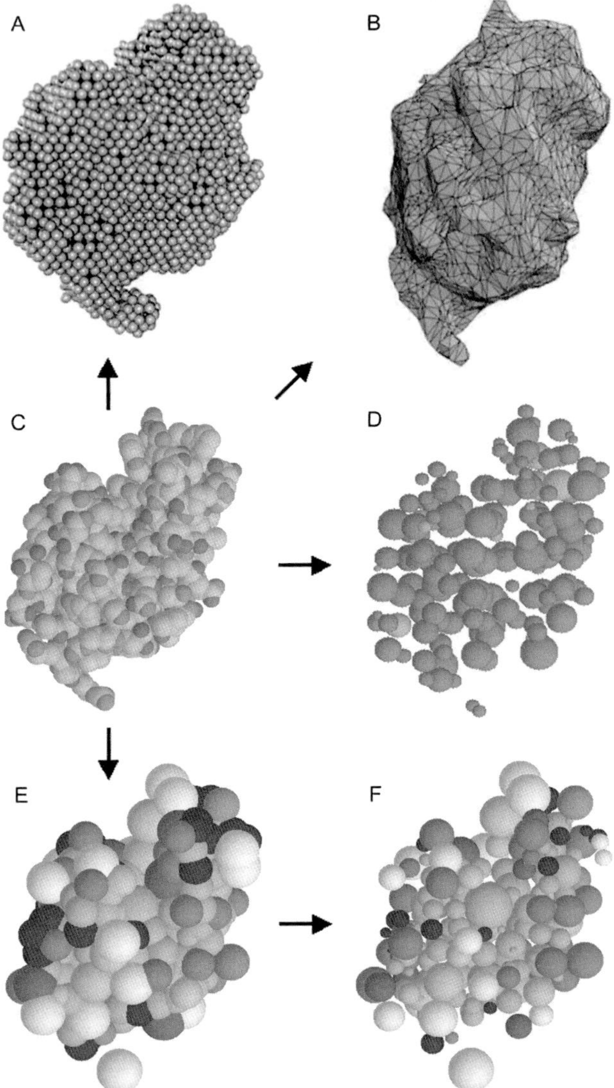

Figure 1 Hydrodynamic models generated by the available modeling programs starting from an atomic resolution structure ((C) lysozyme 6lyz.pdb). (A) A bead-shell model generated by HYDROPRO. (B) A tessellated model generated by BEST. (D) A 5 Å grid bead model generated by the AtoB method in US SOMO. (E) A direct correspondence bead model, with overlaps, generated by the SoMo method in US SOMO. (F) Same as in (E) after overlap removal. In (D), red and orange are the exposed and buried beads, respectively. In (E) and (F), the color coding is: blue, main-chain; cyan, hydrophobic; magenta, non-polar; red, polar; yellow, basic; green, acidic; white, fused beads; orange, buried beads. (See the color plate.)

5.2.2 BEST (as Implemented Within US-SOMO)

Inputs	PDB file (no checks are performed in the original program; in the US-SOMO implementation, checks are automatically performed but the results will not affect BEST execution). In its actual implementation, BEST does not directly compute *s*. A job package created within the Batch/Cluster operations module in US-SOMO is then submitted to a remote supercomputer cluster for processing. Upon completion, the job package is retrieved and extracted for local analysis.
Process	**1.** MSROLL (Connolly, 1993) produces a finely triangulated surface of the input structure using the standard atomic radii, to which (for proteins) 1.1 Å has been added to account for a uniform hydration layer (based on the comparison between experimental and computed data for a series of test proteins (Aragon, 2004)). In US-SOMO, they are stored in a file called *best.radii*. **2.** The BEST module COALESCE generates a number of less finely triangulated surfaces for BE analysis (a typical model is shown in Fig. 1B). **3.** BEST then computes the hydrodynamic parameters for these triangulated models and stores them in a comma-separated values (csv) file. **4.** The *.csv file can be uploaded to the US-SOMO BEST analysis module where a plot of each single hydrodynamic parameter type versus 1/(number of triangular plates) is shown together with a linear regression analysis. A Q-test analysis (Dean & Dixon, 1951) can be performed to objectively find a single outlier (manual discarding is also available, but not recommended). The updated extrapolation(s) can be saved in another *.csv file for further processing.
Options	MSROLL probe radius (default = 1.5 Å), starting finesse angle and maximum output of triangles; choice between an automatic (heuristic, based on *M* of the macromolecule) or manual determination of the minimum and maximum number of triangular plates produced by the COALESCE module and of the number of BE models produced (6 are recommended); manual input of *M*; expansion (or shrinkage) of the radii by an user-entered factor; possibility of entering a different, user-provided atomic radii file.

5.2.3 AtoB in US-SOMO

Inputs	PDB file, with properly defined atoms and residues in the *somo.atom* and *somo.residue* conversion files. Checks are performed to verify this correspondence and the completeness of coded residues (approximate methods are offered by default if residues are not coded or if missing atoms are detected within a coded residue). M and $\bar{v}_{(20,\,\mathrm{w})}$ are automatically computed using the *somo.residue* file.
Process	**1.** A cubic grid with user-defined size (default = 5 Å) is superimposed on the structure. Atoms found within each "cubelet" are replaced by a bead whose volume is computed from the sum of volumes of the atoms it represents, plus that of the theoretically bound water of hydration associated with each atom (currently defined for proteins and carbohydrates only; see Section 3). Beads are centered either on the center of mass of the represented atoms (default) or the cubelet center to generate a primary bead model in which many of the beads will overlap with one another. **2.** An accessible surface area (ASA) screen is performed on this primary bead model. Overlaps between exposed beads are first removed using a synchronous procedure in which the radii of all overlapping beads are reduced in steps by a fixed percentage and at the same time their centers are moved toward the molecular surface by the same amount along a line connecting them to the center of mass of the ensemble ("outward translation," OT), in an attempt to preserve its original contour surface. The procedure is repeated until no overlaps remain between the originally exposed beads. It is then applied to the originally buried beads, but without OT. Finally, another ASA screen is performed to reclassify exposed and buried beads after overlap removal. A resulting model is shown in Fig. 1D. **3.** Hydrodynamic calculations are performed by the supermatrix inversion method utilizing only the exposed beads (Zeno can also be employed).
Options	Grid size; all parameters and choices pertinent to the various steps described earlier are under complete user control (e.g., ASA probe radius and threshold values; choice between synchronous and hierarchical overlap removal methods; overlap tolerance threshold; exclusion/inclusion of buried beads and other features in hydrodynamic computations).

5.2.4 SoMo in US-SOMO

Inputs	As for AtoB, above.
Process	**1.** Based on the *somo.residue* definitions, each residue is approximated with one or more beads. For protein residues, by default one bead, appropriately positioned, is used for each main-chain and side-chain segment; one bead is used for carbohydrate residues, while for nucleic acids residues three beads are used (for phosphate, sugar, and bases, respectively); lipids, prosthetic groups, etc. are typically represented by more beads. The volume of each bead is calculated from the anhydrous volume of the atoms it represents, plus that of the theoretically "bound" water of hydration (see Section 3). **2.** An ASA screen is first performed on the original structure and beads are positioned in the following order according to whether they represent (*i*) exposed side-chains, (*ii*) exposed main-chain segments, or (*iii*) buried side- and main-chain segments. This results in a "primary" model with many overlaps, as shown in Fig. 1E. **3.** Hydrodynamic computations can be performed on this primary SoMo bead model using the *Zeno* method. **4.** For computations using the supermatrix inversion method, overlaps need to be removed as follows: (*i*) exposed beads whose overlap exceeds a predetermined threshold are fused together; (*ii*) remaining overlaps are removed using either the synchronous procedure described earlier for the AtoB models or a hierarchical procedure (overlaps between bead pairs are found, ranked in descending order then removed one at a time). This is done first for exposed side-chain beads, using OT, then for main-chain beads, without OT, and finally for buried beads, again without OT. At the end, another ASA screen is performed to reclassify exposed and buried beads after overlap removal. A final model is shown in Fig. 1F. **5.** Hydrodynamic calculations are performed with the supermatrix inversion method utilizing only the exposed beads (*Zeno* can also be employed).
Options	As for AtoB, above, minus the grid size, plus residues-to-bead conversion rules and overlapping bead fusion threshold.

6. COMPARING THE PERFORMANCE OF VARIOUS METHODS

We have recently performed an in-depth comparison of the performance of the various hydrodynamic computation methods described in this chapter (Rocco & Byron, 2015). This involved selecting some 27 proteins ranging in mass from 12 to 466 kDa whose crystallographic structures are in the PDB and for which reliable experimental values for $D^0_{t(20,w)}$ and $s^0_{(20,w)}$ are found in the literature. These values were carefully evaluated for consistency, proper correction to standard conditions, and extrapolation to infinite dilution. The PDB structures were checked and completed where necessary to ensure that all residues/atoms present in the samples used for the $D^0_{t(20,w)}$ and $s^0_{(20,w)}$ determinations were also present in the structures used for the computation. The structures were then input to US-SOMO for analysis with the methods implemented therein (AtoB with 5 and 2 Å grids, SoMo models without overlaps for which solution parameters were computed with both the supermatrix inversion and *Zeno* methods, SoMo models with overlaps for which solution parameters were computed with *Zeno*, and BEST using both the manual and heuristic approaches to determine the minimum and maximum number of triangular plates in six models), and, separately, to the Windows version of HYDROPRO (Ortega et al., 2011), using the default settings. A graphical summary of the main results of this study is presented in Fig. 2 where the % deviation (Δ%) of calculated values for $D^0_{t(20,w)}$ (A) and $s^0_{(20,w)}$ (B) from their experimental counterparts is plotted as a function of molecular weight (the dotted horizontal lines in both panels delimit the average experimental SD).

The first observation (Fig. 2) is that notwithstanding the better experimental accuracy by which $s^0_{(20,w)}$ can be determined over $D^0_{t(20,w)}$, the Δ% between experimental and computed values is greater for the former than for the latter, irrespective of the computational method used. Since most HM methods compute $f^0_{t(20,w)}$, from which $D^0_{t(20,w)}$ is directly calculated (Eq. 1) while $s^0_{(20,w)}$ requires the knowledge of M and $\bar{v}_{(20,\,w)}$ (Eq. 2), this most likely is due to the usage of inaccurate $\bar{v}_{(20,\,w)}$ values, which are rarely accurately experimentally determined, and thus have had to be calculated from macromolecular composition data. In fact, even if substantial progress has been made in this respect (Durchschlag & Zipper, 2005, 2008), it remains a problematic issue, especially since the dependence of $\bar{v}_{(20,\,w)}$ on solvent composition has not so far been extensively addressed. Thus,

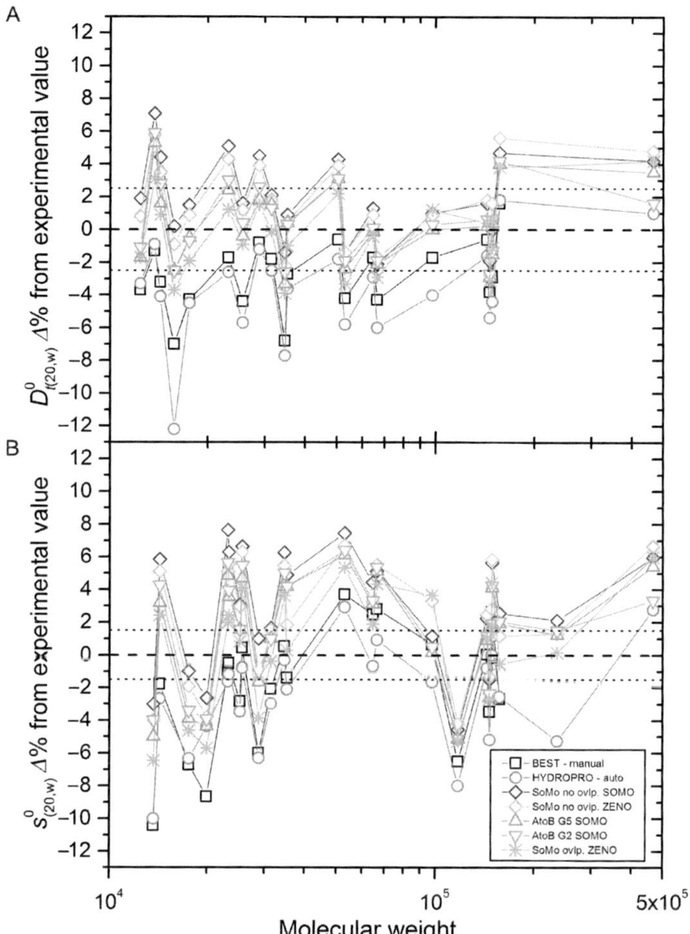

Figure 2 Plots of the percent difference (Δ%) between calculated and experimental values of $D^0_{t(20,w)}$ (A) and $s^0_{(20,w)}$ (B) as a function of their molecular weight for 22 proteins. The correspondence between symbols and modeling/computational methods utilized is shown in the inset to (B). The dotted lines above and below the horizontal Δ% = 0 dashed lines represent the average standard deviation of the experimental values. *Data taken from Rocco and Byron (2015).*

the first take-home message is that whenever possible, the determination of $D^0_{t(20,w)}$ by AUC experimental and data analytical methods not requiring the knowledge of $\bar{\nu}_{(20,\,w)}$ should be sought (DLS being a valid alternative experimental technique).

The performances of the various methods can be now examined. By looking at the plots in Fig. 2A, it is apparent that on average both BEST and HYDROPRO tend to underestimate $D^0_{t(20,w)}$ (by 3–4%), the SoMo models without overlaps (evaluated by either supermatrix inversion or *Zeno*) tend to overestimate them slightly less ($\sim+2\%$) and the AtoB models perform even better ($\sim+1\%$). We have interpreted these results as an indication of excessive "hydration" by BEST and HYDROPRO, and surface shrinking, notwithstanding the OT, in the SoMo models without overlaps; the larger number of smaller beads used by AtoB with grids ≤5 Å most probably attenuates this problem (Rocco & Byron, 2015). Strikingly, when SoMo models with overlaps were input to *Zeno* the best agreement was reached, with an average Δ% of 0% (range −4 to +4%), indicating that probably the preferential hydration scheme employed in SoMo and AtoB models presents some clear advantages (Rocco & Byron, 2015).

7. A STEP-BY-STEP GUIDE TO PERFORMING CORRECT HM USING ATOMIC COORDINATES

1. We assume here that a suitable high-resolution set of coordinates exists for the biomacromolecule for which you have measured some hydrodynamic parameters. The structure could come either from direct structural studies, or be a homology model generated with programs such as SWISS-MODEL (Arnold, Bordoli, Kopp, & Schwede, 2006; Biasini et al., 2014; Guex, Peitsch, & Schwede, 2009; Kiefer, Arnold, Kuenzli, Bordoli, & Schwede, 2009) (http://swissmodel.expasy.org/), I-TASSER (Roy, Kucukural, & Zhang, 2010; Yang et al., 2015; Zhang, 2008) (http://zhanglab.ccmb.med.umich.edu/I-TASSER/), or Phyre2 (Kelley & Sternberg, 2009) (http://www.sbg.bio.ic.ac.uk/phyre2/html/page.cgi?id=index).
2. Check if your structure is complete, i.e., it has all residues and all atoms within each residue. If not, there are a number of programs for completing structures (e.g., WHAT IF (Vriend, 1990) (http://swift.cmbi.ru.nl/whatif/) for completing amino acid residues or MODELLER (Fiser, Do, & Sali, 2000; Marti-Renom et al., 2000; Sali & Blundell, 1993; Webb & Sali, 2014) (https://salilab.org/modeller/) to insert missing residues). More complicated is the case when prosthetic groups (e.g., carbohydrates or detergents) are missing. In this instance, the PDB can be searched for structures with the same or similar prosthetic group which can then be appended to your structure, or modified manually as appropriate, and perhaps subjected to some refinement via, e.g., energy

minimization (using, e.g., Swiss-PdbViewer (Guex & Peitsch, 1997) (http://spdbv.vital-it.ch/)). Although software like HYDROPRO and BEST will take as input even incomplete structures, this is not what you want: it is better to have some incorrectly modeled than missing parts in your structure, since they contribute to the hydrodynamics in solution!

3. For an initial quick evaluation of the hydrodynamic properties, HYDROPRO can then be used.
4. Is your structure rigid or does it have regions that incorporate significant flexibility? Tools are available for a quick evaluation based on primary sequence properties (e.g., PROFbval (Schlessinger, Yachdav, & Rost, 2006) (https://rostlab.org/owiki/index.php/PROFbval) or Predyflexy (de Brevern, Bornot, Craveur, Etchebest, & Gelly, 2012) (http://www.dsimb.inserm.fr/dsimb_tools/predyflexy/)). Starting from a structure, normal modes analysis can provide evidence of hinge regions (e.g., elNémo; Suhre & Sanejouand, 2004) (http://www.sciences.univ-nantes.fr/elnemo/) or FlexServ (Camps et al., 2009) (http://mmb.irbbarcelona.org/FlexServ/). Finally, a DMD or full MD analysis can tell you more, and eventually produce an extended ensemble for hydrodynamic analysis.
5. For sufficiently rigid proteins, and if only $s^0_{(20,\mathrm{w})}$ and $D^0_{\mathrm{t}(20,\mathrm{w})}$ are sought, AtoB with a suitable grid size (5 Å is a good initial choice) will produce accurate results and with reasonable computing times. SoMo models with overlaps followed by *Zeno* computations are even more accurate and will also produce reliable intrinsic viscosity values. While presently *Zeno* is slower than the supermatrix inversion method, this should change in the near future with the implementation of a ~100-fold faster version. BEST needs a reevaluation of the hydration issue before it can be recommended over the other methods, and, since it requires significantly greater computing power, would be advantageous only when the rotational diffusion parameters are sought (HBM is inaccurate in this respect, and *Zeno* presently does not compute rotational diffusion).
6. If flexibility is an issue, a variety of options exists, from full Brownian dynamics simulations with SIMUFLEX (García de la Torre et al., 2009) to DMD/MD methods followed by ensemble averaging after computing the hydrodynamics with either AtoB or SoMo. If preferred conformations are to be searched for, the Model Classifier utility in US-SOMO can rank models according to their closeness to experimentally measured parameters.

8. SELECTED LITERATURE EXAMPLES

As by far the most historically utilized HM program, HYDROPRO has featured in many excellent papers, a few of which will be highlighted here. Doucette et al. (2004) used HYDROPRO to predict $s^0_{(20,w)}$ for the monomer (1.8 S) and dimer (2.9 S) form of human superoxide dismutase (SOD1) from atomic coordinates so that they could determine the association state of species observed by AUC (with $s^0_{(20,\,w)} = 1.8$ and 3.1 S, respectively). Costenaro, Grossmann, Ebel, and Maxwell (2005) used the program to calculate $s_{6,b}$ (i.e., s at 6 °C in a given buffer) for dummy residue models that were generated from SAXS data in order to demonstrate consistency between the dummy atom models ($s_{6,b} = 5.4 \pm 0.1$ S) and the AUC data ($s_{6,b} = 5.25 \pm 0.1$ S) for full-length *E. coli* DNA gyrase A. A similar approach was taken by Vunnam, Flint, Balbo, Schuck, and Pedigo (2011) who elected instead to use BEST to compute $s^0_{(20,w)}$ for the monomer and dimer form (2.06 and 3.21 S, respectively) of murine NCAD12, the first two extracellular domains of N-cadherin, in order to make a comparison with the experimental values (2.10 and 3.27 S, respectively). AUC (especially when used in combination with SAXS and SANS) sometimes reveals that a biomacromolecule adopts a conformation in solution that differs from that observed in crystalline forms. This is the case for the human telomere G-quadruplex for which there are NMR (in the presence of Na^+) and crystal (in the presence of K^+) structures. $s^0_{(20,w)}$ determined by Li, Correia, Wang, Trent, and Chaires (2005) in both Na^+ and K^+ was compared with that computed using HYDROPRO for the NMR and crystal structures. $s^0_{(20,w)}$ computed for the NMR structure (1.86 ± 0.02 S) agreed with the experimentally determined value (1.92 ± 0.01 S) while the value computed from the crystal structure (1.69 ± 0.02 S compared with 2.11 ± 0.01 S determined by AUC) did not, suggesting (when taken together with other data) that the biologically relevant structure of the human telomere G-quadruplex (i.e., in the presence of K^+) is very different from that in the crystal structure. Mavaddat et al. (2000) used AtoB to generate hydrodynamic models of the extracellular region of SLAM for evaluation with SOLPRO (García de la Torre, Carrasco, & Harding, 1997), an ancillary program of HYDRO, in order to interpret experimentally determined values of $s^0_{(20,w)}$. In particular, they were able to establish an end-to-end (rather than side-to-side) geometry for the dimer species but noted that their modeling was inexact because they had not accounted for the probable flexibility of the (significant) sugar moieties.

The combination of AtoB and SOLPRO was also used (by Harkiolaki, Gilbert, Jones, and Feller et al. (2006)) to demonstrate that the first three species observed in the *c*(*s*) size distribution analysis of sedimentation velocity data ($s^0_{(20,\,w)} = 1.4$, 1.9, and 3.0 S, respectively) for the SH3 domain of Crk-like (CRKL) corresponded to monomer, dimer, and tetramer (Fig. 3A). As with their earlier publications, the authors reported values of $s^0_{(20,w)}$ computed for a range of hydration levels (achieved through multiplication of the anhydrous $s^0_{(20,w)}$ by a factor) ($s^0_{(20,\,w)} = 1.31 - 1.56$, 1.80–2.04, and 2.84–3.22 S, respectively). In a very different application of AtoB, Schluttig et al. (2008) used the method to provide one of three levels of coarse-graining of protein surfaces for use in a Langevin equation approach to modeling the formation of protein–protein encounter complexes (Fig. 3B). AtoB was of particular utility in this study as it provided the authors with complete control over the resolution (i.e., detail) of the protein surface, a variable they wanted to assess.

One of the strengths of SoMo is its ability to generate precise models not only for proteins but also other biomacromolecules. This is well illustrated by Rosano and Rocco (2010) in their study of integrin $\alpha_{IIb}\beta_3$ which includes models for the glycoprotein either solubilized in octyl glucoside or embedded in MSP1D1-DMPC/DMPG-nanodiscs (Fig. 3C) and which allowed the authors to conclude that only an extended/closed (as opposed to bent/closed or partially extended/closed) model of the protein satisfactorily agrees with the experimentally determined $D^0_{t(20,w)}$ and $s^0_{(20,w)}$. Pang et al. (2010) used SoMo to assign the two distinct species observed in sedimentation velocity studies of pre-T-cell antigen receptor (pre-TCR) to monomer ($s^0_{(20,\,w)} = 3.1$ S) and dimer ($s^0_{(20,\,w)} = 5.0$S). Nishio et al. (2010) used the program to compute $s^0_{(20,w)}$ for the luminal carbohydrate recognition domain of the transmembrane lectin ERGIC-53 (2.93 S), the soluble calcium-binding protein MCFD2 (1.67 S), and their complex (3.39 or 3.43 S, depending on the contact used to form the complex). The computed values agreed very well with the experimental counterparts (1.73, 2.70, and 3.24 S, respectively) confirming a 1:1 stoichiometry and supporting the contact 1-mode of interaction in solution (for which $s^0_{(20,w)}$ was computed to be 3.39 S, marginally closer to the experimental value than the contact 2 value (3.43 S)). Finally Moeller et al. (2012) provide a great example of the use of

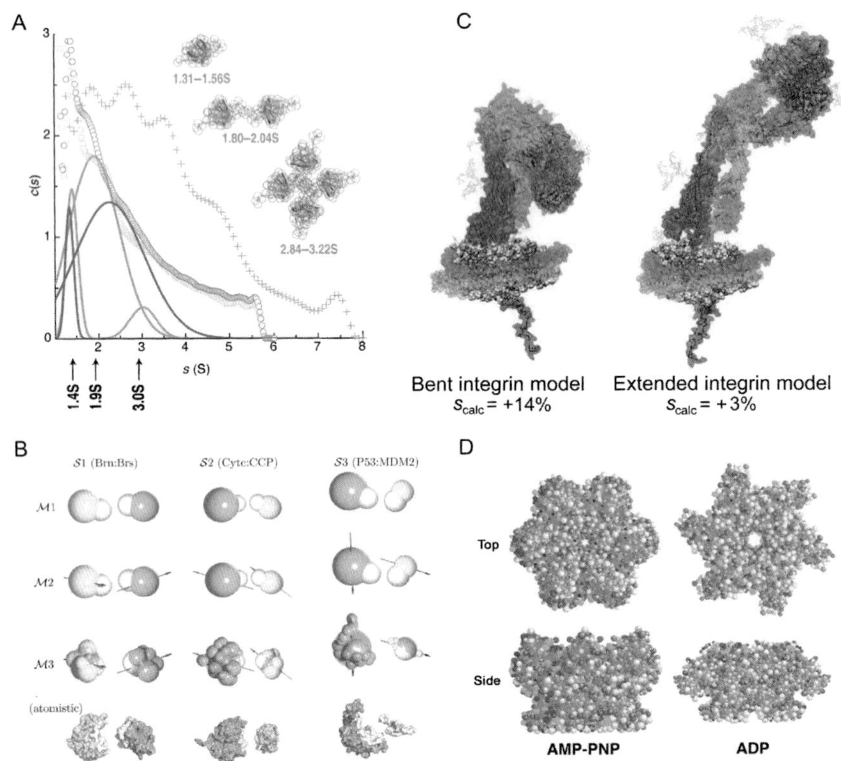

Figure 3 Examples of the use of AtoB (A, B) and SoMo (C, D) drawn from the literature. AtoB was used to (A) generate models for monomer, dimer, and tetramer for the SH3 domain of CRKL for which values of $s^0_{(20,w)}$ were computed (by SOLPRO) for a range of hydration levels and compared with $s^0_{(20,w)}$ determined by *c*(*s*) size distribution analysis of sedimentation velocity data and (B) provide one (M3) of three levels (M1–3) of coarse-graining of protein surfaces for use in a Langevin equation approach to modeling the formation of protein–protein encounter complexes. SoMo was used to generate (C) nanodisc-embedded integrin $\alpha_{IIb}\beta_3$ bent/closed (left) and extended/closed (right) models (their % deviation from experimental $s^0_{(20,w)}$ values is indicated), and (D) models of *N*-ethylmaleimide sensitive factor (NSF) in ADP- and AMP-PNP-bound states. *(A) Adapted from Harkiolaki et al. (2006) with permission. (B) Adapted from Schluttig, Alamanova, Helms, and Schwarz (2008) with permission. (C) Adapted from Rosano and Rocco (2010) with permission. (D) Adapted from Moeller et al. (2012) with permission.* (See the color plate.)

SoMo for computation of $s^0_{(20,w)}$ for the wild-type and mutants of *N*-ethylmaleimide sensitive factor in ADP- and AMP-PNP-bound states (Fig 3D) in order to define the structures underlying conformational changes observed during the ATPase cycle.

9. CONCLUDING REMARKS

HM is a mature technique that has already proven its worth on many occasions, for instance, the examples presented in the preceding section. That said, there is always room for improvement. At the computational level, speeding up by parallelization or implementation on high-performance hardware (such as the CUDA graphic cards used to boost performance in video games) is already taking place for several of the programs mentioned herein. Web servers might become an attractive option, offering high-performance computations. Better integration with other software used in multiresolution modeling is another important line of development, as illustrated by the US-SOMO suite that already comprises advanced small-angle scattering data analysis tools. As general users become more confident with the use of HM methods, we predict a bright future for hydrodynamic modeling!

REFERENCES

Ackerman, C. J., Harnett, M. M., Harnett, W., Svergun, D. I., & Byron, O. (2003). 19 Å solution structure of the filarial nematode immunomodulatory protein, ES-62. *Biophysical Journal, 84*(1), 489–500.

Aragon, S. (2004). A precise boundary element method for macromolecular transport properties. *Journal of Computational Chemistry, 25*(9), 1191–1205.

Aragon, S. R. (2011). Recent advances in macromolecular hydrodynamic modeling. *Methods, 54*(1), 101–114.

Aragon, S. R. (2015). Accurate hydrodynamic modeling with the boundary element method. In S. Uchiyama, F. Arisaka, W. F. Stafford, III, & T. M. Laue (Eds.), *Analytical ultracentrifugation: Instrumentation, software and application*. New York: Springer.

Aragon, S., & Hahn, D. K. (2006). Precise boundary element computation of protein transport properties: Diffusion tensors, specific volume, and hydration. *Biophysical Journal, 91*(5), 1591–1603.

Arnold, K., Bordoli, L., Kopp, J., & Schwede, T. (2006). The SWISS-MODEL workspace: A web-based environment for protein structure homology modelling. *Bioinformatics, 22*(2), 195–201.

Berman, H. M., Westbrook, J., Feng, Z., Gilliland, G., Bhat, T. N., Weissig, H., et al. (2000). The protein data bank. *Nucleic Acids Research, 28*(1), 235–242.

Biasini, M., Bienert, S., Waterhouse, A., Arnold, K., Studer, G., Schmidt, T., et al. (2014). SWISS-MODEL: Modelling protein tertiary and quaternary structure using evolutionary information. *Nucleic Acids Research, 42*(W1), W252–W258.

Boutet, S., Lomb, L., Williams, G. J., Barends, T. R., Aquila, A., Doak, R. B., et al. (2012). High-resolution protein structure determination by serial femtosecond crystallography. *Science, 337*(6092), 362–364.

Broersma, S. (1960). Rotational diffusion coefficient of a cylindrical particle. *Journal of Chemical Physics, 32*(6), 1626–1631.

Brookes, E., Demeler, B., & Rocco, M. (2010). Developments in the US-SOMO bead modeling suite: New features in the direct residue-to-bead method, improved grid routines, and influence of accessible surface area screening. *Macromolecular Bioscience, 10*(7), 746–753.

Brookes, E., Demeler, B., Rosano, C., & Rocco, M. (2010). The implementation of SOMO (SOlution MOdeller) in the UltraScan analytical ultracentrifugation data analysis suite: Enhanced capabilities allow the reliable hydrodynamic modeling of virtually any kind of biomacromolecule. *European Biophysics Journal with Biophysics Letters, 39*(3), 423–435.

Brookes, E., & Rocco, M. (2015). Calculation of hydrodynamic parameters - US-SOMO. In S. Uchiyama, F. Arisaka, W. F. Stafford, III, & T. M. Laue (Eds.), *Analytical ultracentrifugation: Instrumentation, software and application*. New York: Springer.

Burgers, J. M. (1938). On the motion of small particles of elongated form, suspended in a viscous liquid. In J. M. Burgers, F. M. Jaeger, R. Houwink, C. J. Van Nieuwenberg, & R. N. J. Saal (Eds.), *Second report on viscosity and plasticity* (p. 209). Amsterdam: Nordemann.

Byron, O. (1997). Construction of hydrodynamic bead models from high-resolution X-ray crystallographic or nuclear magnetic resonance data. *Biophysical Journal, 72*(1), 408–415.

Byron, O. (2015). Introduction—Calculation of hydrodynamic parameters. In S. Uchiyama, F. Arisaka, W. F. Stafford, III, & T. M. Laue (Eds.), *Analytical ultracentrifugation: Instrumentation, software and application*. New York: Springer.

Camps, J., Carrillo, O., Emperador, A., Orellana, L., Hospital, A., Rueda, M., et al. (2009). FlexServ: An integrated tool for the analysis of protein flexibility. *Bioinformatics, 25*(13), 1709–1710.

Connolly, M. L. (1993). The molecular surface package. *Journal of Molecular Graphics, 11*(2), 139–141.

Costenaro, L., Grossmann, J. G., Ebel, C., & Maxwell, A. (2005). Small-angle X-ray scattering reveals the solution structure of the full-length DNA gyrase A subunit. *Structure, 13*(2), 287–296.

de Brevern, A. G., Bornot, A., Craveur, P., Etchebest, C., & Gelly, J.-C. (2012). PredyFlexy: Flexibility and local structure prediction from sequence. *Nucleic Acids Research, 40*(W1), W317–W322.

Dean, R. B., & Dixon, W. J. (1951). Simplified statistics for small numbers of observations. *Analytical Chemistry, 23*(4), 636–638.

Denisov, V. P., & Halle, B. (1995). Protein hydration dynamics in aqueous-solution—A comparison of bovine pancreatic trypsin-inhibitor and ubiquitin by O-17 spin relaxation dispersion. *Journal of Molecular Biology, 245*(5), 682–697.

Doucette, P. A., Whitson, L. J., Cao, X., Schirf, V., Demeler, B., Valentine, J. S., et al. (2004). Dissociation of human copper-zinc superoxide dismutase dimers using chaotrope and reductant: Insights into the molecular basis for dimer stability. *Journal of Biological Chemistry, 279*(52), 54558–54566.

Durchschlag, H., & Jaenicke, R. (1982). Partial specific changes of proteins densimetric studies. *Biochemical and Biophysical Research Communications, 108*(3), 1074–1079.

Durchschlag, H., & Zipper, P. (1997). Calculation of partial specific volumes and other volumetric properties of small molecules and polymers. *Journal of Applied Crystallography, 30*(2), 803–807.

Durchschlag, H., & Zipper, P. (2005). Calculation of volume, surface, and hydration properties of biopolymers. In D. J. Scott, S. E. Harding, & A. J. Rowe (Eds.), *Analytical ultracentrifugation: Techniques and methods* (pp. 389–431). Cambridge: Royal Society of Chemistry.

Durchschlag, H., & Zipper, P. (2008). Volume, surface and hydration properties of proteins. *Progress in Colloid and Polymer Science, 134*, 19–29.

Fiser, A., Do, R. K. G., & Sali, A. (2000). Modeling of loops in protein structures. *Protein Science, 9*(9), 1753–1773.

Franke, D., & Svergun, D. I. (2009). DAMMIF, a program for rapid ab-initio shape determination in small-angle scattering. *Journal of Applied Crystallography*, *42*(2), 342–346.

García de la Torre, J. (2001). Building hydrodynamic bead-shell models for rigid bioparticles of arbitrary shape. *Biophysical Chemistry*, *94*(3), 265–274.

García de la Torre, J., & Bloomfield, V. A. (1977). Hydrodynamic properties of macromolecular complexes. I Translation. *Biopolymers*, *16*(8), 1747–1763.

García de la Torre, J. G., & Bloomfield, V. A. (1981). Hydrodynamic properties of complex, rigid, biological macromolecules: Theory and applications. *Quarterly Reviews of Biophysics*, *14*(1), 81–139.

García de la Torre, J., Carrasco, B., & Harding, S. E. (1997). SOLPRO: Theory and computer program for the prediction of SOLution PROperties of rigid macromolecules and bioparticles. *European Biophysics Journal*, *25*(5/6), 361–372.

García de la Torre, J., del Rio, G., & Ortega, A. (2007). Improved calculation of rotational diffusion and intrinsic viscosity of bead models for macromolecules and nanoparticles. *Journal of Physical Chemistry B*, *111*(5), 955–961.

García de la Torre, J., Hernández Cifre, J. G., Ortega, Á., Rodríguez Schmidt, R., Fernandes, M. X., Pérez Sánchez, H. E., et al. (2009). SIMUFLEX: Algorithms and tools for simulation of the conformation and dynamics of flexible molecules and nanoparticles in dilute solution. *Journal of Chemical Theory and Computation*, *5*(10), 2606–2618.

García de la Torre, J., Huertas, M. L., & Carrasco, B. (2000). Calculation of hydrodynamic properties of globular proteins from their atomic-level structure. *Biophysical Journal*, *78*(2), 719–730.

García de la Torre, J., Llorca, O., Carrascosa, J. L., & Valpuesta, J. M. (2001). HYDROMIC: Prediction of hydrodynamic properties of rigid macromolecular structures obtained from electron microscopy images. *European Biophysics Journal*, *30*(6), 457–462.

García de la Torre, J., Ortega, A., Pérez Sánchez, H. E., & Hernández Cifre, J. G. (2005). MULTIHYDRO and MONTEHYDRO: Conformational search and Monte Carlo calculation of solution properties of rigid or flexible bead models. *Biophysical Chemistry*, *116*(2), 121–128.

García de la Torre, J., & Rodes, V. (1983). Effects from bead size and hydrodynamic interactions on the translational and rotational coefficients of macromolecular bead models. *Journal of Chemical Physics*, *79*(5), 2454–2460.

Goldstein, R. F. (1985). Macromolecular diffusion constants: A calculational strategy. *The Journal of Chemical Physics*, *83*(5), 2390–2397.

Gregory, L., Davis, K. G., Sheth, B., Boyd, J., Jefferis, R., Nave, C., et al. (1987). The solution conformations of the subclasses of human IgG deduced from sedimentation and small angle X-ray scattering studies. *Molecular Immunology*, *24*(8), 821–829.

Guex, N., & Peitsch, M. C. (1997). SWISS-MODEL and the Swiss-Pdb Viewer: An environment for comparative protein modelling. *Electrophoresis*, *18*(15), 2714–2723.

Guex, N., Peitsch, M. C., & Schwede, T. (2009). Automated comparative protein structure modeling with SWISS-MODEL and Swiss-PdbViewer: A historical perspective. *Electrophoresis*, *30*(1), S162–S173.

Halle, B., & Davidovic, M. (2003). Biomolecular hydration: From water dynamics to hydrodynamics. *Proceedings of the National Academy of Sciences of the United States of America*, *100*(21), 12135–12140.

Harkiolaki, M., Gilbert, R. J. C., Jones, E. Y., & Feller, S. M. (2006). The C-terminal SH3 domain of CRKL as a dynamic dimerization module transiently exposing a nuclear export signal. *Structure*, *14*(12), 1741–1753.

Henchman, R. H., & McCammon, J. A. (2002). Structural and dynamic properties of water around acetylcholinesterase. *Protein Science*, *11*(9), 2080–2090.

Hurton, T., Wright, A., Deubler, G., Bashir, B., Hayes, D. B., Laue, T. M., et al. (2012). SEDNTERP (http://sednterp.unh.edu/).

Kang, E.-H., Mansfield, M. L., & Douglas, J. F. (2004). Numerical path integration technique for the calculation of transport properties of proteins. *Physical Review E, 69*(3), 031918.

Kelley, L. A., & Sternberg, M. J. E. (2009). Protein structure prediction on the Web: A case study using the Phyre server. *Nature Protocols, 4*(3), 363–371.

Kiefer, F., Arnold, K., Kuenzli, M., Bordoli, L., & Schwede, T. (2009). The SWISS-MODEL Repository and associated resources. *Nucleic Acids Research, 37*(Database issue), D387–D392.

Kirkwood, J. G. (1949). The statistical mechanical theory of irreversible processes in solutions of macromolecules (visco-elastic behaviour). *The Journal of Chemical Physics, 68*(7), 649–660.

Kirkwood, J. G. (1954). The general theory of irreversible processes in solutions of macromolecules. *Journal of Polymer Science, 12*(1), 1–14.

Kratky, O., Leopold, H., & Stabinger, H. (1973). The determination of the partial specific volume of proteins by the mechanical oscillator technique. *Methods in Enzymology, 27*, 98–110.

Kreiner, M., Byron, O., Domingues, D., & van der Walle, C. (2009). Oligomerisation and thermal stability of polyvalent integrin α5β1 ligands. *Biophysical Chemistry, 142*(1–3), 34–39.

Kuntz, I. D., & Kauzmann, W. (1974). Hydration of proteins and polypeptides. In C. B. Anfinsen, J. T. Edsall, & F. M. Richards (Eds.), *Advances in protein chemistry: Vol 28.* (pp. 239–345). Waltham, Massachusetts: Academic Press.

Laue, T. M., Shah, D. D., Ridgeway, T. M., & Pelletier, S. L. (1992). Computer-aided interpretation of analytical sedimentation data for proteins. In S. E. Harding, A. J. Rowe, & J. C. Horton (Eds.), *Analytical ultracentrifugation in biochemistry and polymer science* (pp. 90–125). Cambridge, UK: Royal Society of Chemistry.

Li, J., Correia, J. J., Wang, L., Trent, J. O., & Chaires, J. B. (2005). Not so crystal clear: The structure of the human telomere G-quadruplex in solution differs from that present in a crystal. *Nucleic Acids Research, 33*(14), 4649–4659.

Makarov, V. A., Andrews, B. K., Smith, P. E., & Pettitt, B. M. (2000). Residence times of water molecules in the hydration sites of myoglobin. *Biophysical Journal, 79*(6), 2966–2974.

Mansfield, M. L., & Douglas, J. F. (2008). Improved path integration method for estimating the intrinsic viscosity of arbitrarily shaped particles. *Physical Review E, 78*(4).

Mansfield, M. L., Douglas, J. F., & Garboczi, E. J. (2001). Intrinsic viscosity and the electrical polarizability of arbitrarily shaped objects. *Physical Review E, 64*(6).

Marti-Renom, M. A., Stuart, A. C., Fiser, A., Sanchez, R., Melo, F., & Sali, A. (2000). Comparative protein structure modeling of genes and genomes. *Annual Review of Biophysics and Biomolecular Structure, 29*, 291–325.

Mavaddat, N., Mason, D. W., Atkinson, P. D., Evans, E. J., Gilbert, R. J., Stuart, D. I., et al. (2000). Signaling lymphocytic activation molecule (CDw150) is homophilic but self-associates with very low affinity. *Journal of Biological Chemistry, 275*(36), 28100–28109.

Moeller, A., Zhao, C., Fried, M. G., Wilson-Kubalek, E. M., Carragher, B., & Whiteheart, S. W. (2012). Nucleotide-dependent conformational changes in the N-Ethylmaleimide Sensitive Factor (NSF) and their potential role in SNARE complex disassembly. *Journal of Structural Biology, 177*(2), 335–343.

Nakasako, M., Fujisawa, T., Adachi, S.-i, Kudo, T., & Higuchi, S. (2001). Large-scale domain movements and hydration structure changes in the active-site cleft of unligated glutamate dehydrogenase from Thermococcus profundus studied by cryogenic X-ray crystal structure analysis and small-angle X-ray scattering. *Biochemistry, 40*(10), 3069–3079.

Nishio, M., Kamiya, Y., Mizushima, T., Wakatsuki, S., Sasakawa, H., Yamamoto, K., et al. (2010). Structural basis for the cooperative interplay between the two causative gene products of combined factor V and factor VIII deficiency. *Proceedings of the National Academy of Sciences of the United States of America, 107*(9), 4034–4039.

Ortega, A., Amorós, D., & García de la Torre, J. (2011). Prediction of hydrodynamic and other solution properties of rigid proteins from atomic- and residue-level models. *Biophysical Journal, 101*(4), 892–898.

Oseen, C. W. (1927). Neure Methoden und Ergebnisse in der Hydrodynamik. In E. Hilb (Ed.), *Mathematik und ihre Anwendungen in Monographien und Lehrbuchern* (p. 337). Leipzig: Academisches Verlagsgellschaft.

Pang, S. S., Berry, R., Chen, Z., Kjer-Nielsen, L., Perugini, M. A., King, G. F., et al. (2010). The structural basis for autonomous dimerization of the pre-T-cell antigen receptor. *Nature, 467*(7317), 844–848.

Perkins, S. J. (1986). Protein volumes and hydration effects: The calculation of partial specific volumes, neutron scattering matchpoints and 280 nm absorption coefficients for proteins and glycoproteins from amino acid sequences. *European Journal of Biochemistry, 157*(1), 169–180.

Perrin, J. (1936). Mouvement Brownien d'un ellipsoide. II. Rotation libre et depolarisation des fluorescences. Translation et diffusion de molecules ellipsoidales. *Journal de Physique et le Radium*, 7(1), 1–11.

Rai, N., Nöllmann, M., Spotorno, B., Tassara, G., Byron, O., & Rocco, M. (2005). *SOMO* (*SO*lution *MO*deler): Differences between x-ray and NMR-derived bead models suggest a role for side chain flexibility in protein hydrodynamics. *Structure, 13*(5), 723–734.

Rocco, M., & Byron, O. (2015). Computing translational diffusion and sedimentation coefficients: An evaluation of experimental data and programs. *European Biophysics Journal.* http://dx.doi.org/10.1007/s00249-015-1042-9, (in press).

Rosano, C., & Rocco, M. (2010). Solution properties of full-length integrin $\alpha_{IIb}\beta_3$ refined models suggest environment-dependent induction of alternative bent/extended resting states. *FEBS Journal, 277*(15), 3190–3202.

Rotne, J., & Prager, S. (1969). Variational treatment of hydrodynamic interaction in polymers. *Journal of Chemical Physics, 50*(11), 4831–4837.

Roy, A., Kucukural, A., & Zhang, Y. (2010). I-TASSER: A unified platform for automated protein structure and function prediction. *Nature Protocols, 5*(4), 725–738.

Sali, A., & Blundell, T. L. (1993). Comparative protein modeling by satisfaction of spatial restraints. *Journal of Molecular Biology, 234*(3), 779–815.

Schlessinger, A., Yachdav, G., & Rost, B. (2006). PROFbval: Predict flexible and rigid residues in proteins. *Bioinformatics, 22*(7), 891–893.

Schluttig, J., Alamanova, D., Helms, V., & Schwarz, U. S. (2008). Dynamics of protein-protein encounter: A Langevin equation approach with reaction patches. *Journal of Chemical Physics, 129*(15).

Smolle, M., Prior, A. E., Brown, A. E., Cooper, A., Byron, O., & Lindsay, J. G. (2006). A new level of architectural complexity in the human pyruvate dehydrogenase complex. *Journal of Biological Chemistry, 281*(28), 19772–19780.

Spotorno, B., Piccinini, L., Tassara, G., Ruggiero, C., Nardini, M., Molina, F., et al. (1997). BEAMS (BEAds Modelling System): A set of computer programs for the generation, the visualisation and the computation of the hydrodynamic and conformational properties of bead models of proteins. *European Biophysics Journal, 25*(5/6), 373–384.

Squire, P. G., & Himmel, M. E. (1979). Hydrodynamics and protein hydration. *Archives of Biochemistry and Biophysics, 196*(1), 165–177.

Suhre, K., & Sanejouand, Y. H. (2004). ElNémo: A normal mode web server for protein movement analysis and the generation of templates for molecular replacement. *Nucleic Acids Research*, *32*, W610–W614.

Swanson, E., Teller, D. C., & de Haën, C. (1978). The low Reynolds number translational friction of ellipsoids, cylinders, dumbbells, and hollow spherical caps. Numerical testing of the validity of the modified Oseen tensor in computing the friction of objects modeled as beads on a shell. *Journal of Chemical Physics*, *68*(11), 5097–5102.

Trewhella, J., Carlson, V. A. P., Curtis, E. H., & Heidorn, D. B. (1988). Differences in the solution structures of oxidised and reduced cytochrome c measured by small-angle X-ray scattering. *Biochemistry*, *27*(4), 1121–1125.

Vestergaard, B., Sanyal, S., Roessle, M., Mora, L., Buckingham, R. H., Kastrup, J. S., et al. (2005). The SAXS solution structure of RF1 differs from its crystal structure and is similar to its ribosome bound cryo-EM structure. *Molecular Cell*, *20*(6), 929–938.

Vriend, G. (1990). WHAT IF: A molecular modeling and drug design program. *Journal of Molecular Graphics*, *8*(1), 52–56.

Vunnam, N., Flint, J., Balbo, A., Schuck, P., & Pedigo, S. (2011). Dimeric states of neural- and epithelial-cadherins are distinguished by the rate of disassembly. *Biochemistry*, *50*(14), 2951–2961.

Webb, B., & Sali, A. (2014). Comparative protein structure modeling using MODELLER. *Current Protocols in Bioinformatics*, *47*, 5.6.1–5.6.32.

Wegener, W. A. (1982). Bead models of segmentally flexible macromolecules. *The Journal of Chemical Physics*, *76*(12), 6425–6430.

Wegener, W. A., Dowben, R. M., & Koester, V. J. (1980). Diffusion coefficients for segmentally flexible macromolecules: General formalism and application to rotational behaviour of a body with two segments. *The Journal of Chemical Physics*, *73*(8), 4086–4097.

Yamakawa, H. (1970). Transport properties of polymer chains in dilute solution: Hydrodynamic interaction. *Journal of Chemical Physics*, *53*(1), 435–443.

Yang, J., Yan, R., Roy, A., Xu, D., Poisson, J., & Zhang, Y. (2015). The I-TASSER Suite: Protein structure and function prediction. *Nature Methods*, *12*(1), 7–8.

Youngren, G. K., & Acrivos, A. (1975). Stokes flow past a particle of arbitrary shape: A numerical-method of solution. *Journal of Fluid Mechanics*, *69*(2), 377–403.

Zhang, Y. (2008). I-TASSER server for protein 3D structure prediction. *BMC Bioinformatics*, *9*(40), 8.

Zimm, B. H. (1982). Sedimentation of asymmetric elastic dumbbells and the rigid-body approximation in the hydrodynamics of chains. *Macromolecules*, *15*(2), 520–525.

CHAPTER FIVE

Calculations and Publication-Quality Illustrations for Analytical Ultracentrifugation Data

Chad A. Brautigam[1]

Department of Biophysics, The University of Texas Southwestern Medical Center, Dallas, Texas, USA
[1]Corresponding author: e-mail address: chad.brautigam@utsouthwestern.edu

Contents

Abstract

The analysis of analytical ultracentrifugation (AUC) data has been greatly facilitated by the advances accumulated in recent years. These improvements include refinements in AUC-based binding isotherms, advances in the fitting of both sedimentation velocity (SV) and sedimentation equilibrium (SE) data, and innovations in calculations related to posttranslationally modified proteins and to proteins with a large amount of associated cosolute, e.g., detergents. To capitalize on these advances, the experimenter often must prepare and collate multiple data sets and parameters for subsequent analyses; these tasks can be cumbersome and unclear, especially for new users. Examples are the sorting of concentration-profile scans for SE data, the integration of sedimentation velocity distributions ($c(s)$) to arrive at weighted-average binding isotherms, and the calculations to determine the oligomeric state of glycoproteins and membrane proteins. The significant organizational and logistical hurdles presented by these approaches are streamlined by the software described herein, called GUSSI. GUSSI also creates publication-quality graphics for documenting and illustrating AUC and other

Methods in Enzymology, Volume 562
ISSN 0076-6879
http://dx.doi.org/10.1016/bs.mie.2015.05.001

biophysical experiments with minimal effort on the user's part. The program contains three main modules, allowing for plotting and calculations on *c*(*s*) distributions, SV signal versus radius data, and general data/fit/residual plots.

1. INTRODUCTION

Analytical ultracentrifugation (AUC) is a first-principles technique that is increasingly utilized by researchers in the biological sciences. The method can be used to hydrodynamically characterize macromolecules in solution and to study the energetics of their interactions. Currently, two experimental modes comprise the majority of AUC experiments. The first, called "sedimentation velocity" (SV), examines the transport of macromolecules in a high centrifugal field (ca. 100,000–200,000 × *g*), providing high-resolution hydrodynamic information (Schuck, 2000). Moreover, SV can be used to study self- and hetero-interactions, providing information such as stoichiometry (Balbo et al., 2005) and association constants (Brautigam, 2011; Brown, Balbo, & Schuck, 2008; Stafford & Sherwood, 2004). The data obtained from SV, i.e., radial concentration profiles of the sedimenting species captured at various points in time, are described by the Lamm equation (Lamm, 1929), a partial differential equation with no known exact analytical solutions; it is approachable with numerical solutions (Claverie, Dreux, & Cohen, 1975). The other AUC mode commonly used is called "sedimentation equilibrium" (SE), in which a lower centrifugal field (ca. 3000–30,000 × *g*) is applied to the solution, resulting in a final equilibrium concentration gradient. These data may be rigorously analyzed to yield properties such as molar mass and virial coefficients (Casassa & Eisenberg, 1964). SE is also widely used to characterize interacting systems (Ghirlando, 2011). In the following, familiarity with AUC and the attendant data formats is assumed. Those inexperienced in AUC theory and practice are referred to the abovementioned citations, to earlier chapters in this volume (Correia & Stafford, 2015), and to other general treatments (Laue, 1999; Stafford, 2003; Zhao, Brautigam, Ghirlando, & Schuck, 2013) for introductions to SV and SE.

In recent years, significant improvements have been made in the AUC field. For example, advances in the calculation of numerical solutions to the Lamm equation (Brown & Schuck, 2008; Cao & Demeler, 2008; Stafford & Sherwood, 2004) facilitated the extraction of hydrodynamic parameters directly from SV data and fitting diffusion-deconvoluted size distributions (Schuck, 2000) to these data. In addition, deriving thermodynamic quantities for interacting species was eased by advancements in (1) calculating Lamm-equation solutions coupled to thermodynamic/kinetic parameters (Cao & Demeler, 2008; Correia & Stafford, 2009; Dam, Velikovsky, Mariuzza, Urbanke, & Schuck, 2005), (2) isotherm analysis of SV data (Dam & Schuck, 2005; Schuck, 2010), (3) the modeling of SE data (Gillis et al., 2013; Vistica et al., 2004), among others.

Whereas these approaches are powerful, they can increase the demands on the experimenter. For example, isotherm analysis of SV data requires the researcher to integrate several differential ($c(s)$, ls-$g^*(s)$, or $g(s^*)$) distributions, record the signal population and the weighted-average sedimentation coefficients, assemble these values into an electronic file, and load this file into the analysis program. Additionally, taking advantage of SE analytical advances can require the user to sort through many data scans according to sample identity, rotor speed, and wavelength of acquisition, then assembling the scans into the proper file formats. Furthermore, once preliminary analysis of AUC data is complete, other calculations can be performed. Examples of such calculations include the determination of the oligomeric states of glycoproteins and membrane proteins; these calculations can require assembly and collation of a large number of data sets, parameters, and equations. Also, AUC and other biophysical data sets can be large and/or difficult to illustrate. Commercial graphing packages are often either ill-suited to these data types or cumbersome to use, requiring the user to perform actions repetitively or to learn specialized scripting protocols.

In this chapter, a software program called GUSSI (*G*rapher that *U*nderstands data from *S*V, *S*E, and *I*sotherm analyses) is introduced. This program contains utilities that drastically reduce the data-preparation procedures needed to apply advanced analytical techniques to the user's data. GUSSI also solves the aforementioned challenges to plotting several biophysical data types through its ability to automatically parse and present the attendant data formats. The program is designed to be flexible enough for advanced users, but to have an intuitive interface that is inviting to novices.

2. GENERAL FEATURES OF GUSSI

The analytical software programs SEDFIT and SEDPHAT (Zhao, Brautigam, et al., 2013) have plotting functions that conform to GUSSI's formatting requirements. For this reason, this chapter will often reference the interaction of SEDFIT, SEDPHAT, and GUSSI. However, data from any program may be imported into GUSSI as long as they adhere to GUSSI's formatting requirements, which are fully documented in its accompanying manual. Indeed, it has been observed that the power and flexibility of GUSSI's user interface and the quality of the output encourage researchers to use it for purposes afield from the biophysical data formats described herein. The program operates only under the Windows operating system.

GUSSI may be invoked from menu commands in SEDFIT/SEDPHAT, or as a stand-alone executable. In the former case, the software provides a data file to GUSSI; in the latter, the user employs GUSSI menu commands to open a properly formatted file. Next, GUSSI automatically parses and presents the data (Fig. 1). Its graphical user interface shows the graph, and the user can easily manipulate the plot's features, such as line color and width, marker existence/appearance, labeling text, and axes appearances. Changes are automatically updated on the graph, giving the user instantaneous feedback. Menu functions allow writing figures in one of four popular file formats, as well as saving the program's "state," which allows the

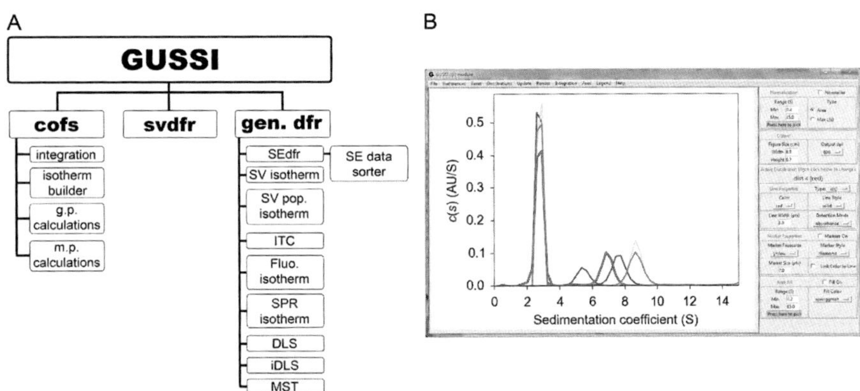

Figure 1 The organization and appearance of GUSSI. (A) The overall organization of the program. The hierarchical relationships between the main modules ("cofs" for *c*(*s*) plots, "svdfr" for SV data/fit/residual plots, and "gen. dfr" for general data/fit/residual plots) are shown. The abbreviations are: g.p., glycoprotein; m.p., membrane proteins; SE dfr, SE data/fit/residuals; pop., population; ITC, isothermal titration calorimetry; Fluo., fluorescence; SPR, surface plasmon resonance; DLS, dynamic light scattering, field autocorrelation mode; iDLS, dynamic light scattering, intensity autocorrelation mode; and MST, microscale thermophoresis. (B) A screenshot of the program's user interface. On the left is the subject graph, and the panel on the right is the "Control Panel," which houses various oft-used controls affecting the appearance of the graph. Calculations performed by GUSSI are accessed from the top menu bar.

figure to be recalled later in GUSSI for examination or revision. This state file is particularly useful because it is completely encapsulated, i.e., it contains all of the information, even the data, needed to reproduce the graph. This represents a convenient means to document analytical results in a publication-ready format. Also, GUSSI state files may be easily shared among collaborators wishing to perfect a graph or illustrate points to one another. GUSSI features three different "modules": (1) a data/fit/residual (dfr) module for sedimentation velocity (svdfr), (2) a $c(s)$ distribution module (cofs), and (3) a dfr module for all other data structures (general dfr) (Fig. 1A).

2.1 The svdfr Module

Following the analysis of SV data, the data points are often presented as markers along with a line representing the fit of the model to the data. It is also good practice to include a plot of the residuals, i.e., the deviations between the data points and the fit lines. There are usually 30–400 "scans" of data that represent the radial concentration profiles of the sedimenting species at a given point in time. When svdfr data are presented to GUSSI, several defaults are automatically applied. For example, because of the high information density of SV data, GUSSI automatically shows only every third scan/fit pair, as well as only every third data point. Importantly, the program does not discard the hidden data/fits. On the contrary, scan and data point sampling are user-adjustable parameters. Second, GUSSI colors the scans in progressive rainbow colors. The colors of the data points are inextricably tied to those of the fits and the residuals. The color scheme can be adjusted; several formats are available, including black-and-white and some that respond to user input (Fig. 2). The residuals can be shown as lines or as markers.

2.2 The cofs Module

The cofs module of GUSSI is by far the most feature-rich, reflecting the popularity of the $c(s)$ analysis as well as the many variations on the analysis that have been introduced over the years. Thus, this module features many integration tools and plot types; those that present the most unique and time-saving features are presented here. Others are also presented in the documentation file that accompanies the program.

A powerful feature of GUSSI is the ease with which multiple distributions may be rendered simultaneously in the same plot (Fig. 2B). After

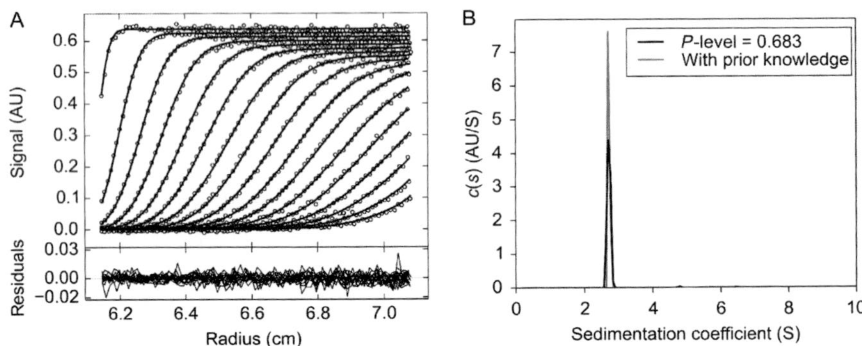

Figure 2 The GUSSI svdfr and *c*(*s*) plots. (A) The default svdfr plot. In the upper panel, the individual data points are circles, and the fits to those data are shown as lines. The lower panel displays the residuals. Time-invariant noise features are subtracted from the data and fit lines. (B) A *c*(*s*) plot. The *c*(*s*) analysis for the data shown in (A) is displayed with the usual regularization (*P*-level = 0.683) along with the distribution after the application of prior knowledge (Brown, Balbo, & Schuck, 2007), i.e., that the species sedimenting at 2.7 S is a single, discrete species ("with prior knowledge"). (See the color plate.)

invocation, additional distributions can be copied (e.g., from SEDFIT) and pasted into GUSSI. A legend can be shown, and the legend's location is user-adjustable. The properties of individual lines can be manipulated by making them "active," which can be accomplished in multiple ways; the easiest is to point and click on the desired distribution on the graph. GUSSI can present the output from discrete/continuous analyses and multisignal SV (Padrick et al., 2010) analyses as well, and the sedimentation coefficients of *c*(*s*) distributions can be converted to *s*-values ($s_{20,w}$) that are corrected to standard conditions, i.e., water at 20 °C. Further, in a single mouse-click, the displayed distributions can be normalized by the integrated area or by the maximum *c*(*s*) value in a user-provided *s*-range. This feature is extremely useful, allowing the simultaneous display of distributions with widely varying total signals.

Arguably, the most important information that can come from a *c*(*s*) analysis is obtained by integrating the individual peaks. GUSSI features a tool for the integration of the distributions that gives the user the weighted-average *s*-values and the area under the distribution (in signal units) within the integrated range for a single distribution or for all distributions simultaneously. An integral form of the *c*(*s*) distribution can be superscribed over the normalized distributions, allowing the user to easily identify the overall and peak signal magnitudes in any given distribution.

2.3 The General dfr Module

GUSSI is capable of displaying the results of analyses of many other biophysical experiments (Fig. 1A). For example, the results of a single-speed or multispeed SE analysis can be graphed. Also, many different kinds of isotherms can be plotted, including those derived from SV. The "general dfr" module of GUSSI is responsible for handling such results (Figs. 1 and 3). Although these disparate biophysical results are plotted under a single computational rubric, GUSSI recognizes that the respective data structures require special features. For example, for long-column SE data, there can be a large number of data points, and thus data downsampling is enabled (Fig. 3A). In SV isotherms, the abscissa is scaled logarithmically (Fig. 3B). GUSSI is well adapted to displaying other biophysical data types, such as those from dynamic light scattering, fluorescence spectroscopy, and microscale thermophoresis. The program is particularly useful in displaying isothermal titration calorimetry data and fits that result from the powerful analytical combination of the programs NITPIC (Keller et al., 2012; Scheuermann & Brautigam, 2015) and SEDPHAT (Zhao, Piszczek, & Schuck, 2015).

In all general dfr plots, legends are available. Also, the residual portion of the plot may be toggled off and on. Error bars can also be added to the data points.

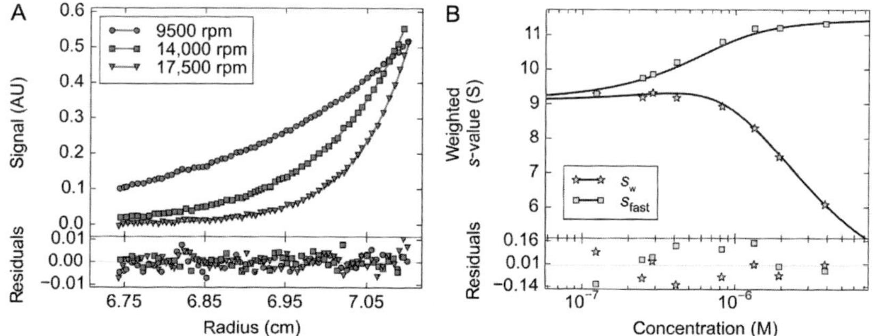

Figure 3 General dfr plots from GUSSI. (A) SE data. As in Fig. 2A, the markers represent data points, and the lines fits to them. The residuals are shown as respective markers. A legend has been activated, indicating the rotor speeds of the respective concentration profiles. Only every third data point is shown. (B) SV isotherm. Globally analyzed s_w (stars) and s_{fast} (i.e., the sedimentation coefficient of the fast-sedimenting species; squares) isotherms are shown, along with fits to those data using effective-particle theory (Schuck, 2010).

3. UTILITY FUNCTIONS OF GUSSI

GUSSI contains three utility functions that significantly simplify data-processing steps required for SV isotherm analysis, SE analysis, and SV analysis of glycoproteins and membrane proteins. The first two prepare the data for subsequent analyses, and the third provides a means to determine the oligomeric state of the subject protein.

3.1 Assembling SV-Based Isotherms

Isotherms derived from $c(s)$ distributions have recently proved very useful to evaluate the association constants of both self- and hetero-association of macromolecules (Ayaz et al., 2014; Dam & Schuck, 2005; Zhao et al., 2012). Ordinarily, the experimenter fits the SV data from several (usually at least five) separate experiments using the $c(s)$ model. The resulting distributions are integrated, and the experimenter records the signal population of the observed peaks and their signal-weighted sedimentation coefficients (i.e., the "weighted-average" s-value, or "s_w"). Once this is accomplished, these values, along with the concentrations of the sedimenting species, are assembled into an electronic file for subsequent analysis. GUSSI streamlines and improves this process considerably. In the GUSSI workflow, the user loads all of the distributions into the cofs module. Then, the "Isotherm Maker" is initiated. After integration limits are chosen, a table appears bearing all of the information that was integrated and prompting the user for the species' concentrations. Optionally, parameters destined for use in SEDPHAT can be entered in this table. Once the user completes the table, the isotherm and any associated files are saved for subsequent analysis. This workflow offers several improvements over the previous one. Importantly, the integration of all isotherms is accomplished at the same time with identical integration limits, thus strictly enforcing consistency across all distributions. Also, the tedious task of assembling the isotherm file is now accomplished automatically and without typographical errors. Isotherm construction based on the mass transport of all species, on effective-particle theory, and on the species' populations are supported (Dam & Schuck, 2005; Schuck, 1998, 2010). Further, a framework for excluding contributions from contaminating species is available.

This latter point is here illustrated in a brief example. Consider a simulated system (Fig. 4A) in which protein A (2.0 S) interacts with protein B (8.0 S) to form an 8.65-S complex with a dissociation constant (K_D) of

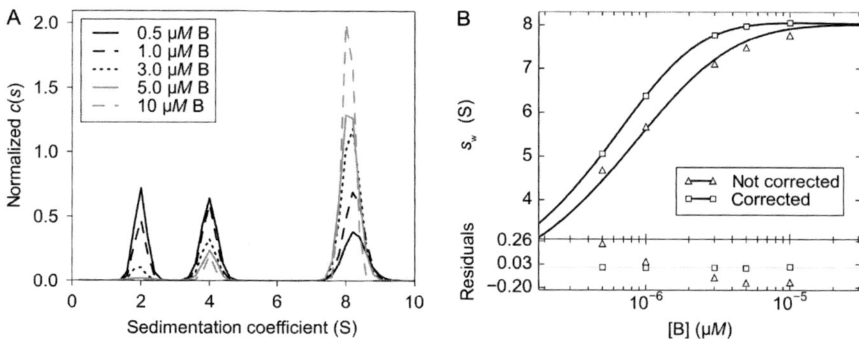

Figure 4 Correction of s_w isotherms in GUSSI. The simulation assumes that species A - associates with species B to form a 1:1 AB complex with a dissociation constant of 1 μ*M*. Also, the sedimentation coefficient of A (s_A) is 2.0 S, s_B is 8.0 S, and s_{AB} is 8.65 S. The preparation of A is contaminated with a 4.0 S, inert species. The [A] was held constant at 1 μ*M* in the titration, while [B] was varied as described in the inset to part (A). (A) The *c*(*s*) distributions for the simulation. The simulated data were analyzed in SEDFIT. The distributions are normalized, and thus the contaminant at 4 S appears to have varying signal amplitudes. (B) Isotherm fitted to the data with (squares) or without (triangles) the correction applied as described in the text. Without the correction, K_D refined to 4.1 μ*M*, whereas the corrected data yielded the simulated value when evaluated, i.e., 1.0 μ*M*. (See the color plate.)

1 μ*M*. However, the protein A preparation is contaminated with a 4.0-S species that does not participate in the interaction. A titration is performed in which [A] is held constant at 1 μ*M* (thus, the signal of the inert contaminant is constant), but [B] is varied from 0.5 to 10 μ*M*. Obviously, simply integrating the distributions over an appropriate range (1–10 S) and analyzing the s_w isotherm will result in an inaccurate K_D (Fig. 4B, triangles; the fitted K_D was 4.1 μ*M*). However, in GUSSI, the user can define an "Exclusion Zone," i.e., an *s*-range that is excluded from the calculation. The corrected s_w values, here termed $s_{w,corr}$, are given by

$$s_{w,corr} = \frac{s_w c_{tot} - s_{w,e} c_{tot,e}}{c_{tot} - c_{tot,e}} \quad (1)$$

where c_{tot} is the total integrated signal over the entire integration region, and $s_{w,e}$ and $c_{tot,e}$ are, respectively, the weighted-average *s*-value and the total integrated signal derived from the Exclusion Zone (Zhao, Lomash, et al., 2013). This simple correction to the simulated system described above results in the isotherm shown in Fig. 4B (squares; the Exclusion Zone was from 3 to 5 S), and fitting this isotherm results in the correct, simulated K_D of 1.0 μ*M*.

3.2 Sorting SE Data

Recent innovations in the analysis of SE data are best utilized when the data are acquired at multiple rotor speeds and wavelengths (Cole, 2004; Vistica et al., 2004). However, these strategies present several bookkeeping hurdles. Hundreds of data files can be acquired, but their respective rotor speeds and wavelengths of acquisition are encoded not in the filenames but in a terse file header. Furthermore, the achievement of thermodynamic and mechanical equilibrium, which is necessary for the validity of the analysis, may be open to question. Finally, the meniscus and bottom of the solution column must be identified and sensible fitting limits chosen. Thus, the experimenter may face a formidable postexperimental sorting and evaluation chore.

The GUSSI SE Data Sorter efficiently handles all of these tasks. The Sorter is invoked from the general dfr module of GUSSI (Fig. 1A). The user is prompted to identify the directory(ies) in which the SE data reside. Next, the program reads the all of the data files, parsing them according to cell number, wavelength, and rotor speed. It then performs a series of analyses. First, the meniscus and bottom of each scan are located. Next, the program evaluates the achievement of equilibrium using an algorithm similar to that employed by the program WinMATCH (similar algorithms are used for this purpose in SEDFIT and HeteroAnalysis (Cole, 2004)). In short, the scans are examined as a function of time and compared to the final scan; when only small differences are consistently calculated, equilibrium is judged to have been achieved. Next, a table is presented that displays the results of the finding and matching routines. Graphs detailing these choices are available, and unsatisfactory scans may be excluded from further consideration. The bottom part of the table prompts the user for information that is optional but useful in subsequent analyses in SEDPHAT. After the user has made all necessary exclusions and adjustments, the program "sorts" the data. The final scan in each cell/wavelength/rotor-speed category is written to a user-selected directory with succinct but informative filenames (e.g., "Eq9k_250.RA3" for absorbance data from Cell 3 acquired at 250 nm at a rotor speed of 9000 rpm). Experimental files needed for analysis of the data in SEDPHAT are also (optionally) written at this time, as is a comprehensive log file.

3.3 Determining the Oligomeric State of Glycoproteins and Membrane Proteins

Finally, routines are present in the GUSSI cofs module that can help the user to establish the oligomeric state of a glycoprotein or membrane protein

(Fig. 1A). Whereas the molar mass (and therefore the oligomeric state) of a protein can be simply derived from an SV experiment (Schuck, Perugini, Gonzales, Howlett, & Schubert, 2002), this task is more difficult for glycoproteins and membrane proteins because the partial specific volume ($\bar{v}$) of the sedimenting species can be indeterminate. For glycoproteins, this complication arises from the possibility that the extent and composition of the modifying carbohydrates may not be known. In the case of membrane proteins, this indeterminacy can be due to unknown detergent content in the sedimenting, protein-containing micelle.

GUSSI's approach to these problems is graphical. It calculates the molar mass (or f/f_0) of the sedimenting species as a function of a hypothesized value, e.g., oligomeric state and extent of carbohydrate modification. It then displays graphs depicting these calculations. In many cases, this allows the user to make an unambiguous determination of the oligomeric state of the protein. In the following, contributions from nondetergent cosolutes such as ions or cosmotropes are considered to be negligible; also, it is assumed that a single association state is populated at all concentrations under investigation.

3.3.1 Protein, Solution, and Chemical Information

In performing these calculations, the user must know some information regarding the protein. These values include the monomeric molar mass, $M_{\mathrm{P,mono}}$, the partial specific volume, $\bar{v}_{\mathrm{P}}$, and the mass-based extinction coefficient (at a given wavelength), ε_{P}. For unmodified proteins, these parameters can conveniently be estimated from the amino acid sequence. SEDFIT (Zhao, Brautigam, et al., 2013), SEDNTERP (Laue, Shah, Ridgeway, & Pelletier, 1992), and UltraScan (Demeler, 2005) are among the platforms that can be used for this purpose; the needed values are obtained after pasting the one-letter amino acid sequence into the respective program's user interface (and providing the experimental temperature). SEDNTERP is also very widely used to calculate the density (ρ) and viscosity (η) of the solution. For this calculation, the user provides the concentrations of solution constituents that are chosen from a table of chemicals with known physical properties. If the solution components are not tabulated in SEDNTERP, ρ and η can be measured using densimetry and viscometry, respectively. Often, $\mathrm{d}n/\mathrm{d}c_{\mathrm{P}}$, the protein's refractive-index increment, is estimated at 0.187 cm^3/g, as most proteins have values close to that. However, there are notable excursions from this value (Zhao, Brown, & Schuck, 2011), and a calculator available in SEDFIT supplies an estimate of the value based on actual amino acid composition. The properties of the cosedimenting chemicals, i.e., carbohydrates and detergents in glycoproteins and membrane proteins, respectively, must be known also as well as possible. With glycoproteins, the identities of the carbohydrate (noted with the subscript C) adducts are often not certain, and thus an approach to determining their oligomerization states taking this uncertainty into account is warranted. With membrane proteins, the identity of the cosedimenting detergent (noted with subscript D) is known, and thus $\bar{v}_D$ and dn/dc_D can usually be obtained from databases or the manufacturer. Therefore, AUC experiments in which the amount of cosedimenting detergent can be firmly established are possible, enabling several powerful and distinct approaches to gleaning oligomeric state. These differences between glycoproteins and membrane proteins inform the distinct strategies outlined below.

3.3.2 Glycoproteins

When studying a glycoprotein, it is possible that the identity of the carbohydrate moieties, extent of glycosylation, and oligomeric state are unknown. SV studies can be carried out to estimate the latter two quantities by calculating the molar mass of the sedimenting species ($M_{GP,s}$) via the Svedberg equation:

$$M_{GP,s} = \frac{s_{GP}RT}{D_{GP}\left(1 - \bar{v}_{GP,h}\rho\right)}, \tag{2}$$

where s_{GP} is the sedimentation coefficient, R is the gas constant, T is the absolute temperature, D_{GP} is the translational diffusion coefficient, and $\bar{v}_{GP,h}$ is the hypothetical partial specific volume ("GP" refers to the glycoprotein). The $\bar{v}_{GP,h}$ must be hypothetical because the identity and extent of glycosylation is unknown in this scenario. Conveniently, the partial specific volumes of most protein-associating carbohydrate moieties ($\bar{v}_C$) fall in a narrow range: 0.58–0.68 cm^3/g (Lewis & Junghans, 2000; Perkins, 1986), allowing the user to make an educated guess regarding the range of $\bar{v}_{GP,h}$ to be investigated. The approach to calculating $\bar{v}_{GP,h}$ is similar to that employed before in the context of SE (Ghirlando et al., 1995; Lewis & Junghans, 2000; Shire, 1992). $M_{P,mono}$ is known from the amino acid sequence. Together with a user-supplied guess regarding the oligomeric state (n, an integer) and an assumed probable mass percentage of the carbohydrate conjugated to the species ($100 \times q$), a hypothetical mass of the glycoprotein can be obtained:

$$M_{\mathrm{GP,h}} = \frac{nM_{\mathrm{P,mono}}}{(1-q)} \tag{3}$$

An assumption is made that $\bar{v}_{\mathrm{GP,h}}$ can be calculated as the weight-average of $\bar{v}_{\mathrm{P}}$ and a hypothesized partial specific volume of the carbohydrate, $\bar{v}_{\mathrm{C,h}}$:

$$\bar{v}_{\mathrm{GP,h}} = \frac{nM_{\mathrm{P,mono}}\bar{v}_{\mathrm{P}} + M_{\mathrm{C,h}}\bar{v}_{\mathrm{C,h}}}{M_{\mathrm{GP,h}}}, \tag{4}$$

or, cast in terms of q, this reduces to

$$\bar{v}_{\mathrm{GP,h}} = q\bar{v}_{\mathrm{C,h}} + \bar{v}_{\mathrm{P}}(1-q). \tag{5}$$

With these quantities determined, the only two obstacles to using Eq. (2) are determining s_{GP} and D_{GP}. The former can be arrived at by integrating $c(s)$ distributions calculated by SEDFIT, defining s_{GP} as the s_{w} of the relevant peak. D_{GP} can be calculated from the SEDFIT-refined frictional ratio, $(f/f_0)_{\mathrm{r}}$, using that program's scaling law (Brown, Balbo, & Schuck, 2007):

$$D_{\mathrm{GP}} = \frac{\sqrt{2}}{18\pi} kTs_{\mathrm{GP}}{}^{-1/2} \left(\eta (f/f_0)_{\mathrm{r}}\right)^{-3/2} \left(\frac{\bar{v}_{\mathrm{u}}}{1-\bar{v}_{\mathrm{u}}\rho}\right)^{-1/2}, \tag{6}$$

where k is the Boltzmann constant and $\bar{v}_{\mathrm{u}}$ is the user-supplied $\bar{v}$ employed in the SEDFIT analysis. Thus, any range of $M_{\mathrm{GP,h}}$ and $M_{\mathrm{GP,s}}$ may be calculated for any combination of hypothesized q and $\bar{v}_{\mathrm{C,h}}$.

To utilize this feature of GUSSI, the user analyzes the SV data from one or more concentrations of a glycoprotein, noting the parameters used (uniformly for all data sets) and the $(f/f_0)_{\mathrm{r}}$'s obtained. Then, the resulting differential distributions (usually $c(s)$ from SEDFIT) are loaded into GUSSI, and the glycoprotein routine is engaged. The user is prompted to select the s-range of interest, after which GUSSI determines the s_{GP}-values together with the standard error of these values. Next, the user inputs the necessary solution, protein, $(f/f_0)_{\mathrm{r}}$, and carbohydrate parameters. Upon actuation, the program calculates and plots $M_{\mathrm{GP,h}}$ and $M_{\mathrm{GP,s}}$ for 1000 q-values from 0.05 to 0.8 (the latter represents a glycoprotein in which the carbohydrate has four times the mass of the protein, a very unusual situation). The plot of $M_{\mathrm{GP,h}}$ versus q yields a single line, while that of $M_{\mathrm{GP,s}}$ results in a swath of probable values taking into consideration the errors in s_{GP}, D_{GP}, and the user-inputted range of $\bar{v}_{\mathrm{C,h}}$ (Fig. 5A, Table 1). The upper and lower bounds are calculated by inputting the limits of these values that result in the highest and lowest (respectively) $M_{\mathrm{GP,s}}$, resulting in conservative error estimates. This

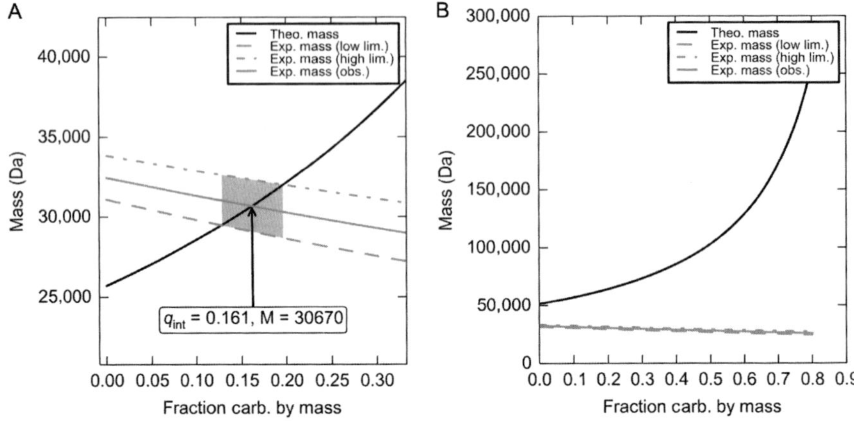

Figure 5 Glycoprotein calculations. (A) The GUSSI calculation on NPC1-NTD with the hypothesis of a monomer. The black, positively trending line represents the hypothetical mass based on a monomer of the protein and carbohydrate composition, and the other lines indicate the mass and limits based on s_{GP} and D_{GP}. The shaded area represents the most likely molar mass and carbohydrate content, and q_{int} (Eq. 7) is indicated, along with the corresponding molar mass. (B) The same calculation, assuming that the protein is a dimer. The value of q_{int} does not appear on this graph because it is negative.

Table 1 Parameters for the Carbohydrate Analysis

Analysis Parameters	
$\bar{\nu}_u$ (cm^3/g)	0.7212
ρ (g/cm^3)	1.00058
η (Poise)	0.01009
T (K)	293
Frictional ratios from analyses	
Exp. 1	1.339
Exp. 2	1.354
Exp. 3	1.400
Protein parameters	
$\bar{\nu}_P$ (cm^3/g)	0.7212
ε_P (L/g cm)	1.109
$M_{P,mono}$ (Da)	25,722
Carbohydrate parameters	
$\bar{\nu}_C$ range (cm^3/g)	0.602–0.64

bounding strategy is recapitulated in all calculations described in this chapter. $M_{GP,h}$ and $M_{GP,s}$ have opposite trends with increasing q; if the two intersect at positive values of q, this indicates that the hypothesized oligomerization state is consistent with the data and that the probable extent of glycosylation may be read from the abscissa of the intersection, while the molar mass of the sedimenting species is derived from the ordinate. Indeed, it can be shown that there is an exact analytical expression for the q-value of intersection, q_{int}:

$$q_{\text{int}} = \frac{1 - [\psi(1 - \bar{\nu}_{P}\rho)]}{1 + [\psi(\bar{\nu}_{P} - \bar{\nu}_{C})]}, \tag{7}$$

with

$$\psi = \frac{nM_{P,\text{mono}}D_{GP}}{s_{GP}RT}. \tag{8}$$

Thus, by inputting the best values of s_{GP}, D_{GP}, and $\bar{\nu}_{C}$ (taken, respectively, in some cases as the mean experimental s and D, and the mean of the user-inputted range of $\bar{\nu}_{C,h}$), then these values may be inserted into Eqs. (7) and (8), allowing the exact calculation of q_{int}. GUSSI calculates q_{int} using these assumptions and displays it in the resulting plot (Fig. 5A). Alternative oligomerization states may be hypothesized and graphed (Fig. 5B). Importantly, there are cases in which two or more hypothesized oligomerization states will lead to intersections of the $M_{GP,h}$ and $M_{GP,s}$ lines; i.e., q_{int} has more than one solution that leads to positive values of q. In such cases, GUSSI detects this issue and warns the user, and all experimental knowledge must be weighed to ascertain which value of q_{int} is the most rational.

Two critical assumptions are made in the above calculations. First, there must be no microheterogeneity in the sedimenting glycoprotein. This is important because this flaw would cause an underestimate of $(f/f_0)_r$ (thus an overestimate of D_{GP}), invalidating the subsequent calculations. However, this assumption can be conveniently checked in SEDFIT through the use of prior information in the $c(s)$ distribution (Brown et al., 2007). In essence, this approach tests the hypothesis that the glycoprotein can be treated as a single species, suggesting no detectable microheterogeneity. The second assumption is that the signal from the glycoprotein comprises the vast majority of the total signal in the experiment. If this condition is not met, the refinement of the $(f/f_0)_r$ may be skewed. There are advanced approaches that may relax this requirement, such as bimodal or size-and-shape distributions (Brown & Schuck, 2006), but they are beyond the scope of this chapter.

As an illustration of the power of this approach, the data (exemplified in Fig. 2A) for a glycosylation mutant of the amino-terminal domain of the Niemann–Pick type C protein 1 (hereafter called NPC1-NTD) (Kwon et al., 2009) were reevaluated. Kwon et al. used a strategy similar to that delineated above, but did a separate SEDFIT analysis for each hypothesized q; only one such analysis is required by GUSSI. Three $c(s)$ distributions for a concentration range of NPC1-NTD were obtained by analysis in SEDFIT, yielding very similar $(f/f_0)_r$'s (Table 1). When all the experimental information was input into GUSSI, the program produced Fig. 5A, showing an intersection of $M_{GP,h}$ and $M_{GP,s}$ at about 16% carbohydrate ($q_{int}=0.161$) and a total molar mass of about 31 kDa, consistent with a monomer of NPC1-NTD. If the hypothesized oligomer was changed to a dimer, the two mass lines did not intersect at positive values of q ($q_{int}=-0.373$; Fig. 5B), indicating that a dimer of NPC1-NTD is very unlikely. Importantly, the hypothesis of little or no microheterogeneity was tested using a prior assumption (Brown et al., 2007) that the main species could be represented by a delta function (i.e., a homogeneous single species) at the s-value of the majority species. Violation of the prior would result in shoulder peaks outside the main peak, but none were observed (Fig. 2B). Thus, the data were consistent with the prior assumption, and the $(f/f_0)_r$ values obtained from the analysis were used without hesitation.

3.3.3 Membrane Proteins

The goal of the membrane-protein routines in GUSSI is essentially the same as that in the glycoprotein routine, i.e., to use SV to determine the oligomeric state of the protein. In this case, the sedimenting species is a protein-containing micelle of uncertain composition. GUSSI's routines follow closely the protocols elaborated by M. le Maire, C. Ebel, and coworkers (Le Maire et al., 2008; Le Roy et al., 2015; Salvay, Santamaria, le Maire, & Ebel, 2008). In all of the following, an assumption is made that the detergent has no absorbance at the protein-detection wavelength. Further, contributions from lipids are neglected below, but they can be included in GUSSI.

There are three strategies to address the oligomeric state of a membrane protein using the tools incorporated into GUSSI. All of them require some knowledge about the protein, the detergent, the protein-containing micelle, and the experimental setting: $M_{P,mono}$, $\bar{\nu}_P$, $\bar{\nu}_D$, δ_D (the amount of detergent bound to the protein in units of g/g), ρ, and η, and T. The value of δ_D is probably not known by the user, but may be measured in an SV experiment in which the evolution of the concentration profiles is monitored by both laser interferometry and absorbance optics. The magnitude of the protein's signal in the interference data can be predicted by knowing its signal magnitude in the absorbance data along with ε_P, dn/dc_P, and the wavelength of the laser used in the interference optics (λ). Any interferometric signal in excess of that expected from the protein is attributed to the detergent. If we denote the species' signal magnitude from the absorbance optics as a_a and that from the interference optics (using the same centrifugation cell) as a_i, then

$$\delta_D = \frac{a_i \varepsilon_P \lambda}{a_a \cdot dn/dc_D} - \frac{dn/dc_P}{dn/dc_D}. \tag{9}$$

With δ_D in hand, the first calculation that can be accomplished is that of a hypothetical f/f_0 for the sedimenting species, given a hypothetical molar mass for the protein oligomer, $M_H = nM_{P,mono}$. This is undertaken by realizing that $f/f_0 = R_S/R_{min}$, where R_S is the Stokes radius of the species and R_{min} is the minimum possible radius of the particle given its mass. In this context,

$$R_{min} = \left(\frac{3M_H(\bar{v}_P + \delta_D \bar{v}_D)}{4\pi N_A}\right)^{1/3}, \tag{10}$$

where N_A is Avogadro's number, and

$$R_S = \frac{M_H(1 - \phi' \rho)}{6\pi \eta s N_A}, \tag{11}$$

where

$$\phi' = \bar{v}_P + \delta_D \bar{v}_D - \frac{\delta_D}{\rho}. \tag{12}$$

The appearance of δ_D in Eq. (12) explicitly makes ϕ' dependent on the relative absorbance and interference signals (see Eq. 9). The buoyant molar mass of the protein equals the product of M_H and $(1 - \phi' \rho)$, as formulated in the trailblazing work of Casassa and Eisenberg (1964) and Tanford and Reynolds (1976). If the f/f_0 derived from this analysis is not in the typical range of 1.1–1.5, the hypothesis of M_H is likely incorrect, and a new calculation with a different hypothesized oligomeric state (n) is warranted.

The second two strategies for testing the likelihood of putative protein oligomers use a modification of the Svedberg equation to calculate the molar mass of the protein component of the protein/detergent complex, M_P, and then to compare it to M_H. This modification makes use of ϕ':

$$M_P = \frac{sRT}{D(1-\phi'\rho)}. \tag{13}$$

These two methods differ only in how they obtain information on D. In the first, D is calculated from the SV result by using Eq. (6), with the caveats pointed out above (Section 3.3.2) for using $(f/f_0)_r$ holding here as well. The second method relies on external information regarding R_S obtained from another experiment, e.g., size-exclusion chromatography. In this case, the Stokes–Einstein equation is used:

$$D = \frac{kT}{6\pi\eta R_S}. \tag{14}$$

In practice, the user would conduct one or more SV experiments, collecting both interference and absorbance data (Le Maire et al., 2008). These data sets would be loaded into SEDFIT and analyzed using the $c(s)$ model, with the user inputting the same analysis parameters for all sets and noting the $(f/f_0)_r$'s for each set. All of the distributions can then be imported into the cofs module of GUSSI, and the membrane-protein routine invoked. After the user selects the peaks of interest, GUSSI determines the s_w values and signal magnitudes associated with the peaks (δ_D is calculated from the latter; see Eq. 9). Then, the user is prompted to input the needed information. Once this is accomplished, the user may toggle between the three types of calculations, supplying hypothetical oligomers and observing how well they conform to the SV data.

To illustrate these functions, SV data for the Ca^{2+}-ATPase SERCA1a (Salvay et al., 2008) were reevaluated.[1] This membrane protein (termed "Ca^{2+}-ATPase" hereafter), solubilized in dodecyl-β-D-maltoside (DM), has a monomeric molar mass of approximately 110 kDa based on its amino acid sequence. Both interference and absorbance data were acquired; because the reference buffer contained no detergent, free micelles of DM were detected in the interference data (Fig. 6A). In sum, six data sets were evaluated (absorbance and interference for each sample): (1) ~0.8 mg/mL Ca^{2+}-ATPase in buffer with 1 mg/mL DM, (2) ~0.3 mg/mL Ca^{2+}-ATPase

[1] Data originally used in the Journal of Biological Physics, "Analytical ultracentrifugation sedimentation velocity for the chracterization of detergent-solubilized membrane proteins Ca^{++}-ATPase and ExbB.", vol. 33, 2007, pages 399–419, Salvay, A.Gl, Santamaria, M. le Maire, M., Ebel, C. Copyright Springer Science + Business B.V. 2008.

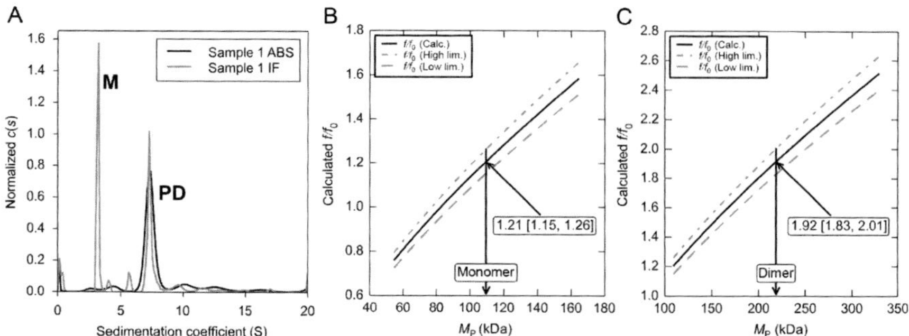

Figure 6 The Ca^{2+}-ATPase distributions and the membrane-protein f/f_0 calculation. (A) Exemplary $c(s)$ distributions for the Ca^{2+}-ATPase/DM data. Shown are the distributions resulting from the analysis of the interference and absorbance data for sample (1) in the text, i.e., 0.8 mg/mL Ca^{2+}-ATPase in a solution containing 1 mg/mL DM. "M" denotes the free detergent micelle, and "PD" the protein/detergent complex. In panels (B) and (C), the lines represent the high, optimal, and low limits for f/f_0 for the protein/micelle species given the parameters. In panel (B), the user has hypothesized a monomer, and the projection from the middle line to the X-axis shows this molar mass. The f/f_0 is shown with an error interval. Panel (C) shows the same calculation with hypothesized dimer.

in buffer with 1 mg/mL DM, and (3) ~0.3 mg/mL Ca^{2+}-ATPase in buffer with 0.3 mg/mL DM. The data were analyzed using the $c(s)$ model in SEDFIT (Table 2, Fig. 6A). After loading all of the $c(s)$ distributions into GUSSI and initiating the membrane-protein routine, the program first calculated that the f/f_0 for a hypothetical monomer was 1.21 (Fig. 6B), well within the range expected for a protein-containing micelle. GUSSI also takes into account the measurement errors in s and δ_D to arrive at an estimate of the standard error in this quantity (Fig. 6B). If the hypothesized oligomer was changed to a dimer, the calculated f/f_0 was 1.92, a value that is physically unlikely (Fig. 6C). This test pointed to a monomeric protein.

The second test was to use s from SV and R_S from chromatography to calculate M_P (Eqs. 13 and 14). This quantity is compared with M_H to assess how well the data fit the hypothesis. The experimentally determined R_S (5.5 nm (Salvay et al., 2008)) was used for this calculation (Table 2, Fig. 7). This resulted in $M_P = 135$ kDa, close to the hypothetical monomer, and well away from the hypothetical dimer. Therefore, this second test was also most consistent with the hypothesis of monomeric Ca^{2+}-ATPase.

The final test was to calculate M_P based on s and D from the SV experiment. Calculating D using Eq. (6) and then inputting that into Eq. (13) yielded $M_P = 111$ kDa (Fig. 8; a range of likely M_P values given the errors

Table 2 Parameters for the Membrane-Protein Analysis

Analysis/Experimental Parameters	
$\bar{\nu}_u$ (cm^3/g)	0.7425
ρ (g/cm^3)	1.004
η (Poise)	0.0100
T (K)	293
λ (nm)	675
Frictional ratios from analyses (absorbance scans only)	
Exp. 1	1.246
Exp. 2	1.324
Protein parameters	
$\bar{\nu}_P$ (cm^3/g)	0.7425
ε_P (L/g cm)	0.966
$M_{P,mono}$ (Da)	109,490
dn/dc_P (cm^3/g)	0.187
R_S (nm)	5.5
Detergent parameters	
$\bar{\nu}_D$ (cm^3/g)	0.82
dn/dc_P (cm^3/g)	0.143
Calculated by GUSSI	
δ_D (g/g)	0.47

in s and D is displayed, 101–121 kDa in this example). The hypothesis of the monomer is again closest to the calculated mass of the protein. Critically, the $(f/f_0)_r$ values from the absorbance data only were used in this calculation because these values from the interference data were probably skewed by the strong signals from the empty detergent micelles (Fig. 6A, Table 2). Also, Cell 3 was excluded from this test because the homogeneity criterion did not appear to hold for this sample. All three tests conclusively pointed to the monomer of this protein being the entity present in the sedimenting protein–detergent complex; hypotheses of $n > 1$ may be safely rejected.

On a final note, the δ_D calculated for Ca^{2+}-ATPase (0.5 g/g) in this work was substantially lower than those reported earlier (0.8–0.9 g/g) using the same data (Le Maire et al., 2008; Salvay et al., 2008). However, these other

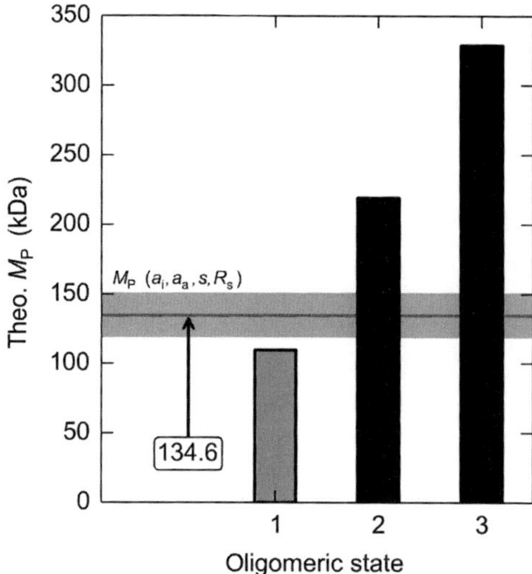

Figure 7 The R_S membrane-protein calculation. The bars represent the M_H given the supplied molar mass of the protein and the hypothesized oligomeric state (the highlighted bar; neighboring bars are shown for reference). A shaded band with an optimal value shows the result of the molar-mass calculation when D is calculated from a supplied R_S. GUSSI notes the exact M_P with an arrow, and terms this mass "$M_P(a_i, a_a, s, R_S)$" to emphasize that the mass is a function of the measured signal amplitudes, the sedimentation coefficient, and the Stokes radius.

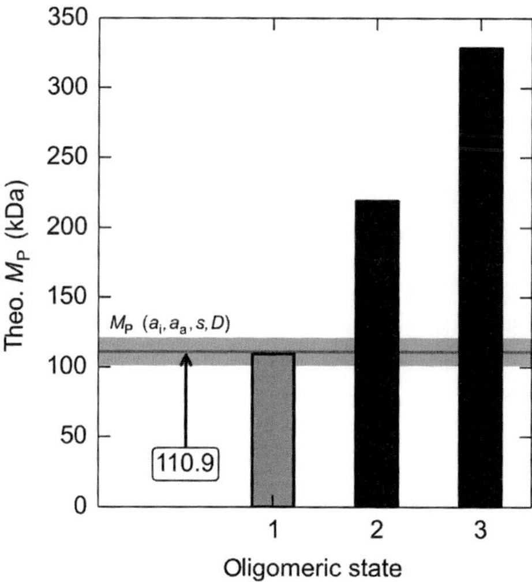

Figure 8 The "fitted f/f_0" membrane-protein calculation. This figure has the same format as Fig. 7. The band represents the molar-mass calculation with D calculated from s and $(f/f_0)_r$ (the latter are taken from the absorbance data only). In this case, the mass is noted as "$M_P(a_i, a_a, s, D)$" to emphasize the contributions of the signal amplitudes, and the sedimentation and diffusion coefficients to the calculation.

authors also came to the conclusion that Ca^{2+}-ATPase is a monomer. It is likely that the variation in the reported δ_D resulted from different integration limits in the *c*(*s*) distributions derived from the absorbance data (i.e., variation in a_i in Eq. 9). These results demonstrate that the determination of oligomeric state can be somewhat insensitive to the exact value of δ_D. Also, the value δ_D calculated herein is similar to that gleaned from a density-variation SV experiment on Ca^{2+}-ATPase if a frictional ratio of 1.25 is assumed (Le Roy et al., 2015), buttressing the validity of the value presented above.

4. SUMMARY

The new software presented above, GUSSI, enables the straightforward graphing of analyses from SEDFIT and SEDPHAT in a publishable form. The graphs can be obtained in just a few mouse clicks. These graphs are highly user-customizable, and the program offers innovative plot formats. The program also provides a convenient platform for simple operations on AUC data. GUSSI is undergoing active development, and it currently encompasses many more functions than detailed in this chapter. These are fully documented in a manual that accompanies the program. The software may be obtained at http://biophysics.swmed.edu/MBR/software.html.

ACKNOWLEDGMENTS

The author gratefully acknowledges the advice, suggestions, and testing efforts of Drs. Patrick Brown, Christine Ebel, Rodolfo Ghirlando, Shae Padrick, Grzegorz Piszczek, Peter Schuck, and Huaying Zhao. The author also thanks Dr. Schuck for adding GUSSI plotting menu items to SEDFIT and SEDPHAT. Drs. Hyock Joo Kwon and Johann Deisenhofer are thanked for providing the NPC1-NTD data, as are Drs. Aline Le Roy and Ebel for graciously supplying the Ca^{2+}-ATPase data. Dr. Thomas Scheuermann and Drs. Padrick, Zhao, Ghirlando, and Ebel are thanked for providing comments on a draft of this chapter. Comprehensive citation of the expansive AUC literature was not possible due to format and space requirements; the author apologizes for any oversights.

REFERENCES

Ayaz, P., Munyoki, S., Geyer, E. A., Piedra, F.-A., Vu, E. S., Bromberg, R., et al. (2014). A tethered delivery mechanism explains the catalytic action of a microtubule polymerase. *ELife*, *3*, e03069.

Balbo, A., Minor, K. H., Velikovsky, C. A., Mariuzza, R. A., Peterson, C. B., & Schuck, P. (2005). Studying multiprotein complexes by multisignal sedimentation velocity analytical ultracentrifugation. *Proceedings of the National Academy of Sciences of the United States of America*, *102*, 81–86.

Brautigam, C. A. (2011). Using Lamm-equation modeling of sedimentation velocity data to determine the kinetic and thermodynamic properties of macromolecular interactions. *Methods*, *54*, 4–15.
Brown, P. H., Balbo, A., & Schuck, P. (2007). Using prior knowledge in the determination of macromolecular size-distributions by analytical ultracentrifugation. *Biomacromolecules*, *8*, 2011–2024.
Brown, P. H., Balbo, A., & Schuck, P. (2008). Characterizing protein-protein interactions by sedimentation velocity analytical ultracentrifugation. In *Current protocols in immunology* (pp. 18.15.1–18.15.39): John Wiley & Sons.
Brown, P. H., & Schuck, P. (2006). Macromolecular size-and-shape distributions by sedimentation velocity analytical ultracentrifugation. *Biophysical Journal*, *90*, 4651–4661.
Brown, P. H., & Schuck, P. (2008). A new adaptive grid-size algorithm for the simulation of sedimentation velocity profiles in analytical ultracentrifugation. *Computer Physics Communications*, *178*, 105–120.
Cao, W., & Demeler, B. (2008). Modeling analytical ultracentrifugation experiments with an adaptive space-time finite element solution for multicomponent reacting systems. *Biophysical Journal*, *95*(1), 54–65.
Casassa, E. F., & Eisenberg, H. (1964). Thermodynamic analysis of multicomponent solutions. *Advances in Experimental Medicine and Biology*, *19*, 287–395.
Claverie, J.-M., Dreux, H., & Cohen, R. (1975). Sedimentation of generalized systems of interacting particles. I. Solution of systems of complete Lamm equations. *Biopolymers*, *14*, 1685–1700.
Cole, J. L. (2004). Analysis of heterogeneous interactions. *Methods in Enzymology*, *384*, 212–232.
Correia, J. J., & Stafford, W. F. (2009). Extracting equilibrium constants from kinetically limited reacting systems. *Methods in Enzymology*, *455*, 419–446.
Correia, J. J., & Stafford, W. F. (2015). Sedimentation velocity: A classical perspective. *Methods in Enzymology*, *562*, 49–80.
Dam, J., & Schuck, P. (2005). Sedimentation velocity analysis of heterogeneous protein-protein interactions: Sedimentation coefficient distributions c(s) and asymptotic boundary profiles from Gilbert-Jenkins theory. *Biophysical Journal*, *89*, 651–666.
Dam, J., Velikovsky, C. A., Mariuzza, R. A., Urbanke, C., & Schuck, P. (2005). Sedimentation velocity analysis of heterogeneous protein-protein interactions: Lamm equation modeling and sedimentation coefficient distributions c(s). *Biophysical Journal*, *89*, 619–634.
Demeler, B. (2005). UltraScan: A comprehensive data analysis software package for analytical ultracentrifugation experiments. In D. J. Scott, S. E. Harding, & A. J. Rowe (Eds.), *Modern analytical ultracentrifugation: Techniques and methods* (pp. 210–229). Cambridge, UK: Royal Society of Chemistry.
Ghirlando, R. (2011). The analysis of macromolecular interactions by sedimentation equilibrium. *Methods*, *54*, 145–156.
Ghirlando, R., Keown, M. B., Mackay, G. A., Lewis, M. S., Unkeless, J. C., & Gould, H. J. (1995). Stoichiometry and thermodynamics of the interaction between the Fc fragment of human IgGl and its low-affinity receptor FcγRIII. *Biochemistry*, *34*, 13320–13327.
Gillis, R. B., Adams, G. G., Heinze, T., Nikolajski, M., Harding, S. E., & Rowe, A. J. (2013). MultiSig: A new high-precision approach to the analysis of complex biomolecular systems. *European Biophysics Journal*, *42*, 777–786.
Keller, S., Vargas, C., Zhao, H., Piszczek, G., Brautigam, C. A., & Schuck, P. (2012). High-precision isothermal titration calorimetry with automated peak-shape analysis. *Analytical Chemistry*, *84*, 5066–5073.
Kwon, H. J., Abi-Mosleh, L., Wang, M. L., Deisenhofer, J., Goldstein, J. L., Brown, M. S., et al. (2009). Structure of N-terminal domain of NPC1 reveals distinct subdomains for binding and transfer of cholesterol. *Cell*, *137*, 1213–1224.

Lamm, O. (1929). Die Differentialgleichung der Ultrazentrifugierung. *Arkiv för Matematik, Astronomi Och Fysik, 21B*, 1–4.
Laue, T. M. (1999). Analytical centrifugation: Equilibrium approach. *Current Protocols in Protein Science, 18*, 20.3.1–20.3.13.
Laue, T. M., Shah, B. D., Ridgeway, R. M., & Pelletier, S. L. (1992). Computer-aided interpretation of analytical sedimentation data for proteins. In S. E. Harding, A. J. Rowe, & J. C. Horton (Eds.), *Analytical ultracentrifugation in biochemistry and polymer science* (pp. 90–125). Cambridge, UK: The Royal Society of Chemistry.
Le Maire, M., Arnou, B., Olesen, C., Georgin, D., Ebel, C., & Moller, J. V. (2008). Gel chromatography and analytical ultracentrifugation to determine the extent of detergent binding and aggregation, and Stokes radius of membrane proteins using sarcoplasmic reticulum Ca^{2+}-ATPase as an example. *Nature Protocols, 3*, 1782–1795.
Le Roy, A., Wang, K., Schaak, B., Schuck, P., Breyton, C., & Ebel, C. (2015). AUC and small-angle scattering for membrane proteins. *Methods in Enzymology, 562*, 257–286.
Lewis, M. S., & Junghans, R. P. (2000). Ultracentrifugal analysis of molecular mass of glycoproteins of unknown or ill-defined carbohydrate composition. *Methods in Enzymology, 321*, 136–149.
Padrick, S. B., Deka, R. K., Chuang, J. L., Wynn, R. M., Chuang, D. T., Norgard, M. V., et al. (2010). Determination of protein complex stoichiometry through multisignal sedimentation velocity experiments. *Analytical Biochemistry, 407*, 89–103.
Perkins, S. J. (1986). Protein volumes and hydration effects: The calculations of partial specific volumes, neutron scattering matchpoints and 280-nm absorption coefficients for proteins and glycoproteins from amino acid sequences. *European Journal of Biochemistry, 157*, 169–180.
Salvay, A. G., Santamaria, M., le Maire, M., & Ebel, C. (2008). Analytical ultracentrifugation sedimentation velocity for the characterization of detergent-solubilized membrane proteins Ca^{++}-ATPase and ExbB. *Journal of Biological Physics, 33*, 399–419.
Scheuermann, T. H., & Brautigam, C. A. (2015). High-precision, automated integration of multiple isothermal titration calorimetric thermograms: New features of NITPIC. *Methods, 76*, 87–98.
Schuck, P. (1998). Sedimentation analysis of noninteracting and self-associating solutes using numerical solutions to the Lamm Equation. *Biophysical Journal, 75*, 1503–1512.
Schuck, P. (2000). Size distribution analysis of macromolecules by sedimentation velocity ultracentrifugation and Lamm equation modeling. *Biophysical Journal, 78*, 1606–1619.
Schuck, P. (2010). Sedimentation patterns of rapidly reversible protein interactions. *Biophysical Journal, 98*, 2005–2013.
Schuck, P., Perugini, M. A., Gonzales, N. R., Howlett, G. J., & Schubert, D. (2002). Size-distribution analysis of proteins by analytical ultracentrifugation: Strategies and application to model systems. *Biophysical Journal, 82*, 1096–1111.
Shire, S. J. (1992). *Determination of molecular weight of glycoproteins by analytical ultracentrifugation: Application note DS-837*. Palo Alto, CA: Beckman-Coulter, Inc.
Stafford, W. F. (2003). Analytical ultracentrifugation: Sedimentation velocity analysis. *Current Protocols in Protein Science, 31*, 20.7.1–20.7.11.
Stafford, W. F., & Sherwood, P. J. (2004). Analysis of heterologous interacting systems by sedimentation velocity: Curve fitting algorithms for estimation of sedimentation coefficients, equilibrium and kinetic constants. *Biophysical Chemistry, 108*, 231–243.
Tanford, C., & Reynolds, J. A. (1976). Characterization of membrane proteins in detergent solutions. *Biochimica et Biophysica Acta, 457*, 133–170.
Vistica, J., Dam, J., Balbo, A., Yikilmaz, E., Mariuzza, R. A., Rouault, T. A., et al. (2004). Sedimentation equilibrium analysis of protein interactions with global implicit mass conservation constraints and systematic noise decomposition. *Analytical Biochemistry, 326*, 234–256.

Zhao, H., Berger, A. J., Brown, P. H., Kumar, J., Balbo, A., May, C. A., et al. (2012). Analysis of high-affinity assembly for AMPA receptor amino-terminal domains. *Journal of General Physiology*, *139*, 371–388.

Zhao, H., Brautigam, C. A., Ghirlando, R., & Schuck, P. (2013). Overview of current methods in sedimentation velocity and sedimentation equilibrium analytical ultracentrifugation. *Current Protocols in Protein Science*, *71*, 20.12.1–20.12.49.

Zhao, H., Brown, P., & Schuck, P. (2011). On the distribution of protein refractive index increments. *Biophysical Journal*, *100*, 2309–2317.

Zhao, H., Lomash, S., Glasser, C., Mayer, M. L., & Schuck, P. (2013). Analysis of high affinity self-association by fluorescence optical sedimentation velocity analytical ultracentrifugation of labeled proteins: Opportunities and limitations. *PLoS One*, *8*, e83439.

Zhao, H., Piszczek, G., & Schuck, P. (2015). SEDPHAT—A platform for global ITC analysis and global multi-method analysis of molecular interactions. *Methods*, *76*, 137–148.

CHAPTER SIX

Sedimentation Equilibrium Analysis of ClpB Self-Association in Diluted and Crowded Solutions

Carlos Alfonso*, Urko del Castillo†, Ianire Martín†, Arturo Muga†,1, Germán Rivas*,1

*Centro de Investigaciones Biológicas, Consejo Superior de Investigaciones Científicas (CSIC), Madrid, Spain
†Unidad de Biofísica (CSIC/UPV-EHU), Departamento de Bioquímica y Biología Molecular, Universidad de País Vasco-Euskal Herriko Unibertsitatea (UPV-EHU), Bilbao, Biscay, Spain
[1]Corresponding authors: e-mail address: arturo.muga@ehu.eus; grivas@cib.csic.es

Contents

Abstract

ClpB belongs to the Hsp100 family of ring-forming heat-shock proteins involved in degradation of unfolded/misfolded proteins and in reactivation of protein aggregates. ClpB monomers reversibly associate to form the hexameric molecular chaperone that, together with the DnaK system, has the ability to disaggregate stress-denatured proteins. Here, we summarize the use of sedimentation equilibrium approaches,

Methods in Enzymology, Volume 562
ISSN 0076-6879
http://dx.doi.org/10.1016/bs.mie.2015.04.007

complemented with sedimentation velocity and composition-gradient static light scattering measurements, to study the self-association properties of ClpB in dilute and crowded solutions. As the functional unit of ClpB is the hexamer, we study the effect of environmental factors, i.e., ionic strength and natural ligands, in the association equilibrium of ClpB as well as the role of the flexible N-terminal and M domains of the protein in the self-association process. The application of the nonideal sedimentation equilibrium technique to measure the effects of volume exclusion, reproducing in part the natural crowded conditions inside a cell, on the self-association and on the stability of the oligomeric species of the disaggregase will be described. Finally, the biochemical and physiological implications of these studies and future experimental challenges to eventually reconstitute minimal disaggregating machineries will be discussed.

1. INTRODUCTION

Most biological processes, as protein homeostasis, are carried out by multiprotein assemblies whose elements reversibly interact to form the functional entities active in maintaining the proper levels and stability of proteins in cells, which is essential for cell viability and growth. Proteome integrity is maintained by the coordinated action of molecular chaperones and the proteolytic machinery (the ubiquitin proteasome system, UPS, and the lysosomal and autophagy systems (Arias & Cuervo, 2011; Hipp, Park, & Hartl, 2014; Kim, Hipp, Bracher, Hayer-Hartl, & Hartl, 2013; Pickart & Cohen, 2004)), which together with other proteostasis network components implicated in protein expression or transport prevent the accumulation of aberrant conformational states. An age-dependent (Ben-Zvi, Miller, & Morimoto, 2009; Cohen et al., 2009) or a stress-induced imbalance in the cellular proteostasis network could exceed its regulatory activity, leading to the accumulation of toxic protein aggregates that cause the loss of protein functions and eventually cell death. Molecular chaperones are key players in this process, as they are involved in *de novo* protein folding, prevent protein aggregation, promote disaggregation and refolding of stress-denatured proteins, and mediate macromolecular complex assembly, protein trafficking, and degradation. Hsp100 chaperones are unique due to their ability to direct protein aggregates to degradation in cooperation with proteolytic components or to extract denatured polypeptides from the aggregates to refold and reactivate them in collaboration with the Hsp70 system (Hodson, Marchall, & Burston, 2012; Mogk, Haslberger, Tessarz, & Bukau, 2008). Therefore, understanding at a molecular level these processes requires to characterize quantitatively these reversible macromolecular associations, to determine the number of subunits forming the final assemblies and the strength of the interactions, and to explore how their formation is modulated by environmental conditions and physiological ligands/proteins. Among the large variety of techniques available to study protein interactions (Schuck & Zhao, 2013), we focus on sedimentation equilibrium (SE), a powerful and versatile analytical ultracentrifugation tool to measure the stoichiometry, affinity, and reversibility of macromolecular interactions in solution. This technique is also uniquely adapted to explore excluded volume effects due to natural crowding on these interactions. These SE approaches, complemented by composition-gradient static light scattering (CG-SLS) and sedimentation velocity (SV), have been applied to define the activities, stability, and association properties of ClpB, a disaggregase involved in protein aggregate reactivation.

1.1 ClpB Assembly and Function

Many of the insoluble aggregated substrates are degraded by Hsp100/Clp proteins, which after extracting polypeptides from the aggregate transfer them to the proteolytic chamber of the associated ClpP. However, severe or prolonged stress conditions can stall the protein degradation machinery and overwhelm the capacity of the conventional protective system with severe consequences, including cell death. Under these conditions, the activity of bacterial ClpB and its yeast homologue Hsp104 is essential, as they are able to interact with protein aggregates and reactivate them in collaboration with the Hsp70 system (Sanchez & Lindquist, 1990; Squires, Pedersen, Ross, & Squires, 1991) Thereby, ClpB and Hsp104 chaperones provide an essential alternative nondestructive pathway to avoid the collapse of the proteostasis network, with the advantage that they rescue aggregated proteins restoring the otherwise lost functions.

ClpB monomers are arranged into homohexamers that constitute the protein functional unit. Each monomer is composed by an N-terminal domain that improves the reactivation efficiency of stable protein aggregates, two nucleotide-binding domains (NBD1 and NBD2) that bind and hydrolyze ATP, and a middle M domain that is inserted into the NBD1 and is strictly required for the disaggregase activity of the chaperone (Fig. 1; Lee et al., 2003). Location of the different protein domains within the hexamer is essential to understand their role in protein activity, allosteric communication within the hexamer, and thus the global mechanism of action of this disaggregase. ClpB oligomerizes dynamically in response

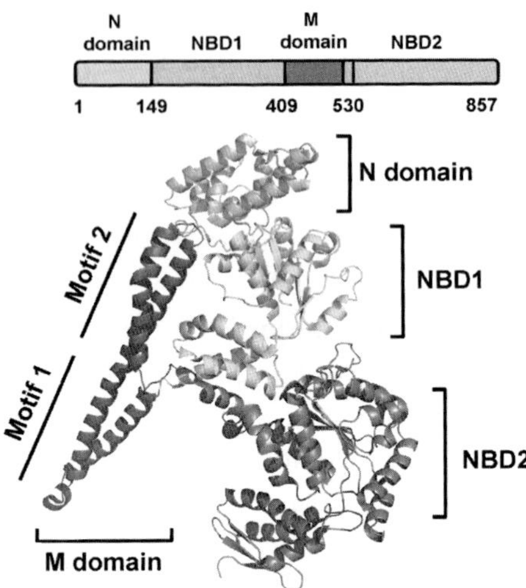

Figure 1 Three-dimensional structure of a ClpB monomer. ClpB from *Thermus thermophilus* was crystallized as a trimer in the presence of AMPPNP (PDB: 1QVR). The different domains of the protein are shown: N-terminal domain (yellow; gray in the print version), NBD1 (blue; light gray color in the print version), NBD2 (gray), and M domain (purple; dark gray in the print version). The four helices that form the M domain are divided into two motifs.

to different stimuli, such as protein concentration (Akoev, Gogol, Barnett, & Zolkiewski, 2004; del Castillo et al., 2011; Zolkiewski, Kessel, Ginsburg, & Maurizi, 1999), nucleotide binding, hydrolysis, and ionic strength (Akoev et al., 2004; del Castillo et al., 2011; Kim et al., 2000; Schlee, Groemping, Herde, Seidel, & Reinstein, 2001; Zolkiewski et al., 1999). As only the hexamer is able to interact with the Hsp70 system and thus to build the bichaperone network active in protein disaggregation, an additional regulatory flexibility is achieved by modulating the oligomerization conditions by these factors. Further, ClpB oligomerization provides the possibility of allosterically coupling different protomers to allow vectorial translocation of unfolded substrates through the central protein channel. Substrate threading most likely needs to productively coordinate the conformational changes brought about by ATP binding and hydrolysis in different subunits within the oligomeric structure. Therefore, the quantitative characterization of the effect of the aforementioned factors on the association equilibrium of this disaggregase will help to understand its mechanism of action.

1.2 Biological Compartments Are Crowded

ClpB has evolved to function in the intracellular environment. This environment is characterized by the presence of high concentrations of proteins and nucleic acids (up to, and in certain environments higher than, 400 g/l), which are present as soluble species and/or structural networks and occupy at least 20–30% of the total volume (Minton, 2006; Rivas, Ferrone, & Herzfeld, 2004). Weak, nonspecific interactions with other soluble macromolecules or structural elements in its immediate vicinity are likely to affect the biochemical reactivity of ClpB. If these interactions are predominantly repulsive, they will lead to preferential exclusion from regions highly volume occupied; on the other hand, if they are predominantly attractive, they will lead to nonspecific associations and adsorption. Theory has predicted and experiments have demonstrated that these nonspecific interactions can have substantial effects on the rates and equilibria of macromolecular reactions (Minton, 1998; Zhou, Minton, & Rivas, 2008).

In this work, we will briefly discuss the background interactions arising from steric repulsion in a volume-occupied cell-like environment, which certainly will have to be taken into account to understand the behavior of ClpB *in vivo*, independently of the presence or absence of other types of interactions. Fractional volume occupancies of the order of those found in the compartments in which ClpB is located (i.e., bacterial cytoplasm) are expected to increase the chemical potential (activity) of all macromolecular species, dilute as well as concentrated, in the solution in a size- and shape-dependent manner. A fundamental chemical consequence of macromolecular crowding is that volume exclusion provides a generalized nonspecific force to facilitate processes leading to a reduction in excluded volume, namely macromolecular compaction and association (Zhou et al., 2008). Volume exclusion is also predicted to result in the acceleration of slow, transition-state-limited association reactions and the deceleration of fast, diffusion-limited association reactions (Zhou et al., 2008). It is furthermore anticipated that crowding will also lead to phase separation phenomena that can affect significantly the spatial arrangement and local distribution of ClpB complexes (Martin et al., 2014). Finally, crowding can contribute to the maintenance of functional activity of essential protein–nucleic acid and protein–protein complexes in *Escherichia coli* against large changes in the osmolarity of its environment (Cayley & Record, 2004).

2. EXPERIMENTAL APPROACHES AND ANALYSIS

A detailed description of the production of wild-type ClpB and deletion mutants $\text{ClpB}_{\Delta(410-455)}$ and $\text{ClpB}_{\Delta(410-520)}$ in which the M domain was partially or totally deleted, respectively, and $\text{ClpB}_{\Delta(1-142)}$, which lacks the N-terminal domain, has already been described (del Castillo et al., 2011;

Martin et al., 2014). All the experiments were done in 50 m*M* Tris–HCl, pH 7.5, 5 m*M* $MgCl_2$ buffer, containing different KCl concentrations (20–500 m*M*), in the absence or in the presence of 1 m*M* nucleotide (ADP or ATP). In the assays containing ATP, an ATP regeneration system (10 m*M* acetylphosphate and 1 U/ml acetyl kinase) was added to maintain a steady-state ATP concentration during the time length of the experiment (2–4 h). Because of this time constraint, SE was not done with protein samples containing ATP.

2.1 Sedimentation Velocity

SV is a hydrodynamic method in which proteins may be fractionated at high centrifugal force on the basis of differences in sedimentation coefficient, which is a function of their buoyant molar mass and shape. Analysis of the time-dependent gradients leads to estimates of the sedimentation coefficient and molar mass of each sedimenting species. The rate of transport of protein species i (J_i) centrifuged at angular velocity ω in a sector-shaped centrifuge cell is described by the following relation:

$$J_i - s_i\omega^2 rS_i - D_i\frac{\mathrm{d}S_i}{\mathrm{d}r} \quad (1)$$

where s_i and D_i are the sedimentation and diffusion coefficients of species i, respectively, ω is the rotor angular velocity, and r the radial position. S_i is a measurable property or signal (S) of the solution that varies linearly with the weight/volume concentration of each protein species at radial position r:

$$S(r) = \sum_i S_i(r) = \sum_i \alpha_i w_i(r) \quad (2)$$

where α_i denotes the signal proportionality constant of species i. Suitable signals are UV–VIS absorbance, or refractive index at a given wavelength, among others.

2.1.1 SV Assays and Analysis

SV experiments were carried out at 40–50 krpm and 18 °C in an Optima XL-I (equipped with UV–VIS and interference detection systems). The SV profiles were registered at given time intervals (5–8 min) using the appropriated optical signal (either UV–VIS or interference) to determine the sedimentation coefficient distribution of ClpB species (monomer and oligomers). The sedimentation coefficient distributions were calculated by least-squares boundary modeling of SV data using the $c(s)$ method

(Schuck, 2000; Schuck, Perugini, Gonzales, Howlett, & Schubert, 2002) as implemented in the SEDFIT program. These *s*-values were corrected to standard conditions (water, 20 °C, and infinite dilution) (van Holde, 1985) using SEDNTERP (Laue, Shah, Ridgeway, & Pelletier, 1992) to obtain the corresponding standard *s*-values ($s_{20,w}$).

2.2 Sedimentation Equilibrium

SE is a method in which a lower centrifugal force is applied to the samples, yielding information about the composition dependence of signal-average buoyant mass. The observed dependence may then be modeled in the context of different association schemes. At SE, no further net transport of species in the centrifuge cell is observed as the sedimenting and diffusing forces equilibrate, which yields the following SE relation:

$$\frac{\mathrm{d}\ln S_i}{\mathrm{d}r^2} = \frac{M^*_{i,\mathrm{app}}\omega^2}{2RT} \tag{3}$$

where $M^*_{i,\mathrm{app}}$ is the apparent buoyant molar mass of species i, which is equal to M^*_i under ideal solution conditions.

If the previous relation is integrated with respect to r^2, the expression for the equilibrium gradient of an ideally sedimenting protein is obtained:

$$S_i(r) = S_i(r_0)\exp\left[\frac{M^*_{i,\mathrm{app}}\omega^2}{2RT}\left(r^2 - r_0^2\right)\right] \tag{4}$$

where r_0 is an arbitrarily selected reference position.

The signal-average buoyant molar mass is given by

$$M^*_s = \frac{\sum_i \alpha_i w_i(r) M^*_i}{\sum_i \alpha_i w_i(r)} \tag{5}$$

The average molecular weight (M_i) of the different proteins can be obtained from the corresponding buoyant values with $M^*_i = M_i d_i$, where d_i denotes the specific density increments of the protein, which can also be expressed as $(1 - v_{\mathrm{bar}} r)$, where v_{bar} is the partial-specific volume of the protein and r is buffer density.

2.2.1 SE Assays and Analysis

Low-speed SE experiments (4000–8000 rpm) using short columns (85 μl) were done to determine the concentration dependence state of association of the ClpB proteins as detailed elsewhere (del Castillo et al., 2011). The absorbance gradients at equilibrium were obtained at the appropriate wavelength by using the UV–VIS detection system. The measured equilibrium concentration (signal) gradients were fit by Eq. (4) to get the whole-cell signal-average buoyant molar masses (M_s^*), from which the corresponding average molecular weights were calculated using 0.737 ml/g as the partial-specific volume of ClpB (calculated from the amino acid composition with SEDNTERP) (Laue et al., 1992).

The experimental approach adopted for the SE measurements yielded shallow protein gradients in which the concentration ratio of all the protein components at the base and meniscus of the solution column was less than 5. Under these conditions, the value of $\mathrm{d}\ln S(r)/\mathrm{d}r^2$ (and $M_{i,\mathrm{app}}^*$) is independent of the radius and $M_{i,\mathrm{app}}^*$ is well described by the solution average molecular weight value, which may be expressed as a function of the loading protein concentration and analyzed employing self-association relationships described elsewhere (Rivas & Minton, 2011; Zorrilla, Jiménez, Lillo, Rivas & Minton, 2004).

To model ClpB self-association in the absence of nucleotides, several association schemes were used to fit the data and that in which ClpB exists as an equilibrium mixture of monomers, hexamers, and dodecamers was the best to describe the dependence of M_s^* with ClpB concentration. This model is defined by the following equilibrium constants:

$$K_6 = w_6/w_1^6 \tag{6}$$

$$K_{12} = w_{12}/w_1^{12} \tag{7}$$

where w_1, w_6, and w_{12} are the weight concentrations of monomers, hexamers, and dodecamers, respectively.

The total concentration is

$$w_{\mathrm{tot}} = w_1 + w_6 + w_{12} \tag{8}$$

Given the values of w_{tot}, M_1, and the two association constants, Eqs. (6)–(8) were solved numerically for the equilibrium value of w_1, and Eqs. (6) and (7) were used to calculate the equilibrium values of w_1, w_6, and w_{12}. Then Eq. (5) was used to calculate M_{w}. All the calculations were performed using MATLAB scripts kindly provided by Dr. Allen Minton (NIH).

2.3 Composition-Gradient Static Light Scattering

Static light (or Rayleigh) scattering measures the intensity of light scattered by protein species in solution, allowing the determination of average molar masses. The methodological approach used in this study (CG-SLS) allows the rapid (order of minutes) acquisition of continuous light scattering data from a solution whose composition varies with time in a controlled manner. The resulting time-dependent scattering and composition data are globally modeled using molecular schemes for composition-dependent scattering (Kameyama & Minton, 2006).

The scattering data are processed to yield the Rayleigh ratio R scaled to an optical constant:

$$K_{\text{opt}} = \frac{4\pi n^2 (\mathrm{d}n/\mathrm{d}w)^2}{\lambda^4 N_A} \tag{9}$$

where n and λ correspond to the refractive index of the solution and the wavelength of incident light (690 nm), respectively, $\mathrm{d}n/\mathrm{d}w$ denotes the specific refractive increment of the scattering species.

If solute species are small relative to the wavelength of incident light, the scaled Rayleigh ratio is independent of scattering angle and depends only upon the protein composition (Attri & Minton, 2005):

$$\frac{R}{K_{\text{opt}}} = \sum_i M_i w_i = \sum_i c_i M_i^2 \tag{10}$$

where M_i, w_i, and c_i, respectively, are the molar mass, w/v concentration, and molar concentration of the ith protein species.

The fractional contribution of a given protein species to the total scattering intensity is then given by

$$f_j = \frac{c_j M_j^2}{\sum_i c_i M_i^2} \tag{11}$$

2.3.1 CG-SLS Assays

Scattering experiments were performed using the Calypso system (Wyatt Technology, Santa Barbara, CA) according to procedures developed by Minton and coworkers (Attri & Minton, 2005). In brief, a programmable three-injector syringe pump was used to generate step gradients of concentration of a protein solution of defined composition in 10% increments into parallel flow cells for concurrent measurement of Rayleigh light scattering at multiple angles (MALLS, using a DAWN-EOS multiangle laser light scattering detector, Wyatt Technology, Santa Barbara, CA) and solute composition (using a Optilab rEX refractive index detector, Wyatt Technology, Santa Barbara, CA).

2.3.2 Scattering Analysis of ClpB

It follows from Eq. (10) that any equilibrium model used to fit the dependence of R/K_{opt} upon w_{tot} may also be used to calculate the dependence of M_w upon w_{tot}, the form in which the SE results are presented in this work. Therefore, the same association schemes described above to model the composition dependence of SE data experiments were used to model the composition dependence of R/K_{opt}. They allow calculating the molar concentration of all significant scattering species and its dependence upon total concentration and the postulated equilibrium association constants. The results of these models were then combined with Eq. (10) to calculate the dependence of total scattering upon solution composition. As in the case of SE data, a monomer–hexamer–dodecamer model was found to be the association scheme that best describes the scattering data in the absence of nucleotide (for details, see del Castillo et al., 2011).

2.4 Studying ClpB Self-Association in Crowded Solutions

The modulation of the association state(s) of ClpB, present at relatively low concentration (9–20 μ*M*) in its naturally crowded intracellular environment, cannot be reliably predicted from the knowledge of the association properties of samples containing these ClpB concentrations, in the absence of background macromolecules present in the cell interior that may interact in a nonspecific manner with ClpB. Among those background interactions, it has already been discussed that excluded volume effects may alter considerably the tendency of a given protein to associate, as illustrated in the following section for ClpB.

2.4.1 Estimation of Crowding Effects on ClpB Self-Association

It is assumed that ClpB (=B) can self-associate to form B_6 according to the following equilibria:

$$6B \rightleftarrows B_6 \quad K=[B_6]/[B]^6$$

The effect of crowding on K can be estimated according to the relation

$$K = K^{\text{o}}\Gamma = K^{\text{o}}\frac{\gamma_1^6}{\gamma_6} \quad (12)$$

where K^{o} is the value of K in an ideal (dilute) solution and γ_i denotes the activity coefficient of species i. For the purpose of estimating the activity coefficients, all species can be represented by spheres of equal density. The crowder is represented by a sphere with mass M_c (in K units), B by a sphere of mass 100 K, and B_6 by spheres with mass 600 K. The scaled particle theory of hard sphere mixtures (Minton, 1998) provides the following simple relation for the activity coefficient of a spherical tracer in a fluid containing a volume fraction ϕ of spherical crowders:

$$\ln \gamma_T = -\ln(1-\phi) + A_1 Q + A_2 Q^2 + A_3 Q^3 \quad (13)$$

where

$$Q = \phi/(1-\phi)$$
$$A_1 = R^3 + 3R^2 + 3R$$
$$A_2 = 3R^3 + 4.5R^2$$
$$A_3 = 3R^3$$

R denotes the ratio of the radius of tracer to that of crowder. For ClpB monomer, $R = (100/M_c)^{1/3}$, and for hexamer, $R = (600/M_c)^{1/3}$. Using Eqs. (12) and (13), log Γ $\left(= \log K_6/K_6^0\right)$ can be calculated as a function of ϕ (Fig. 2). Even if the estimates are only approximate, the monomer–hexamer stoichiometry provides enormous sensitivity of the reaction to crowding effects. Depending upon the value of M_c, a crowder volume fraction of only 0.25 can increase the log Γ value 4–7 units, resulting in an increase in the self-association constant by at least one order of magnitude.

2.4.2 Nonideal Tracer SE

Highly volume-occupied solutions that partially mimic natural crowding in the test tube are usually reconstituted by adding high concentrations of inert unrelated macromolecules (crowders) to the system under study. The analysis of crowding effects on ClpB-related reactions is considerably more challenging than the corresponding studies in dilute solution: (1) the behavior of ClpB species that are diluted has to be detected and measured in the presence of very high concentrations of crowders, leading to the use of tracer methods

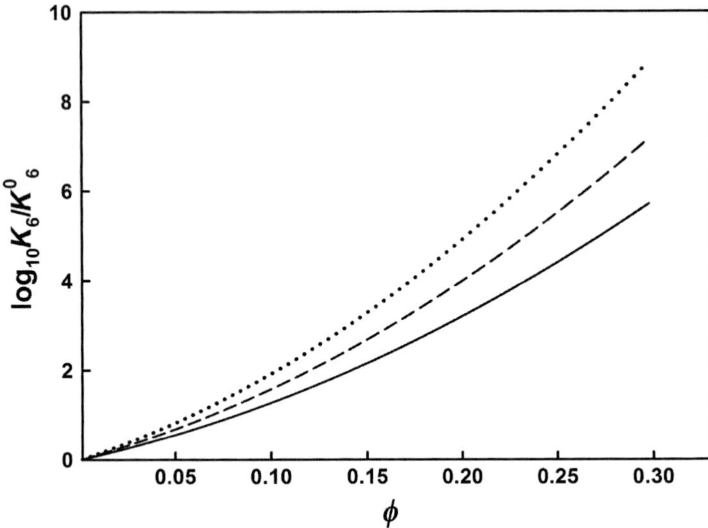

Figure 2 Calculated change in the equilibrium constant for self-association of dilute ClpB protein monomers (100 kDa) into hexamers as a function of the fraction of volume occupied by an inert protein "*c*" of molar mass 30 (solid line), 45 (short-dash line), and 70 (dotted line) kDa. All proteins are represented by equivalent hard spherical particles with respective radii proportional to the cube root of their molar mass. Calculations were performed with the scaled particle theory of hard sphere mixtures as described in Minton (1998).

to track ClpB in such volume-occupied solutions; (2) any observed effect of crowding species on the association properties of ClpB may be caused by other interactions different than excluded volume, as attractive interactions between ClpB and crowding species; and (3) the measurements have to be interpreted taking into account the nonideality of the crowded solution.

In this work, we will describe the application of nonideal tracer SE (NITSE) to study crowding effects on ClpB self-association. NITSE is a variant of SE that is especially well adapted to detect and measure molecular interactions of dilute ClpB protein in solutions containing high concentrations of background macromolecules (Rivas & Minton, 2004). This technique requires that to behave as a tracer, the protein under study (i.e., ClpB) must have an intrinsic or extrinsic (i.e., via fluorescent labeling) signal that can be uniquely detected and is not altered by the presence of the crowder species. The concentration gradients of the tracer protein at SE can be measured independently of the gradients of the background species in solution. Recent advances in the theory and methodology of SE in highly crowded solutions allow interpreting the dependence on protein concentration of signal-average buoyant molar masses in terms of repulsive and attractive interactions between the species in solution, leading to the application of NITSE to study the self- and hetero-associations of tracer proteins in crowded solutions (Rivas & Minton, 2011).

2.4.3 NITSE Assays

Tracer SE was performed to characterize the effect of Ficoll 70 (acting as a crowder) on the composition dependence of the self-association properties of wt and deletion variants of ClpB, using the same experimental conditions previously described in the SE assays. The protein samples contained a fixed amount of ClpB labeled with Alexa 647, acting as a tracer species. Protein labeling did not alter the association properties of the chaperone, as tested by SV. The magnitude of the tracer signal S (absorbance at 650 nm) was followed as a function of the radial position (r) (see Eq. 4). S_r was found to be proportional to the total amount of protein and independent of the concentrations of all unlabeled species. For each solution composition, the radial dependence of the signal at SE was fitted to Eq. (4) to determine the corresponding $M^*_{i,\mathrm{app}}$ values for the different ClpB variants. The molecular weight analysis was done using the EQASSOC and HETEROANALYSIS programs, which yielded the same results within 5% experimental error (Cole, 2004; Minton, 1994).

2.4.4 NITSE Analysis

While in the absence of Ficoll 70, all measurements meet conditions of thermodynamic ideality (high dilution of all macromolecular species) and therefore $M^*_{i,\mathrm{app}} = M^*_i$, this does not apply for the SE measurements done in the presence of Ficoll 70. In this case, to model the dependence of $M^*_{i,\mathrm{app}}$ upon solution composition, it is also necessary to consider the effect that the interaction of ClpB with all the species of the solution mixture might have on its apparent buoyant mass (Zorrilla et al., 2004). The general expression that describes the condition of SE in a solution containing an arbitrary number of macromolecular species at arbitrary concentrations is given by

$$M^*_{i,\mathrm{app}} = M^*_i - \sum_j w_j \left(\frac{\mathrm{d}\ln\gamma_i}{\mathrm{d}w_j}\right) M^*_{i,\mathrm{app}} \tag{14}$$

where w_j denotes the w/v concentration of species j, and γ_i is the activity coefficient of species i. The quantity $(\mathrm{d}\ln\gamma_i/\mathrm{d}w_j)$ is a thermodynamic magnitude measuring the free energy of interactions between species i and j (Rivas & Minton, 2011).

In this work, while the concentration of the dilute protein component (ClpB) changes, the amount of the component present at high concentrations (Ficoll) that significantly contribute to the sum of the right-hand side of Eq. (14) was maintained constant. Under these conditions, the dependence of $M^*_{i,\mathrm{app}}$ with the concentration of the dilute species may be directly modeled by Eq. (4) as in the analysis of the data obtained in the absence of Ficoll 70 (for details, see Rivas & Minton, 2011). Therefore, data in the presence of Ficoll can be analyzed without a previous knowledge of the dependence of the activity coefficients of all species upon crowder concentration. The combined SE data for a given (single) crowder concentration (w_c) can be fitted by a self-association model that makes no assumptions on the nonspecific interactions between dilute protein species and Ficoll 70. It has been found that a self-association scheme, which assumes that ClpB exist as an equilibrium mixture of monomers and hexamers with a thermodynamic equilibrium constant (K_6), was the simplest one that globally described the tracer SE data in the absence and presence of Ficoll 70 with a 95% confidence limit of statistical significance (Saroff, 1989). This model was then fit to the composition-dependent average molecular weight data, $M^*_{i,\mathrm{app}}$, obtained from Eq. (4) using a nonlinear least-squares procedures. The nonlinear modeling procedure (implemented in MATLAB scripts kindly provided by Dr. Allen Minton, NIH) then provides best-fit values of $M^*_{1,\mathrm{app}}$, $M^*_{6,\mathrm{app}}$, and K_6 for the concentration of Ficoll (w_c) at which the molecular weight values were determined.

3. SUMMARY OF EXPERIMENTAL RESULTS

3.1 Environmental Factors That Modulate the Association Equilibrium of ClpB: Ionic Strength

Among the environmental factors that control ClpB self-association, one of the most important is the ionic strength of the medium, as weakening of the electrostatic interactions that stabilize the hexamer favors protein dissociation. Therefore, the association equilibrium of apo-ClpB (no nucleotide bound; 10 μ*M*) was analyzed by SV and equilibrium over a broad range of KCl concentration (50–500 m*M* KCl). Below 150 m*M* KCl, ClpB sedimented as a major single species with standard *s*-value of 18.5 ± 0.5 S, which corresponded with a sedimenting species of average molecular weight 628,000 ± 30,000 (Fig. 3). This value was higher than that obtained from the

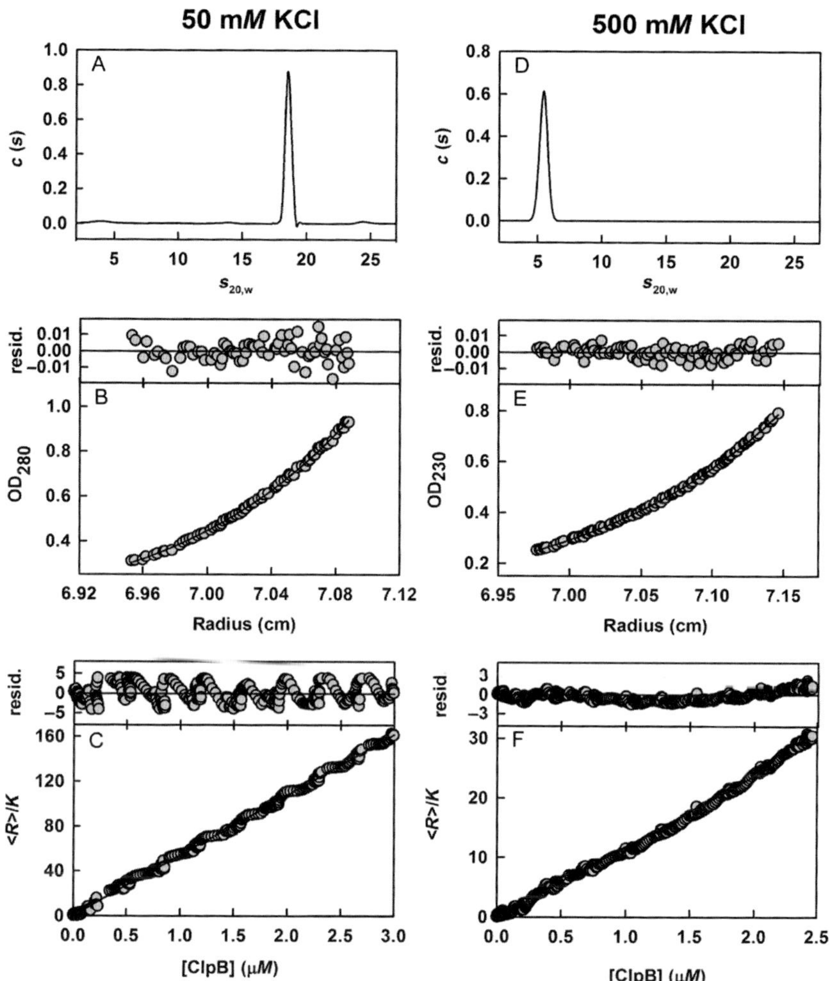

Figure 3 Biophysical characterization of hexameric and monomeric ClpB. Sedimentation coefficient distribution of ClpB (10 μM) in the presence of 50 (A) or 500 mM KCl (D). (B) and (E) Sedimentation equilibrium gradients of ClpB samples at 50 (B) and 500 (E) mM KCl. The circles represent the experimental data and the solid line is the best-fit to single species gradient. (C) and (F) Rayleigh ratio ($\langle R\rangle/K$) as a function of protein concentration. Circles are the experimental data, and the solid line is the best-fit nonassociating model.

sequence of the ClpB hexamer (573,510 Da), suggesting the possible presence of a small proportion of higher-order oligomers. A greater abundance of slower sedimenting species, with lower *s*-values, were found as the KCl concentration was increased, indicating oligomer dissociation. At 500 m*M*

KCl, the protein sedimented as a single species with standard *s*-value of 5.5 ± 0.5 S, and an average molecular weight calculated by SE of $100{,}000 \pm 10{,}000$, corresponding to the monomer mass (Fig. 3). The molecular weight values were confirmed by CG-SLS measurements done in parallel. From these results, we can conclude that the hydrodynamic (SV) and thermodynamic (SE) data obtained for ClpB as a function of salt concentration are semiquantitatively in accord.

To gain further insights on the salt dependence of ClpB oligomer formation, an SE analysis was carried out to define more precisely the dependence of ClpB association state on protein concentration and ionic strength. This study revealed that both parameters (protein and KCl concentration) act coordinately to control, in a linked-process, the association properties of ClpB. While large protein oligomers were found to dissociate at low protein concentration and ionic solution conditions that favored oligomer formation (50 m*M* KCl), species of ClpB with intermediate states of association were observed at high protein concentration under solution conditions favoring oligomer dissociation into monomers (500 m*M* KCl), as can be seen in Fig. 4. CG-SLS experiments aimed to determine the dependence of the excess light scattering of ClpB on KCl concentration confirmed the SE data.

In order to define the association scheme that quantitatively best described these linked association reactions, the complete sets of data obtained at 150 m*M* KCl were analyzed globally. These conditions resulted to be the most appropriate for this modeling exercise as monomeric and oligomeric species were present at significant amounts over the whole range of protein concentrations tested. Following the treatment described in Section 2.2.1, it was found that the model in which ClpB exists in solution as an equilibrium mixture of monomers, hexamers, and higher-order species (tentatively assigned to dodecamers) is the association scheme that best fitted the experimental data. The same model was also found to quantitatively best explain the GC-SLS data. It is important to note that the weight of the largest oligomeric assembly observed in the absence of nucleotides is significantly reduced in the presence of nucleotides (see below), thus suggesting that under physiological conditions, an equilibrium between monomeric and hexameric ClpB species could reasonably account for the experimental data.

3.2 The M Domain Stabilizes ClpB Hexamers

The M domain is the characteristic structural feature of ClpB/Hsp104, as it is absent in the rest of Hsp100 proteins. This domain folds into a helical

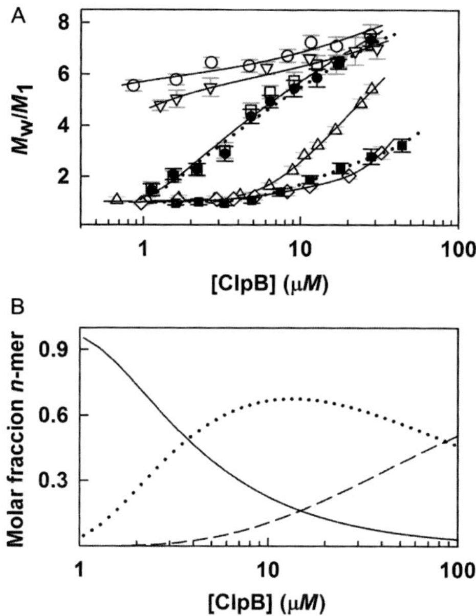

Figure 4 Ionic strength modulates the oligomerization state of wt ClpB and $ClpB_{\Delta(410-455)}$: a sedimentation equilibrium analysis. (A) Dependence of the state of association (M_w/M_1) of wt ClpB and $ClpB_{\Delta(410-455)}$ on salt and protein concentration. Open symbols correspond to sedimentation equilibrium data for wt ClpB at 50 (circles), 100 (inverted triangles), 150 (squares), 300 (triangles), and 500 (diamonds) m*M* KCl. Black symbols correspond to sedimentation equilibrium data for $ClpB_{\Delta(410-455)}$ at 50 (circles) and 150 (squares) m*M* KCl. Solid (wt ClpB) and broken ($ClpB_{\Delta(410-455)}$) lines were calculated using model 1-6-12 with best-fit parameter values obtained for each KCl concentration. (B) Fractional distribution of monomer (solid line), hexamer (dotted line), and dodecamer (dashed line) species obtained from the best-fit of the data corresponding to wt ClpB at 150 m*M* KCl shown in (A).

coiled-coil structure that is inserted in the C-terminal end of the NBD1. It is organized in four antiparallel helices that form two motifs: motif 1 contains helices 1 and 2 and motif 2 helices 2, 3, and 4. Each motif is stabilized by leucine zipper-like hydrophobic interactions (Lee et al., 2003). The role of the M domain on ClpB self-association was studied by SV and SE methods using two M domain deletion mutants that lack motif 1 or the entire domain. SV measurements showed that deletion of the M domain reduces the stability of ClpB, as the dissociation of these truncated protein variants (from the experimental *s*-values) occurred at lower KCl concentrations, as compared with wild-type ClpB. SE data confirmed the stabilization role of the M domain; the association state of the ClpB mutant at 150 m*M* KCl was similar to that of wt ClpB at 300–500 m*M* KCl (Fig. 4). The association free energy of wt ClpB hexamer was 33 and 20 kJ/mol higher than that of $ClpB_{\Delta(410-455)}$ at 50 and 150 m*M* KCl, respectively, indicating that motif 1 of the M domain is involved in intersubunit interactions that modulate oligomer stability. Interestingly, the same interaction also regulates the ATPase activity of the disaggregase that becomes hyperstimulated in the presence of substrates upon elimination of M domain motif 1 (del Castillo et al., 2011). A recent cryo-EM model of the protein hexamer suggests the possibility that an interaction between motif 1 and motif 2 of M domains in adjacent protomers at the surface of the barrel-shaped structure could explain these experimental observations (Carroni et al., 2014).

3.3 Nucleotides Modulate the Stability and Hydrodynamic Behavior of ClpB Hexamers

As mentioned above, the two nucleotide-binding domains are the core motor units of ClpB. The implication of the NBD1 domain in hexamer stability can be explained if, as proposed, the nucleotide-binding sites within the hexamer were located in the interface between adjacent subunits. This has been probed using ClpB variants in which mutations that abolish nucleotide binding to the NBD1 also compromise hexamer stability, whereas the same substitutions in the NBD2 do not affect hexamer formation (Barnett & Zolkiewski, 2002; Mogk et al., 2003). These results suggest that ATP binding to NBD1 but not to NBD2 contributes to protein hexamerization. Despite this large body of experimental data, a quantitative study of the effect of nucleotides on the association equilibrium of ClpB was scarce. To fill this gap, the influence of ADP and ATP on ClpB self-association was determined by CG-SLS and SV measurements and compared with the aforementioned self-association properties of nucleotide-free (apo-form) ClpB. ATP-ClpB studies were done in the presence of an ATP-regenerating system (see Section 2.1.1) to prevent ATP depletion during the timescale of the experiments (1–2 h). This time restriction precluded the use of SE assays, which take much longer to reach equilibrium.

Nucleotides favored formation of oligomeric species, particularly the hexamer, as revealed by CG-SLS measurements. Experiments were performed at 150 m*M* KCl, conditions in which monomeric apo-ClpB self-associates into hexamers and to a lesser extent into dodecamers (del Castillo et al., 2011). The composition-dependent excess scattering R/K (and hence the average molar mass) of nucleotide-bound ClpB was significantly higher than that of the apo-protein, the effect being more pronounced for ATP-ClpB at low protein concentrations (Fig. 5A). The association model that described the oligomerization of the apo-ClpB form was also found to best-fit data obtained in the presence of nucleotides, which

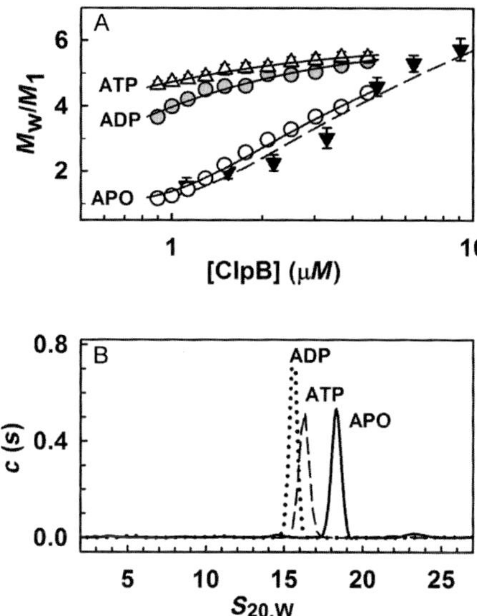

Figure 5 Nucleotides shift the association equilibrium of ClpB toward the hexamer. (A) Effect of nucleotides on the association of wt ClpB as measured by CG-SLS. M_w distribution normalized to the monomeric ClpB molecular weight (95,500 Da), as a function of protein concentration. Experimental data correspond to ClpB (150 m*M* KCl) in the absence of nucleotide (empty circles) and presence of ADP (filled circles) or ATP (empty triangles). For the sake of comparison, sedimentation equilibrium data of the apo-protein is also shown (filled triangles). Solid lines represent the M_w distribution as obtained from the best-fit parameter values shown in Table 1. (B) Modulation of the hydrodynamic properties of wt ClpB by nucleotides. Sedimentation coefficient distribution of wt ClpB (10 m*M*) in the absence of nucleotides (apo-ClpB, black solid line) and in the presence of 1 m*M* ATP (black broken line) or 1 m*M* ADP (black dotted line) in buffer containing 50 m*M* KCl, conditions that favor association.

allows calculating the effect of ADP and ATP on the equilibrium association constant for hexamer formation. The best-fit values of the log K_6 for the ATP and ADP forms of ClpB were higher than the values previously obtained for apo-ClpB (see Table 1). Binding of ADP and ATP to apo-ClpB resulted in a stabilization of the protein hexamer by 19–20 and 28–29 kJ/mol, respectively. Similar nucleotide-induced stabilization effects were found for the hexamer of the M domain deletion mutant $ClpB_{\Delta(410-455)}$, suggesting that they also involve structural elements different from the M domain.

The effect of nucleotide binding on ClpB self-association was studied by SV measurements. SV was employed to follow conformational rearrangements of the protein hexamer upon nucleotide binding. To this aim, experiments were carried out at 20 m*M* KCl as these experimental

Table 1 Effect of Ionic Strength and Nucleotides on the Self-Association of wt ClpB and $ClpB_{\Delta(410-455)}$ as Determined by SE and CG-SLS Measurements[a]

Protein (KCl conc.)	log K'_6[b]	$\Delta G^0{}_6$[c] (kJ/mol)
apo-wt (50 m*M*)	36.0 ± 2.0	−188
apo-wt (500 m*M*)	22.4 ± 2.0	−117
apo-wt (150 m*M*)	27.9 ± 2.0 (28.1)[d]	−156 (−157)[d]
ADP-wt (150 m*M*)	31.3 ± 1.5	−175
ATP-wt (150 m*M*)	33.1 ± 2.0	−185
apo-$ClpB_{\Delta(410-455)}$ (50 m*M*)	27.7 ± 1.0	−155
apo-$ClpB_{\Delta(410-455)}$ (150 m*M*)	24.4 ± 1.0 (23.9)[d]	−137 (−134.0)[d]
ADP-$ClpB_{\Delta(410-455)}$ (150 m*M*)	28.6 ± 2.1	−160
ATP-$ClpB_{\Delta(410-455)}$ (150 m*M*)	30.0 ± 0.4	−168

[a]Model 1-6-12 was used to calculate the self-association parameters.
[b]Association constants are expressed in molar units. They result from the conversion of the the best-fit K_6 values (in weight units) obtained in the self-association analysis of the composition dependence of molecular weights described in the text (see also del Castillo et al., 2011) using the following relation: $K'_n = K_n\, M_1{}^{n-1}/n$.
[c]Standard state free energy change was calculated from the relation $\Delta G^o{}_6 = -2.3\ RT \log K'_6$.
[d]The values in brackets correspond to best-fit parameters obtained from SE.

conditions favor hexamerization of nucleotide-free ClpB. Both the ADP- and ATP-state of the hexameric protein sedimented with lower *s*-values (15.6 and 16.3, respectively) than the apo-form (18.5 S), which may reflect an expansion of the oligomer, greater in the ADP-state than in the ATP-state due to nucleotide-induced conformational changes (Fig 5B).

3.4 Crowding Shifts the Association Equilibrium Toward the Hexameric Species

Excluded volume effects in highly volume-occupied solutions are expected to substantially increase the tendency of ClpB to self-associate, thus counteracting the destabilizing effect of salts and hence extending the range of saline conditions in which the functional ClpB hexamer may exist. In order to determine the impact of volume exclusion on the salt dependence of ClpB self-association, the effect of adding high concentrations of an inert unrelated macromolecule (Ficoll 70) on the tendency of ClpB to form oligomers was studied by NITSE. Experiments were carried out at increasing ClpB concentrations, in the absence and presence of 15% (w/v) Ficoll 70, and then analyzed to obtain the corresponding apparent buoyant molecular weights (Fig. 6).

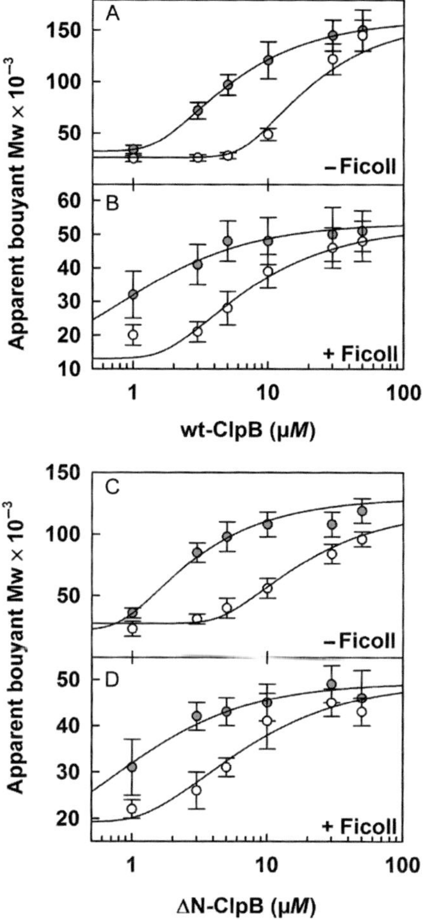

Figure 6 Crowding shifts the association equilibrium of ClpB toward the active hexamer. Sedimentation equilibrium measurements were performed in the absence (A and C) and presence (B and D) of 150 g/l Ficoll 70 to obtain the dependency of the buoyant molecular weight of wt ClpB (A and B) or ΔN-ClpB (C and D). The buffer contained 150 (gray circles) or 300 m*M* KCl (white circles). Solid lines are the best fit of the experimental data to the model described in the Experimental section (Section 3.4; Table 2). Measures are represented as the mean ± SE from three independent experiments.

In the presence of Ficoll, the values of $M^*_{i,\mathrm{app}}$ were significantly lower than those obtained in the absence of crowder, as expected because of the nonideal behavior of the dilute protein in a highly crowded solution (see Eq. 14). The dependence of $M^*_{i,\mathrm{app}}$ upon protein concentration was best described by a two-state monomer–hexamer model, both with and without Ficoll 70 (Fig. 6). The association constant for hexamer formation (K_6) in the

Table 2 Effect of Crowding on the Self-Association of wt ClpB and ΔN-ClpB

Protein	KCl (m*M*)	Ficoll (g/l)	log K_6 (l/g)	ΔG_6^0 (*RT*)
wt ClpB	150	0	2.4 ± 0.4	−61.2
	150	150	6.1 ± 1.2	−70.0
	300	0	−0.8 ± 0.3	−53.8
	300	150	2.3 ± 0.6	−61.0
ΔN-ClpB	150	0	4.3 ± 0.5	−64.5
	150	150	6.6 ± 1.4	−70.0
	300	0	0.1 ± 1.0	−54.8
	300	150	3.2 ± 0.8	−61.9

presence of Ficoll was around 3–4 orders of magnitude higher than the association values in the absence of crowder, both at 150 m*M* and at 300 m*M* KCl (see Table 2), thus confirming the predicted effects of crowding on ClpB self-association. These values corresponded to a decrease in the free energy of hexamer formation of 7 and 9 *RT*, respectively. Similar crowding effects were observed for a ClpB variant lacking the N-terminal domain.

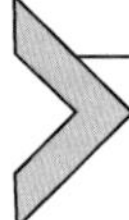

4. ClpB ASSOCIATION EQUILIBRIUM AND DISAGGREGASE ACTIVITY

One of the characteristic features of the ClpB hexamer is its inherent dynamics that seems to be essential to understand the functional mechanism of this disaggregase. The relationship between ClpB dynamics and disaggregase activity remains controversial. It has been shown that cross-linked ClpB hexamers could remodel substrate proteins, and it was therefore proposed that substrate translocation through the central pore did not require hexamer disassembly (Biter, Lee, Sung, & Tsai, 2012). However, the finding that ClpB could solubilize aggregated proteins containing correctly folded domains without unfolding them also suggested that the hexamer must disassembly during aggregate untangling (Haslberger et al., 2008). The work summarized here on the self-association equilibrium of ClpB and a recent FRET study based on the ability of ClpB hexamers to exchange subunits (Aguado, Fernandez-Higuero, Cabrera, Moro, & Muga, 2015) suggest that the ClpB hexamer is able to adjust its dynamics to different environmental stimuli, such as ionic strength and nucleotides, and that salt concentration higher than 150–300 m*M* KCl gradually precludes hexamer assembly. Interestingly, an increase in ClpB dynamics within the KCl concentration range that allows hexamer formation (below 150–300 m*M* KCl, depending on protein concentration) is paralleled by an enhancement of both the ATPase and chaperone activities of the disaggregase, which reached their maxima under conditions that favor a highly dynamic hexamer. This comparison suggests that the hexamer is the active species *in vitro*. This activation might be due to a loosening of subunit packing within the hexamer that could favor the conformational rearrangement associated with the functional cycle of the protein. Higher KCl concentrations progressively hampered formation of the functional hexamer, therefore inhibiting ClpB activity.

The results obtained in the presence of Ficoll demonstrate that excluded volume effects shifted the ClpB association equilibrium toward the functional hexamer at salt conditions that otherwise would promote chaperone dissociation in dilute solutions, and that the N-terminal domain of the protein does not significantly alter this behavior. Shifting of the association equilibrium of ClpB toward the hexamer has important consequences in the activity of this hexameric disaggregase, as it also stimulates the ATPase activity of ClpB and promotes its interaction with substrate proteins and with aggregate-bound DnaK (Martin et al., 2014). These effects strongly enhance protein aggregate reactivation by the DnaK–ClpB network, highlighting the importance of volume exclusion in the regulation of the oligomerization state of ClpB, which in turn modulates its association with DnaK, a requirement to fulfill its biological function. Therefore, these findings illustrate how crowding could act as a nonspecific physiological modulator that can considerably extent the range of intracellular conditions under which a macromolecular reaction may occur in the cell.

5. CONCLUDING REMARKS

We summarize in this review studies aimed to characterize the functionally relevant self-association of ClpB, a chaperone involved in protein aggregate reactivation, by means of SE combined with other complementary techniques, namely SV and composition-gradient light scattering. This experimental strategy has allowed formulating a quantitative description of ClpB association, which exists as an equilibrium mixture of monomers, hexamers, and higher-order oligomers. The relative abundance of these species is modulated by ionic strength and nucleotides, as low salt and the presence of ADP/ATP stabilize the functionally active ClpB hexamer. In addition, the low tendency of ClpB to form hexamers at close to physiological salt concentrations is reversed by the addition of high concentrations of an inert polymer, reproducing the crowded intracellular environment.

This information will contribute to define the conditions that permit assembly of the functional ClpB oligomer and will help to better understand how natural ligands, such as nucleotides, modulate the association reaction. A deep knowledge of these conditions are essential to unravel how the ClpB–DnaK bichaperone complex, responsible for aggregate reactivation, is built and its stability modulated. It is reasonable to imagine that in a near future, using reconstituted systems of increasing complexity, the multiple macromolecular associations involved in protein aggregate untangling could be artificially reproduced and characterized under conditions that mimic the intracellular environment.

ACKNOWLEDGMENTS

We thank J.R. Luque of the Analytical Ultracentrifugation Facility (Centro de Investigaciones Biológicas, CSIC) for help with analytical ultracentrifugation experiments. This work was supported in part by Spanish Government Grants BUF2013-47059 (to A.M.) and BIO2011-28941-C03 (to G.R.) and by Basque Country University and Government Grant IT709-13 (to A.M.).

REFERENCES

Aguado, A., Fernandez-Higuero, J. A., Cabrera, Y., Moro, F., & Muga, A. (2015). ClpB dynamics is driven by its ATPase cycle and regulated by the DnaK system and substrate proteins. *Biochemical Journal, 466*, 561–570.

Akoev, V., Gogol, E. P., Barnett, M. E., & Zolkiewski, M. (2004). Nucleotide-induced switch in oligomerization of the AAA+ ATPase ClpB. *Protein Science, 13*, 567–574.

Arias, E., & Cuervo, A. M. (2011). Chaperone-mediated autophagy in protein quality control. *Current Opinion in Cell Biology, 23*, 184–189.

Attri, A. K., & Minton, A. P. (2005). New methods for measuring macromolecular interactions in solution via static light scattering: Basic methodology and application to non-associating and self-associating proteins. *Analytical Biochemistry, 337*, 103–110.

Barnett, M. E., & Zolkiewski, M. (2002). Site-directed mutagenesis of conserved charged amino acid residues in ClpB from Escherichia coli. *Biochemistry, 41*, 11277–11283.

Ben-Zvi, A., Miller, E. A., & Morimoto, R. I. (2009). Collapse of proteostasis represents an early molecular event in *Caenorhabditis elegans* aging. *Proceedings of the National Academy of Sciences of the United States of America, 106*, 14914–14919.

Biter, A. B., Lee, S., Sung, N., & Tsai, F. T. F. (2012). Structural basis for intersubunit signaling in a protein disaggregating machine. *Proceedings of the National Academy of Sciences of the United States of America, 109*, 12515–12520.

Carroni, M., Kummer, E., Oguchi, Y., Wendler, P., Clare, D. K., Sinning, I., et al. (2014). Head-to-tail interactions of the coiled-coil domains regulate ClpB activity and cooperation with Hsp70 in protein disaggregation. *eLife, 3*, e02481.

Cayley, S., & Record, M. T., Jr. (2004). Large changes in cytoplasmic biopolymer concentration with osmolality indicate that macromolecular crowding may regulate protein-

DNA interactions and growth rate in osmotically stressed Escherichia coli K-12. *Journal of Molecular Recognition*, *17*, 488–496.

Cohen, E., Paulsson, J. F., Blinder, P., Burstyn-Cohen, T., Du, D., Estepa, G., et al. (2009). Reduced IGF-1 signaling delays age-associated proteotoxicity in mice. *Cell*, *139*, 1157–1169.

Cole, J. L. (2004). Analysis of heterogeneous interactions. *Methods in Enzymology*, *384*, 212–232.

del Castillo, U., Alfonso, C., Acebron, S. P., Martos, A., Moro, F., Rivas, G., et al. (2011). A quantitative analysis of the effect of nucleotides and the M domain on the association equilibrium of ClpB. *Biochemistry*, *50*, 1991–2003.

Haslberger, T., Zdanowicz, A., Brand, I., Kirstein, J., Turgay, K., Mogk, A., et al. (2008). Protein disaggregation by the AAA+ chaperone ClpB involves partial threading of looped polypeptide segments. *Nature Structural and Molecular Biology*, *15*, 641–650.

Hipp, M. S., Park, S. H., & Hartl, F. U. (2014). Proteostasis impairment in protein misfolding and aggregation diseases. *Trends in Cell Biology*, *24*, 506–514.

Hodson, S., Marchall, J. J., & Burston, S. G. (2012). Mapping the road to recovery: The ClpB/Hsp104 molecular chaperone. *Journal of Structural Biology*, *179*, 161–171.

Kameyama, K., & Minton, A. P. (2006). Rapid quantitative characterization of protein interactions by composition gradient static light scattering. *Biophysical Journal*, *90*, 2164–2169.

Kim, K. I., Cheong, G. W., Park, S. C., Ha, J. S., Woo, K. M., Choi, S. J., et al. (2000). Heptameric ring structure of the heat-shock protein ClpB, a protein-activated ATPase in Escherichia coli. *Journal of Molecular Biology*, *303*, 655–666.

Kim, Y. E., Hipp, M. S., Bracher, A., Hayer-Hartl, M., & Hartl, F. U. (2013). Molecular chaperone functions in protein folding and proteostasis. *Annual Review of Biochemistry*, *82*, 323–355.

Laue, T. M., Shah, B. D., Ridgeway, T. M., & Pelletier, S. L. (1992). Computer-aided interpretation of analytical sedimentation data for proteins. In S. E. Harding, A. J. Rowe, & J. Horton (Eds.), *Analytical ultracentrifugation in biochemistry and polymer science* (pp. 90–125). Cambridge, UK: Royal Society of Chemistry.

Lee, S., Sowa, M. E., Watanabe, Y. H., Sigler, P. B., Chiu, W., Yoshida, M., et al. (2003). The structure of ClpB: A molecular chaperone that rescues proteins from an aggregated state. *Cell*, *115*, 229–240.

Martin, I., Celaya, G., Alfonso, C., Moro, F., Rivas, G., & Muga, A. (2014). Crowding activates ClpB and enhances its association with DnaK for efficient protein aggregate reactivation. *Biophysical Journal*, *106*, 2017–2027.

Minton, A. P. (1994). Conservation of signal: A new algorithm for the elimination of the reference concentration as an independently variable parameter in the analysis of sedimentation equilibrium. In T. H. Schuster & T. H. Lave (Eds.), *Modern analytical ultracentrifugation* (pp. 81–93). Boston, MA: Birkhauser.

Minton, A. P. (1998). Molecular crowding: Analysis of effects of high concentrations of inert cosolutes on biochemical equilibria and rates in terms of volume exclusion. *Methods in Enzymology*, *295*, 127–149.

Minton, A. P. (2006). How can biochemical reactions within cells differ from those in test tubes? *Journal of Cell Science*, *119*, 2863–2869.

Mogk, A., Haslberger, T., Tessarz, P., & Bukau, B. (2008). Common and specific mechanisms of AAA+ proteins involved in protein quality control. *Biochemical Society Transactions*, *36*, 120–125.

Mogk, A., Schlieker, C., Strub, C., Rist, W., Weibezahn, J., & Bukau, B. (2003). Roles of individual domains and conserved motifs of the AAA+ chaperone ClpB in oligomerization, ATP hydrolysis, and chaperone activity. *The Journal of Biological Chemistry*, *278*, 17615–17624.

Pickart, C. M., & Cohen, R. E. (2004). Proteasomes and their kin: Proteases in the machine age. *Nature Reviews Molecular Cell Biology*, *5*, 177–187.
Rivas, G., Ferrone, F., & Herzfeld, J. (2004). Life in a crowded world. *EMBO Report*, *5*, 23–27.
Rivas, G., & Minton, A. P. (2004). Non-ideal tracer sedimentation equilibrium: A powerful tool for the characterization of macromolecular interactions in crowded solutions. *Journal of Molecular Recognition*, *17*, 362–367.
Rivas, G., & Minton, A. P. (2011). Beyond the second virial coefficient: Sedimentation equilibrium in highly non-ideal solutions. *Methods*, *54*, 167–174.
Sanchez, Y., & Lindquist, S. L. (1990). HSP104 required for induced thermotolerance. *Science*, *248*, 1112–1115.
Saroff, H. A. (1989). Evaluation of uncertainties for parameters in binding studies: The sum-of-squares profile and Monte Carlo estimation. *Analytical Biochemistry*, *176*, 161–169.
Schlee, S., Groemping, Y., Herde, P., Seidel, R., & Reinstein, J. (2001). The chaperone function of ClpB from *Thermus thermophilus* depends on allosteric interactions of its two ATP-binding sites. *Journal of Molecular Biology*, *306*, 889–899.
Schuck, P. (2000). Size-distribution analysis of macromolecules by sedimentation velocity ultracentrifugation and Lamm equation modeling. *Biophysical Journal*, *78*, 1606–1619.
Schuck, P., Perugini, M. A., Gonzales, N. R., Howlett, G. J., & Schubert, D. (2002). Size-distribution analysis of proteins by analytical ultracentrifugation: Strategies and application to model systems. *Biophysical Journal*, *82*, 1096–1111.
Schuck, P., & Zhao, H. (2013). Biophysical methods for the study of protein interactions. *Methods*, *59*, 259–260.
Squires, C. L., Pedersen, S., Ross, B. M., & Squires, C. (1991). ClpB is the Escherichia coli heat shock protein F84.1. *Journal of Bacteriology*, *173*, 4254–4262.
van Holde, K. E. (1985). Sedimentation. In *Physical biochemistry* (2nd ed., pp. 110–136). Englewood Cliffs, NJ: Prentice-Hall.
Zhou, H. X., Minton, A. P., & Rivas, G. (2008). Macromolecular crowding and confinement: Biochemical, biophysical, and potential physiological consequences. *Annual Review of Biophysics*, *37*, 375–397.
Zolkiewski, M., Kessel, M., Ginsburg, A., & Maurizi, M. R. (1999). Nucleotide dependent oligomerization of ClpB from Escherichia coli. *Protein Science*, *8*, 1899–1903.
Zorrilla, S., Jiménez, M., Lillo, P., Rivas, G., & Minton, A. P. (2004). General analysis of sedimentation equilibrium in highly nonideal solutions of associating solutes: Application to ribonuclease. *Biophysical Chemistry*, *108*, 89–100.

CHAPTER SEVEN

Analysis of Linked Equilibria

JiaBei Lin, Aaron L. Lucius[1]
Department of Chemistry, The University of Alabama at Birmingham, Birmingham, Alabama, USA
[1]Corresponding author: e-mail address: allucius@uab.edu

Contents

Abstract

The ATPases associated with diverse cellular activities (AAA+) is a large superfamily of proteins involved in a broad array of biological processes. Many members of this family require nucleotide binding to assemble into their final active hexameric form. We have been studying two example members, *Escherichia coli* ClpA and ClpB. These two enzymes are active as hexameric rings that both require nucleotide binding for assembly. Our studies have shown that they both reside in a monomer, dimer, tetramer, and hexamer equilibrium, and this equilibrium is thermodynamically linked to nucleotide binding. Moreover, we are finding that the kinetics of the assembly reaction are very different for the two enzymes. Here, we present our strategy for determining the self-association constants in the absence of nucleotide to set the stage for the analysis of nucleotide binding from other experimental approaches including analytical ultracentrifugation.

1. INTRODUCTION

Determining a binding constant for a protein–ligand interaction is quite possibly one of the most common experiments in biophysics. Techniques such as ITC, fluorescence titrations, equilibrium dialysis, and many others are commonly used. Analysis of these data is fairly straightforward if

Methods in Enzymology, Volume 562
ISSN 0076-6879
http://dx.doi.org/10.1016/bs.mie.2015.07.003

the protein of interest does not change its oligomeric state as the free concentration of the ligand (chemical potential of the ligand) is increased. However, if the protein does change its oligomeric state as the chemical potential of the ligand is increased, then the analysis becomes substantially more complex.

It is important to recall that a binding constant measured using any of the approaches stated above is an apparent binding constant (Alberty, 2003). Although we typically write down a binding reaction as an association between a macromolecule, M, and a ligand, X, as schematized in Eq. (1), the binding is actually much more complex.

$$M + X \overset{K_{app}}{\rightleftharpoons} MX \tag{1}$$

In fact, the interaction also involves the removal of water and/or ions from the binding pocket as well as from the ligand that is entering the binding pocket. Although there could be many solution condition components, including protons, involved in this interaction, an example scheme that represents only ions and water exchange upon ligand binding is given by Eq. (1a).

$$M \cdot Na_i{}^+ \cdot (H_2O)_j + X \cdot Cl_k{}^- \cdot (H_2O)_l \overset{K}{\rightleftharpoons} MX + iNa^+ + kCl^- + l(H_2O) \tag{1a}$$

These exchanges of water and ions with bulk solvent all contribute to the energetics of the binding interaction. Therefore, the binding equilibrium constant, K_{app}, given in Eq. (1) will exhibit a dependence on solution condition variables such as ion concentration, pH, water concentration. In other words, the binding constant will be thermodynamically linked to all of the components involved in the binding reaction. Thus, linkage analysis can be used to deconvolute the energetics of a "second ligand" (ion, water, protons, etc.) binding. This is accomplished by determining the apparent equilibrium constant for X binding M as a function of, for example, salt concentration. Then, this apparent equilibrium constant can be treated as a signal for a "second ligand" binding event (water, proton, ions, etc.).

With the above in mind, one can recognize that a ligand binding constant for an assembling system may exhibit a protein concentration dependence. Likewise, a protein–protein interaction constant will exhibit a ligand concentration dependence for an assembling system. That is to say, the binding constants are thermodynamically linked to either the protein concentration or ligand concentration.

The ATPases associated with diverse cellular activities (AAA+) is a large superfamily of proteins involved in a broad array of biological processes (Neuwald, Aravind, et al., 1999). Examples include microtubule severing catalyzed by katanin (Roll-Mecak & McNally, 2010); membrane fusion involving *N*-ethylmaleimide-sensitive fusion (NSF) proteins (Yu, Jahn, & Brunger, 1999); morphogenesis and trafficking of endosomes by VPs4p (Babst, Sato, Banta, & Emr, 1997); protein disaggregation by ClpB/Hsp104; and enzyme-catalyzed protein unfolding and translocation by ClpA or ClpX for ATP-dependent proteolysis (Ogura & Wilkinson, 2001; Sauer & Baker, 2011). Another example is human VCP/p97, which has been connected to ubiquitin-dependent reactions, but is implicated in an expanding number of physiological processes (Meyer & Weihl, 2014).

Many of these proteins require nucleoside triphosphate binding to assemble into their final active hexameric form. Thus, ligand-linked assembly is an integral component of the mechanism driving assembly. For many of these examples, the oligomers present in solution will change as the chemical potential of the ligand (nucleotide) is increased.

Analytical ultracentrifugation is the technique of choice for examining the energetics of macromolecular assembly. The technique has been used extensively to do so and a large number of computer applications are available to aid in the analysis of the experimental results, and these have been discussed in many places (Cole, 2004; Demeler, Brookes, Wang, Schirf, & Kim, 2010; Schuck, 1998; Scott, Harding, Rowe, & Royal Society of Chemistry (Great Britain), 2005; Stafford & Sherwood, 2004). Moreover, an enormous body of literature exists on the application of this approach to examine assembling systems (Cole, 1996; Correia & Stafford, 2009; Schuck, 1998, 2003). However, substantially less has been done on examining ligand-linked assembly problems, but some examples can be found (Cole, Correia, & Stafford, 2011; Na & Timasheff, 1985a, 1985b; Streaker, Gupta, & Beckett, 2002; Wong & Lohman, 1995).

Here, we outline our strategy for elucidating the thermodynamic mechanism for an assembling system that exists in a complex, dynamic equilibrium of monomers, dimers, tetramers, and hexamers. This is being done in order to set the stage for an examination of the thermodynamic linkage to the nucleotide-driven hexamer formation for two example AAA+

macromolecular machines, *Escherichia coli* ClpA and ClpB. We have found that both proteins form hexamers and both exist as mixtures of oligomeric states both in the presence and absence of nucleotide (Li, Lin, & Lucius, 2015; Lin & Lucius, 2015; Veronese & Lucius, 2010; Veronese, Stafford, & Lucius, 2009; Veronese, Rajendar, & Lucius, 2011). This is a difficult problem to address for a variety of reasons, some of which will be discussed here. Nevertheless, the number of systems that exhibit ligand-linked assembly is large and there is a pressing need for a set of strategies and approaches to solve the problem.

Oftentimes, without being able to predict the concentration of each species in solution, quantitatively interpreting binding and catalytic data are not possible. However, solving these problems is needed for more than just interpretation of *in vitro* studies. We are finding that the nucleotide binding affinity of these motor proteins is in the range of 10–100 μ*M*, which is at least an order of magnitude below the concentration of nucleotide in the cell (5–10 m*M*). This indicates that the nucleotide binding sites on these macromolecules would likely be saturated in the cell, and thus, nucleotide is not likely to be a regulatory molecule. On the other hand, the concentration of these proteins in the cell (Dougan, Reid, Horwich, & Bukau, 2002; Farrell, Grossman, & Sauer, 2005; Mogk et al., 1999) have been found to be similar to their ligand-linked assembly dissociation equilibrium constant, indicating that the linkage between nucleotide binding and assembly may be an important regulatory component of their function.

To begin to quantitatively address the linkage of ligand binding to macromolecular assembly, we can start by writing down a partition function, Q, that represents the sum of all of the macromolecular states for an arbitrary solution containing monomers (M_1), dimers (M_2), tetramers (M_4), and hexamers (M_6) as follows:

$$Q = [M_1] + [M_2] + [M_4] + [M_6] \tag{2}$$

We can define self-association equilibrium constants for each thermodynamic state. Here, we will use *L* for stoichiometric or overall protein–protein interaction constants and *K* for both stepwise protein–protein interaction constants and for ligand binding constants. For the monomer, dimer, tetramer, and hexamer system, the following three reactions given by Eqs. (3)–(5) can define the equilibria

$$2M_1 \overset{L_{2,0}}{\rightleftharpoons} M_2 \tag{3}$$

$$4M_1 \overset{L_{4,0}}{\rightleftharpoons} M_4 \tag{4}$$

$$6M_1 \overset{L_{6,0}}{\rightleftharpoons} M_6 \tag{5}$$

The three stoichiometric equilibrium constants that result are given by Eqs. (6)–(8)

$$L_{2,0} = \frac{[M_2]}{[M_1]^2} \tag{6}$$

$$L_{4,0} = \frac{[M_4]}{[M_1]^4} \tag{7}$$

$$L_{6,0} = \frac{[M_6]}{[M_1]^6} \tag{8}$$

where the first subscript represents the oligomeric state and the second subscript represents the nucleotide ligation state, in this case the zero represents no ligand bound. The partition function given in Eq. (2) can be simplified by algebraically solving Eqs. (6)–(8) for each oligomer and substituting the solutions into Eq. (2) to yield Eq. (9)

$$Q = [M_1] + L_{2,0}[M_1]^2 + L_{4,0}[M_1]^4 + L_{6,0}[M_1]^6 \tag{9}$$

where the partition function given by Eq. (9) is only a function of the free monomer concentration and the equilibrium constants for each oligomer formed. If each of the oligomers can bind ligand, then each term in the partition function will be multiplied by a partition function for binding of ligand to that particular oligomer to yield Eq. (10)

$$Q = [M_1]P_1 + L_{2,0}[M_1]^2 P_2 + L_{4,0}[M_1]^4 P_4 + L_{6,0}[M_1]^6 P_6 \tag{10}$$

where P_1, P_2, P_4, and P_6 represent the partition functions for the binding of ligand to the monomer, dimer, tetramer, and hexamer, respectively. The steps for deriving Eq. (10) are straightforward and will not be reproduced here; for a review, see Wyman & Gill (1990). For simplicity, if the monomer binds one ligand, X, then the partition function for ligand binding is given by the sum of all of the monomeric states normalized to the unligated state given by Eq. (11)

$$P_1 = \frac{[M_1] + [M_1X]}{[M_1]} \tag{11}$$

If we define an equilibrium constant for nucleotide binding to the monomer as:

$$K_{1,1} = \frac{[M_1X]}{[M_1]} \tag{12}$$

where the first subscript represents the oligomeric state and the second subscript represents the number of ligands bound. We can simplify Eq. (11) to be:

$$P_1 = 1 + K_{1,1}[X] \tag{13}$$

If we assume that the monomer has n-independent and identical binding sites, then the partition function given by Eq. (13) would be expressed as Eq. (14)

$$P_1 = (1 + K_1[X])^{n_1} \tag{14}$$

where n_1 represents the number of binding sites per monomer. If we assume that each oligomer binds n number of ligands independently and identically, then the partition function for the entire system is given by

$$\begin{aligned} Q = {} & [M_1](1 + K_1[X])^{n_1} + L_{2,0}[M_1]^2(1 + K_2[X])^{n_2} + L_{4,0}[M_1]^4(1 + K_4[X])^{n_4} \\ & + L_{6,0}[M_1]^6(1 + K_6[X])^{n_6} \end{aligned} \tag{15}$$

where n_1, n_2, n_4, and n_6 represent the number of binding sites on the monomer, dimer, tetramer, and hexamer, respectively, and K_1, K_2, K_4, and K_6 represent the average binding constant for ligand binding to monomers, dimers, tetramers, and hexamers, respectively, where the average binding constant has been corrected with statistical factors (Wyman & Gill, 1990). Since, in this model, all of the binding sites are assumed to be the same, there is no second subscript on K_x to denote the number of ligands bound.

For any experiment that would seek to examine ligand binding to such a complex system, whether it be ITC, fluorescence titrations, or some other approach, the signal is typically proportional to ligand bound divided by total macromolecule, which is defined as "extent of binding" or $\overline{X}$. The importance of expressing the partition function is that the partition function can now be used to derive an equation that represents the extent of binding, ligand bound over total macromolecule, which could be used to analyze ligand binding data. The extent of binding is given by Eq. (16) (Wyman & Gill, 1990)

$$\frac{[\mathrm{X}]_{\mathrm{Bound}}}{[\mathrm{M}]_{\mathrm{Total}}}=\overline{\mathrm{X}}=\frac{\mathrm{d}Q/\mathrm{d}\ln[\mathrm{X}]}{\mathrm{d}Q/\mathrm{d}\ln[\mathrm{M}]}=\frac{[\mathrm{X}]}{[\mathrm{M}]}\frac{\mathrm{d}Q/\mathrm{d}[\mathrm{X}]}{\mathrm{d}Q/\mathrm{d}[\mathrm{M}]} \tag{16}$$

If the derivatives in Eq. (16) are applied to the partition function given by Eq. (15), then the extent of binding equation is given by Eq. (17)

$$\overline{\mathrm{X}}=\frac{\begin{array}{l}[\mathrm{M}_1]n_1K_1[\mathrm{X}](1+K_1[\mathrm{X}])^{n_1-1}+L_{2,0}[\mathrm{M}_1]^2n_2K_2[\mathrm{X}](1+K_2[\mathrm{X}])^{n_2-1}\\+L_{4,0}[\mathrm{M}_1]^4n_4K_4[\mathrm{X}](1+K_4[\mathrm{X}])^{n_4-1}+L_{6,0}[\mathrm{M}_1]^6n_6K_6[\mathrm{X}](1+K_6[\mathrm{X}])^{n_6-1}\end{array}}{\begin{array}{l}[\mathrm{M}_1](1+K_1[\mathrm{X}])^{n_1}+2L_{2,0}[\mathrm{M}_1]^2(1+K_2[\mathrm{X}])^{n_2}+4L_{4,0}[\mathrm{M}_1]^4(1+K_4[\mathrm{X}])^{n_4}\\+6L_{6,0}[\mathrm{M}_1]^6(1+K_6[\mathrm{X}])^{n_6}\end{array}}. \tag{17}$$

Clearly, Eq. (17) would have entirely too many parameters to apply to a single binding isotherm collected with any technique. However, several important predictions can be made from inspection of Eq. (17) or from inspection of the partition function given in Eq. (15). First, these equations are functions of the self-association constants in the absence of ligands, $L_{n,0}$, the ligand binding constants to each oligomer, K_n, the free monomer concentration $[\mathrm{M}_1]$, and the free ligand concentration [X]. Second, Eq. (17), tells the experimentalist that there is a need to first define the self-association equilibrium constants in the absence of nucleotide, $L_{n,0}$. Third, unlike binding isotherms for simple systems, a binding system that is linked to macromolecular assembly will exhibit a dependence on the free protein concentration. This tells the experimentalist that binding studies will have to be executed over a range of protein concentrations.

Figure 1 shows a series of isotherms simulated using Eq. (17) with several different total macromolecule concentrations ranging from 1 to 10 μM. In this example, $L_{2,0}=1\times10^4\ M^{-1}$, $L_{4,0}=1\times10^{14}\ M^{-3}$, and $L_{6,0}=1\times10^{24}\ M^{-5}$, each monomer is considered to bind one ligand so that $n_1=1$, $n_2=2$, $n_4=4$, and $n_6=6$, the ligand binding constants for the monomers through tetramers are all identical so that $K_1=K_2=K_4=1\times10^5\ M^{-1}$ and the hexamer binding constant is an order of magnitude tighter, $K_6=1\times10^6\ M^{-1}$.

The most salient feature of the binding isotherms shown in Fig. 1 is that there is a shift of the midpoint to lower free ligand concentration as the macromolecule concentration increases. This is the consequence of the fact that as the macromolecule concentration increases, there is a corresponding increase in the concentration of hexamers. This observation is general because the thermodynamic driving force for an assembly reaction is the chemical potential of the free monomer. Thus, there will always be an increase in the

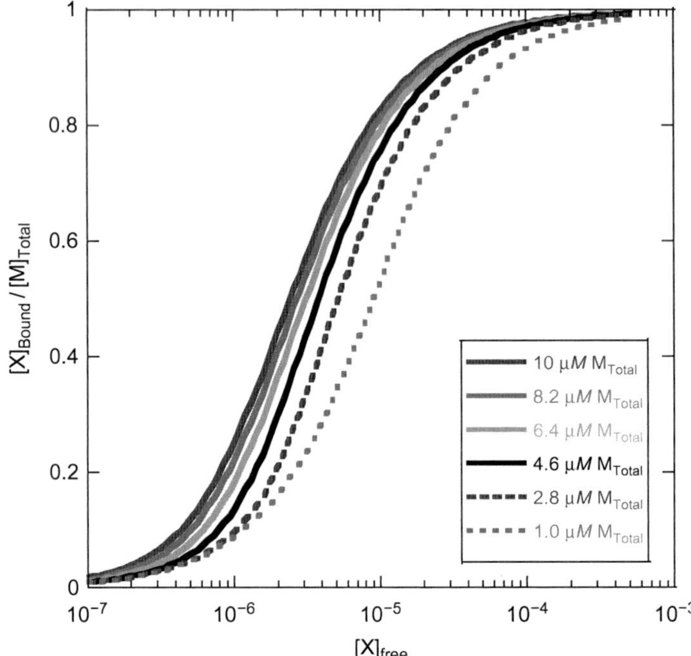

Figure 1 Predicted binding isotherms from Eq. (17) with $L_{2,0}=1\times10^{4}\,M^{-1}$, $L_{4,0}=1\times10^{14}\,M^{-3}$, and $L_{6,0}=1\times10^{24}\,M^{-5}$, each monomer is considered to bind one ligand so that $n_1=1$, $n_2=2$, $n_4=4$, and $n_6=6$, the ligand binding constants for the monomers through tetramers are all identical so that $K_1=K_2=K_4=1\times10^{5}\,M^{-1}$ and the hexamer binding constant is an order of magnitude tighter, $K_6=1\times10^{6}\,M^{-1}$.

population of higher-order oligomers with increasing protein concentration. The apparent increase in ligand binding affinity as the macromolecule concentration is increased is because the ligand binds to the hexamer with an affinity constant one order of magnitude tighter than the other oligomers.

When examining ligand binding to a macromolecule, it is always advisable to perform multiple titrations at several total macromolecule concentrations (Lohman & Mascotti, 1992). If such a strategy is invoked and an apparent change in the affinity constant is observed as a function of macromolecule concentration, then this would be the first indicator that a ligand-linked assembly process is occurring.

2. DETERMINATION OF $L_{n,0}$ FOR AN ASSEMBLING SYSTEM

If assembly is suspected from an experiment such as that illustrated by Fig. 1, then Eq. (17) suggests that the first objective would be to determine the self-association equilibrium constants in the absence of any ligand, $L_{n,0}$, so that this parameter could be constrained in the examination of the titration curves. To determine $L_{n,0}$, we perform sedimentation velocity experiments over a range of protein concentrations.

Sedimentation velocity has been our experiment of choice over sedimentation equilibrium. This is because we have found that the nucleotide bound hexamers are not stable over the timescale required for sedimentation equilibrium, which is often several days. In contrast to a sedimentation equilibrium experiment, a sufficient number of sedimentation boundaries can be collected within 2–3 h in a sedimentation velocity experiment. Further, deconvoluting three or more species from exponential fitting performed on sedimentation equilibrium boundaries can be difficult.

Naturally, one would not know the self-association equilibrium constants, so knowing what concentrations to examine will not be initially certain. However, if a series of isotherms were collected like those shown in Fig. 1, one could judge that the assembly state is making a transition over the range of 1–10 μ*M* and this would be a reasonable starting point.

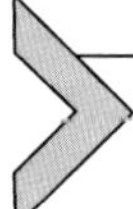

3. GLOBAL FITTING OF SEDIMENTATION VELOCITY DATA AS A FUNCTION OF PROTEIN CONCENTRATION KINETIC CONSIDERATIONS

Sedimentation velocity data are potentially sensitive to the assembly kinetics if the dissociation rate constants for the species are in the range of 10^{-2}–10^{-5} s^{-1} (Correia, Alday, Sherwood, & Stafford, 2009; Dam, Velikovsky, Mariuzza, Urbanke, & Schuck, 2005; Demeler et al., 2010; Stafford & Sherwood, 2004). Consequently, there are two ways to globally analyze sedimentation velocity data. The first is to assume that the system is always at thermodynamic equilibrium. This assumption holds if the dissociation rate constants are faster than 10^{-2} s^{-1}. The second is that the dissociation rate constant is found to be within or slower than the empirical range 10^{-2}–10^{-5} s^{-1}. If the dissociation rate constants are within this empirical range, modeling the reaction kinetics is required. Otherwise, for dissociations that are slower than 10^{-5} s^{-1}, the components can be considered as noninteracting discrete species.

To illustrate the impact of the assembly kinetics, we simulated sedimentation velocity experiments with reverse rate constants of either 1 or 10^{-6} s^{-1}. Dissociation rate constants of $k_r = 1$ or 10^{-6} s^{-1} correspond to half-lives of 0.7 s and 192 h, respectively. It is not difficult to conclude that reaction kinetics occurring with these half-lives would be outside of the detectable range in a sedimentation velocity experiment since the boundaries are typically collected on the minutes timescale. Although one could enhance the temporal resolution with interference, experiments performed on only a single cell since these could be collected every 8 s.

To simulate the sedimentation boundaries using SEDANAL, the concentration of each species needs to be modeled. Here, we will use the language of the Gibbs phase rule. A component is defined as a chemical component and a species is made up of products of reactions between the components. Thus, in an experiment containing a single protein, "M," that reacts to form dimers, tetramers, and hexamers, we define "M" as the component and monomers, dimers, tetramers, and hexamers as species. Therefore, this is a single component—four-species system.

The first step in the simulation is to relate the total loading concentration of the protein to the concentration of each species. This is done by writing down the conservation of mass equation given by Eq. (18)

$$[M_1]_T = [M_1] + 2[M_2] + 4[M_4] + 6[M_6] \tag{18}$$

where $[M_1]_T$ is the total monomer concentration, $[M_1]$, $[M_2]$, $[M_4]$, and $[M_6]$ are the equilibrium concentrations of monomers, dimers, tetramers, and hexamers. The coefficients of 2, 4, and 6 are present because the total concentration is expressed in monomer units, e.g., 2 monomers in a dimer. Equation (18) can be expressed in terms of the equilibrium constants given by Eqs. (6)–(8) and the free monomer concentration, $[M_1]$, to yield Eq. (19)

$$[M_1]_T = [M_1] + 2L_{2,0}[M_1]^2 + 4L_{4,0}[M_1]^4 + 6L_{6,0}[M_1]^6 \tag{19}$$

In a sedimentation velocity experiment, there are two stages of equilibrium that one needs to consider. Those stages are before the force of centrifugation is applied and while the force is present. When the force is present, the equilibrium can be perturbed. However, before the force is applied, the system should be at equilibrium. This "pre-equilibrium" is achieved if the experimentalist has given sufficient time for the system to fully relax to equilibrium after any perturbations, i.e., preparation of samples, dilutions, temperature equilibration. Thus, if an experiment is intended to be performed at, for example, $[M_1]_T = 1\ \mu M$, then the experimentalist would make up this solution in some vessel. In that vessel, the system will distribute itself into free monomers, dimers, tetramers, and hexamers, where the population of each could be defined by Eq. (19) if the equilibrium constants are known. However, an important control would be to allow this sample to incubate for increasing amounts of time before performing the run. Then, the results could be compared after, for example, a 6 h versus 12 h preincubation time. If the system is at equilibrium, one would expect to see results that are independent of incubation time. If incubation time-dependent differences are observed, then the incubation time should be extended until differences are no longer observed.

To determine the concentration of each species for this system, one needs to solve Eq. (19), which is a sixth-order polynomial in the free monomer concentration. However, the free monomer concentration, $[M_1]$, is not known. What is known to the experimentalist is the total monomer concentration, $[M_1]_T$. Thankfully, numerically solving a polynomial is not a difficult task.

In SEDANAL, the roots of Eq. (19) and thus the concentrations of each species are determined by numerical methods, specifically the Newton–Raphson method. In practice, this is achieved in SEDANAL by going to "preferences" choosing "Control extended," "Kinetics/equilibrium control" and under "Initial equilibration" one chooses "No analytic solution" under the Newton–Raphson column. Again, since this model requires solving a sixth-order polynomial, there is no analytic solution and therefore the equation must be solved numerically.

What was just described represents the determination of the concentration of each species in the reaction vessel before the sample is subjected to the force of sedimentation. Since there is no force, there is no perturbation of the equilibrium, and therefore, the concentrations of each species are fixed. The next task at hand is to define the concentrations of each species upon application of force. Since, in a sedimentation velocity experiment, the experimentalist is observing the time-dependent separation of each species, the experiment is potentially sensitive to the reaction kinetics.

To simulate the sedimentation boundaries, one needs to again determine the concentration of each species and then model the sedimentation of each species by passing the determined concentration to the Lamm equation, which defines the movement of the particle under the force of centrifugation and back diffusion. For the monomer, dimer, tetramer, and hexamer reaction, we assume that all species are being formed through bimolecular interactions defined by the following reactions given by Eqs. (20)–(22)

$$2M_1 \underset{k_{r2}}{\overset{k_{f2}}{\rightleftharpoons}} M_2 \tag{20}$$

$$2M_2 \underset{k_{r4}}{\overset{k_{f4}}{\rightleftharpoons}} M_4 \tag{21}$$

$$M_2 + M_4 \underset{k_{r6}}{\overset{k_{f6}}{\rightleftharpoons}} M_6 \tag{22}$$

where $k_{f,n}$ is the bimolecular association rate constant with units of $M^{-1}\ s^{-1}$ and $k_{r,n}$ is the dissociation rate constant with units of s^{-1}. We assume that all species are formed through bimolecular interactions because trimolecular reactions and above are highly improbable. The equilibrium constants for the reactions in Eqs. (20)–(22) are given by Eqs. (23)–(25)

$$K_2 = \frac{k_{f2}}{k_{r2}} = \frac{[M_2]}{[M_1]^2} \tag{23}$$

$$K_4 = \frac{k_{f4}}{k_{r4}} = \frac{[M_4]}{[M_2]^2} \tag{24}$$

$$K_6 = \frac{k_{f6}}{k_{r6}} = \frac{[M_6]}{[M_2][M_4]} \tag{25}$$

It is important to note that if the system is an equilibrium system, then the stepwise equilibrium constants given by Eqs. (23)–(25) can be related to the stoichiometric interaction constants given by Eqs. (6)–(8) as follows:

$$L_{2,0} = K_2 \tag{26}$$

$$L_{4,0} = K_2^2 \cdot K_4 \tag{27}$$

$$L_{6,0} = K_2^3 \cdot K_4 \cdot K_6 \tag{28}$$

Since the equilibrium is potentially being perturbed, determining the concentration of each species is accomplished by numerically solving the following system of coupled differential equation given by Eqs. (29)–(32)

$$\frac{d[M_1]}{dt} = [M_2]k_{r2} - [M_1]^2 k_{f2} = [M_2]k_{r2} - [M_1]^2 K_2 k_{r2} \tag{29}$$

$$\begin{aligned}\frac{d[M_2]}{dt} &= [M_1]^2 k_{f2} - [M_2]k_{r2} - [M_2]^2 k_{f4} + [M_4]k_{r4} - [M]_2[M_4]k_{f6} + [M_6]k_{r6} \\ &= [M_1]^2 K_2 k_{r2} - [M_2]k_{r2} - [M_2]^2 K_4 k_{r4} + [M_4]k_{r4} - [M_2][M_4]K_6 k_{r6} + [M_6]k_{r6}\end{aligned} \tag{30}$$

$$\begin{aligned}\frac{d[M_4]}{dt} &= [M_2]^2 k_{f4} - [M_4]k_{r4} - [M]_2[M_4]k_{f6} + [M_6]k_{r6} \\ &= [M_2]^2 K_4 k_{r4} - [M_4]k_{r4} - [M_2][M_4]K_6 k_{r6} + [M_6]k_{r6}\end{aligned} \tag{31}$$

$$\frac{d[M_6]}{dt} = [M_2][M_4]k_{f6} - [M_6]k_{r6} = [M_2][M_4]K_6 k_{r6} - [M_6]k_{r6}, \tag{32}$$

where the right-hand side of Eqs. (29)–(32) has been simplified to include only the equilibrium constant and reverse rate constants for each reaction. To model the sedimentation of each species, the system of coupled differential equations is numerically integrated to determine the concentration of each species as a function of time based on the values of the equilibrium constants and dissociation rate constants for each reaction.

In the extreme of rapid dissociation, $k_r > 0.01\ \mathrm{s}^{-1}$ ($t_{1/2} < 1.2$ min), the system is considered to be in rapid equilibrium. That is to say, for each infinitely small radial slice of solution the macromolecules that rapidly dissociate within this slice are considered to rapidly reassociate, and therefore, the differential equations given in Eqs. (29)–(32) are all equal to zero, $\mathrm{d}[\mathrm{M}_n]/\mathrm{d}t = 0$. This indicates that the concentration of each oligomer in each infinitely small radial slice is constant and thus at equilibrium.

In contrast, if the dissociation rate constants are on the other extreme, $k_r < 10^{-5}\ \mathrm{s}^{-1}$ ($t_{1/2} > 19$ h), then the differential equations are again equal to zero and the concentrations of each species are again considered to be fixed. However, in this scenario there is no reequilibration at each infinitely small radial slice because no dissociation is occurring during the course of the entire experiment. Thus, the species are being separated when force is applied. In this scenario, the system can be modeled as noninteracting discrete species because they will not appear to react on the timescale of sedimentation.

To illustrate these points, we simulated a series of sedimentation boundaries from the monomer, dimer, tetramer, and hexamer model given by Eqs. (20)–(22) using SEDANAL for three total protein concentrations, $[\mathrm{M}_1]_\mathrm{T} = 1$, 9 and 15 μ*M*. We refer to this model as the "1-2-4-6" model. For the first simulation, all reverse rate constants were considered to be fast relative to sedimentation, $k_{r2} = k_{r4} = k_{r6} = 1\ \mathrm{s}^{-1}$ (see Table 1). In a second simulation, all reverse rate constants were considered to be slow, $k_{r2} = k_{r4} = k_{r6} = 10^{-6}\ \mathrm{s}^{-1}$ (see Table 1). Both extremes of the rate constants were considered to be outside of 10^{-2}–$10^{-5}\ \mathrm{s}^{-1}$, which has been previously reported to be the range over which one would expect to be able to extract meaningful measures of the rate constants from sedimentation velocity data (Dam et al., 2005; Demeler et al., 2010; Stafford & Sherwood, 2004).

Our preferred first level of analysis is to analyze the sedimentation boundaries using SedFit to generate $c(s)$ distributions (Peter Shuck, NIH) (Schuck, 1998). Figure 2A shows the results of a $c(s)$ analysis on the simulated data assuming all dissociation rate constants are $1\ \mathrm{s}^{-1}$. In red is the $c(s)$ distribution from the analysis of the simulation with $[\mathrm{M}_1]_\mathrm{T} = 1$ μ*M* and a single

Table 1 Parameters Used to Simulate the Sedimentation Velocity Data for 1–15 μ*M* Macromolecule Concentration

	Parameters Used for Fast Dissociation		Parameters Used for Slow Dissociation	
Model Used for Simulation	K_n (M^{-1})	$k_{r,n}$ (s^{-1})	K_n (M^{-1})	$k_{r,n}$ (s^{-1})
$2B \underset{k_{r1}}{\overset{k_{f1}}{\rightleftharpoons}} B_2$	6×10^4	1	6×10^4	1×10^{-6}
$2B_2 \underset{k_{r4}}{\overset{k_{f4}}{\rightleftharpoons}} B_4$	1.5×10^5	1	1.5×10^5	1×10^{-6}
$B_2 + B_4 \underset{k_{r6}}{\overset{k_{f6}}{\rightleftharpoons}} B_6$	1.5×10^6	1	1.5×10^6	1×10^{-6}

The stoichiometric equilibrium constants for dimerization, tetramerization, and hexamerization can be calculated using Eqs. (25)–(27) and are: $L_{2,0} = 6 \times 10^4\ M^{-1}$, $L_{4,0} = 5.4 \times 10^{14}\ M^{-3}$, and $L_{6,0} = 4.86 \times 10^{25}\ M^{-5}$. The sedimentation coefficients used for simulations are $s = 3.0$, 5.6, 8.86, and 11.9 for monomer, dimer, tetramer, and hexamer, respectively. The standard deviation of noise added to the data is 0.005 absorbance unit.

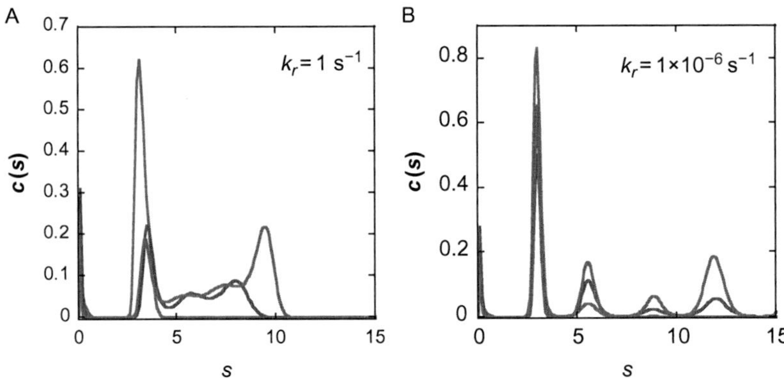

Figure 2 $c(s)$ distributions resulting from $c(s)$ analysis of data simulated from 1, 9, and 15 μ*M* protein. (A) Parameters are given in Table 1 and (B) parameters are given in Table 2. In both cases, the sedimentation coefficients used to simulate the data were $s = 3.0$, 5.6, 8.86, and 11.9 for monomer, dimer, tetramer, and hexamer, respectively. The extinction coefficient used in the simulation is 4.2 $(mg/ml)^{-1}$ cm^{-1} at 230 nm and 0.45 $(mg/ml)^{-1}$ cm^{-1} at 280 nm. Time between simulated scans is 4 min. (See the color plate.)

peak is observed that corresponds to the monomer. In blue is the $c(s)$ analysis for $[M_1]_T = 9$ μ*M*. The peak corresponding to monomer shifts to the right slightly and a broad distribution from ~5 to 8 S emerges. In green is the $c(s)$ distribution from the simulation with $[M_1]_T = 15$ μ*M* and what is observed is a clear shifting of the peaks further to the right. The peak shifting can be taken as the first indication that the system is exhibiting fast reaction kinetics (Dam & Schuck, 2005; Dam et al., 2005). Thus, a preliminary $c(s)$ analysis that shows this type of broad $c(s)$ distribution may serve to be the first indicator that rapid dissociation may be occurring on the timescale of sedimentation, and this hypothesis would warrant further testing.

Figure 2B shows a $c(s)$ analysis resulting from analysis of simulations performed with all dissociation rate constants $k_r = 10^{-6}\ \mathrm{s}^{-1}$. At 1 μ*M* total protein, two peaks are observed that correspond to the sedimentation coefficient of monomers and dimers. In contrast, at 15 μ*M* all four peaks appear that correspond to the monomers, dimers, tetramers, and hexamers, respectively. Under these conditions of slow dissociation, the oligomers sediment as noninteracting discrete species. Under such conditions, the area under each of these peaks would represent the equilibrium concentrations of each species and could be analyzed to yield the equilibrium constants.

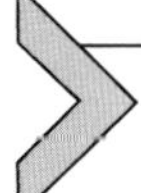

4. GLOBAL ANALYSIS USING THE 1-2-4-6 MODEL WITH RAPID DISSOCIATING OLIGOMERS

If the $c(s)$ plot given by Fig. 2A was experimentally observed, our next step would be to globally fit all protein concentrations by direct boundary analysis in SEDANAL. Again, the goal here is to determine the self-association equilibrium constants, $L_{n,0}$. Since the $c(s)$ plot suggests there may be rapid dissociation, the first strategy in this analysis would be to float each of the reverse rate constants starting with a guess of around $0.01\ \mathrm{s}^{-1}$. Thus, the global floating fitting parameters are the loading concentrations, K_2, K_4, K_6, k_{r2}, k_{r4}, and k_{r6} as given by Eqs. (23)–(25).

As with the simulations described above to fit the data accounting for the kinetics, there are several steps that have to be executed in SEDANAL. We always assume that the system is at equilibrium at the start of the run. In practice, this is achieved in SEDANAL by going to "preferences" choosing "Control extended," "Kinetics/equilibrium control" and under "Initial equilibration" one chooses "No analytic solution" under the Newton–Raphson column. What this accomplishes is numerically solving Eq. (19) to determine the free monomer concentration and thus the concentration of each oligomer is determined based on the initial guesses of the equilibrium constants. Next, under "Equilibration during run," we choose either BulSt or SEulEx under "Kinetic integrator" in the "No analytic solution" row.

The choice of kinetic integrator primarily impacts the speed of executing the fit and the differences have been discussed elsewhere (Stafford & Sherwood, 2004), including the SEDANAL manual. What this accomplishes is numerically solving the system of coupled differential equations given by Eqs. (29)–(32) for the concentrations of each species as a function of time and passes this to the Lamm equation to define the movement of the oligomer in the field.

The other parameters required for this analysis are the molar mass, the sedimentation coefficient, density increment, and mass extinction coefficient. The molar mass would be known from sequence information and thus the molar mass of each oligomer is calculated. Estimates of the sedimentation coefficients are needed and solution conditions can usually be modified to acquire reasonable estimates of the monomer and largest oligomers. One can estimate the intermediate sedimentation coefficients based on the $s_n = s_1(n)^{2/3}$ with the assumption that the frictional ratio of the oligomers are the same as monomer's, where s_n is the sedimentation coefficient for the oligomer containing n protomer units and s_1 is the sedimentation coefficient for the monomer (Correia, 2000). Alternatively, if data on the three-dimensional structure are available, then one can approximate hydrodynamic information using applications like Hydropro (Ortega, Amoros, & Garcia de la Torre, 2011; Chapter "Hydrodynamic Modeling and Its Application in AUC" by Rocco and Byron). The extinction coefficient should be rigorously determined by denaturing the protein in 6 *M* guanidine (Edelhoch, 1967; Gill & von Hippel, 1989; Pace, Vajdos, Fee, Grimsley, & Gray, 1995). The density increments, $d\rho/dc$, can be determined experimentally, but for most cases it is acceptable to substitute $(1-\bar{v}\rho)$, where $\bar{v}$ is the partial specific volume of the protein and ρ is the density of the buffer. These parameters are typically calculated with SednTerp (David Hayes, Magdalen College; Tom Laue, University of New Hampshire; and John Philo, Alliance Protein Laboratories) (Laue, Shah, Ridgeway, & Pelletier, 1992).

Global NLLS analysis of the data simulated from the 1-2-4-6 model with all the dissociation rate constants $k_{r,n} = 1\ s^{-1}$ yields good estimates of the equilibrium constants (compare values used to generate data in Table 1 to fitted values in Table 2, first column). The difference curves and their associated fits are shown in Fig. 3. Values of the rate constants float somewhere between ~0.3 and 2 s^{-1} (see Table 3). We have interpreted this to indicate that the kinetic parameters are unconstrained because the value used to simulate the data of 1 s^{-1} results in a half-life of ~0.7 s. Thus, there is little

Table 2 Examination of Data Simulated with Fast Dissociation Using the "1-2-4-6" Model

	Fit with Kinetic Integrator Allowing $k_{r,n}$ to Float		**Fit with Kinetic Integrator, Constraining $k_{r,n} = 0.01\ s^{-1}$**		**Fit with Newton–Raphson, No $k_{r,n}$ Input**	
	RMSD = 5.003×10^{-3}		**RMSD = 5.929×10^{-3}**		**RMSD = 5.003×10^{-3}**	
Model Used for Fitting	**K_n (M^{-1})**	**$k_{r,n}$ (s^{-1})**	**K_n (M^{-1})**	**$k_{r,n}$ (s^{-1})**	**K_n (M^{-1})**	**$k_{r,n}$ (s^{-1})**
$2B \underset{k_{r1}}{\overset{k_{f1}}{\rightleftharpoons}} B_2$	6.00×10^4	0.29	5.87×10^4	0.01	6.00×10^4	NA
$2B_2 \underset{k_{r4}}{\overset{k_{f4}}{\rightleftharpoons}} B_4$	1.49×10^5	1.52	2.02×10^5	0.01	1.49×10^5	NA
$B_2 + B_4 \underset{k_{r6}}{\overset{k_{f6}}{\rightleftharpoons}} B_6$	1.51×10^6	1.94	1.02×10^6	0.01	1.52×10^6	NA

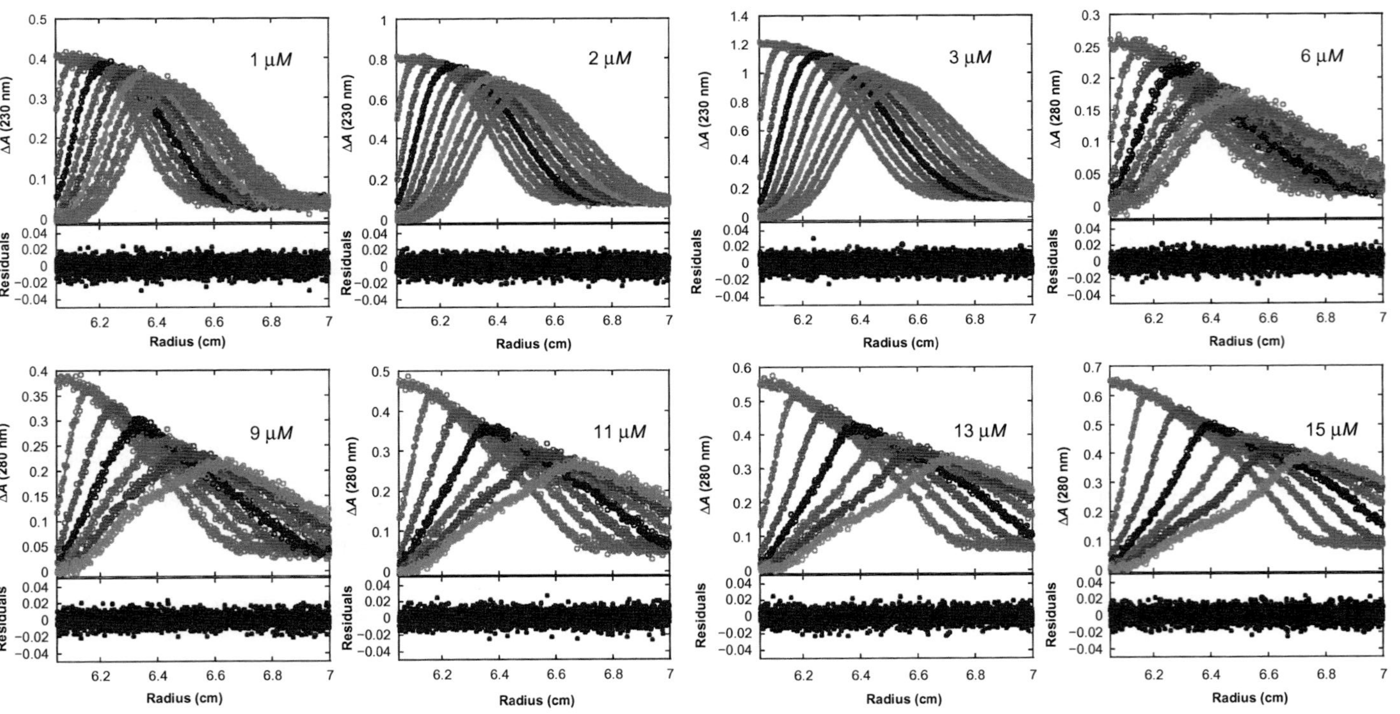

Figure 3 Global analysis of simulated sedimentation velocity data using 1-2-4-6 model by allowing dissociation rate constant, $k_{r,n}$, to float. The data were simulated using parameters presented in Table 1. The fitting results are presented in Table 3. The concentrations of protein are indicated on the plots. Every fourth difference curve is presented.

Table 3 Examination of Data Simulated with Slow Dissociation Using the "1-2-4-6" Model

	Fit with Kinetic Integrator Allowing $k_{r,n}$ to Float		Fit with Kinetic Integrator, Constraining $k_{r,n} = 1 \times 10^{-5}\ s^{-1}$		Fit with Newton–Raphson, No $k_{r,n}$ Input	
	RMSD = 5.000×10^{-3}		RMSD = 5.188×10^{-3}		RMSD = 3.178×10^{-2}	
Model Used for Fitting	K_n (M^{-1})	$k_{r,n}$ (s^{-1})	K_n (M^{-1})	$k_{r,n}$ (s^{-1})	K_n (M^{-1})	$k_{r,n}$ (s^{-1})
$2B \underset{k_{r1}}{\overset{k_{f1}}{\rightleftharpoons}} B_2$	6.00×10^4	1.16×10^{-6}	5.91×10^4	1×10^{-5}	5.89×10^3	NA
$2B_2 \underset{k_{r4}}{\overset{k_{f4}}{\rightleftharpoons}} B_4$	1.52×10^5	4.86×10^{-9}	1.65×10^5	1×10^{-5}	6.45×10^6	NA
$B_2 + B_4 \underset{k_{r6}}{\overset{k_{f6}}{\rightleftharpoons}} B_6$	1.49×10^6	5.13×10^{-7}	1.43×10^6	1×10^{-5}	7.00×10^7	NA

information in the sedimentation boundaries that yield constraints on the values of these rate constants.

The next step in our analysis strategy would be to constrain the dissociation rate constants to a value of 0.01 s^{-1} and repeat the analysis. In doing this on the same simulated data as above, we acquire estimates of the equilibrium constants that have the correct order of magnitude but are as much as 25% higher than the values used to generate the data (Table 2, second column). Moreover, the RMSD is significantly larger than the RMSD determined when allowing the rate constants to float. This cannot simply be the consequence of having three additional floating parameters because the degrees of freedom (DOFs) between the fits are essentially identical. That is to say, since the DOF is the difference between the number of data points and the number of parameters and here the number of data points is ~80,000, the additional three parameters do not influence this number enough to account for the deviation in the RMSD.

Despite the fact that constraining the rate constants to the empirical upper bound for "fast dissociation" leads to an ~25% overestimate of the equilibrium constant, the observation that both fits lead to the correct order of magnitude in the equilibrium constant suggests that the data could be modeled with a purely thermodynamic model. This leads to the suggestion that a path-independent thermodynamic model could be used to describe the data. To test this, the data were fit by choosing Newton–Raphson under the "Equilibration during run" in the "No analytic solution" row. What this accomplishes is numerically solving the sixth-order polynomial in the free monomer concentration to yield the concentration of species, thereby modeling the system under the assumption that each infinitely small slice of radial position is at equilibrium. This analysis yields equilibrium constants that are in good agreement with the values used to generate the data and the RMSD is identical to the value acquired when allowing the rate constants to float as fitting parameters (see Table 2, third column).

5. GLOBAL ANALYSIS USING THE 1-2-4-6 MODEL WITH SLOW DISSOCIATION OF OLIGOMERS

Figure 2B shows a $c(s)$ analysis from sedimentation velocity experiments simulated for the 1-2-4-6 model with all dissociation rate constants of 10^{-6} s^{-1}. Again, our first level of global analysis is to allow both the kinetic parameters, $k_{r,n}$, and the equilibrium constants, K_n, to float as fitting parameters. As seen in Table 3, the values of the equilibrium constants are in good agreement with those used to simulate the data (compare Table 1 for simulated values to Table 3 for fitted values). However, the rate constants are as much as three orders of magnitude slower than the values used to simulate the data. This is not surprising because a dissociation rate constant of 10^{-6} s^{-1} yields a half-life of ~192 h if one assumes a simple first-order reaction. Thus, on the timescale of sedimentation, there is no appreciable dissociation that occurs for the simulated data with a given 10^{-6} s^{-1} rate constant.

When the data are analyzed with the rate constants constrained to 10^{-5} s^{-1}, the values of the equilibrium constants determined are in agreement with the values used to simulate the data within 2–9% errors (see Table 3). However, the RMSD is significantly worse than the value when the rate constants are allowed to float as fitting parameters. This is surprising because with such a slow dissociation rate constant, one does not expect the data to contain any information on these parameters. One possible explanation is that the empirical bound of 10^{-5} s^{-1} may be on the edge of values that would define no dissociation for the simulated system.

To probe this further, the data were analyzed by constraining all $k_{r,n} \leq 10^{-6}$ s^{-1}. As shown by the analysis performed when allowing $k_{r,n}$ to float, $k_{r,n} = 10^{-6}$ s^{-1} is slow enough to indicate no dissociation for the simulated system. Therefore, rate constants smaller than 10^{-6} s^{-1} should be able to represent reactions with no dissociation and describe the simulated data adequately well. Our results show that an RMSD $= 5.000 \times 10^{-3}$ was determined when constraining all $k_{r,n} = 10^{-6}$ s^{-1} and an RMSD $= 5.003 \times 10^{-3}$ was determined when constraining all $k_{r,n} = 10^{-7}$ s^{-1}. Both analyses yield identical equilibrium constants and are identical to the value used to generate the data.

Above all, the analyses suggest that the empirical boundary for defining "no dissociation" is in the range of 10^{-6}–10^{-5} s^{-1}. We suspect that for systems with different levels of complexity and sizes of species, the empirical boundaries for rapid and slow dissociation may be slightly larger than the reported 10^{-2}–10^{-5} s^{-1} range. Based on these observations, we recommend testing the reaction kinetics by allowing the rate constants to float as fitting parameters in the analysis.

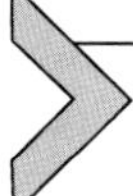

6. GLOBAL ANALYSIS WITH RATE CONSTANTS IN THE DETECTABLE RANGE

The next and most obvious question is how well can the rate constants be determined if they fall between the empirical bound of 10^{-2}–10^{-5} s^{-1}?

Table 4 Parameters Used to Simulate the Sedimentation Velocity Data for 1–15 μ*M* Macromolecule with Rate Constants Indicated in the Table

Model Used for Simulation	Values Used for Simulation		Parameters Resulting from Analysis RMSD = 4.972×10^{-3}	
	K_n (M^{-1})	$k_{r,n}$ (s^{-1})	K_n (M^{-1})	$k_{r,n}$ (s^{-1})
$2B \underset{k_{r1}}{\overset{k_{f1}}{\rightleftharpoons}} B_2$	6×10^4	3×10^{-3}	6.00×10^4	3.04×10^{-3}
$2B_2 \underset{k_{r4}}{\overset{k_{f4}}{\rightleftharpoons}} B_4$	1.5×10^5	4×10^{-4}	1.50×10^5	3.98×10^{-4}
$B_2 + B_4 \underset{k_{r6}}{\overset{k_{f6}}{\rightleftharpoons}} B_6$	1.5×10^6	7×10^{-4}	1.50×10^6	7.01×10^{-4}

The stoichiometric equilibrium constants for dimerization, tetramerization, and hexamerization can be calculated using Eqs. (25)–(27) and are: $L_{2,0} = 6 \times 10^4\ M^{-1}$, $L_{4,0} = 5.4 \times 10^{14}\ M^{-3}$, and $L_{6,0} = 4.86 \times 10^{25}\ M^{-5}$.

To address this question, we simulated data using the 1-2-4-6 model with equilibrium constants and rate constants given in Table 4. As shown in Table 4, the rate constants are in the range of 10^{-3}–10^{-4} s^{-1}. Table 4 shows the results of globally fitting these simulated data to the 1-2-4-6 model. It is not surprising that the data are well described by the correct model and the predicted parameters are well within range of the values used to simulate the data. Although not surprising, one might expect that the information on the kinetic rate constants could be lost due to the complexity of the model. Thus, what this analysis does show is that the global fitting strategy is able to detect these rate constants for this model.

Thermodynamics is path independent. Sedimentation velocity data where the oligomers are either in rapid dissociation or slow dissociation do not contain information about path. Although the specific strategy for doing so is different due to the process of sedimentation, in the two extremes fast and slow kinetics the data can be described by either stoichiometric or stepwise equilibrium constants when allowing the kinetic parameters to float and the results are the same. This is not the case for a system that exhibits dissociation with rate constants in the range of 10^{-2}–10^{-5} s^{-1}. This is shown in Table 5 where the simulated data were analyzed assuming that each oligomer forms in a single step from monomers as given by Eqs. (3)–(8) where monomers form dimers, tetramers, or hexamers in a single kinetic step. The analysis of these data does not accurately predict the equilibrium constants or

Table 5 Analysis of Data Simulated in Table 4 Using "1-2, 1-4, 1-6" Model to Analyze the Data

	Fit with Kinetic Integrator Allow $k_{r,n}$ to Float		
	RMSD = 5.539×10^{-3}		
Model Used for Fitting	$L_{n,0}$	**Calculated K_n (M^{-1})**	$k_{r,n}$ (s^{-1})
$2B \overset{L_{2,0}}{\rightleftharpoons} B_2$	$6.24 \times 10^4\ M^{-1}$	6.24×10^4	1.88×10^{-3}
$4B \overset{L_{4,0}}{\rightleftharpoons} B_4$	$3.91 \times 10^{14}\ M^{-3}$	1.00×10^5	1.16×10^{-4}
$6B \overset{L_{6,0}}{\rightleftharpoons} B_6$	$5.64 \times 10^{25}\ M^{-5}$	2.31×10^6	1.80×10^{-4}

the rate constants, with the exception of dimer formation which is not different for the two models. What this analysis reveals is that when the rate constants are in the detectable range, there is information on the path and the data cannot be modeled by simple path-independent thermodynamic models. Thus, under these conditions care needs to be taken to write down appropriate path-dependent models. Although it may take more computational power to do so, we assert that this should be done by assuming all reactions occur through bimolecular interactions.

7. CONCLUSIONS

Similar simulations and examination of the kinetics as discussed here have also been discussed elsewhere (Correia & Stafford, 2009; Dam et al., 2005). The advance presented here is an analysis of a more complex model. Most previous discussions have centered on a monomer–dimer equilibrium. Here, we have focused on the monomer, dimer, tetramer, and hexamer equilibrium since this is what we are experimentally observing for two AAA+ molecular motors, *E. coli* ClpA (Veronese & Lucius, 2010; Veronese et al., 2011; Veronese et al., 2009) and ClpB (Li et al., 2015; Lin & Lucius, 2015). Most importantly, there are some general principles that have been derived from previous studies on less complex systems that do not seem to generalize to the more complex systems we are studying. That is to say, the empirical bound of $0.01–10^{-5}\ s^{-1}$ may be a bit wider depending on the size of molecules and the rotor speed.

One observation that holds for all values of the kinetic rate constants is that floating the values always seems to lead to acquisition of the values used to simulate the data. So why not always float the kinetic rate constants? The answer is the lack of computational power. Indeed, 20 years ago, globally analyzing upward of 80,000 data points by numerically solving the Lamm equation combined with numerically solving a complex system of coupled differential equations was outside of our computational reach. However, we now have desktop computers that can accomplish these tasks. Nevertheless, fitting to such complex models is very slow and it can be accelerated when limiting assumptions can be made.

One limiting factor, which is true for all NLLS minimization routines, is that each computation is dependent on the outcome of the last. Thus, such approaches are not amenable to parallel computing. In contrast, routines such as the genetic algorithm are gaining interest because of the independence of each calculation and thus the ease with which such routines can be parallelized. Ultrascan is an application for the analysis of sedimentation velocity data that take advantage of the genetic algorithm and thus takes advantage of advances in parallel computing (Demeler et al., 2010).

Sedphat does not allow the user to write their own models for global analysis. Thus, the experimentalist is constrained to use prewritten models that assume, for example, that tetramers would form in a single kinetic step. One could make an argument for trimolecular interactions but anything higher than three bodies colliding in a single kinetic step is unrealistic. Although Ultrascan allows the user to write down their own models, it does not allow the product of one reaction to be the reactant of another. As shown by the simulations presented here, there is a need to be able to write down kinetic models that reduce the reactions to bimolecular steps when kinetic information is present.

To our knowledge, the only software application that allows the experimentalist to construct their own models and directly fit boundaries by numerically solving the Lamm equation is SEDANAL (Stafford & Sherwood, 2004). More importantly, SEDANAL is numerically solving a system of coupled differential equations describing the reaction. Thus, the rate constants that emerge have the potential to contain a great deal more information about the reaction than arbitrary rate constants that are being applied in other applications. We assert that the gravity of what SEDANAL is doing behind the scenes is not fully appreciated. Indeed, sedimentation velocity would not be the experiment of choice if one were trying to thoroughly deconvolute a kinetic mechanism. However, not adequately accounting for the kinetics in the analysis of these data may lead to significant errors in the equilibrium parameters of interest. Moreover, it is clear that sedimentation velocity experiments can lead the experimentalist to pursue more direct kinetic approaches if something interesting is revealed in the modeling.

REFERENCES

Alberty, R. A. (2003). *Thermodynamics of biochemical reactions*. Hoboken, NJ: Wiley Interscience.

Babst, M., Sato, T. K., Banta, L. M., & Emr, S. D. (1997). Endosomal transport function in yeast requires a novel AAA-type ATPase, Vps4p. *The EMBO Journal, 16*(8), 1820–1831.

Cole, J. L. (1996). Characterization of human cytomegalovirus protease dimerization by analytical centrifugation. *Biochemistry, 35*(48), 15601–15610.

Cole, J. L. (2004). Analysis of heterogeneous interactions. *Methods in Enzymology, 384*, 212–232.

Cole, J. L., Correia, J. J., & Stafford, W. F. (2011). The use of analytical sedimentation velocity to extract thermodynamic linkage. *Biophysical Chemistry, 159*(1), 120–128.

Correia, J. J. (2000). Analysis of weight average sedimentation velocity data. *Methods in Enzymology, 321*, 81–100.

Correia, J. J., Alday, P. H., Sherwood, P., & Stafford, W. F. (2009). Effect of kinetics on sedimentation velocity profiles and the role of intermediates. *Methods in Enzymology, 467*, 135–161.

Correia, J. J., & Stafford, W. F. (2009). Extracting equilibrium constants from kinetically limited reacting systems. *Methods in Enzymology, 455*, 419–446.

Dam, J., & Schuck, P. (2005). Sedimentation velocity analysis of heterogeneous protein-protein interactions: Sedimentation coefficient distributions c(s) and asymptotic boundary profiles from Gilbert-Jenkins theory. *Biophysical Journal, 89*(1), 651–666.

Dam, J., Velikovsky, C. A., Mariuzza, R. A., Urbanke, C., & Schuck, P. (2005). Sedimentation velocity analysis of heterogeneous protein-protein interactions: Lamm equation modeling and sedimentation coefficient distributions c(s). *Biophysical Journal, 89*(1), 619–634.

Demeler, B., Brookes, E., Wang, R., Schirf, V., & Kim, C. A. (2010). Characterization of reversible associations by sedimentation velocity with UltraScan. *Macromolecular Bioscience, 10*(7), 775–782.

Dougan, D. A., Reid, B. G., Horwich, A. L., & Bukau, B. (2002). ClpS, a substrate modulator of the ClpAP machine. *Molecular Cell, 9*(3), 673–683.

Edelhoch, H. (1967). Spectroscopic determination of tryptophan and tyrosine in proteins. *Biochemistry, 6*(7), 1948–1954.

Farrell, C. M., Grossman, A. D., & Sauer, R. T. (2005). Cytoplasmic degradation of ssrA-tagged proteins. *Molecular Microbiology, 57*(6), 1750–1761.

Gill, S. C., & von Hippel, P. H. (1989). Calculation of protein extinction coefficients from amino acid sequence data. *Analytical Biochemistry, 182*(2), 319–326.

Laue, T. M., Shah, B. D., Ridgeway, T. M., & Pelletier, S. L. (1992). Computer-aided interpretation of analytical sedimentation data for proteins. In A. J. Rowe, S. E. Harding, & J. C. Horton (Eds.), *Analytical ultracentrifugation in biochemistry and polymer science*. Cambridge: Royal Society of Chemistry.

Li, T., Lin, J., & Lucius, A. L. (2015). Examination of polypeptide substrate specificity for Escherichia coli ClpB. *Proteins, 83*(1), 117–134.

Lin, J., & Lucius, A. L. (2015). Examination of the dynamic equilibrium for *E. coli* ClpB. Submitted to *Proteins*.

Lohman, T. M., & Mascotti, D. P. (1992). Nonspecific ligand-DNA equilibrium binding parameters determined by fluorescence methods. *Methods in Enzymology, 212*, 424–458.

Meyer, H., & Weihl, C. C. (2014). The VCP/p97 system at a glance: Connecting cellular function to disease pathogenesis. *Journal of Cell Science, 127*(Pt. 18), 3877–3883.

Mogk, A., Tomoyasu, T., Goloubinoff, P., Rudiger, S., Roder, D., Langen, H., et al. (1999). Identification of thermolabile Escherichia coli proteins: Prevention and reversion of aggregation by DnaK and ClpB. *The EMBO Journal, 18*(24), 6934–6949.

Na, G. C., & Timasheff, S. N. (1985a). Measurement and analysis of ligand-binding isotherms linked to protein self-associations. *Methods in Enzymology, 117*, 496–519.

Na, G. C., & Timasheff, S. N. (1985b). Velocity sedimentation study of ligand-induced protein self-association. *Methods in Enzymology, 117*, 459–495.

Neuwald, A. F., Aravind, L., Spouge, J. L., & Koonin, E. V. (1999). AAA+: A class of chaperone-like ATPases associated with the assembly, operation, and disassembly of protein complexes. *Genome Research, 9*(1), 27–43.

Ogura, T., & Wilkinson, A. J. (2001). AAA+ superfamily ATPases: Common structure–diverse function. *Genes to Cells, 6*(7), 575–597.

Ortega, A., Amoros, D., & Garcia de la Torre, J. (2011). Prediction of hydrodynamic and other solution properties of rigid proteins from atomic- and residue-level models. *Biophysical Journal, 101*(4), 892–898.

Pace, C. N., Vajdos, F., Fee, L., Grimsley, G., & Gray, T. (1995). How to measure and predict the molar absorption coefficient of a protein. *Protein Science, 4*(11), 2411–2423.

Roll-Mecak, A., & McNally, F. J. (2010). Microtubule-severing enzymes. *Current Opinion in Cell Biology, 22*(1), 96–103.

Sauer, R. T., & Baker, T. A. (2011). AAA+ proteases: ATP-fueled machines of protein destruction. *Annual Review of Biochemistry, 80*, 587–612.

Schuck, P. (1998). Sedimentation analysis of noninteracting and self-associating solutes using numerical solutions to the Lamm equation. *Biophysical Journal, 75*(3), 1503–1512.

Schuck, P. (2003). On the analysis of protein self-association by sedimentation velocity analytical ultracentrifugation. *Analytical Biochemistry, 320*(1), 104–124.

Scott, D. J., Harding, S. E., Rowe, A. J., & Royal Society of Chemistry (Great Britain). (2005). *Analytical ultracentrifugation: Techniques and methods*. Cambridge, UK: RSC Publications.

Stafford, W. F., & Sherwood, P. J. (2004). Analysis of heterologous interacting systems by sedimentation velocity: Curve fitting algorithms for estimation of sedimentation coefficients, equilibrium and kinetic constants. *Biophysical Chemistry, 108*(1–3), 231–243.

Streaker, E. D., Gupta, A., & Beckett, D. (2002). The biotin repressor: Thermodynamic coupling of corepressor binding, protein assembly, and sequence-specific DNA binding. *Biochemistry, 41*(48), 14263–14271.

Veronese, P. K., & Lucius, A. L. (2010). Effect of temperature on the self-assembly of the Escherichia coli ClpA molecular chaperone. *Biochemistry, 49*(45), 9820–9829.

Veronese, P. K., Rajendar, B., & Lucius, A. L. (2011). Activity of Escherichia coli ClpA bound by nucleoside di- and triphosphates. *Journal of Molecular Biology, 409*(3), 333–347.

Veronese, P. K., Stafford, R. P., & Lucius, A. L. (2009). The Escherichia coli ClpA molecular chaperone self-assembles into tetramers. *Biochemistry, 48*(39), 9221–9233.

Wong, I., & Lohman, T. M. (1995). Linkage of protein assembly to protein-DNA binding. *Methods in enzymology, 259*, 95–127.

Wyman, J., & Gill, S. J. (1990). *Binding and linkage: Functional chemistry of biological macromolecules*. Mill Valley, CA: University Science Books.

Yu, R. C., Jahn, R., & Brunger, A. T. (1999). NSF N-terminal domain crystal structure: Models of NSF function. *Molecular Cell, 4*(1), 97–107.

CHAPTER EIGHT

Elucidating Complicated Assembling Systems in Biology Using Size-and-Shape Analysis of Sedimentation Velocity Data

Catherine T. Chaton*, Andrew B. Herr[†,‡,1]

*Graduate Program in Molecular Genetics, Biochemistry & Microbiology, University of Cincinnati College of Medicine, Cincinnati, Ohio, USA

[†]Division of Immunobiology and Center for Systems Immunology, Cincinnati Children's Hospital Medical Center, Cincinnati, Ohio, USA

[‡]Division of Infectious Diseases, Cincinnati Children's Hospital Medical Center, Cincinnati, Ohio, USA

[1]Corresponding author: e-mail address: andrew.herr@cchmc.org

Contents

Abstract

Sedimentation velocity analytical ultracentrifugation (SV-AUC) has seen a resurgence in popularity as a technique for characterizing macromolecules and complexes in solution. SV-AUC is a particularly powerful tool for studying protein conformation, complex stoichiometry, and interacting systems in general. Deconvoluting velocity data to determine a sedimentation coefficient distribution *c*(*s*) allows for the study of either individual proteins or multicomponent mixtures. The standard *c*(*s*) approach estimates molar masses of the sedimenting species based on determination of the frictional ratio (f/f_0) from boundary shapes. The frictional ratio in this case is a weight-averaged parameter, which can lead to distortion of mass estimates and loss of information when attempting to analyze mixtures of macromolecules with different shapes. A two-dimensional extension of the *c*(*s*) analysis approach provides size-and-shape distributions that describe the data in terms of a sedimentation coefficient and frictional ratio grid. This allows for better resolution of species with very distinct shapes that may co-sediment and provides better molar mass determinations for multicomponent

Methods in Enzymology, Volume 562
ISSN 0076-6879
http://dx.doi.org/10.1016/bs.mie.2015.04.004

mixtures. An example case is illustrated using globular and nonglobular proteins of different masses with nearly identical sedimentation coefficients that could only be resolved using the size-and-shape distribution. Other applications of this analytical approach to complex biological systems are presented, focusing on proteins involved in the innate immune response to cytosolic microbial DNA.

1. TECHNIQUES FOR STUDYING MACROMOLECULAR MIXTURES

Characterizing multicomponent macromolecular systems in solution is important in many areas of biology and biotechnology. Many biological systems involve complicated macromolecular assembly or aggregation phenomena; two examples from the field of innate immunity will be presented in this chapter. In biotechnology, quality control studies on biopharmaceutical products often need to resolve the bioactive species from potential contaminants and aggregated species (Arthur, Kendrick, & Gabrielson, 2015). However, distinguishing the different components in a multicomponent mixture, particularly those of widely varying shapes and sizes, can present significant challenges. Some techniques such as mass spectrometry that might be used are not solution-based approaches and require assumptions when extrapolating back to the aqueous sample. Other techniques such as native PAGE restrict analysis to macromolecular species that are small enough to enter the gel and require specialized running buffers.

Several solution-based techniques offer some advantages and the flexibility to analyze the macromolecule(s) of interest in defined buffers, which is often important both for biological function and for testing of biopharmaceuticals in their approved formulation state. Dynamic light scattering (DLS) is one such technique for studying macromolecular molecules in solution. By measuring time-dependent fluctuations in light scattering intensity, it is possible to determine a diffusion coefficient and hydrodynamic radius (Cantor & Schimmel, 1980). For globular proteins, DLS can provide an estimate of the molecular weight, although without a direct determination of the diffusion or frictional coefficient, mass determinations are inaccurate for nonglobular macromolecules. The DLS method has high sensitivity and requires very little sample volume. However, since the raw intensity of scattering is proportional to the square of the protein's molecular weight (i.e., for Rayleigh scattering, in which the particle size is small compared to the wavelength), large complexes and aggregates are overweighted and contribute disproportionately to the overall signal. This increases the uncertainty when measuring relative concentrations of different species within a population if large complexes or aggregates are present. Another popular technique for analyzing macromolecular mixtures, size-exclusion (gel filtration) chromatography (SEC), can distinguish multiple species in solution. In general, SEC suffers from the same limitation as DLS when estimating the molecular weight of nonglobular proteins. Combining data from SEC with experimentally determined diffusion coefficients can provide accurate molecular weights for individual macromolecules (Ackers, 1967a,1967b), but this is not technically feasible for mixtures of macromolecular species with distinct diffusion and frictional coefficients.

Sedimentation velocity (SV) analytical ultracentrifugation (AUC) is a powerful technique for the study of macromolecules and nanoparticles in solution. This approach involves spinning a macromolecular sample in an ultracentrifuge and tracking the sedimentation rate of the macromolecule(s) through solution in real time, providing information on the size and shape of the sedimenting species. The macromolecular migration is independent of matrix interactions (unlike native PAGE or SEC) and it takes place at constant temperature under user-defined solution conditions. Unlike DLS, the signal observed in AUC is proportional to sample concentration, allowing accurate determination of the relative proportions of different species. Over the years, many different methods have been developed to analyze SV data, including van Holde–Weischet analysis (van Holde & Weischet, 1978); direct fitting to forms of the Lamm equation (Behlke & Ristau, 1997; Demeler & Saber, 1998; Philo, 1997); time-derivative approaches (Philo, 2000, 2006; Schuck & Rossmanith, 2000; Stafford, 1992; Stafford & Sherwood, 2004); and more recently, the *c*(*s*) analysis approach that yields a sedimentation coefficient distribution (Schuck, 2000; Schuck, MacPhee, & Howlett, 1998). There are several good review articles that describe and compare these different analysis methods (Cole & Hansen, 1999; Correia, 1998; Howlett, Minton, & Rivas, 2006; Stafford, 2009; Zhao, Brautigam, Ghirlando, & Schuck, 2013). The *c*(*s*) approach determines a sedimentation coefficient distribution for the sedimenting species and provides both hydrodynamic parameters and—particularly for single-component systems—quite accurate molecular weight determinations. However, there are some particular challenges when analyzing multicomponent mixtures or complicated assembling systems. The Sedfit program developed by Schuck and colleagues now includes a two-dimensional extension of the *c*(*s*) approach, allowing the study of complex mixtures of macromolecules by determining a size-and-shape distribution (Brown & Schuck, 2006). Other groups have also introduced similar two-dimensional approaches for analyzing velocity data (Brookes, Cao, & Demeler, 2010; Carney et al., 2011). In this chapter, we will briefly introduce the sedimentation coefficient distribution $c(s)$ and the size-and-shape distribution $c(s,f/f_0)$ and provide examples of how the new analytical approach can provide insight into multicomponent mixtures or complex assembling systems in biology.

2. AN OVERVIEW OF THE STANDARD SEDIMENTATION COEFFICIENT DISTRIBUTION APPROACH

A typical SV experiment measures the concentration gradients formed by a sedimenting macromolecule driven by centrifugal force as it moves through a column of aqueous solvent. As the protein sediments, the boundary between protein and solvent moves from the meniscus (i.e., the inner boundary of the sample column) toward the outer wall of the sample cell. The rate of sedimentation and shape of the boundary give information about the macromolecule's size and shape via hydrodynamic properties such as the sedimentation, frictional, and diffusion coefficients (Cantor & Schimmel, 1980; Laue & Stafford, 1999). These hydrodynamic properties are related by the Svedberg equation, shown in several equivalent forms:

$$\frac{s}{D} = \frac{\left(\frac{M_b}{f}\right)}{\left(\frac{RT}{f}\right)} = \frac{M_b}{RT} = \frac{M(1-\bar{v}\rho)}{RT}$$

where s is the sedimentation coefficient, D is the diffusion coefficient, f is the frictional coefficient, M is the molar mass, M_b is the buoyant mass, $\bar{v}$ is the partial specific volume of the macromolecule, ρ is the solvent density, R is the gas constant, and T is the absolute temperature. One can see that the sedimentation coefficient is proportional to the buoyant mass but inversely proportional to the frictional coefficient, and that knowledge of both s and f or s and D allows determination of the molar mass of the sedimenting macromolecule, after accounting for the partial specific volume of the macromolecule and the buffer density.

Sedimentation of a particle in a sector-shaped column within a centrifugal field is described by the Lamm equation (Lamm, 1929). Written as a partial differential, it is:

$$\frac{dc}{dt} = \frac{1}{r}\frac{d}{dr}\left[rD\frac{dc}{dr} - s\omega^2 r^2 c\right]$$

The concentration distribution $c(r,t)$ of the macromolecule as a function of radial position r and time t depends on the diffusion coefficient D, the strength of the centrifugal field $\omega^2 r$, and the sedimentation coefficient of the species. Measured in Svedberg units (1 S $=10^{-13}$ s), the sedimentation coefficient is a proportionality constant that relates the velocity of the sedimenting particle to the applied force resulting from the centrifugal field. The sedimentation coefficient is dependent on both the size and shape of the particle.

In an idealized case where diffusion is neglected, the Lamm equation can be solved exactly. In this theoretical case, each sedimentation profile would have a sharp boundary created between the bulk macromolecules and the solution as macromolecules are driven toward the outer wall of the sample cell. In a more realistic experiment, diffusion acts to broaden the sedimentation boundary over time. This diffusional spread in the boundary is highly dependent on the frictional coefficient of the macromolecule; in other words, nonglobular particles experience greater drag, which retards their rate of diffusion (and their rate of sedimentation, albeit to a lesser extent). Experimental factors such as radial dilution, nonuniformity in cell loading concentration, the formation of density gradients, and the accumulation of sedimented protein at the outside of the sample cell are several reasons that exact analytical solutions of the Lamm equation are not practical. Approximate analytical solutions do exist to but are sensitive to experimental conditions (Behlke & Ristau, 1997; Philo, 1997). Another potential approach is to use numerical simulation to reproduce the natural progression of the sedimentation boundary (Demeler & Saber, 1998; Schuck et al., 1998; Stafford & Sherwood, 2004). Schuck and colleagues have developed the $c(s)$ analysis approach that directly models the boundary as a distribution of noninteracting particles and describes diffusion of the ensemble of sedimentation species (Dam & Schuck, 2004; Schuck, 2000; Schuck et al., 1998). They have shown that despite being based on equations describing the sedimentation behavior of noninteracting particles, the $c(s)$

analysis can also be applied to systems of interacting macromolecules that produce asymmetric boundaries according to Gilbert–Jenkins theory (Dam, Velikovsky, Mariuzza, Urbanke, & Schuck, 2005; Gilbert & Jenkins, 1959; Schuck, 2010).

Besides the sedimentation coefficient, which is the experimental parameter directly determined in an SV experiment, the major hydrodynamic property fitted for during $c(s)$ analysis is the frictional ratio f/f_0. It is a measure of the frictional coefficient of the macromolecule compared to that of a compact, anhydrous, ideal sphere. Thus, the lower bound of this ratio is 1.0, corresponding to an ideal sphere; the frictional ratio increases for sedimenting macromolecules that are increasingly extended or nonglobular. Because the diffusion rate is highly sensitive to the frictional coefficient, the shape of sedimentation boundaries can be directly fitted to determine the weight-averaged frictional ratio in the $c(s)$ analysis. Thus, for a single-component sample, it is possible to experimentally determine the values of the sedimentation coefficient s and the frictional coefficient f, allowing determination of the mass of the sedimenting species according to the Svedberg equation. Since the frictional ratio determined in this way is a weight-averaged value, accurate mass determination is generally only possible for a single-component system unless additional species happen to have the same frictional ratio. As differences in the frictional ratio between components increase, the distortion in mass estimates will also increase. Despite this fact, the standard $c(s)$ approach is an excellent method for characterizing the number of species present and their relative concentrations in discrete macromolecular mixtures, even if accurate mass determinations are not feasible in all situations.

We will present an experimental situation that illustrates the limits of the standard $c(s)$ approach to distinguish heterogeneous species with similar sedimentation rates. We conducted SV experiments on two noninteracting proteins with very different shapes, each of which sediments near 2.0 S (Fig. 1A–C). For a nonglobular, highly elongated protein, we chose a construct from the accumulation-associated protein (Aap), a cell-surface protein from *Staphylococcus epidermidis*. We have previously demonstrated that Aap is necessary for biofilm formation, as it mediates intercellular adhesion through zinc-induced self-assembly (Conrady et al., 2008). The C-terminal portion of Aap contains 5–17 tandem copies of the "B-repeat," a 128-amino acid sequence comprising two subdomains. The crystal structure of a minimal B-repeat construct, called Brpt1.5, revealed a highly elongated fold without a canonical hydrophobic core, with overall dimensions of 180 Å long by

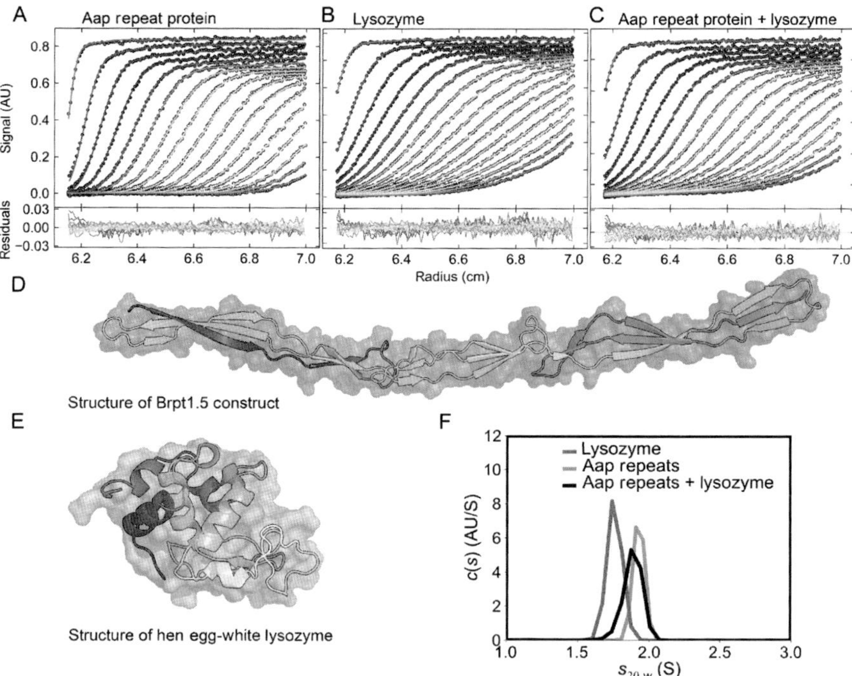

Figure 1 Standard $c(s)$ analysis of test proteins. SV-AUC raw data for (A) the Aap repeat protein, (B) lysozyme, and (C) the ~1:9 molar mixture of Aap:lysozyme. (D) Crystal structures of a Brpt1.5 Aap construct (Conrady, Wilson, & Herr, 2013) (PDB 4FUN) and (E) hen egg-white lysozyme (Vaney, Maignan, Riès-Kautt, & Ducriux, 1996) (PDB 193L). (F) The calculated $c(s)$ distributions for lysozyme, the highly elongated Aap protein, and the mixture. Despite a small shift in $s_{20,w}$, multiple species are not observed in the mixture.

20 Å wide (Fig. 1D; Conrady et al., 2013). Hen egg-white lysozyme (Fig. 1E) was chosen as an example of a globular protein (Vaney et al., 1996). In order to compare the sedimentation behavior of lysozyme and the Aap construct, each protein was sedimented individually, followed by a mixture of the two. The results will illustrate the challenges in resolving two proteins with different shapes but similar sedimentation coefficients.

When sedimented individually, the $c(s)$ distributions of lysozyme and the Aap construct are distinct but overlap (Fig. 1F). The Aap construct sedimented with an $s_{20,w}$ (i.e., sedimentation coefficient corrected to standard conditions of 20 °C in water) of 1.94 S, a frictional ratio of 2.41, and a resulting apparent molecular weight of 47.1 kDa (Fig. 1A). The frictional ratio of the Aap construct is high, indicating a very elongated or nonglobular protein, which is consistent with the crystal structure of the Brpt1.5 Aap construct (Fig. 1D). Lysozyme sedimented with an $s_{20,w}$ of 1.77 S, a much lower frictional ratio of 1.21, and a lower apparent molecular weight of 14.5 kDa (Fig. 1B). This frictional ratio is in the range typically observed for globular proteins (1.2–1.4). Since the buoyant mass is dependent on both s and f ($s = M_b/f$), it is clear that accurate determination of f/f_0 is essential to calculate the mass, since both proteins have nearly identical sedimentation coefficients.

The Aap construct and lysozyme were then mixed together in a 1:9 molar ratio, which corresponded to a slightly higher absorbance of lysozyme relative to the Aap construct at 235 nm (Fig. 1C). Analysis of the mixture yielded a single peak in the $c(s)$ distribution at $s_{20,w} = 1.85$ S, with a weight-averaged frictional ratio of 1.62. Despite the acceptable residuals on all three experiments and RMSDs on all fits of less than 0.005, there is little evidence in the $c(s)$ distribution that would indicate the presence of two distinct species, despite the large difference in mass between the two constructs (Fig. 1F). In order to resolve these two species with very different shapes, a new analytical approach is required: determining the size-and-shape distribution $c(s,f/f_0)$ from the experimental velocity data.

3. TWO-DIMENSIONAL SIZE-AND-SHAPE ANALYSIS

As mentioned earlier, the standard $c(s)$ analysis approach uses the assumption that all sedimenting species can be described by a single frictional ratio, which is determined experimentally from the shape of the boundary. In short, nonglobular or elongated particles with higher frictional coefficients experience greater drag, which reduces the rate of diffusion and therefore reduces boundary spreading. The use of a weight-averaged frictional ratio to estimate molar mass assumes that M scales with s according to a 2/3 power relationship, based on the Svedberg equation and the Stokes–Einstein relationship (Dam & Schuck, 2004). This assumption of increasing M with increasing s is appropriate when dealing with a single species, or with several species having similar frictional ratios. However, the assumption of a single frictional ratio obviously does not hold in the case of multiple heterogeneous species. Cases such as these can instead be analyzed using a relatively new approach that calculates a two-dimensional size-and-shape distribution describing the sedimentation behavior of a sample in terms of both sedimentation coefficient and frictional ratio distributions (Brown & Schuck, 2006).

This $c(s,f/f_0)$ distribution is calculated on a grid of s and f/f_0 values, determining the best-fit value of f/f_0 for each s-value across the grid. The data are fitted to a modified form of the Lamm equation using finite element solutions and combined with Tikhonov–Phillips or maximum entropy regularization to produce the most parsimonious $c(s,f/f_0)$ distribution that is consistent with the input data (Brown & Schuck, 2006). As with the $c(s)$ analysis, both time-invariant and radial-invariant noise are accounted for, improving the signal-to-noise ratio in the data. In some cases, f/f_0 may not be the hydrodynamic parameter of primary interest. This same computational approach allows for transformation into other two-dimensional distributions such as mass-Stokes radius distributions $c(M,R_s)$ or sedimentation coefficient-molar mass distributions $c(s,M)$. Furthermore, integration along the f/f_0 dimension can return the results to a one-dimensional $c(s)$-equivalent distribution without the limitation of using a single weight-averaged frictional ratio. Referred to as the general $c(s,*)$ distribution, this transformation can serve as the final step in an analysis if diffusion information is not required or if the frictional ratio dimension is not well determined.

The ability to resolve distinct frictional ratios for multiple species and therefore to more accurately assign molecular masses of multiple components adds a significant amount of valuable information. Importantly, calculation of these size-and-shape distributions is only marginally more computationally intensive than the standard $c(s)$ analysis, taking only minutes on a standard laptop. Experiments involving heterogeneous mixtures such as proteins of very different shapes, combinations of folded and intrinsically disordered proteins, or multiple assembling species in equilibrium are good candidates for the size-and-shape distribution approach. Returning to our test case data, we reanalyzed the same scans of the Aap protein and lysozyme mixture to determine the size-and-shape distribution $c(s,f/f_0)$. The $c(s,*)$ distribution (Fig. 2A, black line) shows two peaks. Superimposed on this $c(s,*)$ distribution is a heat map illustrating a bird's-eye view of the resolved peaks in the size-and-shape distribution, plotted in terms of $s_{20,w}$ (the corrected sedimentation coefficient) along the x-axis and f/f_0 along the y-axis. The same size-and-shape distribution is illustrated as a 3D plot in Fig. 2B, showing the clear separation of the two peaks along the f/f_0 axis, even though their sedimentation coefficient distributions almost overlap. Comparing the final $c(s,*)$ distribution of the mixture to the standard $c(s)$ distribution reveals additional information about the system and can resolve the second peak corresponding to the Aap construct (Fig. 2C).

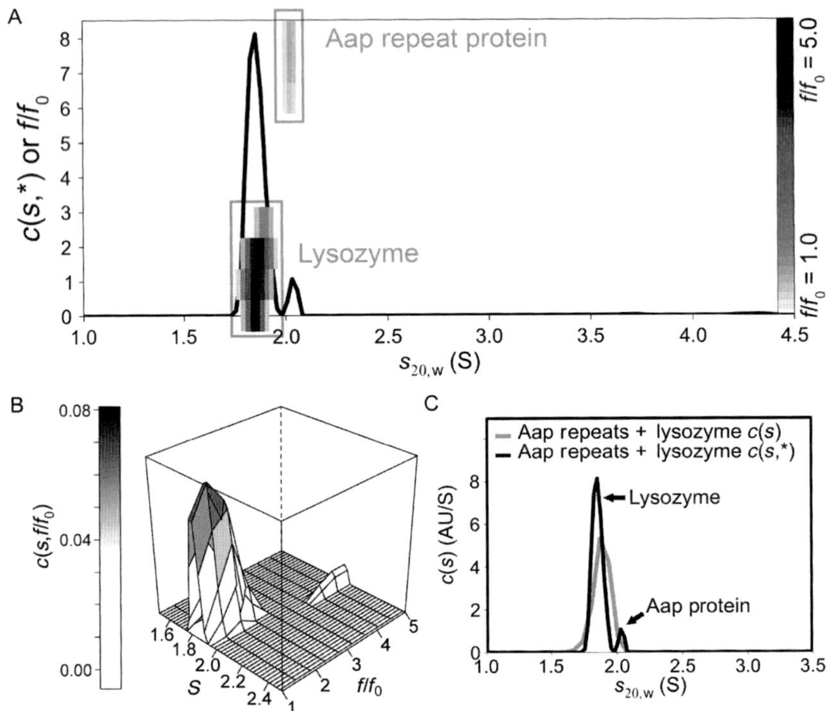

Figure 2 Size-and-shape $c(s, f/f_0)$ analysis of test proteins. (A) The $c(s, f/f_0)$ distribution for the mixture of Aap and lysozyme is able to clearly resolve both species. Lysozyme appears at an $s_{20,w}$ of 1.84 and accounts for 90% of the loading signal. The highly elongated Aap protein sediments with an $s_{20,w}$ of 1.98 and accounts for 9% of the loading signal. (B) 3D surface representation of the $c(s, f/f_0)$ distribution visualized with the lattice package in R. (C) A comparison showing the original $c(s)$ distribution calculated for the mixture of Aap and lysozyme overlaid on the $c(s,*)$ distribution. The presence of the minor species, the elongated Aap protein, is only visible after decoupling the scaling relationship between s and M implicit in the standard $c(s)$ analysis. (See the color plate.)

4. CASE STUDY: CYCLIC GMP-AMP SYNTHASE OLIGOMERIZATION

We present two case studies of proteins involved in cytosolic innate immune responses that undergo multiple assembly events upon activation. In both cases, use of the $c(s, f/f_0)$ analysis approach was essential to understand the system under study. The cGAS-STING pathway connects a cytosolic microbial DNA sensor to the production of type I interferons, forming a critical link in innate antimicrobial immunity (Shu, Li, & Li, 2014). The cyclic GMP-AMP synthase (cGAS), a 60-kDa dsDNA sensor, synthesizes cyclic 2′,5′ cGAMP from ATP and GTP upon activation by microbial dsDNA. cGAMP acts as a secondary messenger, facilitating the dimerization of STING (Ishikawa & Barber, 2008). In addition, STING can be activated directly by the cyclic dinucleotides c-di-GMP or c-di-AMP, both of which are conserved bacterial signaling molecules (Burdette et al., 2011). Upon activation by these second messengers, STING recruits tank-binding kinase (TBK)-1, an ER membrane serine/threonine protein kinase. TBK-1 can phosphorylate interferon regulatory factor (IRF)-3, leading to its assembly and translocation into the nucleus, where it stimulates transcription of interferon-β (Fig. 3).

Several crystal structures of cGAS bound to dsDNA were published in 2013; the initial reports described an apparent 1:1 complex of cGAS:dsDNA (Civril et al., 2013; Gao et al., 2013; Li et al., 2013; Kato et al., 2013; Kranzusch, Lee, Berger, & Doudna, 2013; Zhang et al., 2014). The

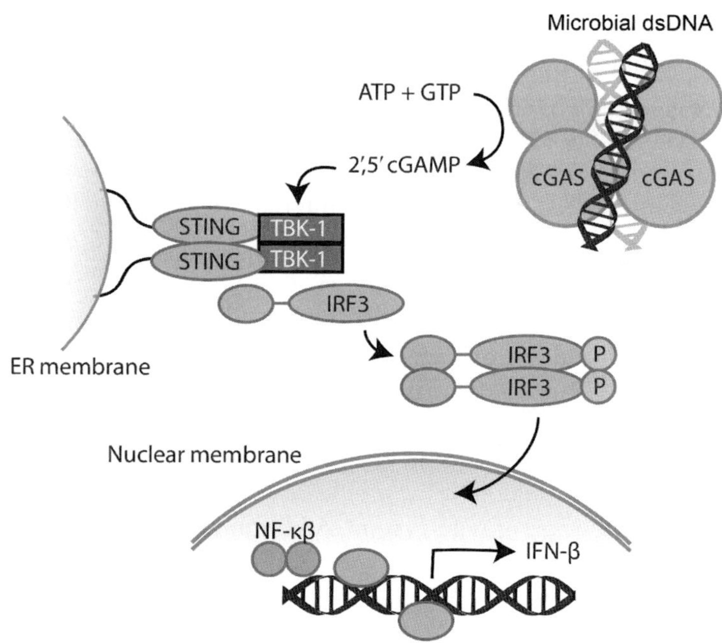

Figure 3 Schematic of innate immune signaling pathways in the cytosol. Microbial dsDNA is recognized by the sensor protein cGAS, which upon activation catalyzes the synthesis of the secondary messenger 2′,5′ cGAMP. cGAMP binds to STING, activating recruitment of the kinase TBK-1. TBK-1 then phosphorylates IRF-3, which translocates to the nucleus and triggers expression of interferon-β. *This figure was adapted from Shu et al. (2014).*

monomer's nucleotidyltransferase (NTase) fold (N-lobe) and five-helix bundle (C-lobe) were found to bind dsDNA down the length of the N-lobe. It was hypothesized that this allowed for a change in conformation of the active site buried in the cleft between both lobes. Our collaborators solved the cGAS:dsDNA structure as well and noticed the formation of a putative 2:2 dimer in the crystal (Li et al., 2013). The N-lobes of each protomer face one another, creating a ~1330 Å patch of buried surface area which is mediated by 15 hydrogen bonds. Each protomer contains two distinct dsDNA-binding sites, both the originally described dsDNA-binding site (site A) next to the dimer interface and a secondary site B on the opposite side of the dimer interface. Site B contributes 34–40% of the total buried surface area between cGAS and complexed dsDNA. Comparison of the free and bound cGAS structures revealed that the B-factors of the N-lobe dropped significantly when dsDNA was bound, suggesting that reducing flexibility of the NTase fold is necessary for formation of a functional active site.

In order to verify that the putative 2:2 cGAS:dsDNA dimer observed in the crystal could form in solution, we analyzed the complex by SV-AUC. cGAS alone sedimented as a monomer, but in the presence of dsDNA cGAS formed several higher-order oligomeric species, as revealed by $c(s)$ analysis (Fig. 4A). The cGAS dimer observed in the crystal revealed an important H-bonding interaction across the dimer interface mediated by K382; mutation of this residue to alanine prevented cGAS oligomerization, although the monomeric K382A mutant still was capable of binding dsDNA (Fig. 4A). Importantly, the K382A mutant revealed a nearly complete loss of cGAS enzymatic activity, indicating that formation of the dsDNA-induced cGAS dimer is essential for catalysis. Mutation of residues at each DNA-binding site (K372E and R158E for site A and R342E and K335E for site B) also prevented cGAS oligomerization, showing the importance of DNA binding for dimer formation (Fig. 4B).

The $c(s)$ distribution of wild-type cGAS bound to dsDNA indicated the presence of multiple species, not simply a dimeric 2:2 complex (Fig. 4A). In order to delineate the species present in this sample, we determined the $c(s,f/f_0)$ distribution of cGAS in the presence of dsDNA (Fig. 4C). While ~36% of the cGAS population formed a compact 2:2 complex, many other oligomeric species were observed. A more elongated complex with apparent 6:6 stoichiometry (25% of protein signal) was prominent along with several highly elongated, putative fiber-like species. Furthermore, the concentration regime over which cGAS assembly occurred was consistent with its affinity for dsDNA. cGAS samples at ~150 μ*M* formed higher-order species, but

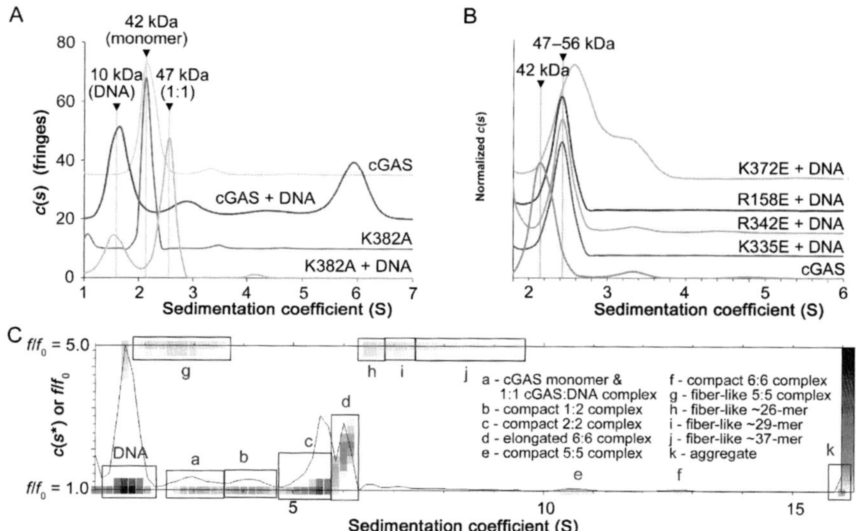

Figure 4 Sedimentation analysis of cGAS interacting with dsDNA. (A) *c*(*s*) distributions of WT and K382A mutants of mouse cGAS in the presence and absence of dsDNA. WT cGAS forms multiple oligomeric species in the presence of dsDNA. The K382A mutant, which disrupts H-bonds at the cGAS dimer interface, forms a 1:1 complex with the DNA but was unable to oligomerize. (B) Normalized *c*(*s*) distributions of cGAS mutants altered in the DNA-binding sites, including A-site mutants (K372E and R158E) and B-site mutants (R34E and K335E), each sedimented in the presence of dsDNA. (C) Size-and-shape $c(s, f/f_0)$ analysis of WT cGAS:dsDNA demonstrates that multiple oligomeric forms of cGAS occur upon dsDNA binding. Both elongated and compact species are seen at a variety of possible stoichiometries, based on mass estimates from the $c(s, f/f_0)$ distribution. *This figure was reprinted from Li et al. (2013) with permission.* (See the color plate.)

dilution of the sample to ~30 μ*M* led to dissociation of the oligomeric species. This agrees with the low affinity (K_D ~20 μ*M*) observed for cGAS binding to dsDNA (Li et al., 2013).

5. CASE STUDY: IRF-3 PHOSPHORYLATION BY TBK-1

As mentioned earlier, cGAS activated by dsDNA-mediated oligomerization catalyzes the synthesis of the second messenger cGAMP, which binds to STING and triggers recruitment of the kinase TBK-1 (Fig. 3). Our collaborators solved the crystal structure of TBK1 and analyzed its phosphorylation of IRF-3; we again used the $c(s, f/f_0)$ analysis of SV data to fully understand the effects of phosphorylation on IRF-3 assembly state (Shu et al., 2013). Once phosphorylated by TBK1 or IκB kinase-ε (IKKε), phospho-IRFs translocate to the nucleus and bind to CREB-binding

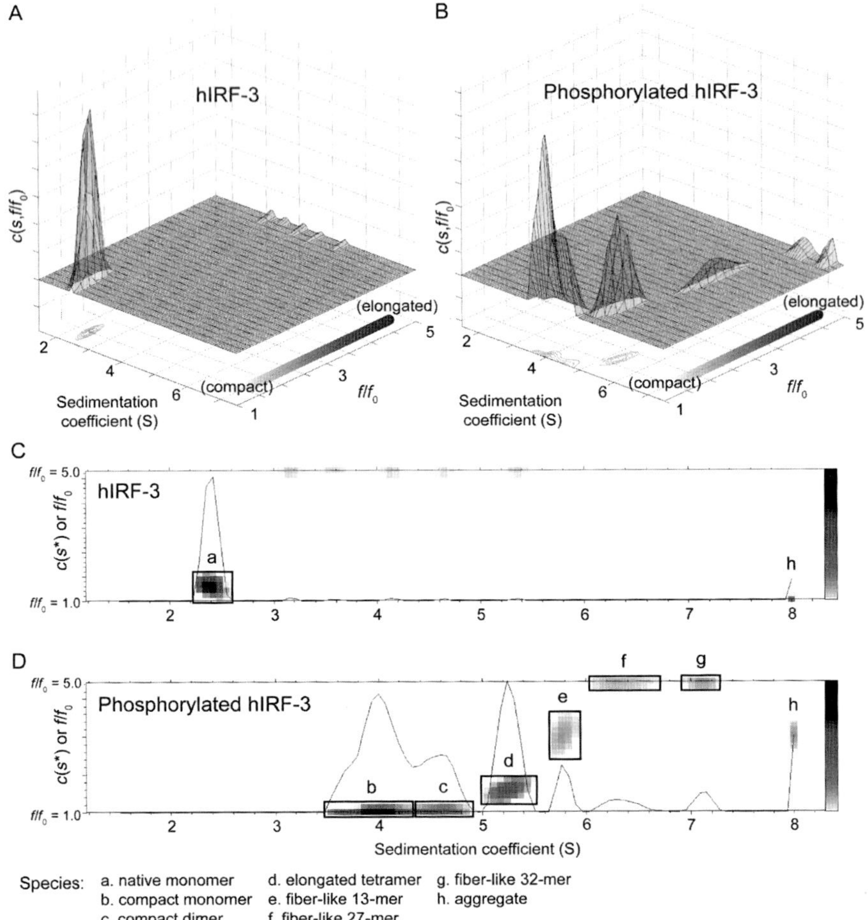

Figure 5 Size-and-shape analysis of hIRF-3 phosphorylated by TBK-1. The $c(s,f/f_0)$ distribution was determined for native IRF-3 (A) and phosphorylated IRF-3 (B). A number of distinct species appear upon phosphorylation, including compact monomer and dimer forms as well as elongated species. Two-dimensional size-and-shape distributions in (C) and (D) show major peaks and the putative oligomeric states based on predicted mass from the $c(s,f/f_0)$ analysis. *This figure was reprinted from Shu et al. (2013) with permission.* (See the color plate.)

protein (CBP) in order to initiate transcription of type I interferons (Fitzgerald et al., 2003; Lin, Heylbroeck, Pitha, & Hiscott, 1998). SEC analysis of *in vitro* phosphorylated human IRF-3 indicated that it migrated as a broad peak with a much higher apparent mass (~300 kDa) than the native monomeric form (~95 kDa). SV analyses of native hIRF-3 revealed a single monomeric species with a frictional ratio of 1.5 (Fig. 5A and C). Upon phosphorylation, the frictional ratio of the hIRF-3 monomer was lowered to near 1.0, indicating a conformational change to a compact structure (Fig. 5B and D). In addition to conformational collapse of the primary species, several discrete higher-order oligomers were observed. These included a potential tetramer with a frictional ratio above 3.0, as well as a series of highly elongated fiber-like assemblies with apparent masses between a 13-mer and 32-mer. Further biochemical experiments showed that mutation of S86 to alanine prevented phosphorylation-dependent assembly of hIRF-3, and that an oligomeric form of phosphorylated hIRF-3 specifically bound to CBP (most likely the dimeric form) (Shu et al., 2013). Similar results were found when IRF-3 was coexpressed with the kinase IKKε, with IRF-3 forming monomers, dimers, and higher-order oligomers (Takahasi et al., 2010). Though not yet fully understood, it is believed that the oligomeric forms of IRF-3 could play important roles in transcription activation due to higher avidity for DNA.

6. CONCLUSIONS

SV analysis of macromolecules in solution is a powerful technique that can provide a wealth of information about particle size, shape, and assembly state, and can elucidate complex stoichiometry or conformational changes in solution. Deconvoluting the data into a $c(s)$ continuous sedimentation coefficient distribution is useful for determining the number of species in a sample and characterizing their relative size, but mass estimates are not accurate for mixtures of particles with varying frictional ratios. In contrast, the size-and-shape $c(s,f/f_0)$ distribution allows the determination of f/f_0 for each s-value on a two-dimensional grid. This improves the mass determination for multiple species and allows the resolution of multiple species of distinct shape, even those that sediment with similar rates.

There are certain limitations to the size-and-shape distribution approach that should be kept in mind. The $c(s)$ approach in general is based on equations for noninteracting particles, and some have raised concerns about the general applicability of this analysis method to interacting systems (Stafford, 2009); however, Schuck and colleagues have shown that $c(s)$ was able to accurately model sedimentation coefficients of interacting species under conditions of rapid equilibrium in a manner consistent with Gilbert–Jenkins theory (Dam et al., 2005; Schuck, 2010). Brown and Schuck do point out that the $c(s,f/f_0)$ distribution was not as successful as $c(s)$ at recreating the asymmetric boundaries from Gilbert–Jenkins theory for interacting systems, probably due to the slightly lower resolution of the $c(s,f/f_0)$ distribution (Brown & Schuck, 2006). Furthermore, the molar mass estimates for low-abundance species in the $c(s,f/f_0)$ distribution were not as accurate, and they suggest that $c(s)$ is better suited to the analysis of trace components (Brown & Schuck, 2006). Thus, one should use caution in interpreting mass estimates from size-and-shape distributions for interacting systems or for low-abundance species. Regardless, the use of the size-and-shape distribution will be very helpful in providing a greater level of detail than is otherwise possible for complicated multicomponent solutions, particularly for species of widely varying shape. We believe that this type of analysis will find an increasing number of applications when studying biological systems that involve complicated assembly mechanisms or multicomponent mixtures of macromolecules.

ACKNOWLEDGMENTS

We thank Peter Schuck and Jim Cole for comments on the manuscript. This work was supported by funding from the National Institute of General Medical Sciences, R01 GM094363 (to A.B.H.). Figure 4 was reprinted from Li et al. (2013) with permission from Elsevier. Figure 5 was reprinted from Shu et al. (2013) with permission from Elsevier.

REFERENCES

Ackers, G. K. (1967a). Molecular sieve studies of interacting protein systems: 1. Equations for transport of associating systems. *The Journal of Biological Chemistry*, *242*, 3026–3034.

Ackers, G. K. (1967b). A new calibration procedure for gel filtration columns. *The Journal of Biological Chemistry*, *242*, 3237–3238.

Arthur, K. K., Kendrick, B. S., & Gabrielson, J. P. (2015). Guidance to achieve accurate aggregate quantitation in biopharmaceuticals by SV-AUC. *Methods in Enzymology*, *562*, 477–500.

Behlke, J., & Ristau, O. (1997). Molecular mass determination by sedimentation velocity experiments and direct fitting of the concentration profiles. *Biophysical Journal*, *72*, 428–434.

Brookes, E., Cao, W., & Demeler, B. (2010). A two-dimensional spectrum analysis for sedimentation velocity experiments of mixtures with heterogeneity in molecular weight and shape. *European Biophysics Journal*, *39*, 405–414.

Brown, P. H., & Schuck, P. (2006). Macromolecular size-and-shape distributions by sedimentation velocity analytical ultracentrifugation. *Biophysical Journal*, *90*, 4651–4661.

Burdette, D. L., Monroe, K., Sotelo-Troha, K., Iwig, J., Eckert, B., Hyodo, M., et al. (2011). STING is a direct innate immune sensor of cyclic di-GMP. *Nature*, *478*, 515–518.

Cantor, C., & Schimmel, P. (1980). *Biophysical chemistry part II: Techniques for the study of biological structure and function*. New York: W.H. Freeman and Company.

Carney, R. P., Kim, J. Y., Qian, H., Jin, R., Mehenni, H., Stellacci, F., et al. (2011). Determination of nanoparticle size distribution together with density or molecular weight by 2D analytical ultracentrifugation. *Nature Communications*, *2*, 335.

Civril, F., Deimling, T., de Oliveira Mann, C., Ablasser, A., Moldt, M., Witte, G., et al. (2013). Structural mechanism of cytosolic DNA sensing by cGAS. *Nature*, *498*, 332–337.

Cole, J., & Hansen, J. (1999). Analytical ultracentrifugation as a contemporary biomolecular research tool. *Journal of Biomolecular Techniques, 10*, 163–176.

Conrady, D. G., Brescia, C., Horii, K., Weiss, A., Hassett, D., & Herr, A. B. (2008). A zinc-dependent adhesion module is responsible for intercellular adhesion in staphylococcal biofilms. *Proceedings of the National Academy of Sciences of the United States of America, 105*, 19456–19461.

Conrady, D. G., Wilson, J., & Herr, A. B. (2013). Structural basis for Zn^{2+}-dependent intercellular adhesion in staphylococcal biofilms. *Proceedings of the National Academy of Sciences of the United States of America, 110*, 202–211.

Correia, J. J. (1998). Sedimentation velocity data analysis methods: What, when and why? *Chemtracts. Biochemistry and Molecular Biology, 11*, 944–948.

Dam, J., & Schuck, P. (2004). Calculating sedimentation coefficient distributions by direct modeling of sedimentation velocity concentration profiles. *Methods in Enzymology, 384*, 185–212.

Dam, J., Velikovsky, C. A., Mariuzza, R. A., Urbanke, C., & Schuck, P. (2005). Sedimentation velocity analysis of heterogeneous protein-protein interactions: Lamm equation modeling and sedimentation coefficient distributions c(s). *Biophysical Journal, 89*, 619–634.

Demeler, B., & Saber, H. (1998). Determination of molecular parameters by fitting sedimentation data to finite-element solutions of the Lamm equation. *Biophysical Journal, 74*(1), 444–454.

Fitzgerald, K. A., McWhirter, S. M., Faia, K. L., Rowe, D. C., Latz, E., Golenbock, D. T., et al. (2003). IKKepsilon and TBK1 are essential components of the IRF3 signaling pathway. *Nature Immunology, 4*, 491–496.

Gao, P., Ascano, M., Zillinger, T., Wang, W., Dai, P., Serganov, A. A., et al. (2013). Structure-function analysis of STING activation by c[G(2',5')pA(3',5')p] and targeting by antiviral DMXAA. *Cell, 154*, 748–762.

Gilbert, G. A., & Jenkins, R. C. (1959). Sedimentation and electrophoresis of interacting substances. II. Asymptotic boundary shape for two substances interacting reversibly. *Proceedings of the Royal Society A, 253*, 420–437.

Howlett, G. J., Minton, A. P., & Rivas, G. (2006). Analytical ultracentrifugation for the study of protein association and assembly. *Current Opinion in Chemical Biology, 10*, 430–436.

Ishikawa, H., & Barber, G. N. (2008). STING is an endoplasmic reticulum adaptor that facilitates innate immune signalling. *Nature, 455*, 674–678.

Kato, K., Ishii, R., Goto, E., Ishitani, R., Tokunaga, F., & Nureki, O. (2013). Structural and functional analyses of DNA-sensing and immune activation by human cGAS. *PLoS One, 8*, e76983.

Kranzusch, P. J., Lee, A. S., Berger, J. M., & Doudna, J. A. (2013). Structure of human cGAS reveals a conserved family of second-messenger enzymes in innate immunity. *Cell Reports, 3*, 1362–1368.

Lamm, O. (1929). Die differentialgleichung der ultrazentrifugierung. *Arkiv för Matematik, Astronomi, och Fysik, 21B*(2), 1–4.

Laue, T. M., & Stafford, W. F. (1999). Modern applications of analytical ultracentrifugation. *Annual Review of Biophysics and Biomolecular Structure, 28*, 75–100.

Li, X., Shu, C., Yi, G., Chaton, C. T., Shelton, C. L., Diao, J., et al. (2013). Cyclic GMP-AMP synthase is activated by double-stranded DNA-induced oligomerization. *Immunity, 39*, 1019–1031.

Lin, R., Heylbroeck, C., Pitha, P. M., & Hiscott, J. (1998). Virus-dependent phosphorylation of the IRF-3 transcription factor regulates nuclear translocation, transactivation potential, and proteasome-mediated degradation. *Molecular and Cellular Biology, 18*, 2986–2996.

Philo, J. S. (1997). An improved function for fitting sedimentation velocity data for low-molecular-weight solutes. *Biophysical Journal, 72*, 435–444.

Philo, J. S. (2000). A method for directly fitting the time derivative of sedimentation velocity data and an alternative algorithm for calculating sedimentation coefficient distribution functions. *Analytical Biochemistry, 279*, 151–163.

Philo, J. S. (2006). Improved methods for fitting sedimentation coefficient distributions derived by time-derivative techniques. *Analytical Biochemistry, 354*, 238–246.

Schuck, P. (2000). Size distribution analysis of macromolecules by sedimentation velocity ultracentrifugation and Lamm equation modeling. *Biophysical Journal, 78*, 1606–1619.

Schuck, P. (2010). Sedimentation patterns of rapidly reversible protein interactions. *Biophysical Journal, 98*, 2005–2013.

Schuck, P., MacPhee, C. E., & Howlett, G. J. (1998). Determination of sedimentation coefficients for small peptides. *Biophysical Journal, 74*, 466–474.

Schuck, P., & Rossmanith, P. (2000). Determination of the sedimentation coefficient distribution by least-squares boundary modeling. *Biopolymers, 54*, 328–341.

Shu, C., Li, X., & Li, P. (2014). The mechanism of double-stranded DNA sensing through the cGAS-STING pathway. *Cytokine & Growth Factor Reviews, 25*, 641–648.

Shu, C., Sankaran, B., Chaton, C. T., Herr, A. B., Mishra, A., Peng, J., et al. (2013). Structural insights into the functions of TBK1 in innate antimicrobial immunity. *Structure, 21*, 1137–1148.

Stafford, W. F. (1992). Boundary analysis in sedimentation transport experiments: A procedure for obtaining sedimentation coefficient distributions using the time derivative of the concentration profile. *Analytical Biochemistry, 203*, 295–301.

Stafford, W. F. (2009). Protein-protein and ligand-protein interactions studied by analytical ultracentrifugation. In K. W. Shriever (Ed.), *Protein structure, stability, and interactions: 490* (pp. 83–113). Totowa, NJ: Humana Press.

Stafford, W. F., & Sherwood, P. J. (2004). Analysis of heterologous interacting systems by sedimentation velocity: Curve fitting algorithms for estimation of sedimentation coefficients, equilibrium and kinetic constants. *Biophysical Chemistry, 108*, 231–243.

Takahasi, K., Horiuchi, M., Fujii, K., Nakamura, S., Noda, N. N., Yoneyama, M., et al. (2010). Ser386 phosphorylation of transcription factor IRF-3 induces dimerization and association with CBP/p300 without overall conformational change. *Genes to Cells, 15*, 901–910.

Vaney, M. C., Maignan, S., Riès-Kautt, M., & Ducriux, A. (1996). High-resolution structure (1.33 A) of a HEW lysozyme tetragonal crystal grown in the APCF apparatus. Data and structural comparison with a crystal grown under microgravity from SpaceHab-01 mission. *Acta Crystallographica. Section D, Biological Crystallography, 52*, 505–517.

van Holde, K. E., & Weischet, W. O. (1978). Boundary analysis of sedimentation velocity experiments with monodisperse and paucidisperse solutes. *Biopolymers, 17*, 1387–1403.

Zhang, X., Wu, J., Du, F., Xu, H., Sun, L., Chen, Z., et al. (2014). The cytosolic DNA sensor cGAS forms an oligomeric complex with DNA and undergoes switch-like conformational changes in the activation loop. *Cell Reports, 6*, 421–430.

Zhao, H., Brautigam, C. A., Ghirlando, R., & Schuck, P. (2013). Current methods in sedimentation velocity and sedimentation equilibrium analytical ultracentrifugation. *Current Protocols in Protein Science, 20*, Unit 20.12. http://dx.doi.org/10.1002/0471140864.ps2012s71.

CHAPTER NINE

Quaternary Structure Analyses of an Essential Oligomeric Enzyme

Tatiana P. Soares da Costa*[,1], Janni B. Christensen*[,1], Sebastien Desbois*[,1], Shane E. Gordon*[,1], Ruchi Gupta*[,1], Campbell J. Hogan*[,1], Tao G. Nelson*[,1], Matthew T. Downton[†], Chamodi K. Gardhi[‡], Belinda M. Abbott[‡], John Wagner[†], Santosh Panjikar[§,¶], Matthew A. Perugini*[,2]

*Department of Biochemistry and Genetics, La Trobe Institute for Molecular Science, La Trobe University, Melbourne, Victoria, Australia
[†]IBM Research Collaboratory for Life Sciences-Melbourne, Victorian Life Sciences Computation Initiative, Carlton, Victoria, Australia
[‡]Department of Chemistry and Physics, La Trobe Institute for Molecular Science, La Trobe University, Melbourne Victoria, Australia
[§]Australian Synchrotron, Clayton, Victoria, Australia
[¶]Department of Biochemistry and Molecular Biology, Monash University, Clayton, Victoria, Australia
[1]Designates equal first authorship
[2]Corresponding author: e-mail address: m.perugini@latrobe.edu.au

Contents

Abstract

Here, we review recent studies aimed at defining the importance of quaternary structure to a model oligomeric enzyme, dihydrodipicolinate synthase. This will illustrate the complementary and synergistic outcomes of coupling the techniques of analytical ultracentrifugation with enzyme kinetics, *in vitro* mutagenesis, macromolecular crystallography, small angle X-ray scattering, and molecular dynamics simulations, to demonstrate the role of subunit self-association in facilitating protein dynamics and enzyme function. This multitechnique approach has yielded new insights into the molecular evolution of protein quaternary structure.

Methods in Enzymology, Volume 562
ISSN 0076-6879
http://dx.doi.org/10.1016/bs.mie.2015.06.020

1. INTRODUCTION

Since the first oligomeric protein structure of hemoglobin was determined by Max Perutz and colleagues (Perutz, Muirhead, Cox, & Goaman, 1968; Perutz et al., 1960), protein biochemists have been interested in understanding the importance of quaternary structure to protein function. In some cases, oligomerization has been shown to be critical for allostery, such as in the case of hemoglobin, where the heterotetramerization of 2 α and 2 β subunits introduces a central cavity to which the allosteric regulator 2,3-bisphosphoglycerate binds (Muirhead & Perutz, 1963; Perutz, 1976). In other cases, oligomerization is simply thought to stabilize the active form of a protein, as demonstrated for the enzyme ornithine carbamoyltransferase from *Pyrococcus furiosus*, which self-associates to form a thermostable dodecamer (Villeret et al., 1998). For the model oligomeric protein of interest here, dihydrodipicolinate synthase (DHDPS) (Fig. 1), one can appreciate why the enzyme needs to be at least a dimer, given that dimerization allows the formation of the catalytic triad in the active site required for enzyme function. Dimerization of DHDPS also introduces allosteric sites to which lysine can bind and feedback inhibit the enzyme. Therefore, it is not surprising to find native dimeric forms of the enzyme, such as *Staphylococcus aureus* DHDPS (Fig. 1B). However, why the enzyme in most species forms a homotetramer is not apparent by simply examining the crystal structure of the typical bacterial enzyme (Fig. 1A). Furthermore, it is also not apparent why two different forms of the DHDPS homotetramers exist in nature, namely the "head-to-head" dimer-of-dimers canonical to bacteria

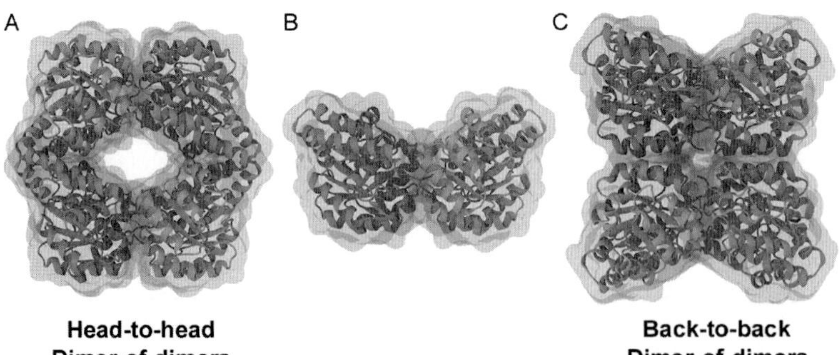

Figure 1 Quaternary structure of (A) *Bacillus anthracis* DHDPS (PDB ID: 3HIJ), (B) *S. aureus* DHDPS (PDB ID: 3DAQ), and (C) *Vitis vinifera* DHDPS (PDB ID: 3TUU). (See the color plate.)

(Fig. 1A), and the "back-to-back" dimer-of-dimers observed for plant orthologs (Fig. 1C).

In this chapter, we will explore how analytical ultracentrifugation, coupled to enzyme kinetics, *in vitro* mutagenesis, macromolecular crystallography, small angle X-ray scattering (SAXS), and molecular dynamics (MD) simulations, has provided unique insights into understanding the molecular evolution of DHDPS quaternary structure.

2. CATALYTIC FUNCTION OF DHDPS

DHDPS (E.C. 4.2.1.52) catalyzes the condensation of pyruvate and (*S*)-aspartate semialdehyde [(*S*)-ASA] to form (4*S*)-4-hydroxy-2,3,4,5-tetrahydro-(2*S*)-dipicolinic acid (HTPA) (Fig. 2A). This is the first committed step in the diaminopimelate (DAP) biosynthesis pathway in bacteria and plants (Fig. 2B).

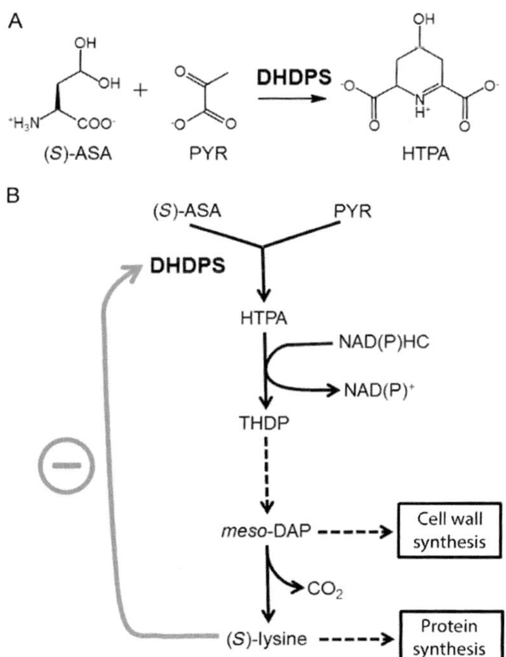

Figure 2 (A) Schematic of the DHDPS-catalyzed condensation reaction. (B) Diaminopimelate biosynthesis pathway, including feedback inhibition of DHDPS by lysine.

The DAP pathway (Fig. 2B) yields *meso*-2,6-diaminopimelate (*meso*-DAP) and (*S*)-lysine, which are important building blocks for the synthesis of housekeeping proteins, virulence factors, and the peptidoglycan cell wall (Atkinson et al., 2012; Dogovski et al., 2009; Hutton, Perugini, & Gerrard, 2007). Given that DHDPS is the product of an essential bacterial gene that is absent in humans, DHDPS represents a promising but as yet unexploited antimicrobial target (Dogovski et al., 2012, 2009, 2013; Hutton et al., 2007). The enzyme is also of interest as a model oligomeric protein, given that several different quaternary structures exist in nature despite the fact that each quaternary form supports the same catalytic function (Fig. 1). Understanding the structure–function relationship of this model, oligomeric enzyme may provide insights into the role of protein quaternary structure in general.

The DHDPS-catalyzed reaction follows a typical ping-pong mechanism, whereby pyruvate binds to the active site first forming a covalent enzyme-substrate intermediate, followed then by the release of a protonated water molecule. (*S*)-ASA then binds in the active site and condenses with the bound pyruvate intermediate to form the heterocyclic product, HTPA (Blickling, Renner, et al., 1997). Specifically, the first step of the mechanism involves Schiff base formation between the keto carbon of pyruvate and a highly conserved lysine residue (Lys161 in *Escherichia coli* DHDPS) via a tetrahedral intermediate (Laber, Gomis-Rüth, Romão, & Huber, 1992). A catalytic triad (Tyr133, Thr44, and Tyr107, *E. coli* numbering) has been proposed to transfer protons between the active site and bulk solvent through a water-filled channel (Dobson, Valegård, & Gerrard, 2004). The Schiff base (imine) is converted to an enamine form, allowing (*S*)-ASA to bind to the active site, where it undergoes an aldol-like condensation reaction, leading to cyclization, and the release of the product, HTPA (Fig. 2).

3. STRUCTURE OF DHDPS

Mirwaldt, Korndorfer, and Huber (1995) first crystallized and determined the three-dimensional structure of the DHDPS enzyme from *E. coli*. There have since been more than 75 DHDPS structures determined from approximately 25 bacterial species and 3 plant species (refer to the Protein Data Bank (PDB)). Based on crystallographic structural studies supported by analytical ultracentrifugation and SAXS experiments in solution, DHDPS exists primarily as a homotetramer (Dogovski et al., 2012, 2009; Griffin et al., 2008). However, DHDPS from the bacterial pathogens,

S. aureus (Fig. 1B) and *Pseudomonas aeruginosa*, exist as homodimers in the crystal state (Burgess et al., 2008; Girish, Sharma, & Gopal, 2008; Kaur et al., 2011). In addition, two different tetrameric architectures of the enzyme are observed in bacteria (Fig. 1A) and plants (Fig. 1C). The importance of quaternary structure will be discussed in detail below, commencing firstly with a description of the subunit structure including the key active site residues.

3.1 Subunit Structure

The *E. coli* DHDPS monomer is comprised of 292 amino acids folded to form two distinct domains (Dobson, Griffin, Jameson, & Gerrard, 2005; Mirwaldt et al., 1995). The amino-terminal domain is a $(\beta/\alpha)_8$ TIM-barrel (residues 1–224) with the active site located at the center of the barrel (Dobson et al., 2005; Mirwaldt et al., 1995). The carboxyl-terminal domain (residues 225–292) is comprised of three α-helices that contain key residues mediating tetramerization (Dobson et al., 2005; Mirwaldt et al., 1995; Fig. 3A).

3.2 Active Site

The crystal structure of *E. coli* DHDPS with pyruvate bound provided insight to the identity of the key active site residues important for catalytic function (Blickling, Renner, et al., 1997). Lys161, involved in Schiff base formation, is present within the β-barrel adjacent to the catalytic triad residues Thr44, Tyr107, and Tyr133 (Fig. 3B). Thr44 forms hydrogen bonds with Tyr133 and Tyr107, and its position in the hydrogen bonding network is critical for Schiff base formation and cyclization (Dobson et al., 2005).

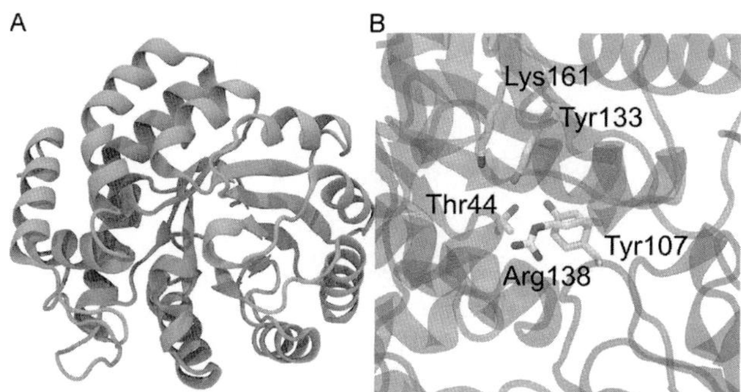

Figure 3 (A) Structure of *E. coli* DHDPS monomer (PDB ID: 1YXC) and (B) Active site of *E. coli* DHDPS, highlighting the catalytic triad residues Thr44, Tyr107, and Tyr133.

Tyr133 is crucial for substrate binding events during catalysis as it acts as a proton donor during Schiff base formation and also accepts a proton while coordinating the amino group of (*S*)-ASA (Dobson et al., 2005; Fig. 2B). Tyr107 interdigitates across the dimerization interface between adjacent subunits to complete the catalytic triad, by forming a hydrogen bond with Thr44. This residue is believed to play a key role in the stability of the enzyme along with shuttling protons to and from the active site (Dobson et al., 2005; Fig. 2B). Arg138, located at the active site entrance, plays a major part in coordinating (*S*)-ASA and stabilizing the catalytic triad through a network of hydrogen bonds. Mutational studies have highlighted the importance of these conserved active site residues in DHDPS (Dobson et al., 2004; Soares da Costa et al., 2010).

3.3 Canonical Bacterial Tetramer

The DHDPS dimer-of-dimers or homotetramer consists of a tight-dimer interface between monomers A & B and C & D, and a weak-dimer interface with hydrogen bonds and noncovalent interactions between the dimers A–B and C–D (Dobson et al., 2005; Dogovski et al., 2012; Griffin, Dobson, Gerrard, & Perugini, 2010; Griffin et al., 2008; Perugini et al., 2005; Fig. 4).

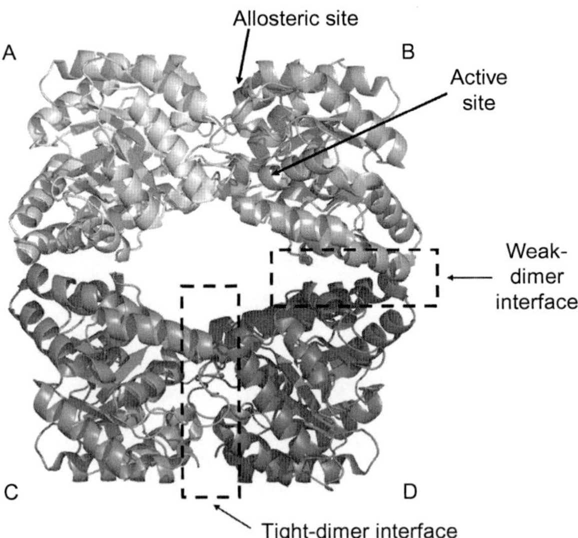

Figure 4 Structure of *E. coli* DHDPS, highlighting the location of the active site, allosteric site (which binds lysine), tight-dimer, and weak-dimer interfaces (PDB ID: 1YXC). *Adapted from Dogovski et al. (2012).*

In the tetrameric structure, each monomer contains one active site at the center of the barrel and an (*S*)-lysine allosteric binding site within the cleft between the subunits in each tight-dimer (Blickling, Renner, et al., 1997; Dogovski et al., 2009; Mirwaldt et al., 1995; Fig. 4). In *E. coli* DHDPS, there are 25 residues from each monomer at the tight-dimer interface, where a number of interactions are observed including hydrogen bonding between Ser111 and Cys141, and hydrophobic interactions between Leu51 and Ala81. Tyr107 of one monomer also coordinates with Tyr106 from the adjacent monomer, forming a strong hydrophobic stack of aromatic rings (Dogovski et al., 2012). The tight-dimers of *E. coli* DHDPS assemble via two isologous interfaces found between corresponding monomers and its formation involves nine residues from each monomer, where a number of hydrophobic contacts between Leu167, Thr168, and Leu197 have been shown to stabilize the weak-dimer interface (Dobson et al., 2004).

Interestingly, even though the tight-dimer interface is similar between bacterial and plant DHDPS, they adopt different tetrameric architectures as the residues in the weak-dimer interface are situated on the opposite end of the monomer (Blickling, Beisel, et al., 1997; Blickling & Knäblein, 1997). In bacteria, the tetramer is formed by the buttressing of the two tight-dimers in a "head-to-head" configuration, where the lysine binding sites are found at the top and bottom of the tetramer (Dobson et al., 2005; Voss et al., 2010; Figs. 1A and 4). However, the plant DHDPS tetramer forms a "back-to-back" configuration, with the lysine allosteric clefts facing each other in the interior portion of the structure (Fig. 1C; Atkinson et al., 2013, 2012; Blickling, Beisel, et al., 1997; Griffin, Billakanti, Wason, Keller, & Mertens, 2012) (also refer to Section 3.5). The weak-dimer interface is less extensive in bacterial DHDPS compared to the plant counterpart, which is reflected in a smaller number of residues making contacts between subunits, resulting in a large water-filled cavity (Dogovski et al., 2012). Despite these differences in configurations, the position and orientation of all active site residues are conserved.

Analytical ultracentrifugation studies in conjunction with mutagenesis, enzyme kinetics, X-ray crystallography, and SAXS have been carried out to investigate the importance of the tetrameric structure and shed light into the molecular evolution of DHDPS. Single point mutants of both *E. coli* and *Bacillus anthracis* DHDPS have been generated, where repulsive forces or steric bulk were introduced to the weak-dimer interface (Griffin et al., 2008; Voss et al., 2010). The role of Leu197 at the weak-dimer interface of *E. coli* DHDPS has been examined by making mutations (L197D and L197Y) that resulted in dimeric enzymes unable to form a tetramer as assessed by analytical ultracentrifugation (Fig. 5A), X-ray crystallography, and size exclusion chromatography (Griffin et al., 2008). The mutant dimeric enzymes displayed attenuated catalytic activity and increased K_M values for pyruvate, but the K_i for (*S*)-lysine was unaltered (Griffin et al., 2008). In the crystal

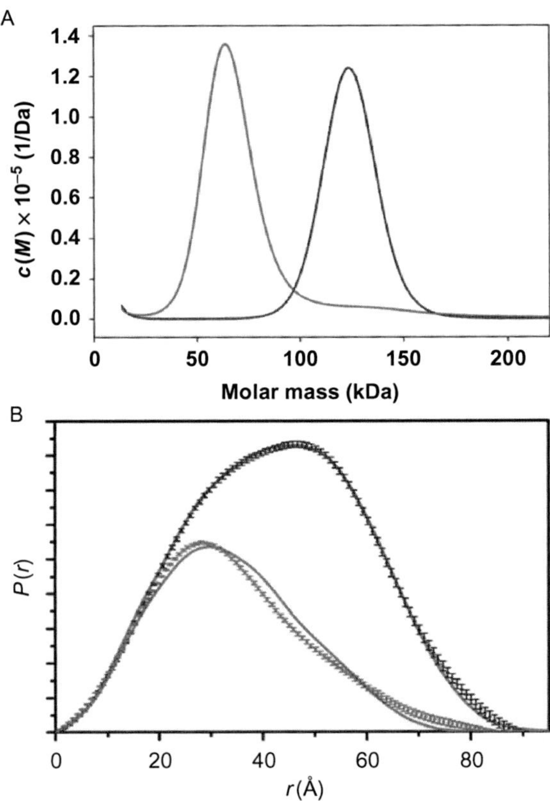

Figure 5 Structural characterization of wild-type *E. coli* DHDPS and L197Y. (A) Sedimentation velocity analyses showing the continuous mass [*c*(*M*)] distribution plotted as a function of molar mass (kDa) for wild-type *E. coli* DHDPS (blue line) and L197Y (red line). The L197Y mutant exists as a dimer (expected molecular mass of 62.5 kDa) compared to the tetrameric structure adopted by the wild-type enzyme (expected molar mass of 125 kDa). (B) Overlay of the interatomic distance distribution functions, *P*(*r*) *versus r*, derived from the scattering profiles for tetrameric wild-type *E. coli* DHDPS (blue symbols) and dimeric L197Y (red symbols) using GNOM (Svergun, 1992) alongside the profiles predicted from their corresponding crystal structures (wild-type, purple line; mutant, green line) using CRYSOL (Svergun, Barberato, & Koch, 1995). *Adapted from Griffin et al. (2008).* (See the color plate.)

structure of both mutants, the substrate analogue α-ketoglutarate, which was not added during the enzyme preparation, was covalently trapped at the highly conserved active site Lys161 residue of the dimer normally involved in Schiff base formation with pyruvate (Griffin et al., 2008). SAXS analysis indicates that L197Y exists as a population of structures in solution, indicating a rearrangement of the monomer units within the dimer. As can be seen in Fig. 5B, the interatomic distance [$P(r)$] profile is narrower at its half height and extends to longer vector lengths in comparison to those derived from the crystal structures, demonstrating that there are differences between the solution and crystal structures (Griffin et al., 2008). It has thus been proposed that the reduction in substrate specificity and the deviations from the crystal structure can be attributed to an increase in "breathing motion" associated with the movement between the subunits in the dimeric enzymes (Griffin et al., 2008).

Additional mutations involving the introduction of electrostatic repulsion (Q196D, Q234D), hydrogen bond removal (D193A, D193Y), and introduction of steric bulk (D193Y) in *E. coli* DHDPS also resulted in a destabilization of the tetramer (Griffin et al., 2010). Based on sedimentation velocity experiments, the mutants Q196D, D193A, and D193Y show two nonresolving peaks at 4.6S and 6.6S, suggesting an equilibrium mixture of dimer and tetramer (Griffin et al., 2010; Perugini et al., 2005), with the Q196D enzyme containing the greatest proportion of dimer. Sedimentation equilibrium experiments show that the data best fits to a dimer–tetramer equilibrium with $K_d^{4\rightarrow2}$ values for Q196D and D193A that are 240- and 310-fold weaker than wild-type enzyme, respectively (Griffin et al., 2010; Table 1). It appears that the position of the mutation has a greater influence on the interface than the nature of the substitution as the introduction of a tyrosine (D193Y) did not significantly weaken the interface more than the alanine mutant (D193A). This is consistent with the analytical ultracentrifugation experiments conducted with the L197 point mutants (Griffin et al., 2008). Kinetic analyses also show that the Q196D, D193A, and D193Y mutant enzymes possess significantly reduced catalytic activity compared to the wild-type tetramer, but lysine inhibition was unchanged (Griffin et al., 2010; Table 1).

Unlike the tight-dimer interface, the weak-dimer interface is not conserved amongst DHDPS enzymes, with a greater number of contacts observed in DHDPS from *B. anthracis* compared to *E. coli* DHDPS. *B. anthracis* DHDPS has been shown to exist in a reversible self-association with sedimentation coefficients of 4.0S and 6.5S at low micromolar

Table 1 Summary of the Hydrodynamic and Kinetic Properties of Wild-Type *E. coli* DHDPS and Mutant Enzymes

		Wild-type	**Q196D**	**D193A**	**D193Y**	**Q234D**
s_w(S)[a]		6.58	6.01	6.12	6.19	6.42
$K_d^{4\rightarrow 2}$(μM)[b]		0.076[c]	18.4	23.7	–	–
k_{cat}(s^{-1})		78 ± 0.06[d]	13.28 ± 0.16[d]	9.15 ± 0.16[d]	–	–
k_{cat}^{app}	Pyruvate	–	–	–	7.2 ± 0.1	11.8 ± 0.2
	(*S*)-ASA	–	–	–	6.5 ± 0.1	12.4 ± 0.4
K_m(mM)	Pyruvate	0.16 ± 0.003[d]	0.32 ± 0.01[d]	0.44 ± 0.02[d]	–	–
	(*S*)-ASA	0.13 ± 0.002[d]	0.15 ± 0.005[d]	0.17 ± 0.0007[d]	–	–
K_m^{app}(mM)	Pyruvate	–	–	–	0.57 ± 0.04	0.46 ± 0.02
	(*S*)-ASA	–	–	–	0.15 ± 0.01	0.15 ± 0.02

[a]Weight average sedimentation coefficient determined by integration of the *c*(*s*) distribution.
[b]Tetramer dissociation constant calculated from a global nonlinear least squares best-fit to a tetramer-dimer equilibrium.
[c]Data was obtained from Perugini et al. (2005).
[d]Data best fitted to the ping-pong model.
Adapted from Griffin et al. (2010).

concentrations (Voss et al., 2010), consistent with the dimer-tetramer equilibrium observed with *E. coli* DHDPS (Perugini et al., 2005). In the presence of pyruvate, the equilibrium shifts toward the tetrameric species, with sedimentation equilibrium experiments conducted at multiple concentrations and rotor speeds resulting in a $K_d^{4\rightarrow2}$ of 0.66 μ*M*, which is three times tighter than in the absence of substrate (Voss et al., 2010). To confirm the importance of the homotetrameric structure of *B. anthracis* DHDPS, a double mutant (L170E/G191E) was designed to prevent association at the weak-dimer interface by the introduction of electrostatic repulsion similar to the aforementioned studies incorporating aspartate mutations in *E. coli* DHDPS (Griffin et al., 2010, 2008). Kinetic analysis of L170E/G191E showed that the mutant possessed <2.5% of the catalytic activity of the wild-type enzyme (Voss et al., 2010). This was not surprising, since sedimentation velocity studies showed that L170E/G191E was a stable dimer in solution, unable to self-associate to form a tetramer (Voss et al., 2010). To further understand substrate-mediated stabilization by pyruvate, a crystal structure of the pyruvate-bound enzyme was determined and compared to the apo enzyme. Upon inspection of the weak dimer interface using PISA (Krissinel & Henrick, 2007) analysis, an increase in the number of inter-residue contacts was observed at the weak-dimer interface resulting in an increase of approximately 90 Å^2 in buried surface area and 12 additional hydrogen bonds in the pyruvate-bound structure (Voss et al., 2010). This provided a structural basis for the substrate-mediated stabilization phenomenon observed for the *B. anthracis* enzyme.

3.4 Dimeric Structure of *S. aureus* DHDPS

In addition to the canonical "head-to-head" dimer-of-dimers observed for most bacterial DHDPS structures to date (Figs. 1A and 4), native dimeric forms of the enzyme have been observed in the crystal structures of DHDPS from *S. aureus* (Burgess et al., 2008; Fig. 1B) and *P. aeruginosa* (Kaur et al., 2011). However, it should be noted that the *P. aeruginosa* ortholog is yet to be confirmed as a dimer in solution. Nevertheless, the discovery that *S. aureus* DHDPS exists as a dimer was first revealed using native polyacrylamide gel electrophoresis, which showed that the *S. aureus* enzyme has greater electrophoretic mobility compared to the canonical tetrameric enzyme from *E. coli* (Burgess et al., 2008). As a first step to providing a quantitative view of *S. aureus* DHDPS dimerization, absorbance-detected sedimentation velocity experiments were conducted in the analytical ultracentrifuge (Burgess et al., 2008). Examination of the shift in sample sedimentation boundary over the time course of the experiment using enhanced van Holde–Weischet analysis (Demeler & van Holde, 2004; van Holde & Weischet, 1978) showed that *S. aureus* DHDPS sedimented as a single component with a $s_{20,\mathrm{w}}$ value of 4.2S (Fig. 6A, open circles). This is a significant decrease from 6.5S for the tetrameric control of *E. coli* DHDPS (Burgess et al., 2008) (Fig. 6A, closed circles). These values agree with those reported for dimeric and tetrameric forms of DHDPS (Perugini et al., 2005), respectively, indicating that *S. aureus* DHDPS exists as a stable dimer in solution.

To further investigate the self-association behavior of *S. aureus* DHDPS, sedimentation velocity experiments were performed using the analytical ultracentrifuge equipped with a fluorescence detection system (AU-FDS). AU-FDS studies of Alexa Fluor 488-labeled *S. aureus* DHDPS at low nanomolar concentrations revealed that the enzyme sediments with a $s_{20,\mathrm{w}}$ value of 3.3S (Fig. 6B, broken line). However, in the presence of the substrate, pyruvate, the modal sedimentation coefficient shifts to a $s_{20,\mathrm{w}}$ value of 4.0S (Fig. 6B, unbroken line), consistent with the DHDPS dimer (Perugini et al., 2005). This effect was subsequently measured using sedimentation equilibrium experiments at multiple protein concentrations

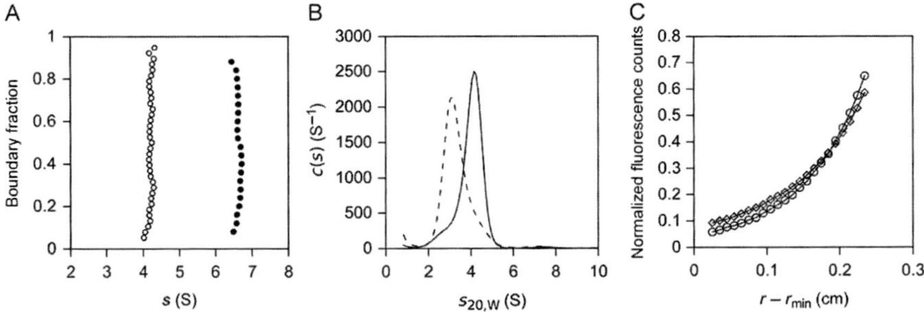

Figure 6 Analytical ultracentrifugation studies of *S. aureus* DHDPS. (A) Enhanced van Holde–Weischet integral distribution plot extrapolated from raw sedimentation velocity data for *S. aureus* DHDPS (open circles) and *E. coli* DHDPS (closed circles) at an initial concentration of 3.1 μ*M*; (B) *c*(*s*) distribution of *S. aureus* DHDPS (25 n*M*) in the absence (broken line) or presence of 2 m*M* pyruvate (unbroken line). (C) Sedimentation equilibrium data for *S. aureus* DHDPS (25 n*M*) in the absence (open diamonds) and presence (open circles) of 2 m*M* pyruvate collected at a rotor speed of 10,000 rpm. Global nonlinear best fits to a monomer–dimer self-association model (global reduced χ^2 values 3.8×10^{-2} and 4.3×10^{-2}, respectively) are shown. *Adapted from Burgess et al. (2008).*

and rotor speeds. As presented in Fig. 6C, addition of pyruvate induces a steeper distribution. This indicates a shift in equilibrium in favor of the higher order dimeric species. Quantification of this phenomenon was achieved by fitting the sedimentation equilibrium data to various self-association models. In agreement with the sedimentation velocity studies, the best-fit was obtained to a monomer–dimer equilibrium model. Other models, such as monomer–trimer and dimer–tetramer, were ruled out given poorer fit statistics. The nonlinear regression best-fit yielded a $K_d^{4\to 2}$ of 33 n*M* for the apo enzyme and 1.6 n*M* in the presence of pyruvate. Thus, the presence of pyruvate shifts the *S. aureus* DHDPS equilibrium approximately 20-fold in favor of the dimeric species (Burgess et al., 2008). Understanding the molecular basis for the unusual dimeric form of *S. aureus* DHDPS warranted atomic-level resolution not accessible using analytical ultracentrifugation. Burgess et al. (2008) therefore determined the X-ray crystal structure of the enzyme. Subsequent computational analysis using PISA software (Krissinel & Henrick, 2007) showed that the tight-dimer interface has a greater proportion of buried surface area and incorporates more noncovalent contacts compared to the tight-dimer interface of tetrameric orthologs (Burgess et al., 2008). This is consistent with a tighter interaction at the dimer interface of *S. aureus* DHDPS, providing a direct connection with the high affinity dimerization calculated in the analytical ultracentrifuge.

3.5 Canonical Structure of Plant DHDPS

The first crystal structure of a plant DHDPS was determined in 1997 from the woodland tobacco species, *Nicotiana sylvestris*, using the *E. coli* DHDPS structure as the molecular replacement model (Blickling, Beisel, et al., 1997). A Patterson search employing the *E. coli* tetramer produced no solution, while (equivalent) solutions were found using the dimer and monomer. When the structure was determined to 2.79 Å, it was clear why the *E. coli* tetramer was unsuccessful in yielding a statistically acceptable model, since the quaternary architecture of the *N. sylvestris* DHDPS tetramer was arranged as a "back-to-back" dimer-of-dimers (Fig. 1C) compared to the "head-to-head" arrangement of *E. coli* DHDPS (Fig. 1A). More specifically, although the tight-dimer unit of *N. sylvestris* DHDPS is identical to the bacterial equivalent, the plant dimers come together enclosing the allosteric clefts that bind lysine on the interior of the tetrameric structure (Fig. 1C), instead of the exterior as observed in the bacterial "head-to-head" structure

(Fig. 1A). However, the *N. sylvestris* DHDPS tetramer was not confirmed in solution and the structural coordinates are not available in the PDB raising questions of whether the "back-to-back" architecture is biologically relevant.

Accordingly, Atkinson and colleagues set out to validate the quaternary architecture of plant DHDPS by characterizing the structure and function of the enzyme from the common grapevine, *Vitis vinifera*. First, size exclusion chromatography measurements were performed on *V. vinifera* DHDPS, which estimated the molar mass of the enzyme to be 131 kDa, consistent with a tetramer (Atkinson, Dogovski, Newman, Dobson, & Perugini, 2011). This was confirmed using sedimentation velocity experiments analyzed using two-dimensional spectrum analysis (2DSA) (Fig. 7A) supported by enhanced van Holde–Weischet analysis, which yielded a $s_{20,\mathrm{W}}$ of 7.3S, molecular weight of 135 kDa and f/f_0 of ~1.4 for a His-tagged construct (Atkinson et al., 2012). The quaternary architecture of *V. vinifera* DHDPS was then determined by X-ray crystallography, which showed the enzyme adopts the "back-to-back" architecture (Atkinson et al., 2013, 2012; Fig. 1C) consistent with *N. sylvestris* DHDPS (Blickling, Beisel, et al., 1997). The "back-to-back" dimer-of-dimers can also be observed in the

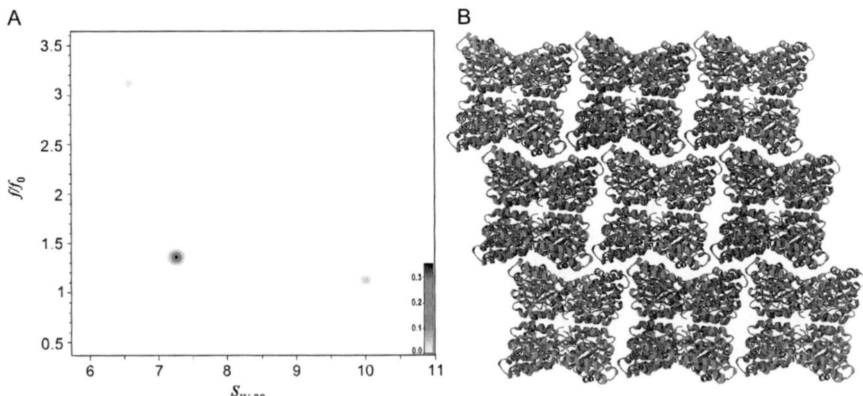

Figure 7 Sedimentation velocity and crystal lattice analyses of *V. vinifera* DHDPS. (A) Pseudo-3D plots of solute distributions derived from 2DSA analysis of sedimentation velocity data. (B) The orientation of the dimeric units in the crystal lattice demonstrates that *V. vinifera* DHDPS (PDB ID: 3TUU) adopts a "back-to-back" assembly, rather than the canonical bacterial "head-to-head" architecture. Lattice generated using the symmetry function of the Visual Molecular Dynamics software package (Humphrey, Dalke & Schulten, 1996). *Panel (A) Figure adapted from Atkinson et al. (2012).*

crystal lattice providing further evidence that this quaternary architecture is biologically relevant (Fig. 7B). SOMO bead modeling (Rai et al., 2005) of the "back-to-back" and "head-to-head" dimer-of-dimers (not including expression tags) was also performed, but the computed frictional ratio obtained for the two structures was almost identical ($f/f_0 \sim 1.2$), indicating that the plant and bacterial architectures could not be differentiated in the analytical ultracentrifuge (Atkinson et al., 2012). Therefore, SAXS analyses were performed to show that the scattering data obtained for *V. vinifera* DHDPS overlayed with the theoretical scattering profile derived from the crystal structure of the enzyme, but fitted poorly to the theoretical scattering profile derived from the bacterial structure of *B. anthracis* DHDPS (Atkinson et al., 2012). This provided unprecedented evidence that the "back-to-back" dimer-of-dimers exists in solution. As a control, SAXS data generated for *B. anthracis* DHDPS was shown to fit well to the theoretical scattering profile of the bacterial "head-to-head" tetramer, but poorly to the "back-to-back" equivalent observed for the plant enzyme (Atkinson et al., 2012). Likewise, this provides evidence that the "head-to-head" architecture canonical to bacteria also presents in solution and cannot be attributed to crystal packing artifacts.

Together, the studies of *N. sylvestris* and *V. vinifera* DHDPS suggest that although the plant tetramer adopts a "back-to-back" dimer-of-dimers compared to the "head-to-head" arrangement canonical to bacterial orthologs, this may represent a rare example of convergent evolution to negate "breathing motion" of the tight-dimer unit. Moreover, it has been proposed by Griffin et al. (2008) that the self-association of the tight-dimer unit to form either a "head-to-head" or "back-to-back" tetramer arrangement achieves the same objective of reducing "breathing motion" of the tight-dimer unit. To support this hypothesis, MD simulations were conducted to compare the *V. vinifera* DHDPS tetramer to a putative dimeric form of the enzyme. It was found that most residues displayed a substantially higher root mean square fluctuation (RMSF) in the dimer compared to the equivalent residues of the tetramer (Fig. 8), indicating that tetramerization does indeed stabilize the functional dimeric unit. This supports the idea that while the dimer forms a complete active site, higher order quaternary structures have evolved to optimize protein dynamics required for maximal catalytic function.

Shortly after the study of *V. vinifera* DHDPS was published, a third plant DHDPS structure was reported from *Arabidopsis thaliana* (Griffin et al., 2012). Similar to *V. vinifera* DHDPS, analytical ultracentrifugation

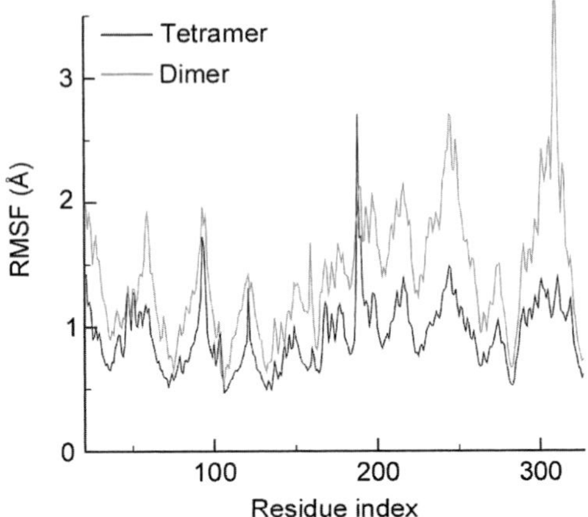

Figure 8 Comparison of the *V. vinifera* DHDPS tetramer with the putative dimer using molecular dynamics simulations. Compared to the tetramer control, the dimeric form displayed substantially higher RMSF values for almost all residues, indicating that tetramerization stabilizes the "breathing motion" of the dimeric unit. *Figure adapted from Atkinson et al. (2012).*

experiments determined that *A. thaliana* DHDPS sediments as a 7.0S species, whereas light scattering/refractive index data analysis showed the enzyme has a molar mass of 140 kDa consistent with a tetramer. In order to determine the quaternary architecture of the tetramer, Griffin et al. (2012) solved the crystal structure of *A. thaliana* DHDPS to 2 Å, which not only confirmed the tetramer observed in solution but also indicated that the *A. thaliana* ortholog adopts a "back-to-back" dimer-of-dimers as observed for DHDPS from *N. sylvestris* and *V. vinifera*.

Together, the studies of bacteria and plant DHDPS enzymes offer new insights into the importance of quaternary structure in optimizing protein dynamics for function. It remains to be seen whether other quaternary forms of DHDPS have also evolved to negate "breathing motion" in addition to the "head-to-head" homotetramer (Fig. 1A), *S. aureus* DHDPS tight-dimer (Fig. 1B), and "back-to-back" dimer-of-dimers (Fig. 1C).

4. CONCLUSIONS

The studies described in this chapter demonstrate how analytical ultracentrifugation in conjunction with SAXS, X-ray crystallography, *in vitro* mutagenesis, enzyme kinetics, and MD simulations, have shed light into the molecular evolution of DHDPS quaternary structure. We propose that

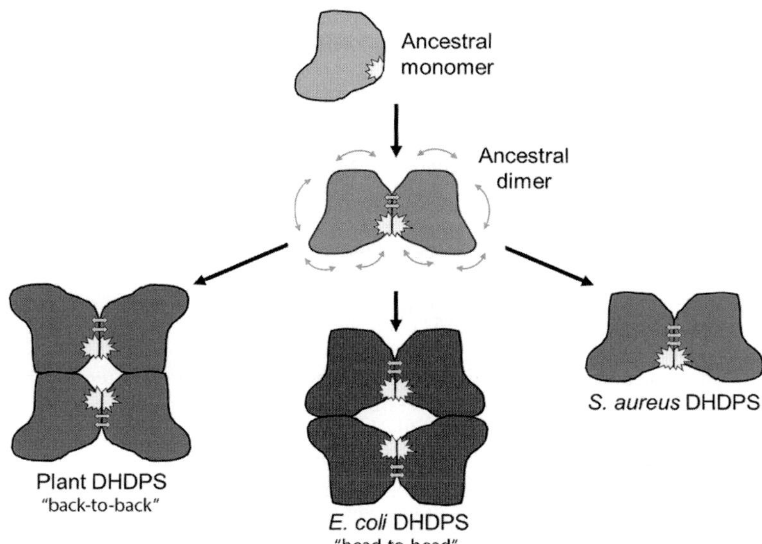

Figure 9 Simplified molecular evolution trajectory of the quaternary structure of bacterial and plant DHDPS. Enzyme active sites are represented by yellow stars (white in the print version), "breathing motion" by doubled headed green arrows (dark gray in the print version) (specifically for the ancestral dimer), and the degree of noncovalent interactions stabilizing the tight-dimer interface of each oligomeric structure represented by light blue double-headed arrows (light gray in the print version) (which are more extensive in the *S. aureus* dimer compared to the other oligomeric structures shown).

the ancestral form of DHDPS was a monomer (Fig. 9) containing a catalytic dyad comprised of Tyr133 and Thr44 (*E. coli* numbering). To enhance catalytic function, the ancestral monomer then evolved into a dimer that included the interdigitation of Tyr107 from the opposite subunit to complete the catalytic triad. However, the ancestral dimer lacked catalytic function due to "breathing motion" at the tight-dimer interface (Fig. 9). As a result, the ancestral dimer evolved in bacteria to form a "head-to-head" dimer-of-dimers as observed for *E. coli* DHDPS, and likewise evolved independently in plants to adopt a "back-to-back" dimer-of-dimers (Fig. 9). By contrast, the enzyme from *S. aureus* evolved to form a tight-dimer structure with enhanced buried surface area at the dimer interface (Fig. 9). Despite the obvious differences in quaternary structure, the three terminal oligomeric forms depicted in Fig. 9 are proposed to have evolved to optimize protein dynamics for catalytic function. Accordingly, the molecular evolution of DHDPS quaternary structure (Fig. 9) may represent a common trajectory for self-associating proteins in general.

REFERENCES

Atkinson, S. C., Dogovski, C., Downton, M. T., Czabotar, P. E., Dobson, R. C. J., Gerrard, J. A., et al. (2013). Structural, kinetic and computational investigation of *Vitis vinifera* DHDPS reveals new insight into the mechanism of lysine-mediated allosteric inhibition. *Plant Molecular Biology*, *81*, 431–446.

Atkinson, S. C., Dogovski, C., Downton, M. T., Pearce, F. G., Reboul, C. F., Buckle, A. M., et al. (2012). Crystal, solution and *in silico* structural studies of dihydrodipicolinate synthase from the common grapevine. *PloS One*, 7, e38318.

Atkinson, S. C., Dogovski, C., Newman, J., Dobson, R. C. J., & Perugini, M. A. (2011). Cloning, expression, purification and crystallization of dihydrodipicolinate synthase from the grapevine *Vitis vinifera*. *Acta crystallographica. Section F, Structural biology and crystallization communications*, *67*(12), 1537–1541.

Blickling, S., Beisel, H. G., Bozic, D., Knäblein, J., Laber, B., & Huber, R. (1997). Structure of dihydrodipicolinate synthase of *Nicotiana sylvestris* reveals novel quaternary structure. *Journal of Molecular Biology*, *274*(4), 608–621.

Blickling, S., & Knäblein, J. (1997). Feedback inhibition of dihydrodipicolinate synthase enzymes by *L*-lysine. *Biological Chemistry*, *378*(3–4), 207–210.

Blickling, S., Renner, C., Laber, B., Pohlenz, H., Holak, T. A., & Huber, R. (1997). Reaction mechanism of *Escherichia coli* dihydrodipicolinate synthase investigated by X-ray crystallography and NMR spectroscopy. *Biochemistry*, *36*(1), 24–33.

Burgess, B. R., Dobson, R. C. J., Bailey, M. F., Atkinson, S. C., Griffin, M. D. W., Jameson, G. B., et al. (2008). Structure and evolution of a novel dimeric enzyme from a clinically important bacterial pathogen. *Journal of Biological Chemistry*, *283*(41), 27598–27603.

Demeler, B., & van Holde, K. E. (2004). Sedimentation velocity analysis of highly heterogeneous systems. *Analytical Biochemistry*, *335*(2), 279–288.

Dobson, R. C. J., Griffin, M. D., Jameson, G. B., & Gerrard, J. A. (2005). The crystal structures of native and (*S*)-lysine-bound dihydrodipicolinate synthase from *Escherichia coli* with improved resolution show new features of biological significance. *Acta Crystallographica. Section D, Biological Crystallography*, *61*(8), 116–1124.

Dobson, R. C. J., Valegård, K., & Gerrard, J. A. (2004). The crystal structure of three site-directed mutants of *Escherichia coli* dihydrodipicolinate synthase: Further evidence for a catalytic triad. *Journal of Molecular Biology*, *338*(2), 329–339.

Dogovski, C., Atkinson, S. C., Dommaraju, S. R., Downton, M., Hor, L., Moore, S., et al. (2012). Enzymology of bacterial lysine biosynthesis. In D. Ekinci (Ed.), *Biochemistry* (pp. 225–262). Rijeka: InTech Open Access Publisher.

Dogovski, C., Atkinson, S. C., Dommaraju, S. R., Hor, L., Dobson, R. C. J., Hutton, C. A., et al. (2009). Lysine biosynthesis in bacteria: An unchartered pathway for novel antibiotic design. In H. W. Doelle & S. Rokem (Eds.), *Biotechnology Part I:* Vol. 11. *Encyclopedia of life support systems (EOLLS)* (pp. 116–136). Oxford: Eolls Publishers.

Dogovski, C., Gorman, M. A., Ketaren, N. E., Praszkier, J., Zammit, L. M., Mertens, H. D., et al. (2013). From knock-out phenotype to three-dimensional structure of a promising antibiotic target from *Streptococcus pneumoniae*. *PloS One*, *8*(12), e83419.

Girish, T. S., Sharma, E., & Gopal, B. (2008). Structural and functional characterization of *Staphylococcus aureus* dihydrodipicolinate synthase. *FEBS Letters*, *582*(19), 2923–2930.

Griffin, M. D. W., Billakanti, J. M., Wason, A., Keller, S., & Mertens, H. D. T. (2012). Characterisation of lysine biosynthetic enzymes in *Arabidopsis thaliana*. *PloS One*, 7, e40318.

Griffin, M. D. W., Dobson, R. C. J., Gerrard, J. A., & Perugini, M. A. (2010). Exploring the dimer-dimer interface of the dihydrodipicolinate synthase tetramer: How resilient is the interface? *Archives of Biochemistry and Biophysics*, *494*(1), 58–63.

Griffin, M. D. W., Dobson, R. C. J., Pearce, F. G., Antonio, L., Whitten, A. E., Liew, C. K., et al. (2008). Evolution of quaternary structure in a homotetrameric protein. *Journal of Molecular Biology*, *380*(4), 691–703.

Humphrey, W., Dalke, A., & Schulten, K. (1996). VMD - Visual molecular dynamics. *Journal of Molecular Graphics*, *14*(1), 33–38.

Hutton, C. A., Perugini, M. A., & Gerrard, J. A. (2007). Inhibition of lysine biosynthesis: An emerging antibiotic strategy. *Molecular BioSystems*, *3*(7), 458–465.

Kaur, N., Gautam, A., Kumar, S., Singh, A., Singh, N., Sharma, S., et al. (2011). Biochemical studies and crystal structure determination of dihydrodipicolinate synthase from *Pseudomonas aeruginosa*. *International Journal of Biological Macromolecules*, *48*(5), 779–787.

Krissinel, E., & Henrick, K. (2007). Inference of macromolecular assemblies from crystalline state. *Journal of Molecular Biology*, *372*(3), 774–797.

Laber, B., Gomis-Rüth, F. X., Romão, M. J., & Huber, R. (1992). *Escherichia coli* dihydrodipicolinate synthase. Identification of the active site and crystallization. *The Biochemical Journal*, *288*(Pt. 2), 691–695.

Mirwaldt, C., Korndorfer, I., & Huber, R. (1995). The crystal structure of dihydrodipicolinate synthase from *Escherichia coli* at 2.5Å resolution. *Journal of Molecular Biology*, *246*(1), 227–239.

Muirhead, H., & Perutz, M. F. (1963). Structure of haemoglobin: A three-dimensional Fourier synthesis of reduced human haemoglobin at 5.5 Å resolution. *Nature*, *17*, 633–638.

Perugini, M. A., Griffin, M. D. W., Smith, B. J., Webb, L. E., Davis, A. J., Handman, E., et al. (2005). Insight into the self-association of key enzymes from pathogenic species. *European Biophysics Journal*, *34*(5), 469–476.

Perutz, M. F. (1976). Structure and mechanism of haemoglobin. *British Medical Bulletin*, *32*, 195–208.

Perutz, M. F., Muirhead, H., Cox, J. M., & Goaman, L. C. (1968). Three-dimensional Fourier synthesis of horse oxyhaemoglobin at 2.8 Å resolution: The atomic model. *Nature*, *219*, 131–139.

Perutz, M. F., Rossmann, M. G., Cullis, A. F., Muirhead, H., Will, G., & North, A. C. (1960). Structure of haemoglobin: a three-dimensional Fourier synthesis at 5.5 Å resolution, obtained by X-ray analysis. *Nature*, *185*(4711), 416–422.

Rai, N., Nöllmann, M., Spotorno, B., Tassara, G., Byron, O., & Rocco, M. (2005). SOMO (SOlution MOdeler): Differences between X-Ray- and NMR-derived bead models suggest a role for side chain flexibility in protein hydrodynamics. *Structure*, *13*(5), 723–734.

Soares da Costa, T. P., Muscroft-Taylor, A. C., Dobson, R. C. J., Devenish, S. R. A., Jameson, G. B., & Gerrard, J. A. (2010). How essential is the 'essential' active-site lysine in dihydrodipicolinate synthase? *Biochimie*, *92*(7), 837–845.

Svergun, D. I. (1992). Determination of the regularization parameter in indirect-transform methods using perpetual criteria. *Journal of Applied Crystallography*, *25*, 495–503.

Svergun, D. I., Barberato, C., & Koch, M. H. J. (1995). CRYSOL—A program to evaluate X-ray solution scattering of biological macromolecules from atomic coordinates. *Journal of Applied Crystallography*, *28*, 768–773.

van Holde, K. E., & Weischet, W. O. (1978). Boundary analysis of sedimentation-velocity experiments with monodisperse and paucidisperse solutes. *Biopolymers*, *17*(6), 1387–1403.

Villeret, V., Clantin, B., Tricot, C., Legrain, C., Roovers, M., Stalon, V., et al. (1998). The crystal structure of *Pyrococcus furiosus* ornithine carbamoyltransferase reveals a key role for oligomerization in enzyme stability at extremely high temperatures. *Proceedings of the National Academy of Sciences of the United States of America*, *95*, 2801–2806.

Voss, J. E., Scally, S. W., Taylor, N. L., Atkinson, S. C., Griffin, M. D. W., Hutton, C. A., et al. (2010). Substrate-mediated stabilization of a tetrameric drug target reveals Achilles heel in anthrax. *Journal of Biological Chemistry*, *285*(8), 5188–5195.

CHAPTER TEN

Characterization of Intrinsically Disordered Proteins by Analytical Ultracentrifugation

David J. Scott[*,†,1], **Donald J. Winzor**[‡]
[*]National Center for Macromolecular Hydrodynamics, School of Biosciences, University of Nottingham, Nottingham, United Kingdom
[†]ISIS Spallation Neutron and Muon Source and Research Complex at Harwell, Rutherford-Appleton Laboratory, Oxford, United Kingdom
[‡]School of Chemistry and Molecular Biosciences, University of Queensland, Brisbane, Queensland, Australia
[1]Corresponding author: e-mail address: david.scott@nottingham.ac.uk

Contents

Abstract

Intrinsically disordered proteins have traditionally been largely neglected by structural biologists because a lack of rigid structure precludes their study by X-ray crystallography. Structural information must therefore be inferred from physicochemical studies of their solution behavior. Analytical ultracentrifugation yields important information about the gross conformation of an intrinsically disordered protein. Sedimentation velocity studies provide estimates of the weight-average sedimentation and diffusion coefficients of a given macromolecular state of the protein.

Methods in Enzymology, Volume 562
ISSN 0076-6879
http://dx.doi.org/10.1016/bs.mie.2015.06.034

1. INTRODUCTION

Intrinsically disordered proteins differ from their globular counterparts by virtue of their failure to exhibit tertiary structure and hence rigid three-dimensional structure (Tompa, 2010; Uversky & Longhi, 2010). For some proteins, this nonrigid structure occurs throughout the entire molecule; but in others, e.g., the plasmid partition protein KorB (Rajasekar et al., 2010) there are intrinsically disordered regions amidst sections exhibiting the compact structure that is amenable to delineation by X-ray crystallography. Historically, intrinsically disordered proteins have been much ignored in the literature, as the drive has always been to isolate globular proteins for structural and functional studies. However, it has become apparent that these proteins are very important to many processes, such as cell signaling, due to their propensity to form tight, but not specific, interactions with of proteins. As such they are often found at the nodes of interaction pathways (Wright & Dyson, 1999).

The absence of a rigid three-dimensional shape precludes use of the classical X-ray crystallographic procedures for structure determination. Reliance must therefore be placed on solution techniques to characterize the array of conformations adopted by intrinsically disordered proteins (Tompa, 2010; Uversky & Longhi, 2010). This chapter assesses the extent to which analytical centrifugation can contribute to that characterization. There are two ultracentrifuge techniques to consider: (i) sedimentation velocity, where the rotor is spun at sufficiently high speed to allow measurement of the rate of migration of a sedimenting boundary and (ii) sedimentation equilibrium, where the rotor speed is lower to permit the attainment of equilibrium between the opposing sedimentation and diffusional forces. Whereas the sedimentation coefficient (s) derived from the former technique is a function of size and shape of the macromolecular solute, the parameter obtained by sedimentation equilibrium is the buoyant molecular mass, $M(1-\bar{v}\rho_s)$, the product of molecular mass (M), and a buoyancy term in which $\bar{v}$ is the protein partial-specific volume and ρ_s is the solvent density—two parameters for which programs are available (Laue, Shah, Ridgeway, & Pelletier, 1992) to determine their magnitudes.

2. ULTRACENTRIFUGAL PROCEDURES

2.1 Sedimentation Equilibrium

A sedimentation equilibrium distribution comprises a relatively featureless, monotonic increase in solute concentration with radial distance (r) that extends from the air–liquid meniscus to the outer extremity of the centrifuge cell. Studies of intrinsically disordered proteins have invariably employed very low solute concentrations, and hence the absorption optical system of a Beckman XL-A (or XL-I) analytical ultracentrifuge to record the $A(r)-r$ distribution. For a species with molecular mass M, the equilibrium distribution at angular velocity ω and absolute temperature T (recognized by its time-independence) is described by

$$A(r) = A(r_F)\exp\left[M(1-\bar{v}\rho_s)\omega^2\left(r^2 - r_F^2\right)/(2RT)\right] \quad (1)$$

where $A(r_F)$ is the absorbance at a selected radial distance r_F and R is the universal gas constant. In as much as the ensemble of random conformations of an intrinsically disordered protein has the same molecular mass, the analysis of the sedimentation equilibrium distribution is relatively straightforward. Indeed, programs are available for this purpose. Such analysis of the distribution (Sánchez-Puig, Veprintsev, & Fersht, 2005) for securin (or pituitary tumor gene 1 product) obtained at 45,000 rpm and 10 °C is shown (•) in Fig. 1A, together with its best-fit description in terms of Eq. (1): a molecular mass of 18,600 (±1000) Da was derived therefrom (Sánchez-Puig et al., 2005).

In the event that a single exponential fails to provide an adequate description of an experimental distribution, Eq. (1) requires modification to the form

$$A(r) = \sum_i A_i(r_F)\exp\left[M_i(1-\bar{v}\rho_s)\omega^2\left(r^2 - r_F^2\right)/(2RT)\right] \quad (2)$$

to account for the presence of oligomeric states of the intrinsically disordered protein: $i=1$ for monomer, $i=2$ for dimer, etc. Such action was required to account for the sedimentation equilibrium distribution obtained (Batra-Safferling et al., 2006) at 20,000 rpm and 4 °C for GARP2 (a glutamic acid-rich protein of rod photoreceptors)—a distribution shown (•) in Fig. 1B, together with the best-fit descriptions in terms of a 39.1 kDa monomer (— — —), dimer (.......), and tetramer (—.—.—) present in proportions 0.47:0.16:0.37.

In summary, sedimentation equilibrium experiments provide an unequivocal means for determining the molecular mass and proportion of each oligomeric state of an intrinsically disordered protein. However, the technique provides no information on the ensemble of conformations (isomers) comprising each oligomeric state; or, indeed, the rate of interconversion between states.

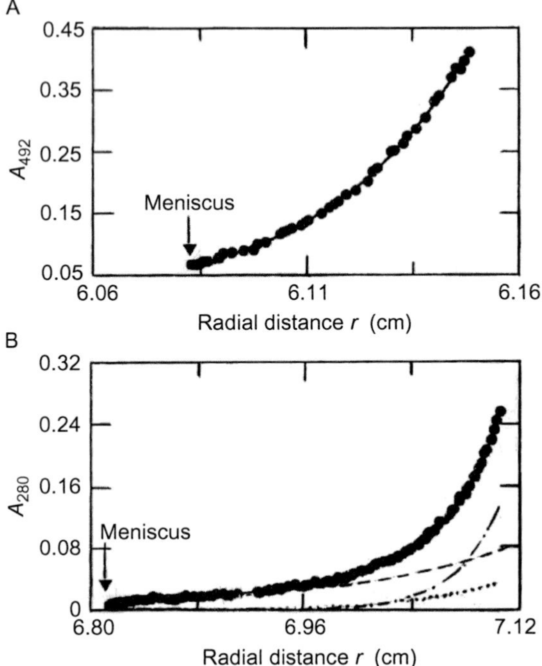

Figure 1 Sedimentation equilibrium distributions for intrinsically disordered proteins. (A) Distribution for fluorescently labeled securin at 45,000 rpm and 10 °C, together with the best-fit description for a single species with $M=18{,}600\ (\pm 1000)$ Da. (B) Equilibrium distribution for GARP2 at 20,000 rpm and 4 °C, which was resolved into separate distributions for 31.9 kDa monomer (— — —), dimer (·····), and tetramer (—.—.—) in proportions 0.48:0.15:0.37. *Panel (A): Data taken from Fig. 1B of Sánchez-Puig et al. (2005); Panel (B): Data taken from Fig. 5B of Batra-Safferling et al. (2006).*

2.2 Sedimentation Velocity

The migration parameter derived from sedimentation velocity experiments is the sedimentation coefficient (s), the rate of boundary movement per unit centrifugal force. Thus,

$$s=\frac{1}{\omega^2 r}\frac{\mathrm{d}r_b}{\mathrm{d}t}=\frac{1}{\omega^2}\frac{\mathrm{dln}r_b}{\mathrm{d}t} \tag{3}$$

where the boundary position r_b is defined by the square root of the second moment of the boundary (Goldberg, 1953; Trautman & Schumaker, 1954). For a symmetrical boundary, its position may be identified with the radial distance at which the concentration is half of that in the solution plateau, the concentration of which decreases progressively because of radial dilution in the sector-shaped cell. This radial dilution effect is evident in Fig. 2, which summarizes the migration process for a solution (0.3 mg/mL) of the cytoplasmic domain of neuroligin 3 (hNL3-cyt) subjected to centrifugation at 50,000 rpm and 20 °C (Paz et al., 2008).

Although *s* may, in principle, be obtained by the application of Eq. (3), more information can be obtained by employing numerical simulation to solve the Lamm equation,

$$\left(\frac{\partial c}{\partial r}\right)_r = -\frac{1}{r}\frac{\partial}{\partial r}\left(r\left(cs\omega^2 r - D\left(\frac{\partial c}{\partial r}\right)_t\right)\right) \tag{4}$$

for *s* and *D*, the diffusion coefficient that takes into account the spreading of the boundary. In studies of intrinsically disordered proteins the value of the curve-fitting parameter also includes any additional spreading as the result of variation in *s* across the ensemble of species comprising a given oligomeric state. Of the various software packages available for numerical solution of the Lamm equation (BPCFIT, LAMM, SEDANAL, SEDFIT, SEDPHAT, SVEDBERG, ULTRASCAN), the SEDFIT program (Schuck, 1998, 2000) has been the procedure of choice for studies of intrinsically disordered proteins (Batra-Safferling et al., 2006; Paz et al., 2008; Rajasekar et al., 2010). Its application to the neurologin 3 distributions presented in Fig. 2

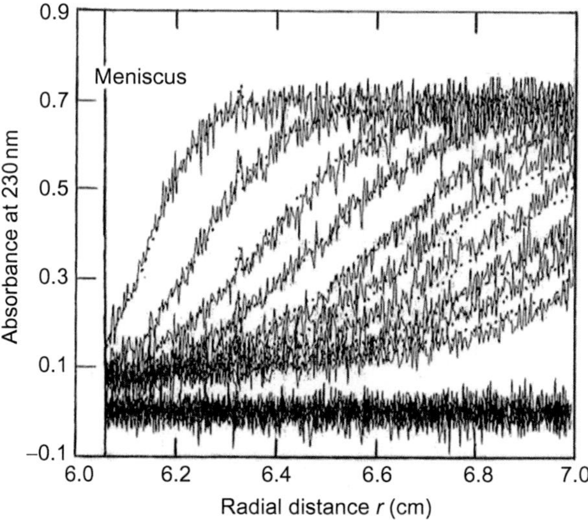

Figure 2 Sedimentation velocity distributions obtained for the cytoplasmic domain of neuroligin 3 at 50,000 rpm and 20 °C. *Data taken from Fig. 1A of Paz et al. (2008).*

is summarized in Fig. 3A, which signifies the presence of a principal species (or ensemble of species) with $s = 1.33$ S, and a smaller proportion of oligomer with $s \approx 2$ S (Paz et al., 2008). Even greater heterogeneity with respect to sedimentation coefficient is observed (Batra-Safferling et al., 2006) for GARP2 (Fig. 3B), the intrinsically disordered protein for which description of the sedimentation equilibrium distribution required the introduction of dimeric and tetrameric forms (Fig. 1B). A point of interest to note from Fig. 3B is the clear resolution of $c(s)$ peaks, a feature which signifies a slow

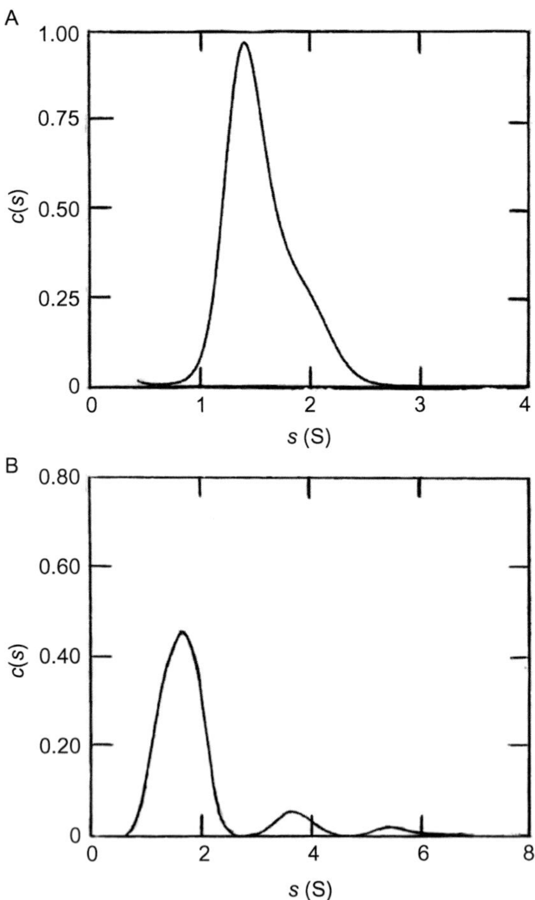

Figure 3 Analysis of sedimentation velocity distributions for intrinsically disordered proteins by the SEDFIT procedure (Schuck, 1998, 2000). (A) $c(s) - s$ distributions for the cytoplasmic domain of neuroligin 3. (B) Corresponding distribution for a glutamic acid-rich protein, GARP 2. *Panel (A): Data taken from Fig. 1B of Paz et al. (2008); Panel (B): Data taken from Fig. 1A of Batra-Safferling et al. (2006).*

rate of oligomer interconversion relative to that of oligomer separation. This is at variance with other measurements of on IDPs, where a fast interconversion rate leaves only a single sedimenting boundary; the theoretical basis for this is enumerated below.

2.3 Sucrose Density Gradient Centrifugation

Sedimentation coefficients for intrinsically disordered proteins have also been obtained by sucrose density gradient centrifugation (Denning, Patel, Uversky, Fink, & Rexach, 2003; Kalthoff, Alves, Urbanke, Knorr, & Ungewickell, 2002), a technique in which a small zone of sample solution is layered over a linear 5–20% sucrose solution in the same buffer to maintain gravitational stability. This particular gradient was selected (Martin & Ames, 1961) to allow the calculation of $s_{20,w}$, the sedimentation coefficient in a medium with the density and viscosity characteristics of water, by comparing the position of the solute zone after sedimentation in a swinging-bucket rotor (SW-39) of a preparative ultracentrifuge with those for a set of standard proteins with known sedimentation coefficients ($s_{20,w}$): a Beckman SW-55 rotor has now supplanted its predecessor. The rationale of this procedure is evident from the definition of the standardized sedimentation coefficient, namely,

$$s_{20,w} = \left(\frac{\eta_s}{\eta_{20,w}}\right)\left(\frac{1-\bar{\upsilon}\rho_{20,w}}{1-\bar{\upsilon}\rho_s}\right) \tag{5}$$

where η_s and $\eta_{20,w}$ are the respective viscosities of buffer (at the temperature of the experiment) and water at 20 °C. Incorporation of Eq. (3) for s and rearrangement then gives

$$\frac{\mathrm{d}r}{\mathrm{d}t} = s_{20,w}\left(\frac{r\omega^2\eta_{20,w}(1-\bar{\upsilon}\rho_s)}{\eta_s(1-\bar{\upsilon}\rho_{20,w})}\right) \tag{6}$$

where the bracketed term on the right-hand side is effectively constant for a protein ($\bar{\upsilon} = 0.71 - 0.75\,\mathrm{g/mL}$) in a linear 5–20% sucrose gradient. On the grounds that the radial distance moved by the sample zone from its initial position (r_0) after centrifugation for time t is therefore,

$$(r_t - r_0) = ks_{20,w}t \tag{7}$$

the sedimentation coefficient for the solute of interest may be obtained from its position in the gradient relative to that of a protein with known $s_{20,w}$.

An obvious advantage of this procedure is its compatibility with the determination of $s_{20,w}$ for a protein of interest in a crude tissue extract—a potential demonstrated in studies of epsin 1, an endocytic accessory protein (Kalthoff et al., 2002). Pig brain cytosol or protein standard (200 μL) was layered onto the top of a sucrose gradient formed in a 13 × 51 mm centrifuge tube and subjected to centrifugation in the SW-55 Ti rotor at 40,000 rpm for 18 h, after which a micropipette was used to fractionate the tube contents (5 mL) from the meniscus downwards in 220 μL steps. Fractions were analyzed by SDS-PAGE and the proteins visualized either by staining with Coomassie Blue (protein standards) or Western blotting. From the resulting zonal distribution for epsin 1 (Fig. 4A), the peak concentration is located in tube 5, which on comparison with corresponding results for a range of protein standards (Fig. 4B) signifies a sedimentation coefficient ($s_{20,w}$) of 2.4 S for this 60 kDa cytosolic protein (Kalthoff et al., 2002).

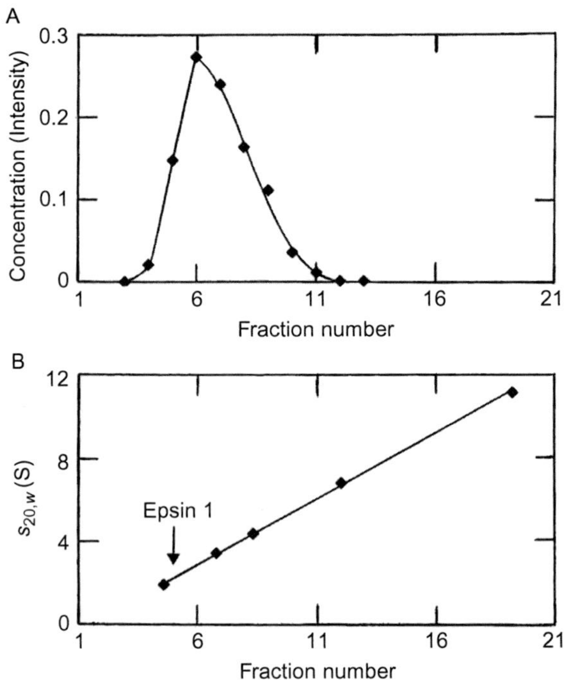

Figure 4 Density gradient centrifugation of pig brain cytosol. (A) Distribution of epsin 1 (detected by Western blotting) in a linear 5–20% sucrose gradient after centrifugation at 40,000 rpm in a Beckman SW-55 rotor for 18 h. (B) Calibration plot from which $s_{20,w}$ for epsin 1 has been obtained. *Data in (A) and (B) taken from Fig. 3 of Kalthoff et al. (2002).*

This procedure is certainly far more labor-intensive than a conventional sedimentation velocity experiment; but it does allow determination of the sedimentation coefficient in a crude tissue extract without the requirement of an XL-A or XL-I analytical ultracentrifuge.

3. EXTENT OF STRUCTURAL INFORMATION OBTAINED

The foregoing considerations of sedimentation equilibrium and sedimentation velocity have detailed the basic information available from the application of these techniques to intrinsically disordered proteins. Sedimentation equilibrium provides an unequivocal estimate of the buoyant molecular mass of a given oligomeric state, while the sedimentation coefficient obtained from velocity experiments provides information on its size and shape. The extent of that information obtained from the sedimentation coefficient is the weighted average of those for the species comprising the ensemble of isomeric conformations (Scott & Winzor, 2012).

3.1 Stokes Radius and the Frictional Ratio

Combination of the buoyant molecular mass with the sedimentation coefficient allows some assessment of the size of the molecule from the expressions

$$s = \frac{M(1-\bar{v}\rho_s)}{N_A f} = \frac{M(1-\bar{v}\rho_s)}{N_A 6\pi\eta_s R_h} \tag{8}$$

where N_A is Avogadro's number, and where f, the frictional ratio, has then been expressed in terms of the Stokes radius R_h (the radius of the equivalent hydrodynamic sphere). In studies of intrinsically disordered proteins, this solvated radius has routinely been obtained (Adkins & Lumb, 2002; Batra-Safferling et al., 2006; Denning et al., 2003; Kalthoff et al., 2002; Paz et al., 2008; Rajasekar et al., 2010; Sánchez-Puig et al., 2005) from size-exclusion chromatography experiments by comparing its elution characteristics with those for globular proteins with defined Stokes radii (Laurent & Killander, 1964; Siegel & Monty, 1966).

In order to obtain a structure parameter that is independent of molecular mass, the frictional ratio, f/f_0, has been introduced as the ratio of the Stokes radius to its unsolvated counterpart (R_u). Thus,

$$\frac{f}{f_0} = \frac{R_h}{R_u} = \frac{R_h}{\sqrt[3]{\frac{3M\bar{v}}{4\pi N_A}}} \tag{9}$$

which clearly has a value of unity for a spherical unsolvated solute. In that situation a larger value of f/f_0 is diagnostic of protein asymmetry. However, because proteins are invariably solvated (hydrated) to some extent, the Stokes radius reflects the additional contribution to solute volume. Specifically,

$$R_h = \sqrt[3]{\frac{3M\bar{v}}{4\pi N_A}\left(1 + \frac{\delta}{\rho_s\bar{v}}\right)} \tag{10}$$

where δ, the extent of solvation, is typically 0.3–0.4 g/g protein for globular proteins (Tanford, 1961)—an extent that leads to a frictional ratio of about 1.1 for spherical proteins. The observation of larger f/f_0 values for globular proteins is therefore attributed to asymmetry.

Values of the frictional ration for intrinsically disordered proteins are relatively high: 2.8 for nucleoporin (Denning et al., 2003), 2.1 for securin (Sánchez-Puig et al., 2005), 1.9 for KorB (Rajasekar et al., 2010), and 1.6 for neuroligin (Paz et al., 2008). Whether these higher values of f/f_0 reflect asymmetry of a compact protein or a larger value of δ for an unstructured species cannot be determined by analytical ultracentrifugation. Instead, spectral techniques such as circular dichroism, nuclear magnetic resonance, and small-angle X-ray scattering must be used to obtain evidence of any regular and rigid structure. Failure to detect such characteristics of detailed structure has thus provided the major criterion for classifying the macromolecules under consideration as intrinsically disordered proteins (Tompa, 2010; Uversky & Longhi, 2010).

3.2 Analysis of Boundary Spreading: Theoretical Aspects

Attention thus far has centered on interpretation of sedimentation velocity distributions in terms of the Lamm equation, Eq. (4), which pertains to a single migrating species with sedimentation coefficient s and diffusion coefficient D: the resulting analysis thus yields a weight-average value for each parameter in the event that a given boundary reflects an ensemble of isomers. The point at issue is whether expanding the Lamm equation to encompass the existence of multiple isomeric states can lead to a better description of the ensemble (Scott & Winzor, 2012).

For a mixture of n independent isomeric species, Eq. (4) requires modification to the form (Cann, 1970; Fujita, 1962)

$$\sum_{i=1}^{i=n}\left(\frac{\partial c_i}{\partial r}\right)_r = -\left(\frac{1}{r}\right)\sum_{i=1}^{i=n}\left(\frac{\partial}{\partial r}\right)\left(r\left(c_i s_i \omega^2 r - D_i\left(\frac{\partial c_i}{\partial r}\right)_t\right)\right)_t \tag{11}$$

However, the existence of intrinsically disordered proteins in multiple conformations that interconvert in solution (Tompa, 2010) dictates the inclusion of kinetic terms into the continuity equation for each conformer to accommodate the rates of these interconversions. For illustrative purposes, consideration is restricted to a mixture of two isomeric states, i and j, a simplification that still provides insight into the effect of isomeric interconversions. As noted by Cann (1970), the Lamm equations for the two isomeric species then become

$$\left(\frac{\partial c_i}{\partial r}\right)_r = -\left(\frac{1}{r}\right)\left(\frac{\partial}{\partial r}\right)\left(r\left(c_i s_i \omega^2 r - D_i\left(\frac{\partial c_i}{\partial r}\right)_t\right)\right)_t - k_1 c_i + k_2 c_j \tag{12a}$$

$$\left(\frac{\partial c_j}{\partial r}\right)_r = -\left(\frac{1}{r}\right)\left(\frac{\partial}{\partial r}\right)\left(r\left(c_j s_j \omega^2 r - D_j\left(\frac{\partial c_j}{\partial r}\right)_t\right)\right)_t - k_2 c_j + k_1 c_i \tag{12b}$$

where k_1 is the rate constant for the conversion of i to j and k_2 is the corresponding rate constant for the reverse reaction. There are five possible rate constant combinations that need to be considered (Longsworth & MacInnes, 1942). In situations where either $k_1 \gg k_2$ or $k_2 \gg k_1$, Eq. (4) continues to apply because of the effective elimination of one isomeric state. On the other hand, continuity equations for both species need to be retained in the event that k_1 and k_2 are sufficiently small for essentially no interconversion of species to occur on the timescale of species separation. Both continuity equations also need to be retained for systems in which the rates of species interconversion are of similar magnitude to the rate of species separation; and the two continuity equations need to be summed (Fujita, 1962). For slow interconversion between two isomeric states, the sedimentation velocity distribution comprises separate boundaries for each isomer, the relative sizes of which depend upon the ratio of rate constants for the forward and reverse reactions, i.e., upon the thermodynamic isomerization constant (Cann, Kirkwood, & Brown, 1957).

The fifth potential situation recognized by Longsworth and MacInnes (1942) is that in which the forward and reverse rate constants are both rapid, in which case there is rapidly established equilibrium between isomeric states. Because the relative concentrations of isomers coexisting in rapidly established equilibrium are independent of total concentration, the rates of sedimentation and diffusional spreading are invariant across the migrating boundary system (Cann, 1970, Fujita, 1962). The concentration distribution thus comprises a single symmetrical boundary whose migration is described by Eq. (4) with c identified as the total concentration at radial distance r, but with s and D identified as weight-average quantities (Cann, 1970, Fujita, 1962).

3.3 Analysis of Boundary Spreading: Practical Applications

The analysis of boundary spreading in sedimentation velocity distributions for intrinsically disordered proteins clearly depends upon the rates of interconversion between the various members of the isomeric ensemble. For illustrative purposes only the two extreme situations need to be addressed: that in which the rates of interconversion are sufficiently slow for the ensemble to comprise essentially a noninteracting mixture over the time-course of the sedimentation velocity experiment; and that in which interconversion rates are sufficiently rapid for essentially instantaneous establishment of equilibrium. As noted in relation to Fig. 3B, the observation of completely resolved peaks in the $c(s) - s$ distributions for GARP-2 (Batra-Safferling et al., 2006) signifies a lack of equilibrium readjustment between oligomeric states (the former extreme situation). However, intrinsically disordered proteins are considered to have very shallow free energy profiles—the characteristic that allows them to adopt a multiplicity of isomeric conformational states in solution (Fisher & Stultz, 2011). The fact that there is then little difference between rate constants for the formation and depletion of a given isomer increases greatly the opportunity for the coexistence of multiple iso-energetic states in rapidly established equilibrium (the other extreme situation). Experimental support for this interpretation of sedimentation velocity distributions is seemingly provided by the finding (Paz et al., 2008) that combination of the diffusion and sedimentation coefficients obtained by SEDFIT analysis according to Eq. (4) signify a molecular mass of 15,200 ($\pm$300) Da for neuroligin 3, which encompasses the value of 15,378 that is calculated from the amino acid sequence.

Although this observation is certainly consistent with migration governed by an invariant sedimentation coefficient, no information about the ensemble of isomers responsible for the $c(s) - s$ distribution has been obtained. Indeed, it cannot be established unequivocally whether the weight-average values of s and D obtained from the analysis refer to an ensemble of conformers that are in rapidly established or slowly attained equilibrium because the distribution of s and D values are likely to be very narrow for isomeric systems. Furthermore, it is unlikely that elimination of the consequences of diffusion to obtain the asymptotic form of the sedimentation velocity pattern (Van Holde & Weischet, 1978; Winzor, Tellam, & Nichol, 1977) would assist in that regard because of experimental difficulties in establishing whether the ordinate intercepts of dependencies of s^*, the apparent sedimentation coefficient determined from different regions of the boundary, upon $1/\sqrt{t}$ describe a unique value or a very narrow range of values for the sedimentation coefficient. As noted previously (Scott & Winzor, 2012), the information to be obtained from sedimentation velocity distributions for an intrinsically disordered protein is the weight-average sedimentation coefficient of the isomeric ensemble. That conclusion is now modified by adding the corresponding weight-average diffusion coefficient in studies where numerical simulation procedures are employed to solve the Lamm equation for both s and D.

4. CONCLUDING REMARKS

Intrinsically disordered proteins are of immense interest due to both their diverse biological roles and also their apparent abrogation of the structure-equals-function paradigm. As such many diverse biophysical techniques can be applied to their analysis in the view that a new application of an established method will have great impact in the literature. Analytical ultracentrifugation sedimentation velocity has immense potential to probe the gross conformation of proteins in solution in their native conformations. Careful consideration of the fundamental equations of sedimentation shows that the amount of information that can be extracted from such an analysis for intrinsically disordered proteins is that the molecules are extended in solution: the determined sedimentation and diffusion coefficients represent weighted averages of those for each oligomeric species present. The analytical ultracentrifuge provides unequivocal characterization of the macromolecular state of an IDP, which is clearly valuable information in the lead up to analysis by other solution methods such as NMR and small-angle scattering.

REFERENCES

Adkins, J. N., & Lumb, K. J. (2002). Intrinsic structural disorder and sequence features of the cell cycle inhibitor p57^{Kip2}. *Proteins, 46*, 1–7.

Batra-Safferling, R., Abarca-Heidemann, K., Körschen, H. G., Tziatzios, C., Stoldt, M., Budyak, I., et al. (2006). Glutamic acid-rich proteins of rod photoreceptors are natively unfolded. *The Journal of Biological Chemistry, 281*, 1449–1460.

Cann, J. R. (1970). *Interacting macromolecules: The theory and practice of their electrophoresis, ultracentrifugation and chromatography*. NewYork: Academic Press.

Cann, J. R., Kirkwood, J. G., & Brown, R. A. (1957). Theory of isomerization equilibrium in electrophoresis. *Archives of Biochemistry and Biophysics, 72*, 37–41.

Denning, D. P., Patel, S. S., Uversky, V., Fink, A. L., & Rexach, M. (2003). Disorder in the nuclear pore complex: The FG repeat regions of nucleoporins are natively unfolded. *Proceedings of the National Academy of Sciences of the United States of America, 100*, 2450–2455.

Fisher, C. K., & Stultz, C. M. (2011). Constructing ensembles for intrinsically disordered proteins. *Current Opinion in Structural Biology, 21*, 426–431.

Fujita, H. (1962). *Mathematical theory of sedimentation analysis*. New York: Academic Press.

Goldberg, R. J. (1953). Sedimentation in the ultracentrifuge. *The Journal of Physical Chemistry, 57*, 194–202.

Kalthoff, C., Alves, J., Urbanke, C., Knorr, R., & Ungewickell, E. J. (2002). Unusual structural organization of the endocytic proteins AP180 and epsin 1. *The Journal of Biological Chemistry, 277*, 8209–8216.

Laue, T. M., Shah, B. D., Ridgeway, T. M., & Pelletier, S. L. (1992). Computer-aided interpretation of analytical sedimentation data for proteins. In S. E. Harding, A. J. Rowe, & J. C. Horton (Eds.), *Analytical ultracentrifugation in biochemistry and polymer science* (pp. 90–125). Cambridge, UK: Royal Society of Chemistry.

Laurent, T. C., & Killander, J. (1964). A theory of gel filtration and its experimental verification. *Journal of Chromatography, 14*, 317–330.

Longsworth, L. G., & MacInnes, D. A. (1942). An electrophoretic study of mixtures of ovalbumin and yeast nucleic acid. *The Journal of General Physiology, 25*, 507–516.

Martin, R. G., & Ames, B. N. (1961). A method for determining the sedimentation behavior of enzymes: Application to protein mixtures. *The Journal of Biological Chemistry, 236*, 1372–1379.

Paz, A., Zeiv-Ben-Mordehai, T., Lundqvist, M., Sherman, E., Mylonas, E., Weiner, L., et al. (2008). Biophysical characterization of the unstructured cytoplasmic domain of the human neuronal adhesion protein neuroligin 3. *Biophysical Journal, 95*, 1928–1944.

Rajasekar, K., Muntaha, S. T., Tame, J. R. H., Kommareddy, S., Morris, G., Wharton, C. W., et al. (2010). Order and disorder in the domain organization of the plasmid partition protein KorB. *The Journal of Biological Chemistry, 285*, 15440–15449.

Sánchez-Puig, N., Veprintsev, D. B., & Fersht, A. R. (2005). Human full-length securin is a natively unfolded protein. *Protein Science, 14*, 1410–1418.

Schuck, P. (1998). Sedimentation analysis of noninteracting and self-associating solutes using numerical solutions to the Lamm equation. *Biophysical Journal, 75*, 1503–1512.

Schuck, P. (2000). Size-distribution analysis of macromolecules by sedimentation velocity ultracentrifugation and Lamm equation modeling. *Biophysical Journal, 78*, 1606–1619.

Scott, D. J., & Winzor, D. J. (2012). Sedimentation velocity of intrinsically disordered proteins: What information can we actually obtain? *Molecular BioSystems, 8*, 378–380.

Siegel, L. M., & Monty, K. J. (1966). Determination of the molecular weights and frictional ratios of proteins in impure systems by use of gel filtration and density gradient centrifugation: Application to crude preparations of sulfite and hydroxylamine reductases. *Biochimica et Biophysica Acta, 112*, 346–362.

Tanford, C. (1961). *Physical chemistry of macromolecules*. New York: Wiley.
Tompa, P. (2010). *Structure and function of intrinsically disordered proteins*. Boca Raton, FL: CRC Press.
Trautman, R., & Schumaker, V. N. (1954). Generalization of the radial dilution squared law in ultracentrifugation. *The Journal of Chemical Physics*, *22*, 551–554.
Uversky, V., & Longhi, S. (2010). *Instrumental analysis of intrinsically disordered proteins*. Hoboken, NJ: Wiley.
Van Holde, K. E., & Weischet, W. O. (1978). Boundary analysis of sedimentation velocity experiments with monodisperse and paucidisperse systems. *Biopolymers*, *17*, 1387–1403.
Winzor, D. J., Tellam, R., & Nichol, L. W. (1977). Determination of the asymptotic shape of sedimentation velocity patterns for reversibly polymerizing solutes. *Archives of Biochemistry and Biophysics*, *178*, 327–332.
Wright, P. E., & Dyson, H. J. (1999). Intrinsically unstructured proteins: Re-assessing the protein structure function paradigm. *Journal of Molecular Biology*, *293*, 321–331.

CHAPTER ELEVEN

Sedimentation Velocity Analysis of the Size Distribution of Amyloid Oligomers and Fibrils

Yee-Foong Mok, Geoffrey J. Howlett[1], Michael D.W. Griffin[1]
Department of Biochemistry and Molecular Biology, Bio21 Molecular Science and Biotechnology Institute, University of Melbourne, Parkville, Victoria, Australia
[1]Corresponding authors: e-mail address: ghowlett@unimelb.edu.au; mgriffin@unimelb.edu.au

Contents

Abstract

Amyloid fibrils result from the self-assembly of proteins into large aggregates with fibrillar morphology and common structural features. These fibrils form the major component of amyloid plaques that are associated with a number of common and debilitating diseases, including Alzheimer's disease. While a range of unrelated proteins and peptides are known to form amyloid fibrils, a common feature is the formation of aggregates of various sizes, including mature fibrils of differing length and/or structural morphology, small oligomeric precursors, and other less well-understood forms such as amorphous aggregates. These various species can possess distinct biochemical, biophysical, and pathological properties. Sedimentation velocity analysis can characterize amyloid fibril formation in exceptional detail, providing a particularly useful method for resolving the complex heterogeneity present in amyloid systems. In this chapter, we describe analytical methods for accurate quantification of both total amyloid fibril formation and the formation of distinct amyloid structures based on differential

Methods in Enzymology, Volume 562
ISSN 0076-6879
http://dx.doi.org/10.1016/bs.mie.2015.06.024

sedimentation properties. We also detail modern analytical ultracentrifugation methods to determine the size distribution of amyloid aggregates. We illustrate examples of the use of these techniques to provide biophysical and structural information on amyloid systems that would otherwise be difficult to obtain.

1. INTRODUCTION

Amyloid fibrils form by the self-association of proteins or peptides into large, fibrillar structures with a common cross-β secondary structure core. Over 20 unrelated proteins form amyloid fibrils that are associated with a range of major diseases including Alzheimer's, and Parkinson's disease and atherosclerosis (Antoni et al., 2013; Sipe & Cohen, 2000; Zerovnik, 2002). It has become apparent that under different conditions, amyloid proteins aggregate into a multitude of species that vary in size, morphology, and solution properties. For instance, aggregates formed by the β2-microglobulin protein are comprised of a heterogeneous mixture of amorphous masses and two types of fibrils with either worm-like or long-straight morphology (Gosal et al., 2005). Studies of how the size and morphologies of β2-microglobulin fibrils change with pH, salt, and temperature conditions provide valuable insights into the competing pathways of amyloid fibril formation. Furthermore, it appears that distinct aggregates formed by specific amyloid-forming proteins affect different cellular processes. Emerging evidence suggests, for example, that small, amyloid oligomers of Aβ are cytotoxic in Alzheimer's disease neurons, whereas other structural forms of Aβ fibrils are more protective (Petkova et al., 2005). Thus, detailed study of the size distributions of fibrils and aggregates formed by amyloidogenic proteins are vital to understanding the underlying mechanisms of amyloid disease development. Yet, the isolation and biochemical analysis of different amyloid species remain challenging due to their size and relatively insoluble nature. Tackling this problem has led to a range of techniques to characterize the heterogeneity of amyloid fibrils (Bruggink, Muller, Kuiperij, & Verbeek, 2012). Over the last decades, centrifugation technologies have emerged as a robust way to separate and analyze many types of amyloid aggregates. In this chapter, we describe preparative ultracentrifuge methods to isolate specific amyloid aggregates, as well as analytical ultracentrifuge methods for quantifying both total amyloid fibril formation and the size distribution of amyloid fibrils and aggregates. We also illustrate examples of the use of fluorescence detection coupled with analytical ultracentrifugation to provide biophysical and structural information on the interaction of nonfibrillar components with amyloid fibrils.

2. PREPARATIVE ULTRACENTRIFUGATION

While centrifugation provides the potential to globally examine the size distributions of amyloid aggregates in solution, in some cases, it can be particularly useful to isolate an amyloid species for further study. The size and shape differences between species in a heterogeneous amyloid fibril sample enable the use of defined centrifugal conditions to fractionate various amyloid species in a preparative centrifuge. The theory and factors governing the sedimentation of protein molecules in a centrifugal field have been covered extensively in an earlier review (Mok & Howlett, 2006). In general, the amyloid field has used relatively short periods of low-speed centrifugation around 14,000–16,000 × *g* and aqueous buffers to pellet insoluble fibrils and large aggregates of the major amyloid proteins. Smaller amyloid species (e.g., short fibril fragments or oligomers) can be sedimented using longer periods of high-speed centrifugation (>50,000 × *g*) (Mok & Howlett, 2006). In recent years, we have developed protocols to isolate distinct species within complex amyloid mixtures and present, in this section, three preparative centrifugation regimes used to isolate different sized aggregates formed by apolipoprotein C-II and huntingtin.

2.1 Separation of Fibrils that Differ in Morphology

Human apolipoprotein (apo) C-II is an exchangeable apolipoprotein that binds reversibly to the polar lipid surface of plasma lipoprotein particles *in vivo* and associates with a range of natural and synthetic lipid surfaces *in vitro*. In the absence of lipid, apoC-II is largely unstructured and readily self-assembles into amyloid fibrils with a characteristic flexible ribbon morphology (Fig. 1A). These fibrils are 11–12 nm in width, micrometers in length, and appear to have recurrent constrictions along their contour length every 30–70 nm (Hatters, MacPhee, Lawrence, Sawyer, & Howlett, 2000). In contrast, at low concentrations of phospholipid micelles, apoC-II aggregates via a two-phase process (Griffin et al., 2008), with aggregate morphology after the second stage resembling straight, inflexible rods (Fig. 1B). These straight fibrils are 13–14 nm wide, with no apparent large-scale constrictions, and appear to be up to several micrometers long (Griffin et al., 2008). Straight fibrils sediment more rapidly than their flexible ribbon counterparts, a property that was exploited to separate straight

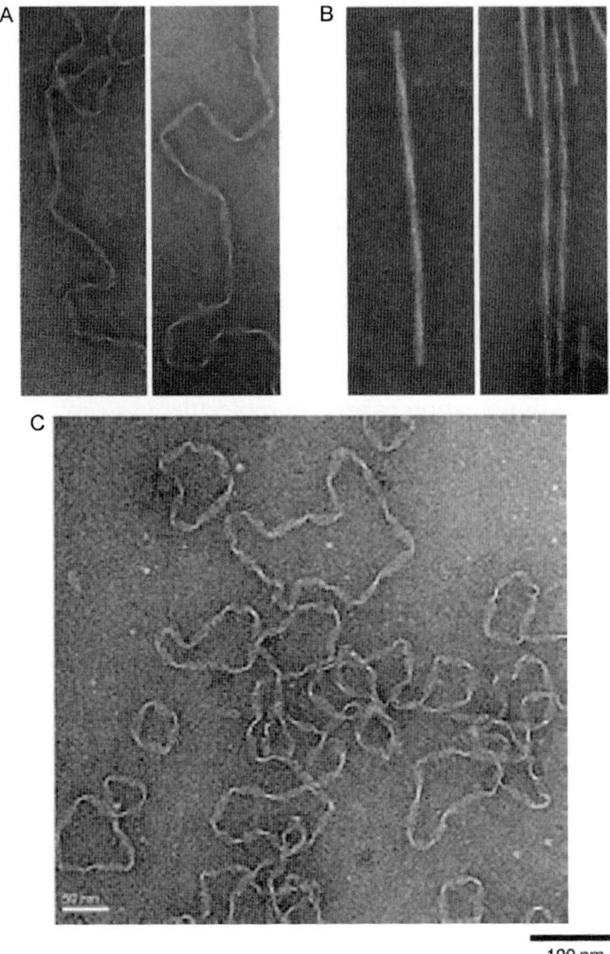

Figure 1 Transmission electron micrographs of apoC-II amyloid fibrils. (A) ApoC-II fibrils alone formed at 1 mg/mL, showing flexible ribbon morphology. (B) ApoC-II fibrils formed in the presence of 500 μ*M* 1,2-di-octanoyl-*sn*-glycero-3-phosphocholine, showing straight-rod morphology. (C) ApoC-II closed loops purified from preformed fibrils using preparative centrifugation. *Panels (A) and (B) adapted with permission from Griffin et al. (2008) and panel (C) adapted with permission from Yang, Griffin, Binger, Schuck, and Howlett (2012).*

fibrils from ribbon-type fibrils for further study: fibril samples formed at 1.0 mg/mL (112.1 μ*M*) apoC-II in the presence of 500 μ*M* phospholipid were centrifuged for 2 min at relatively low speeds of 14,500 × *g* and the supernatant removed. The resulting straight fibrils in the pellet were then resuspended in an equal volume of 100 m*M* phosphate, pH 7.4. The process was repeated a total of three times and the isolated straight fibrils in the pellet fraction pooled. Under the centrifugal conditions, the bulk of ribbon-type fibrils remained in the supernatant and was separately pelleted by centrifuging at 350,000 × *g* for 10 min.

2.2 Separation of Linear and Closed-Loop Fibrils

Both sedimentation velocity analysis and electron microscope images of ribbon-type apoC-II fibrils have consistently revealed a subpopulation of fibrils that appear as closed-loop structures (Hatters et al., 2000). Similar loop or "ring" structures have been observed in amyloid fibrils formed by apoC-III (de Messieres, Huang, He, & Lee, 2014), κ-casein (Thorn et al., 2005), and α-synuclein (Conway et al., 2000), with the rings in the latter hypothesized to be on-pathway for mature fibril formation. ApoC-II fibril formation proceeds via a reversible model that includes elongation, dissociation, fibril breaking, and fibril rejoining (Binger et al., 2008). At equilibrium, closed loops comprise a small proportion of apoC-II ribbon-type fibrils (Yang et al., 2012) and are thought to form by annealing of the free ends of individual, short fibrils. ApoC-II loops appear similar in width and morphology to flexible ribbons, but have a much shorter average contour length of approximately 292 nm (Yang et al., 2012). To prepare these loop structures for study, it was necessary to first enrich the population of ring structures. ApoC-II ribbon-type fibrils were formed at a protein concentration of 0.3 mg/mL by incubation at 25 °C for 7 days. Mature fibrils (1 mL aliquots) were then freeze–thawed 10 times by submerging fibril samples in liquid nitrogen for 25 s followed by rapid thawing in water at 42 °C. This process serves to fragment large fibrils into shorter lengths that form ring structures more readily. After reannealing at 25 °C for 24 h, the samples were centrifuged at 47,500 × *g* for 8 min using an Optima Max centrifuge and a TL-100.1 rotor (Beckman Coulter, Fullerton, CA) resulting in sedimentation of large linear fibrils and significant enrichment of closed-loop fibrils in the supernatant fraction (Fig. 1C). The centrifugal speed and time used was intermediate between those used for pelleting straight fibrils and those for pelleting flexible ribbon fibrils, and demonstrates the utility of preparative ultracentrifugation for separating distinct amyloid aggregates (Yang et al., 2012).

2.3 Isolation of Huntingtin Oligomers and Aggregates

The flexibility of preparative centrifugation is illustrated in a final example of a protocol for huntingtin aggregates. The mutant form of huntingtin exon 1 protein is associated with the fatal neurological disorder, Huntington's disease and accumulates as massive inclusion bodies within neurons. Neuronal cell lysates containing mutant huntingtin exon 1 contain heterogeneous populations of inclusions, fibrillar aggregates, monomers, and an invariant pool of oligomers (Olshina et al., 2010; Ormsby, Ramdzan, Mok, Jovanoski, & Hatters, 2013). Isolation of huntingtin oligomers is of interest due to hypotheses that oligomers are the toxic agents of amyloid disease (Slow, Graham, & Hayden, 2006). However, it has been technically difficult to fractionate small populations of oligomers from bulk monomers and inclusions using preparative centrifugation alone. Thus, the following protocol was developed (Ormsby et al., 2013). Neuronal cell lysate containing mutant huntingtin aggregates was centrifuged at $14{,}000 \times g$ for 10 min, which pellets inclusions, oligomers, and cell debris. The supernatant, containing monomers, was decanted and the pellet was resuspended in aqueous buffer. Desalting resin from prepacked 0.8 mL Pierce Zeba centrifuge columns (Thermo Scientific, Waltham, MA) was manually removed and replaced with 300 μL of Sephacryl S1000 gel filtration resin (GE Life Sciences, Pittsburgh, PA). The columns were placed on top of empty microfuge tubes, and the resuspended pellet loaded on to the columns. The columns were centrifuged at $1500 \times g$ for 1 min, whereby oligomers passed through the resin and collected in the microfuge tubes, while inclusions and large aggregates remained trapped at the top of the columns. The procedure was repeated three times, and the accumulated oligomers pooled. Here, creative refinement of centrifugal procedures enabled the isolation of a smaller entity within a highly complex mixture of aggregates and cell debris.

3. ANALYTICAL ULTRACENTRIFUGATION

The analytical ultracentrifuge has been widely used in sedimentation equilibrium and velocity studies to characterize the size and interactions of macromolecules in solution. Sedimentation equilibrium methods offer the advantage that the system at equilibrium yields molecular weight information directly. However, the very large size of amyloid fibrils precludes this approach, since it is generally not possible to centrifuge at sufficiently low speeds to obtain equilibrium distributions. Nevertheless, sedimentation equilibrium is useful in defining the molecular weights of subunits that form amyloid fibrils (Hammarstrom, Jiang, Deechongkit, & Kelly, 2001; Lashuel et al., 2002; Lashuel, Lai, & Kelly, 1998). On the other hand, sedimentation velocity analysis, which measures the rate of movement of fibrils in a centrifugal field, offers a very powerful method to characterize the size and heterogeneity of amyloid fibrils. Typical sedimentation velocity experiments on amyloid fibrils take 3–5 h to complete, permitting the time course of slow amyloid formation to be determined (Binger et al., 2008). Recent studies include analysis of the size distributions of Aβ oligomers (Nagel-Steger et al., 2010), ultrasonication effects on β2-microglobulin (Chatani et al., 2009), and the analysis of huntingtin aggregates (Ormsby et al., 2013).

We have previously reviewed some of the basic information pertaining to the analysis of sedimentation velocity data for amyloid fibrils (Mok & Howlett, 2006; Pham, Mok, & Howlett, 2009). This review established that, at the lowest speed achievable in the analytical ultracentrifuge (about 1500 rpm or $164 \times g$), fibrils with sedimentation coefficients as large as 10,000 can be characterized. Note, however, that some fibrillar aggregates become colloidal, sedimenting rapidly even in very low centrifugal fields, and therefore prohibiting sedimentation velocity analysis.

3.1 Sedimentation Velocity Theory

The evolution of the concentration distribution, $c(r,t)$, of a single species of diffusing particles in a spinning rotor is given by the Lamm equation (Fujita, 1962):

$$\frac{\mathrm{d}c}{\mathrm{d}t} = \frac{1}{r}\frac{\mathrm{d}}{\mathrm{d}r}\left[rD\frac{\mathrm{d}c}{\mathrm{d}r} - s\omega^2 r^2 c\right] \tag{1}$$

where r is the radial position, t is the time, c is the concentration, s is the sedimentation coefficient, ω is the angular velocity, and D is the translational diffusion coefficient. The development of rapid desktop computers allowed the use of numeric solutions to the Lamm equation as fitting functions, to fit experimental data to models comprising sedimentation coefficient distributions ($c(s)$) or several noninteracting components (Demeler & Saber, 1998; Schuck, 1998, 2000). Equation (1) forms the basis for the analysis of the sedimentation velocity of amyloid fibrils. One approach that simplifies data analysis is based on the assertion that the large size of amyloid fibrils reduces their rate of diffusion. Using the assumption that the diffusion coefficient of the fibrils is negligible, the evolution of the concentration distribution of identical nondiffusing particles with a uniform initial distribution is described by a step function:

$$c(r,t) = \mathrm{e}^{-2\omega^2 st} \times \begin{cases} 0 & \text{for}\, r < r_{\mathrm{m}}\mathrm{e}^{\omega^2 st} \\ 1 & \text{else} \end{cases} \tag{2}$$

where r_m is the position of the meniscus. A distribution of nondiffusing and noninteracting particles can be analyzed using an approach referred to as ls-$g^*(s)$, where experimental concentration distributions are modeled in a least-squares fashion by a summation of step functions weighted according to a variable distribution of sedimentation coefficients (Schuck, 2000). A computer program called SEDFIT for analyzing sedimentation data includes an ls-$g^*(s)$ option and is freely available from http://www.analyticalultracentrifugation.com.

For many amyloid fibril systems, electron micrograph images indicate either worm-like or straight rod-like structures (Fig. 1). In the case of worm-like fibrils, explicit equations are available to relate the size of fibrils to their respective diffusion coefficients. The results in Fig. 2 show the dependences of the sedimentation coefficient and diffusion coefficient for apoC-II amyloid fibrils based on their mass per unit length and subunit molecular weights (MacRaild et al., 2003). The results show a strong dependence of the diffusion coefficient for fibrils in the range of 1–2000 subunits reaching a plateau level for larger fibrils. The ability to relate sedimentation coefficients to the molecular weights and diffusion coefficients of fibrils

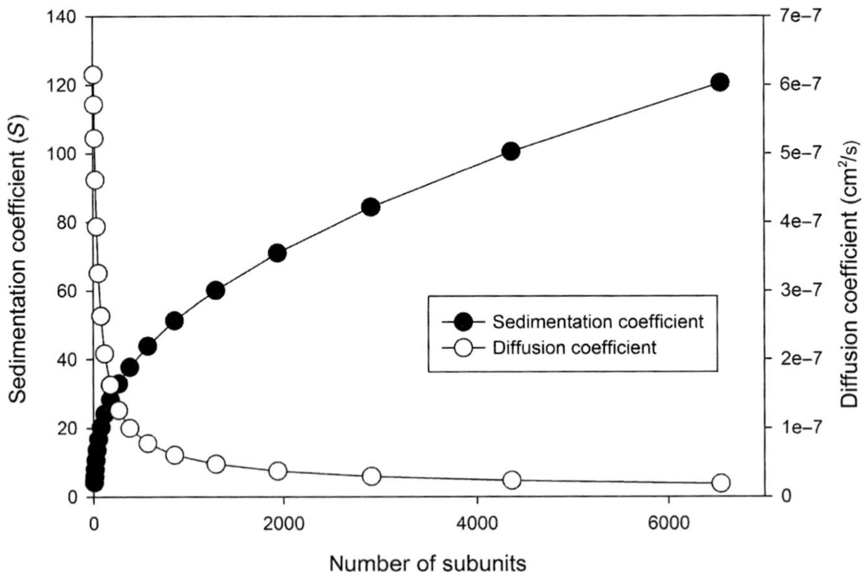

Figure 2 Dependence of sedimentation coefficient and diffusion coefficient on the number of subunits within a worm-like chain. Parameters used for the worm-like chain are based on the values pertaining to human apoC-II amyloid fibrils.

permits the use of the program SEDFIT and $c(s)$ analysis to determine the size distribution of diffusing amyloid fibrils (Schuck, 1998, 2000). An example of the use of $c(s)$ for the analysis of apoC-II fibrils is presented in Fig. 3, which shows the fits obtained to experimental data and the resulting size

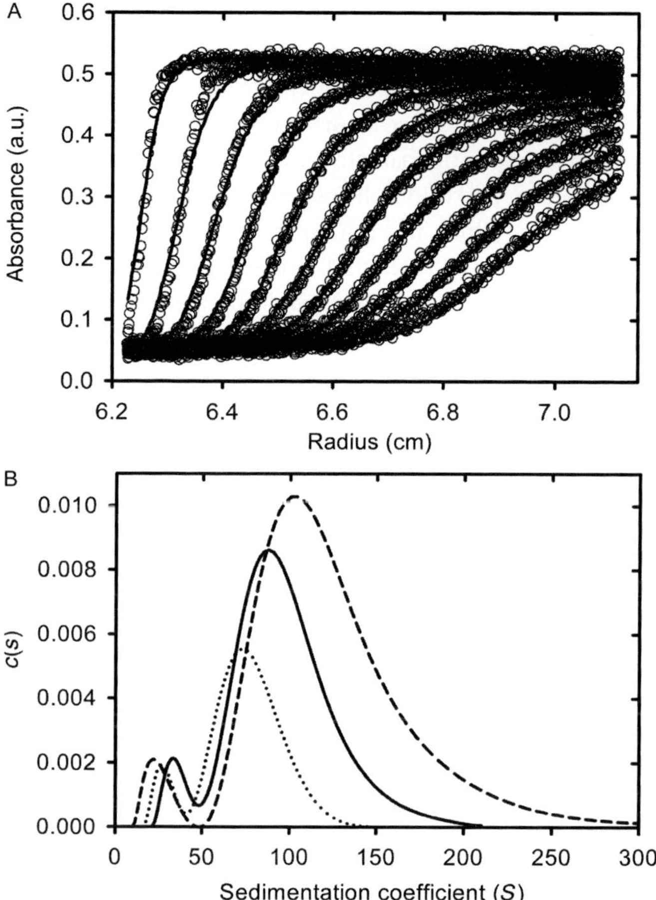

Figure 3 Sedimentation velocity analysis of apoC-II fibrils formed at different apoC-II concentrations. ApoC-II samples were prepared by incubating apoC-II in refolding buffer (100 m*M* sodium phosphate, 0.05% sodium azide, pH 7.4) at 25 °C for 14 days prior analytical ultracentrifugation at 8000 rpm at 20 °C with radial scans taken at 8 min intervals. (A) Radial scans for apoC-II sample (0.3 mg/mL), monitored at 280 nm (open circles). The fit of the data to a $c(s)$ model (Schuck, 2000) is also shown (solid lines). (B) Sedimentation coefficient distributions obtained by $c(s)$ analysis for fibril samples formed at 0.1 mg/mL (dotted line), 0.3 mg/mL (solid line), and 0.5 mg/mL (dashed line). *Adapted with permission from Yang et al. (2012).*

distributions for fibrils formed at different starting concentrations. These data formed the basis for the development of an isodesmic self-association model coupled to closed-loop formation to describe the formation of apoC-II fibrils (Yang et al., 2012).

3.2 Sedimentation Velocity Procedures

Procedures for the analysis of sedimentation data for amyloid fibrils have previously been reviewed (Mok & Howlett, 2006; Pham, Mok, & Howlett, 2009). Briefly, there are a number of considerations. It is important to ensure, when comparing data for amyloid fibrils, that sedimentation experiments are performed under the same conditions of temperature, buffer density, and viscosity. This can be particularly relevant for amyloid fibrils, which often have different structural and solution properties at different temperatures. Experimentally, rotors should be preequilibrated with ultracentrifuge cells assembled ahead of time. The rotor and cells can be subjected to up to 2 h of temperature equilibration in the ultracentrifuge prior to analysis. A recent benchmarking and validation exercise has confirmed the stability of the temperature within the cells when the ultracentrifuge chamber temperature is at 20 °C (Zhao et al., 2014). A further consideration is that large aggregates, such as amyloid fibrils and inclusions, may sediment significantly at 3000 rpm. Thus, minimizing the setup period where the rotor is spinning but data not yet being collected is particularly important.

Sedimentation data can be collected using either optical density, interference, or the fluorescence detection system (FDS). For optical density measurements, a wavelength of 280 nm is commonly used although increased sensitivity can be achieved at a wavelength of 230 nm. At this wavelength, advantage is taken of the high absorbance of peptide bonds and high intensity of the light source. Measurements can also be made using the interference optics available on the model XL-I. Interference optics offer the advantage that scans can be collected quickly (<10 s) with high accuracy and sensitivity and are useful for solution conditions where there is interference from strongly absorbing ligands or buffer components. The recent development by AVIV Biomedical of the FDS for the Beckman XL-A analytical ultracentrifuge has greatly increased the versatility and sensitivity of AUC in a broad range of contexts. In particular, the FDS increases the accessible range of analyte concentrations, allowing accurate detection of sedimenting molecules in the picomolar concentration range, and provides specific detection of fluorescent molecules within complex mixtures (Kroe & Laue, 2009). Using this detection system, initiation, laser lock, and setup of the emission gain and experiment settings require the rotor to be spinning at 3000 rpm, processes that can take up to 10–15 min to complete. In order to minimize the time of this setup period where fibrils may already start to sediment, fluorescence gains may be set using prior knowledge of the approximate fluorescence yields of samples. Previously saved experimental method conditions may also be used. In addition, firing of the laser requires a centrifuge chamber vacuum below 150 μm. It is thus recommended that the vacuum be allowed to equilibrate to below this value while the rotor is stationary.

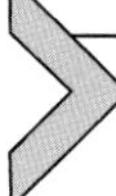

4. HETEROGENEOUS SYSTEMS

4.1 Effects of αB-Crystallin on Mature ApoC-II Amyloid Fibrils

Continuous sedimentation coefficient ($c(s)$) distribution analysis can be extended to determine the effects of extrinsic molecules and proteins on the interactions between fibrils. We studied the interaction of the small heat-shock protein, αB-crystallin, with mature apoC-II amyloid fibrils and showed that αB-crystallin binds directly to apoC-II fibrils with an apparent K_d of 5.4 μM (Binger et al., 2013). αB-crystallin exists in solution as a polydisperse and dynamic population of oligomeric complexes, including high-molecular-weight oligomers consisting of more than 50 subunits (Baldwin, Lioe, Robinson, Kay, & Benesch, 2011). Thus, we used analytical ultracentrifugation to investigate the possibility that αB-crystallin oligomers can bind multiple apoC-II fibrils simultaneously thereby mediating their lateral association. Mature apoC-II fibrils formed at an apoC-II concentration of 33 μM were incubated in the presence of increasing concentrations of αB-crystallin up to 15 μM for 72 h and subjected to sedimentation velocity analysis. Sedimentation velocity data (Fig. 4) revealed a significant increase in fibril sedimentation rate in the presence of αB-crystallin at a concentration of 5 μM or greater, indicated by the increased distance between radial scans in these samples. Analysis of the sedimentation velocity data using the $c(s)$ model and conversion of the resulting distributions to weight-average sedimentation coefficients (S_{av}) confirmed this large increase in sedimentation rate. The calculated weight-average sedimentation coefficient in the presence of 15 μM αB-crystallin was 14-fold that of apoC-II fibrils alone, suggesting that the observed changes in sedimentation behavior were not simply due to binding of αB-crystallin to individual fibrils. Transmission electron microscopy confirmed the presence of large aggregates consisting of many fibrils that were not seen in samples of apoC-II alone, indicating that αB-crystallin induces the lateral association and tangling of apoC-II fibrils.

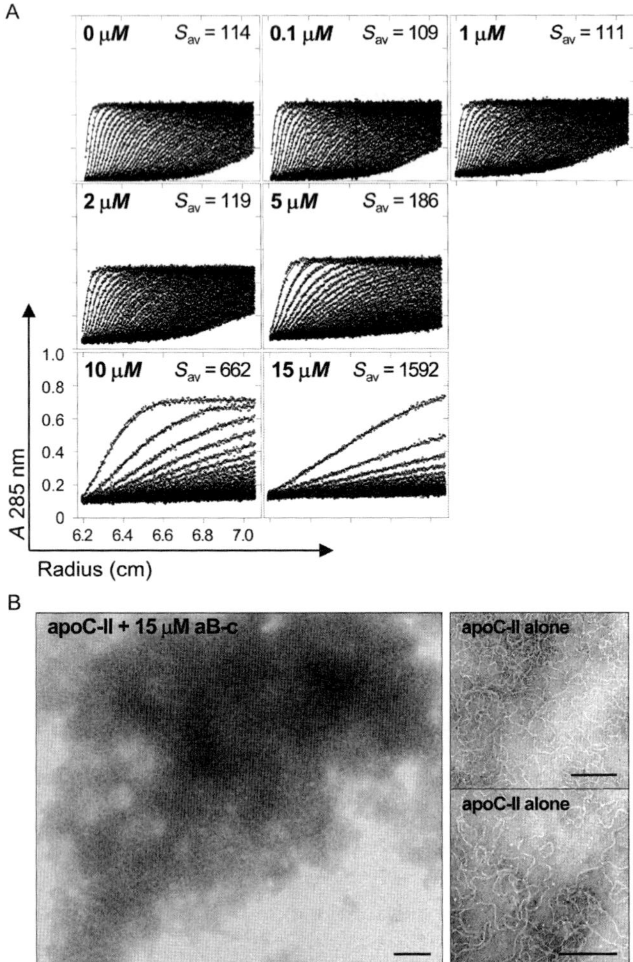

Figure 4 Incubation of preformed fibrils with different concentrations of αB-crystallin. Fibrils were incubated with increasing concentrations of αB-crystallin for 72 h and examined by sedimentation velocity and TEM. (A) Radial absorbance scans at 285 nm (circles) of preformed apoC-II fibrils incubated with 0, 0.1, 1, 2, 5, 10, or 15 μ*M* αB-crystallin for 72 h. Fit of the data to the *c*(*s*) model is also shown (solid lines). Inset: S_{av} for each sample. (B) TEM of preformed apoC-II fibrils incubated with 15 μ*M* αB-crystallin for 72 h. Also shown are control samples of apoC-II fibrils incubated for the same period without αB-crystallin (smaller panels). Scale bars correspond to 200 nm. *Adapted with permission from Binger et al. (2013).*

4.2 The Use of Fluorescence Detection

We have recently reviewed the applications of fluorescence detected (FD) AUC to the analysis of amyloid systems, including investigation of amyloid-forming proteins labeled with small-molecule fluorophores or green

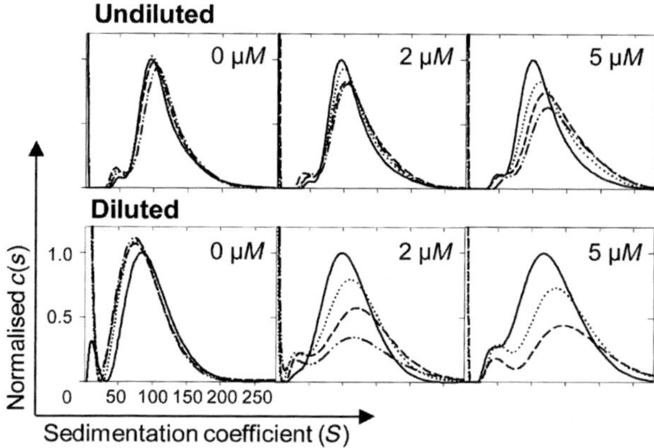

Figure 5 Fluorescence-detected sedimentation analysis of apoC-II amyloid fibrils under dilution in the presence and absence of αB-crystallin. (A) Continuous sedimentation coefficient distributions of apoC-II fibrils undiluted or diluted 1:10 and incubated with 0, 2, or 5 μ*M* αB-crystallin for 0 h (solid), 8 h (dotted), 24 h (dashed), or 72 h (dash–dot–dot). Total integrated signal for each distribution was normalized with respect to the corresponding $t=0$ sample. *Adapted with permission from Binger et al. (2013).*

fluorescent protein, and the use of extrinsic fluorophores and fluorescently labeled lipids that bind to amyloid aggregates and intermediates (Mok et al., 2011).

We further exploited the ability of FD-AUC to detect the sedimentation of fluorescent components of multicomponent protein mixtures to accurately determine the effect of αB-crystallin on the size distribution of apoC-II fibrils after dilution. Fluorescent apoC-II was prepared by labeling of the protein using Alexa-488 maleimide at an engineered cysteine residue (Ryan, Howlett, & Bailey, 2008), and fibrils were formed from a mixture of 2% Alexa-488-apoC-II and 98% unlabeled, wild-type apoC-II at a total apoC-II concentration of 33 μ*M*. Fluorescent fibrils were then incubated with 0, 2, or 5 μ*M* αB-crystallin and subsequently diluted 1:10 in buffer containing the appropriate concentration of αB-crystallin. After this dilution, the apoC-II concentration was approximately 3.3 μ*M* and, thus, αB-crystallin in the samples comprised a large proportion of the total protein concentration. As high-molecular-weight αB-crystallin oligomers sediment under the centrifugal fields used for analysis of apoC-II fibrils, it was critical to monitor only the fibrillar component of the mixture in order to avoid interference from αB-crystallin and to obtain accurate size distributions and weight-average sedimentation coefficients for the apoC-II fibrils. Analysis of the FD sedimentation velocity data (Fig. 5) showed that dilution of the fibrils in the absence of αB-crystallin resulted in a reduction in the weight-average sedimentation coefficient corresponding to fragmentation of fibrils (Binger et al., 2008) and consistent with our model of apoC-II fibril formation including fibril breakage and rejoining (Binger et al., 2008). The presence of αB-crystallin during and after dilution protected fibrils against fragmentation. Interestingly, the sedimentation velocity data also showed a time-dependent reduction in fluorescence signal sedimenting in the presence of αB-crystallin suggesting dissociation of fibrils. Subsequent fibril pelleting assays, performed as described in our earlier review (Mok & Howlett, 2006), confirmed that αB-crystallin mediated the dissociation of fibrils resulting in monomeric apoC-II.

5. SUMMARY

The preparative ultracentrifuge provides a useful method for the fractionation of amyloid oligomers, closed loop and linear fibrils, as well as fibrils with different morphologies. Fibril aggregates with sedimentation coefficients up to 10,000 S can be fractionated using this approach. The analytical ultracentrifuge offers much higher precision in the characterization of the size distributions of amyloid fibrils. For fibrils where there is prior knowledge of the relationship between the molecular weight of the fibrils and their diffusion coefficients, *c*(*s*) analysis allows the size dependence of fibril diffusion coefficients to be included in the analysis. The recent development of the FDS for the analytical ultracentrifuge has greatly extended the ability to analyze heterogeneous systems and the effects of specific reagents on the sedimentation behavior of amyloid fibrils.

ACKNOWLEDGMENTS

We thank Danny Hatters and Angelique Ormsby for permission to include details of the preparation of huntingtin aggregates. This work was supported by the Australian Research Council (projects DP0877800 and DP0984565). M.D.W.G. is the recipient of the C.R. Roper Fellowship and an Australian Research Council Future Fellowship (project number FT140100544).

REFERENCES

Antoni, G., Lubberink, M., Estrada, S., Axelsson, J., Carlson, K., Lindsjo, L., et al. (2013). In vivo visualization of amyloid deposits in the heart with 11C-PIB and PET. *Journal of Nuclear Medicine: Official Publication, Society of Nuclear Medicine*, *54*, 213–220.

Baldwin, A. J., Lioe, H., Robinson, C. V., Kay, L. E., & Benesch, J. L. (2011). alphaB-crystallin polydispersity is a consequence of unbiased quaternary dynamics. *Journal of Molecular Biology*, *413*, 297–309.

Binger, K. J., Ecroyd, H., Yang, S., Carver, J. A., Howlett, G. J., & Griffin, M. D. (2013). Avoiding the oligomeric state: AlphaB-crystallin inhibits fragmentation and induces dissociation of apolipoprotein C-II amyloid fibrils. *FASEB Journal: Official Publication of the Federation of American Societies for Experimental Biology*, *27*, 1214–1222.

Binger, K. J., Pham, C. L., Wilson, L. M., Bailey, M. F., Lawrence, L. J., Schuck, P., et al. (2008). Apolipoprotein C-II amyloid fibrils assemble via a reversible pathway that includes fibril breaking and rejoining. *Journal of Molecular Biology*, *376*, 1116–1129.

Bruggink, K. A., Muller, M., Kuiperij, H. B., & Verbeek, M. M. (2012). Methods for analysis of amyloid-beta aggregates. *Journal of Alzheimer's Disease: JAD*, *28*, 735–758.

Chatani, E., Lee, Y. H., Yagi, H., Yoshimura, Y., Naiki, H., & Goto, Y. (2009). Ultrasonication-dependent production and breakdown lead to minimum-sized amyloid fibrils. *Proceedings of the National Academy of Sciences of the United States of America*, *106*, 11119–11124.

Conway, K. A., Lee, S. J., Rochet, J. C., Ding, T. T., Williamson, R. E., & Lansbury, P. T., Jr. (2000). Acceleration of oligomerization, not fibrillization, is a shared property of both alpha-synuclein mutations linked to early-onset Parkinson's disease: Implications for pathogenesis and therapy. *Proceedings of the National Academy of Sciences of the United States of America*, *97*, 571–576.

de Messieres, M., Huang, R. K., He, Y., & Lee, J. C. (2014). Amyloid triangles, squares, and loops of apolipoprotein C-III. *Biochemistry*, *53*, 3261–3263.

Demeler, B., & Saber, H. (1998). Determination of molecular parameters by fitting sedimentation data to finite-element solutions of the Lamm equation. *Biophysical Journal*, *74*, 444–454.

Fujita, H. (1962). *Mathematical theory of sedimentation analysis*. New York: Academic Press.

Gosal, W. S., Morten, I. J., Hewitt, E. W., Smith, D. A., Thomson, N. H., & Radford, S. E. (2005). Competing pathways determine fibril morphology in the self-assembly of beta2-microglobulin into amyloid. *Journal of Molecular Biology*, *351*, 850–864.

Griffin, M. D., Mok, M. L., Wilson, L. M., Pham, C. L., Waddington, L. J., Perugini, M. A., et al. (2008). Phospholipid interaction induces molecular-level polymorphism in apolipoprotein C-II amyloid fibrils via alternative assembly pathways. *Journal of Molecular Biology*, *375*, 240–256.

Hammarstrom, P., Jiang, X., Deechongkit, S., & Kelly, J. W. (2001). Anion shielding of electrostatic repulsions in transthyretin modulates stability and amyloidosis: Insight into the chaotrope unfolding dichotomy. *Biochemistry*, *40*, 11453–11459.

Hatters, D. M., MacPhee, C. E., Lawrence, L. J., Sawyer, W. H., & Howlett, G. J. (2000). Human apolipoprotein C-II forms twisted amyloid ribbons and closed loops. *Biochemistry*, *39*, 8276–8283.

Kroe, R. R., & Laue, T. M. (2009). NUTS and BOLTS: applications of fluorescence-detected sedimentation. *Analytical Biochemistry*, *390*, 1–13.

Lashuel, H. A., Hartley, D. M., Balakhaneh, D., Aggarwal, A., Teichberg, S., & Callaway, D. J. (2002). New class of inhibitors of amyloid-beta fibril formation. Implications for the mechanism of pathogenesis in Alzheimer's disease. *The Journal of Biological Chemistry*, *277*, 42881–42890.

Lashuel, H. A., Lai, Z., & Kelly, J. W. (1998). Characterization of the transthyretin acid denaturation pathways by analytical ultracentrifugation: Implications for wild-type, V30M, and L55P amyloid fibril formation. *Biochemistry*, *37*, 17851–17864.

MacRaild, C. A., Hatters, D. M., Lawrence, L. J., & Howlett, G. J. (2003). Sedimentation velocity analysis of flexible macromolecules: Self-association and tangling of amyloid fibrils. *Biophysical Journal*, *84*, 2562–2569.

Mok, Y. F., & Howlett, G. J. (2006). Sedimentation velocity analysis of amyloid oligomers and fibrils. *Methods in Enzymology*, *413*, 199–217.

Mok, Y. F., Ryan, T. M., Yang, S., Hatters, D. M., Howlett, G. J., & Griffin, M. D. (2011). Sedimentation velocity analysis of amyloid oligomers and fibrils using fluorescence detection. *Methods*, *54*, 67–75.

Nagel-Steger, L., Demeler, B., Meyer-Zaika, W., Hochdorffer, K., Schrader, T., & Willbold, D. (2010). Modulation of aggregate size- and shape-distributions of the amyloid-beta peptide by a designed beta-sheet breaker. *European Biophysics Journal: EBJ*, *39*, 415–422.

Olshina, M. A., Angley, L. M., Ramdzan, Y. M., Tang, J., Bailey, M. F., Hill, A. F., et al. (2010). Tracking mutant huntingtin aggregation kinetics in cells reveals three major populations that include an invariant oligomer pool. *The Journal of Biological Chemistry*, *285*, 21807–21816.

Ormsby, A. R., Ramdzan, Y. M., Mok, Y. F., Jovanoski, K. D., & Hatters, D. M. (2013). A platform to view huntingtin exon 1 aggregation flux in the cell reveals divergent influences from chaperones hsp40 and hsp70. *The Journal of Biological Chemistry*, *288*, 37192–37203.

Petkova, A. T., Leapman, R. D., Guo, Z., Yau, W. M., Mattson, M. P., & Tycko, R. (2005). Self-propagating, molecular-level polymorphism in Alzheimer's beta-amyloid fibrils. *Science*, *307*, 262–265.

Pham, C., Mok, Y.-F., & Howlett, G. J. (2009). Sedimentation velocity analysis of amyloid fibrils. *Methods in Molecular Biology*, *752*, 179–196.

Ryan, T. M., Howlett, G. J., & Bailey, M. F. (2008). Fluorescence detection of a lipid-induced tetrameric intermediate in amyloid fibril formation by apolipoprotein C-II. *The Journal of Biological Chemistry*, *283*, 35118–35128.

Schuck, P. (1998). Sedimentation analysis of noninteracting and self-associating solutes using numerical solutions to the Lamm equation. *Biophysical Journal*, *75*, 1503–1512.

Schuck, P. (2000). Size-distribution analysis of macromolecules by sedimentation velocity ultracentrifugation and Lamm equation modeling. *Biophysical Journal*, *78*, 1606–1619.

Sipe, J. D., & Cohen, A. S. (2000). Review: History of the amyloid fibril. *Journal of Structural Biology*, *130*, 88–98.

Slow, E. J., Graham, R. K., & Hayden, M. R. (2006). To be or not to be toxic: Aggregations in Huntington and Alzheimer disease. *Trends in Genetics: TIG*, *22*, 408–411.

Thorn, D. C., Meehan, S., Sunde, M., Rekas, A., Gras, S. L., MacPhee, C. E., et al. (2005). Amyloid fibril formation by bovine milk kappa-casein and its inhibition by the molecular chaperones alphaS- and beta-casein. *Biochemistry*, *44*, 17027–17036.

Yang, S., Griffin, M. D., Binger, K. J., Schuck, P., & Howlett, G. J. (2012). An equilibrium model for linear and closed-loop amyloid fibril formation. *Journal of Molecular Biology*, *421*, 364–377.

Zerovnik, E. (2002). Amyloid-fibril formation. Proposed mechanisms and relevance to conformational disease. *European Journal of Biochemistry/FEBS*, *269*, 3362–3371.

Zhao, H., Balbo, A., Metger, H., Clary, R., Ghirlando, R., & Schuck, P. (2014). Improved measurement of the rotor temperature in analytical ultracentrifugation. *Analytical Biochemistry*, *451*, 69–75.

CHAPTER TWELVE

AUC and Small-Angle Scattering for Membrane Proteins

Aline Le Roy*,†,‡, Kai Wang*,†,‡, Béatrice Schaack*,†,‡, Peter Schuck§, Cécile Breyton*,†,‡, Christine Ebel*,†,‡,1

*Université Grenoble Alpes, IBS, Grenoble, France
†CNRS, IBS, Grenoble France
‡CEA, IBS, Grenoble, France
§Dynamics of Macromolecular Assembly Section, Laboratory of Cellular Imaging and Macromolecular Biophysics, National Institute of Biomedical Imaging and Bioengineering, National Institutes of Health, Bethesda, Maryland, USA
[1]Corresponding author: e-mail address: christine.ebel@ibs.fr

Contents

Methods in Enzymology, Volume 562
ISSN 0076-6879
http://dx.doi.org/10.1016/bs.mie.2015.06.010

Abstract

Analytical ultracentrifugation is a key tool to assess homogeneity of membrane protein samples, to determine protein association state and detergent concentration, and to characterize protein–protein equilibrium. Combining absorbance and interference detections gives information on the amount of the detergent and lipid bound to proteins. Changing the solvent density affects specifically the buoyancy of each of the different components, and can also be used to gain information on particle composition and interaction. We will present the related tools, recently implemented in the softwares Sedphat (sedfitsedphat.nibib.nih.gov/software) and Gussi (http://biophysics.swmed.edu/MBR/software.html), which help to measure the amount of detergent bound to the protein, and ascertain the protein association state within the protein–detergent complex. In addition, fluorescence detection allows focusing specifically on a labeled component within a complex mixture. We present two examples of sedimentation velocity experiments, allowing on one hand to evidence complex formation between an unpurified GFP-labeled protein and a membrane protein, and on the other hand to characterize fluorescent lipid vesicles. Small-angle X-ray and neutron scattering are techniques that give insights into the structure and conformation of macromolecules in solution. However, the detergents used to purify membrane protein are often imperfectly masked due to their amphipathic character. Particular strategies addressing membrane proteins were recently proposed, which are shortly presented.

1. INTRODUCTION

Membrane proteins account for about one-third of the proteins encoded in all genomes (Fagerberg, Jonasson, von Heijne, Uhlen, & Berglund, 2010) and about half of drug targets (Overington, Al-Lazikani, & Hopkins, 2006; Terstappen & Reggiani, 2001). Integral membrane proteins are physiologically embedded in a lipid bilayer, which confers a hydrophobic environment to their hydrophobic transmembrane surface. Detergents used to extract membrane proteins from the lipid bilayer allow their solubilization, purification, and characterization. Proteins, however, are usually less stable in a detergent environment when compared to their native lipidic one. This hampers their biochemical, biophysical, and structural study: only a relatively low number of high-resolution structures, for example, are available to date.

We address here the characterization of the structure and interactions of membrane proteins in solution. They represent a multicomponent system, in which different components (protein, detergent, lipid, water...) interact to form the different types of sedimenting and scattering particles (species). Detergents are characterized by their critical micelle concentration (cmc) above which detergent monomers autoassemble to form small, compact aggregates, i.e., the micelle. Above the cmc, detergents can bind the hydrophobic transmembrane domain of the proteins and solubilize them. Thus, the protein–detergent (pd) complex(es), the detergent micelles—often in unknown concentration—and detergent monomers are present in solution. Additionally, very often, lipids—or other hydrophobic cofactors—copurify with the pd complex and/or are present in the detergent micelles in ill-defined amounts. Finally, the membrane protein is possibly glycosylated.

Analytical ultracentrifugation (AUC) is a key tool to assess sample homogeneity, determine the protein association state and the detergent concentration, and characterize the protein–protein equilibrium and the particle composition. The different types of particles in solution will be distinguished by their sedimentation coefficients in sedimentation velocity (SV), or their buoyant molar masses in sedimentation equilibrium. Combining absorbance and interference detections provides information on particle composition. Changing the solvent density, which affects differentially the buoyancy of the various components, can also be used to get information on particle composition and interaction. Reference works or reviews in the field of membrane protein are, e.g., Burgess, Stanley, and Fleming (2008), Ebel, Møller, and le Maire (2007), Ravaud et al. (2006), Rosenbusch, Lustig, Grabo, Zulauf, and Regenass (2001), and Tanford, Nozaki, Reynolds, and Makino (1974). We already published related protocols dealing with membrane proteins and AUC (Ebel, 2011; le Maire et al., 2008; Le Roy et al., 2013; Salvay, Santamaria, le Maire, & Ebel, 2007). These publications have helped to implement specific tools, which will be presented here, for the analysis of membrane proteins, in the softwares Sedphat (sedfitsedphat.nibib.nih.gov/software) and Gussi (http://biophysics.swmed.edu/MBR/software.html), described in Chapter 5 (Calculations and Publication-Quality Illustrations for Analytical Ultracentrifugation Data by Chad Brautigam). In addition, the fluorescence detection allows focusing specifically on a labeled component within a complex mixture. We present two examples of SV experiments allowing to evidence the formation of a membrane protein complex with an unpurified GFP-labeled soluble protein, and to characterize fluorescently labeled lipid vesicles.

Small-angle X-ray scattering (SAXS) and small-angle neutron scattering (SANS) are techniques that give an insight into the structure and the conformation of macromolecules in solution. Particular strategies to address the multicomponent nature of membrane proteins were recently proposed, which will be also shortly presented.

2. MEMBRANE PROTEINS IN AUC

2.1 On the General Principles of AUC and Analysis for Membrane Proteins

The theory of centrifugation, described in Chapter 3 (Sedimentation Velocity: Fundamental Principles by Walter Stafford and John J. Correia), is general and stands for membrane proteins. The dynamic exchange of detergent between the monomer, the pd complex, and the detergent micelle has in practice no consequence, and these three types of molecules or assemblies will be described as noninteracting particles (Salvay et al., 2007), if there is no equilibrium between different protein–detergent complexes (see Section 2.4). The programs of analysis will be the same as for nonmembrane proteins, e.g., Sedfit/Sedphat developed by Peter Schuck (sedfitsedphat.nibib.nih.gov/software), Ultrascan by Borris Demeler (http://www.ultrascan3.uthscsa.edu), and Sedanal by Walter Stafford (http://www.sedanal.org/). Basically, the sedimentation depends on the sedimentation and diffusion coefficients (s and D, respectively) of noninteracting species, plus eventually, in the case of chemical equilibrium, of the equilibrium and kinetic constants of association and dissociation. The specificity of studying membrane proteins comes from the obligatory heterogeneity of the sample, and of the more complex description of the particle.

2.2 Expression of the Buoyant Molar Mass for Membrane Proteins

For a pd complex, we label protein, detergent, and lipid, p, d, and l, respectively. Possible glycosylation can be taken into account in the proteic component p or described as an individual component. Hydration (label w) has also to be considered, when the solvent density differs from $1/\bar{v}_w$, i.e., contains a significant amount of another component (e.g., sucrose). M_b is the buoyant molar mass of the pd complex, M_p the protein molar mass, δ_d and δ_l the amount of bound detergent and lipid, in gram per gram of protein, $\rho°$ the solvent density, and $\bar{v}_i$ the partial specific volumes of the components (i = p, d, l). M_b can be expressed as a function of the pd complex molar mass, M_{pd}, and related partial specific volume, $\bar{v}_{pd}$:

$$M_b = M_{pd}\left(1 - \rho^\circ \bar{v}_{pd}\right) \quad (1)$$

$$M_{pd} = M_p(1 + \delta_d + \delta_l + \delta_w) \quad (2)$$

$$\bar{v}_{pd} = \left(\bar{v}_p + \delta_d\bar{v}_d + \delta_l\bar{v}_l + \delta_w\bar{v}_w\right)/(1 + \delta_d + \delta_l + \delta_w) \quad (3)$$

M_b can be conveniently expressed as a function of protein molar mass (Casassa & Eisenberg, 1964; Eisenberg, 1976; Tanford & Reynolds, 1976), with $(\partial\rho/\partial c_p)_\mu$ being the density increment at constant chemical potential μ of the solvent components:

$$M_b = M_p\left(\partial\rho/\partial c_p\right)_\mu \quad (4)$$

In the following, for simplicity, we will express the lipid and detergent contribution together within component d. $(\partial\rho/\partial c_p)_\mu$ can be expressed as:

$$\left(\partial\rho/\partial c_p\right)_\mu = \left[\left(1 - \rho^\circ\bar{v}_p\right) + \delta_d(1 - \rho^\circ\bar{v}_d)\right] \quad (5)$$

$(\partial\rho/\partial c_p)_\mu$ can be measured in density experiments performed with appropriately prepared samples, given the protein concentration, c_p, is precisely determined. By analogy with a two-component system (protein + solvent), some authors use the operational partial specific volume ϕ', which is not the partial specific volume of the particle:

$$\left(\partial\rho/\partial c_p\right)_\mu = \left(1 - \rho^\circ\phi'\right) \quad (6)$$

Partial specific volumes of the protein can be precisely estimated from tabulated data according to amino acid composition. The partial specific volume of lipid is often close to 1, making its contribution often not considered in the buoyant mass. However, the value for specific lipids can deviate significantly from this value (Tanford et al., 1974). Partial specific volumes of detergent are highly variable and depend on the detergent type (le Maire, Champeil, & Møller, 2000; Tanford et al., 1974): they can be measured by precise density measurements, or by measurement of sedimentation in solvents of different densities (see, e.g., Salvay & Ebel, 2006), or from tabulated volume of chemical groups (Durchschlag & Zipper, 1997).

When performing the analysis, in the programs or calculations, attention will be paid to the fact that the determined molar mass is mathematically related to the input values of the solvent density, $\rho^\circ{}_{input}$, and of the partial specific volume for the pd complex, $\bar{v}_{input}$, the latter being in general an approximate value. The molar mass provided by the programs is thus an apparent molar mass (M_{app}). In most case, it is judicious to recalculate the buoyant molar mass, $M_b = M_{app}\left(1 - \rho^\circ{}_{input}\bar{v}_{input}\right)$, prior further analysis.

2.3 Sedimentation Equilibrium Is Difficult and Restricted to Favorable Cases

Heterogeneity is difficult to solve by sedimentation equilibrium. In the case of membrane proteins, the intrinsic heterogeneity of the sample (at least three type of particles: detergent monomer and micelle, and pd complex) may not be detected using absorbance optics if the detergent does not absorb (which should be checked, since it often dissolves impurities that absorb). In the favorable former situation, measurement using absorbance will focus on the pd complex. Interpretation of the buoyant molar mass in terms of protein molar mass, and equilibrium constants in the case of protein association–dissociation reactions, will be done either by measuring $(\partial\rho/\partial c_p)_\mu$ (see, e.g., Butler, Ubarretxena-Belandia, Warne, & Tate, 2004) or by estimating it from the composition in terms of bound detergent and lipids (see, e.g., le Maire et al., 2008; le Maire et al., 2000; Ravaud et al., 2006), or by working at the detergent neutral buoyancy, i.e., with the condition $(1-\rho^\circ\bar{v}_d)=0$. Using sucrose or other solvent components for changing solvent density (Lustig, Engel, Tsiotis, Landau, & Baschong, 2000) is more complicated because it makes hydration contributes to the pd complex buoyant mass. Some detergents, however, have nearly the same density as water and can be matched, in the buffer or using an appropriate mixture of H_2O and D_2O. The latter approach was developed by Reynolds and Tanford (1976) and applied, e.g., by Fleming et al. (2004) for successfully deciphering the energetics of membrane protein association. Principle and comments on the methods are reported in details in, e.g., Burgess et al. (2008). Note that, as will be presented in Section 2.7, when using D_2O, the exchange of labile hydrogens atoms in the protein and detergent molecules for deuterium has to be taken into account in the analysis of the buoyant molar mass, a feature that is not experienced when using $H_2{}^{18}O$.

2.4 SV—General Considerations

Our standard protocol is to perform SV at different concentrations. Because association equilibrium can be modulated by detergent—which takes part in this equilibrium—it may be appropriate to investigate different protein and detergent concentrations, in particular when the detergent concentration is close to the cmc. Analysis of SV experiments in terms of a distribution of sedimentation coefficients, $c(s)$, is the best way to detect concentration effects. Changes in the sedimentation coefficients will indicate macromolecular equilibrium. As an example, the sedimentation coefficient of the membrane protein HUPON increases, when, for a given detergent concentration, protein concentration increases, indicating an equilibrium of association between proteins. It increases also when decreasing the detergent concentration for a given concentration of protein, which was rationalized considering that the protein monomer and dimer bind detergent in different ways (Josse et al., 2002). The $c(s)$ analysis also allows one to characterize the pd complex within a heterogeneous sample. Analysis in terms of non-interacting species will provide M_b values that should be interpreted with caution, since heterogeneity or weak signal over noise ratio makes the values of D defined with a poor precision. The values of s, as the absorbance and interference signals, are determined in a robust way from the $c(s)$ analysis and will be analyzed with confidence. Assuming that the boundary represents a defined species, the Svedberg equation can be expressed as:

$$s = M_b / N_A 6\pi\eta^\circ R_s, \tag{7}$$

where the sedimentation coefficient, s, is related to the particle buoyant molar and Stokes radius, M_b and R_s, the solvent viscosity, η°, with N_A Avogadro's number. Details and comments on M_b are given in Section 2.1.

The frictional ratios f/f_{min} are usually defined as the ratio of the hydrodynamic radius to R_{min}, the radius of the anhydrous volume $\left(M_p\bar{\nu}_p + \delta_d M_p \bar{\nu}_d\right)$ of the pd complex (and not of the protein). The typical values of f/f_{min} do not differ from those of soluble proteins. A value of 1.2–1.25 indicates a globular compact species, >1.5 values correspond to elongated or/and glycosylated particles.

The calculation of the normalized s_{20w} needs an input value for the partial specific volume of the pd complex. Because this value is entailed with uncertainty, we sometimes prefer to fix it considering an arbitrary hypothesis, e.g., considering the mean value of the partial specific volumes of the protein and detergent.

2.5 SV in a Given Buffer Condition

We generally use a simple protocol, detailed previously in, e.g., Ebel (2011), le Maire et al. (2008), and Salvay et al. (2007). SV profiles are measured at typically three different protein or detergent concentrations, at 280 nm and using interference. In general, we measure an AUC cell filled with the buffer containing only the detergent. We usually fill the reference channel with buffer without detergent. From the superposition of the $c(s)$ distributions, which we obtain with the program Sedfit (Schuck, 2000) and analyze now with the program Gussi (see below), we determine the contributions arising from the free detergent micelles and for the main pd complex(es). We build a table with their *s*- and signal values. The free detergent micelle contribution is easily identified because of its low or null absorbance, and/or by comparison with the detergent-only sample. The free micelle concentration can be determined from the interference signal. If the *s*-value measured for the dp complex within the different samples is constant, we consider that the system is noninteracting.

From the signals (absorbance and interference), and inputs of the extinction coefficients and refractive index increments, $(\partial n/\partial c)$, for the protein (which may be glycosylated) and the detergent (which may comprise lipid), we obtain an estimate of the amount of bound detergent δ_d. A more sophisticated "multiwavelength analysis" in the program Sedphat (Balbo et al., 2005) may be useful in specific cases. It allows, from pairs of SV sets obtained in absorbance and interference, to derive detergent and protein distributions (in μM units) within the investigated *s*-space (Salvay et al., 2007). Further calculations are based on the estimates for the pd complex of *s* and δ_d, and other knowledge we have on the solvent (density and viscosity) and components (partial specific volumes): (1) considering successively different discrete values for the protein molar mass, defined by different assemblages of the constituting polypeptide chain(s): what would be the frictional ratio and R_s of the pd complex? (2) Having an independent estimate of R_s (e.g., from size calibrated size-exclusion chromatography (SEC)): what would be the protein molar mass? (3) Using the fitted value of the frictional ratio in Sedfit: what would be the protein molar mass? This sequence of calculations is now implemented in the program Gussi, as described in Chapter 5 (Calculations and Publication-Quality Illustrations for Analytical Ultracentrifugation Data by Chad Brautigam). Briefly, the $c(s)$ have to be opened in the order interference and then absorbance; the menu "Integration" and "Membrane Protein" are used; options 1, 2, and 3 described above correspond to the calculation types "$f/f°$," "R_s," and "Fitted $f/f°$," respectively. When the sample contains a limited number of species, the results are compared to the independent estimates of M_b and R_s obtained from noninteracting species analysis.

2.6 Complementarity with Size-Exclusion Chromatography Coupled to Light Scattering (SEC/MALS)

We have detailed the complementarity between SV and SEC/MALS in Le Roy, Breyton, and Ebel (2014) and will mention here briefly the main features. SEC/MALS instrumentation for characterizing protein assemblies not only comprises SEC coupled to static light scattering detection but also refractive index and absorbance detectors (and optionally dynamic light scattering, DLS). Measurements are made along the elution profile. Performing the experiment and analysis requires limited expertise when compared to AUC. AUC is more versatile for changing buffer composition, SEC/MALS leading also to the uncontrolled dilution of the sample during the experiment. Both techniques are separative, the quality of the results being dependent on the separation efficiency. AUC is, however, intrinsically better, since the separation of particles of same shape and composition depends on $M^{2/3}$ for AUC compared to $M^{1/3}$ for SEC/MALS. In AUC, particle separation is also modulated by difference in the particle and solvent densities (e.g., LAPAO-lipid micelles do not sediment in dilute aqueous buffer; Nury et al., 2008). In SEC/MALS, analysis of the pd complex mass is made first from the combination of refractive index and absorbance signals, allowing to derive the concentrations of the "detergent" and "protein" components (and thus δ_d). This step is totally similar to that described above for SV, since the same physical characteristics are measured (interference measures variation in the refractive index). In addition, the molar masses of the "detergent" and "protein" components within the complex are obtained from the additional static light scattering signal. The common limit of SEC/MALS and SV analysis, in the discrimination of the detergent and protein compounds, is mainly the uncertainty on the extinction coefficients of the different components.

2.7 SV in Two Buffers, One Containing Isotopically Labeled Water

Varying solvent density in SV gives independent information on the particle density, thus composition. Using isotopically labeled water is a good strategy to change solvent density, since adding an additional compound in the solvent would induce preferential hydration, i.e., add to the complexity of the system.

Global analysis in Sedphat of a set of SV experiments performed for the same sample, in different mixtures of H_2O and isotopically enriched $H_2{}^{18}O$ buffers, was shown to allow the determination, with high precision, of a common partial specific volume and a common *c*(*s*) distribution, for soluble proteins (Brown, Balbo, Zhao, Ebel, & Schuck, 2011). This strategy can be adapted to the study of membrane proteins, which is complicated by the heterogeneity of the partial specific volume within the sample (it is different for the detergent micelles and the pd complex), and therefore requires assignment of different partial specific volumes in different sedimentation coefficient ranges (Brown et al., 2011). We describe below a different approach that additionally allows to exploit the narrow range of frictional ratios of pd complexes to reach directly pd masses and composition.

Using D_2O for varying density contrast is quite common (particularly when characterizing the samples by SANS, see below). We recently published a step-by-step protocol for setting up, realizing, and analyzing SV experiments in hydrogenated and deuterated solvents, in the context of the characterization of membrane protein, in terms of homogeneity, association state, and amount of bound detergent (Le Roy et al., 2013). Due to D–H exchange of the labile hydrogens of the protein and detergent, the molar mass of the component increases in the deuterated buffer, but not the molar volume. In consequence, the buoyant factors $(1-\rho^{\circ}\bar{v}_i)$ should be expressed as $\left((M_D/M)_i - \rho^{\circ}\bar{v}_i\right)$, where $(M_D/M)_i$ is the ratio of the molar mass in D_2O to that in H_2O for component i. With ρ° being named ρ_H and ρ_D in hydrogenated and deuterated buffers, respectively:

$$s_H = M_p\left[\left(1-\rho_H\bar{v}_p\right) + \delta_d\left(1-\rho_H\bar{v}_d\right)\right]/N_A 6\pi\eta_H R_s; \tag{8}$$

$$s_D = M_p\left[\left((M_D/M)_p - \rho_D\bar{v}_p\right) + \delta_d\left((M_D/M)_d - \rho_D\bar{v}_d\right)\right]/N_A 6\pi\eta_D R_s. \tag{9}$$

We keep in mind that R_s is related to particle volume (defined by M_p and δ_d, $\bar{v}_p$, and $\bar{v}_d$) as well as f/f_{min}.

A graphical presentation was proposed (Dach et al., 2012; Le Roy et al., 2013; Nury et al., 2008) to interpret the s_H/s_D results and ascertain if a given protein molecular mass can be compatible with the two sedimentation coefficients measured in the hydrogenated and isotopically labeled buffers. Some parameters ($\bar{v}_p$, $(M_D/M)_p$ and $(M_D/M)_d$, η_H, η_D, ρ_H, and ρ_D) are indeed known with reasonable confidence. Various hypotheses regarding M_p (plausible protein association states) and f/f_{min} (reasonable shape factor, since f/f_{min} is typically in the range 1.25–1.75) will be investigated successively. Any value of the sedimentation coefficient is compatible with a variety of couples of mathematical solutions $(\delta_d; \bar{v}_d)$. The aim is to determine the $(\delta_d; \bar{v}_d)$ ensemble that is compatible with the experimental values of s_H and s_D. We aim to consider also the uncertainty on the sedimentation coefficients. For this purpose, we plot four $((\delta_d; \bar{v}_d)$ line solutions, corresponding to the minimum and maximum (determined by the uncertainty) values of

s_H and s_D. It allows one to determine the (δ_d; $\bar{v}_d$) that are mathematical solutions for s_H and s_D. Then the physical relevance of the (δ_d; $\bar{v}_d$) pairs are examined: δ_d, for example, should be in line with values determined from the signals (see above), the corresponding R_s may be compared to that from other experimental work; $\bar{v}_d$ should be in between values for individual bound compounds (e.g., detergent and lipid). In our previously published works (Dach et al., 2012; Le Roy et al., 2013; Nury et al., 2008), we derived implicit functions $\delta_d = f(\bar{v}_d)$ using the software Maple. The calculation was recently implemented in Sedphat (version 12.02 and later, sedfitsedphat.nibib.nih.gov/software), allowing easy use.

As a demonstration, we use here the sedimentation coefficients determined for the sarcoplasmic reticulum Ca^{2+}-ATPase solubilized in dodecyl-β-D-maltoside (DDM) and delipidated (Salvay et al., 2007). Numerical values are for (hydrogenated) buffer H: $\eta_H = 1.00$ cp and $\rho_H = 1.004\,\mathrm{g\,mL^{-1}}$; for (deuterated) buffer D: $\eta_D = 1.23\,\mathrm{cp}$ and $\rho_D = 1.109\,\mathrm{g\,mL^{-1}}$. The sedimentation coefficients are $s_H = 7.13\,\mathrm{S}$ (we consider as minimum and maximum values 7.03 and 7.23 S) and $s_D = 4.1\,\mathrm{S}$ (we consider as minimum and maximum values 4.0 and 4.2 S). For Ca^{2+}-ATPase: $\bar{v}_p = 0.7425\,\mathrm{mL\,g^{-1}}$, $M_{p\ \mathrm{monomer}} = 109{,}490\,\mathrm{g\,mol^{-1}}$, and $(M_D/M)_p - 1.015$. For DDM, $\bar{v}_d = 0.82\,\mathrm{mL\,g^{-1}}$ and $(M_D/M)_d = 1.014$. For the plots, the partial specific volume of bound component (DDM here) is to be drawn between 0.72 or 0.75 and 0.85 mL g^{-1}; and its bound amount δ_d between 0 and 2.5 g g^{-1}, for a range of frictional ratio comprised between 1 and 2.

The numerical values are introduced in the subprogram "Ebel B-v Plot of detergent binding from Density contrast SV" localized in Sedphat, in the submenu localized in "Options" and "Interaction Calculator." Considering Ca^{2+}-ATPase monomer, a red (dark gray in the print version) zone delimits the values of (δ_d; $\bar{v}_d$) compatible with the experimental values of the sedimentation coefficients. A screenshot is presented for $f/f_{min} = 1.5$ (Fig. 1A). Clicking on the plot leads to a window (Fig. 1B), providing $\bar{v}_d$ and δ_d values corresponding to the position of the arrow, and the related values for R_{min}, and s and M_b in the two buffers. Estimates for the uncertainty of $\bar{v}_d$ and δ_d are easily determined by clicking on the corners of the red (dark gray in the print version) area delimitating the limits of the possible mathematical solutions. Similar plots drawn with $f/f_{min} = 1.25$ and 1.75 are easily obtained by changing the f/f_{min} slider position on the Sedphat windows. When considering the molar mass of a dimer, there is clearly no solution in the—reasonable—range of $\bar{v}_d$ and δ_d considered above; mathematical solutions are found changing

A

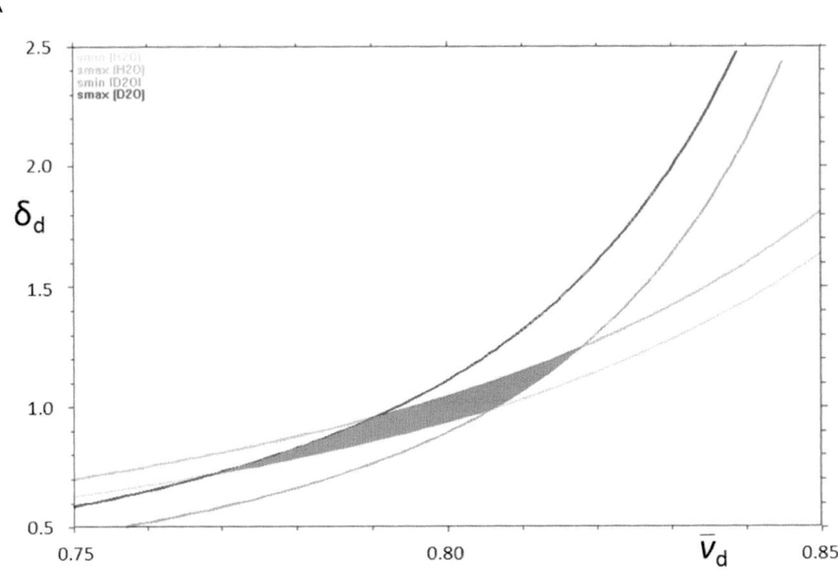

B

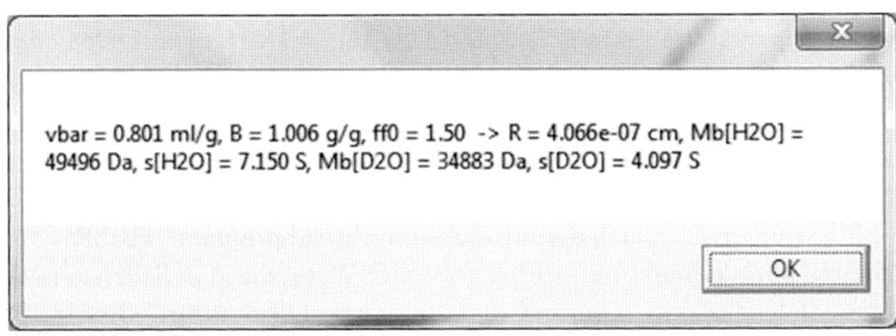

C

Input			Mathematical solution								Conclusion
N_{agg}	M_p (Da)	f/f_{min}	v_{bar} (mL/g)	±	B_{det} (g/g)	±	R_{min}	±	R_s	±	
1	109,490	1.75	0.79	0.2	1.5	0.3	4.4	0.3	7.7	0.5	Excluded
1	109,490	1.5	0.8	0.2	1	0.25	4.0	0.2	6.0	0.3	Possible
1	109,490	1.25	0.83	0.2	0.5	0.2	3.7	0.2	4.6	0.3	Possible
2	218,980	1.75	0.88	0.2	0.25	0.2					Excluded
2	218,980	1.5	≥1.13		≤0.13						Excluded
2	218,980	1.25	>2		<0.05						Excluded

Figure 1 Graphical analysis using Sedphat of s_H and s_D of Ca^{2+}-ATPase. (A) Screenshot from Sedphat program, considering Ca^{2+}-ATPase as a monomer with $f/f_{min} = 1.5$; input parameters for calculation and drawing have been erased. The light green (light gray in the print version), green (gray in the print version), light blue (dark gray in the print version), and blue (black in the print version) lines give $(\bar{v}_d; \delta_d)$ solutions compatible with

the scale of the graphs. The (δ_d; $\bar{v}_d$) solutions and derived R_s values are reported in Fig. 1C and considered in view of what we know about the Ca^{2+}-ATPase complex: Ca^{2+}-ATPase here is delipidated, the $\bar{v}_d$ is thus expected to be close to that of pure DDM; information about the amount of protein-bound detergent and remaining lipid is derived from the analysis of the interference and absorbance signals in AUC; and the Stokes radius is estimated from SEC. The s_H/s_D graphical representation definitively eliminates the possibility of a dimer and of a very elongated shape for the monomer.

To conclude, this method considers the solubilized membrane proteins as a two-component system composed of (possibly glycosylated) protein and bound compounds (detergent + lipid + ...). It is particularly appropriate for ill-defined pd complexes binding lipids or unknown components in undefined amounts. It is based on the fact that proteins are well defined and can associate only into discrete oligomeric states. The method allows to explore rapidly if a given protein association state is compatible or not with s_H and s_D values. It does not consider the optical properties of the components and is complementary to the analysis based on the absorbance and interference signals of the boundaries.

2.8 Following Complex Formation in Cellular Extracts Using the Fluorescence Detection

The fluorescence detection system in AUC has a high selectivity and excellent sensitivity (MacGregor, Anderson, & Laue, 2004) that allows to follow the sedimentation of labeled proteins at very low concentration (p*M*) sample in complex background solution (i.e., cytosol or serum). The available model (from Aviv Biomedical, Inc.), given excitation wavelength is fixed to 488 nm and emission range is between 505 and 565 nm, is suitable for

the minimum and maximum experimental values of s_H and s_D. The red (dark gray in the print version) zone gives the ($\bar{v}_d$; δ_d) solutions compatible with the experimental sedimentation coefficients given their uncertainties. (B) Screenshot providing, from a particular $\bar{v}_d$;δ_d position of the mouse arrow on the plot, here nearly in the center of the red (dark gray in the print version) area, values of $\bar{v}_d$;δ_d; f/f_{min}; minimum radius R_{min}; M_b in buffer H, s_H, M_b in buffer D, s_D. (C) Table reporting the results for different hypothesis. R_s is R_{min} f/f_{min}. Given $\bar{v}_d$ is expected to be close to 0.82 mL g^{-1}, given δ_d was estimated to be in the range 0.5–0.9 from the signals in interference and absorbance, given R_s = 5.5 nm estimated from size-exclusion chromatography, and the dimeric association state and elongated shape are excluded.

use with fluorescein-like (FITC) dyes or GFP-labeled proteins. Due to the short exposure time in the AUC, no significant photoconversion occurs in the detection of conventional dyes, in contrast to photoswitchable fluorescent proteins (Zhao, Ma, et al., 2014). Fluorescent ligand for His-tagged proteins was also described (Hellman et al., 2011). We have checked with FITC-labeled BSA that the sedimentation coefficients measured with fluorescence optics, $s = 4.28 \pm 0.12$ S, agree with those determined using absorbance or interference (Fig. 2A), close to literature value (Ghirlando et al., 2013), in a concentration range from 1 n*M* (without adjuvant), or less than 3 p*M* (with adjuvant), to 15 μ*M*. Adjuvant (here unlabeled BSA at 0.5 mg mL^{-1}) avoids protein at low concentration sticking onto plastic and quartz surfaces (Fig. 2B). At 30 μ*M*, *s* decreases due to optical effects (Zhao, Casillas, Shroff, Patterson, & Schuck, 2013). The low detection

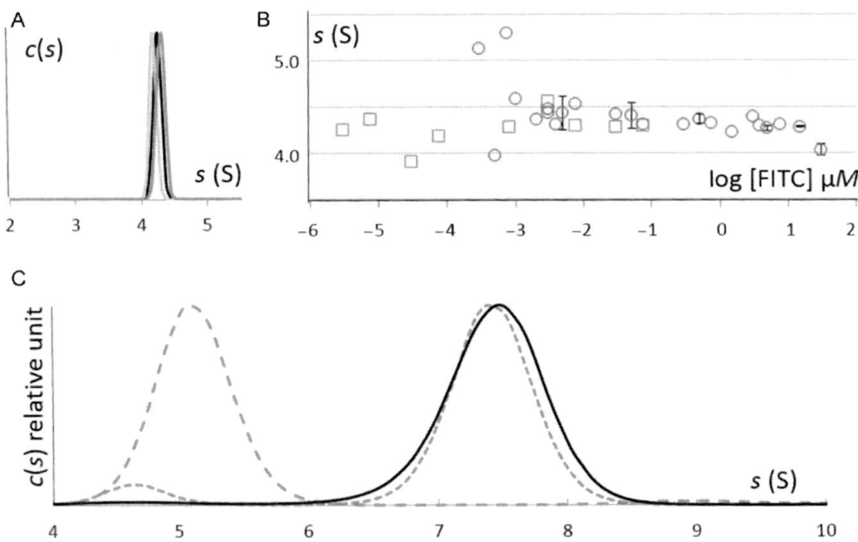

Figure 2 Sedimentation of labeled proteins using fluorescence optics. (A) *c*(*s*) superposition, from 280 nm absorbance, interference, and fluorescence data, for FITC-labeled BSA (12 and 14 μ*M*, respectively) in 30 m*M* phosphate pH 7.4, 100 m*M* NaCl. (B) *s* measured for BSA-FITC at different concentrations, with (red square) or without (purple) circle) 0.5 mg mL^{-1} (8 μ*M*) unlabeled BSA. (C) *c*(*s*) superposition for 2.5 μL GFP-pb5 crude extract, diluted at 0.5% for a final concentration of 8 n*M*, without (red dashed line) or with FhuA at a molar ratio FhuA/GFP-pb5 of 1 (green dotted line) and 5 (black, continuous line), in 50 m*M* Tris–HCl pH 8, 150 m*M* NaCl supplemented with 30 m*M* octyl glucoside. The AUC experiments were performed at 20 °C and 42,000 rpm using 12 mm path length cells. (See the color plate.)

limits offer the potential, for example, to study very high-affinity interactions (Zhao, Lomash, Glasser, Mayer, & Schuck, 2013; Zhao, Mayer, & Schuck, 2014).

Selectivity of fluorescence detection allows investigating the behavior of fluorescent macromolecules by SV-AUC even in very complex medium. We investigated whether the association state and interactions of a fluorescent protein can be studied in crude cellular extracts. We have studied the phage T5 receptor-binding protein pb5, which forms a very stable complex with its membrane receptor FhuA (Flayhan, Wien, Paternostre, Boulanger, & Breyton, 2012; Plançon et al., 2002). pb5 was fused at its N-terminal with GFP and the fusion protein was overexpressed in *Escherichia coli*. Cells were chemically broken with Bugbuster (Novagen) and centrifuged at 16,000 × *g* for 20 min at 4 °C. Aliquots of 2.5 μL of supernatant, considered as the crude extract, were diluted to 500 μL with buffer containing or not the separately purified membrane protein FhuA. The concentration of pb5 was estimated from fluorescence and that of FhuA from absorbance at 280 nm. The $c(s)$ profiles obtained from SV-AUC with fluorescence detection (Fig. 2C) show, in the absence of FhuA, at 5–5.5 S, the presence of the pb5-GFP monomer (M=99 kDa). In the presence of FhuA, pb5-GFP is no more detected, and the pb5-GFP:FhuA complex (M=181 kDa + bound detergent) sediments at 7.4 S. At 2.5 S (not shown), free GFP from proteolysis of pb5-GFP was observed. The nondetected unlabeled FhuA is known to sediment at 6.2 S (not shown). In conclusion, our results showed the possibility of studying complex formation with unpurified GFP-labeled proteins in diluted cellular extracts.

2.9 Preliminary Investigation of Lipid Vesicles Sedimentation Using Fluorescent Detection

Membrane protein are very often studied after reconstitution in lipid vesicles, a more native environment, in which most often the protein is more stable, and specific activities like transport can be measured. As a preliminary step before more complex studies, we have investigated the behavior of pure lipid vesicles of 1,2-diphytanoyl-*sn*-glycero-3-phosphocholine (DPhPC) (4ME 16:0 PC) (from Avanti Polar Lipids, 25 mg mL^{-1} stock solution in $CHCl_3$) in SV-AUC. Two labeling strategies were assayed: (1) with a fluorescent lipid, 18:1–12:0 NBD-PC (obtained from Avanti Polar Lipids) which was dried—from a $CHCl_3$ solution at 1 mg mL^{-1}—with DPhPC before vesicle formation, and (2) with a soluble-labeled amphiphilic probe,

5-dodecanoyl aminofluorescein (DAF, Invitrogen), which was added to diluted preformed vesicles prior AUC.

Results in terms of size distributions using DLS and SV-AUC did not depend significantly on the labeling strategies (NBD-PC or DAF). Further investigations were made with DAF to investigate the effect of lipid/probe ratios. DPhPC vesicles were preformed at 5 mg mL^{-1} with a first protocol consisting of 19 extrusions through a polycarbonate membrane with a nominal pore diameter of 100 nm. From DLS, $R_s = 100$ nm, with a polydispersity index of 40%. Liposomes were diluted to 0.1 mg mL^{-1} (0.12 m*M*) and labeled with DAF at 1 n*M*, 10 n*M* 100 n*M*, 1 μ*M*, or 5 μ*M* (lipid/probe ratios from 120,000 to 24). With overnight incubations prior AUC, maximum *s*-values of −24, −30, −31, −26, and −16S were measured, respectively. In view of the rather large heterogeneity of these samples, since the value for DAF of 1 n*M* is poorly precise, only the 5 μ*M* DAF sample has a significantly decreased value possibly related to the high-fluorescence intensity level. With shorter incubation times (labeling about 30 min before SV experiments), AUC results did not significantly differ. In conclusion, satisfying signal/noise in SV was obtained for lipid at 0.1–1 mg mL^{-1} and DAF at 0.1 μ*M*, corresponding to a ratio lipid/probe of 1180–11,800. These conditions allow the characterization of aliquots of *a posteriori* labeled vesicles or proteoliposomes, and without long incubation times.

DLS and AUC results for a sample prepared with a late protocol with an additional extrusion step are shown in Fig. 3. DLS showed one main population (99% intensity) with $R_s = 85$ nm, and a polydispersity index of 22%. SV-AUC showed a large distribution of negative sedimentation coefficients up to −160 S, with a maximum at −20 S. Flotation is expected in view of the values of the solvent density and lipid partial specific volume. The maximum *s*-value is within the expected order magnitude. We indeed calculate $s = -14$ S for an 85-nm R_s DPhPC liposome, considering $f/f_{min} = 1$.

The calculations show different interesting features: on the one hand, a variation of R_s between 50 and 120 nm leads to moderate variation in *s* (±4 S). On the other hand, even slight differences between the internal or external solvent would have a huge effect. We calculate, e.g., the effect of adding 5% sucrose (density: 1.0175 g mL^{-1} and viscosity: 1.145 cp) outside or inside the liposome. *s* would reach values below −100 S (sucrose only outside) or above 100 S (sucrose only inside). The parameters used were $M_l = 846.25$ Da; $A_l = 69$ A^2, the molecular area (Pownall, Pao, Brockman, & Massey, 1987); and $e_l = 5$ nm the lipid thickness (Zaccai, Blasie, & Schoenborn, 1975). $R_{min} = R_s/(f/f_{min})$ is used to calculate the

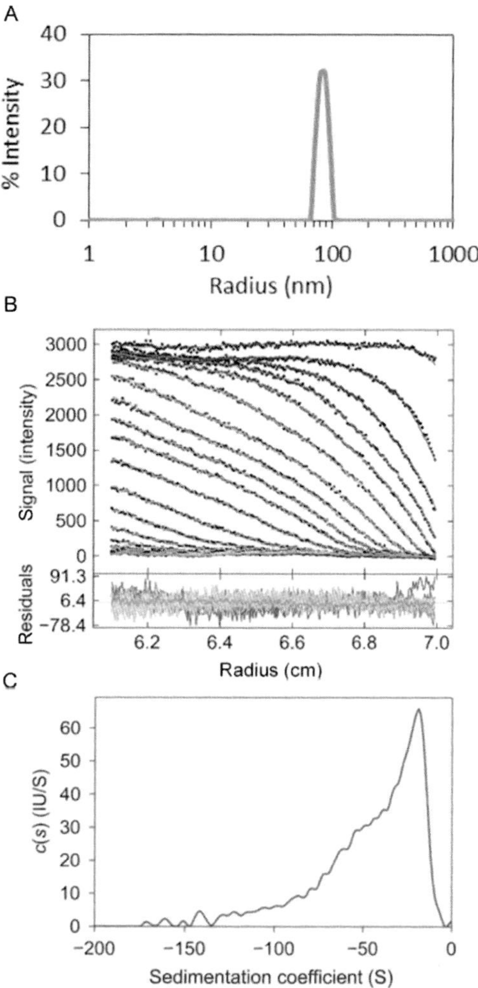

Figure 3 Size distributions of DAF-labeled DPhPC liposome. Final concentration of DAF and DPhPC are 0.1 μ*M* and 0.5 mg mL^{-1}, respectively. R_s-distribution from DLS (A); superposition of experimental and fitted sedimentation velocity profiles (top B), and their differences (bottom B); for clarity one over two acquisitions is shown corresponding to about 12-min time intervals. Sedimentation coefficient distributions (C). Liposomes were prepared at 10 mg mL^{-1}. $CHCl_3$ was evaporated, using a weak N_2 beam then vacuum, from the appropriate volume of lipid solution, and the dry lipid film was hydrated using 200 μL of buffer (10 m*M* pipes, 150 mM KCl, pH 7.4). After resuspension, the solution was first extruded five times through a 200-nm-pore-diameter polycarbonate membrane filter using a miniextruder (Avanti Polar Lipids, Inc.). This first extrusion step leads to more homogeneous distributions of liposome (polydispersity index in DLS decreasing from 40% to 20%). The vesicle suspension was then extruded 19 times through a polycarbonate membrane (Avanti Polar Lipids, Inc.) with a nominal pore diameter of 100 nm. (*Continued*)

outer and inner liposome surfaces, $S_{outer} = \pi R_{min}^2$ and $S_{inner} = \pi(R_{min} - e_l/2)^2$; the lipid number, $N_l = (S_{outer} + S_{inner})/A_l$; and buoyant mass, $M_{b,l} = M_l N_l(1 - \rho^\circ \bar{v}_l)$, leading to $s = M_{b,l}/N_A 6\pi\eta R_s$. If the internal solvent has a different density, ρ_{int}, the internal radius $R_{int} = R_{min} - e_l$, volume ($V_{int} = 4/3\pi R_{int}^3$), mass ($m_{int} = V_{int}\ \rho_{int}$), and buoyant mass $m_{b,\ int} = m_{int} - V_{int}\ \rho$ are calculated to derive $s = (M_{b,l}/N_A + m_{bint})/6\pi\eta R_s$.

3. MEMBRANE PROTEINS IN SAXS AND SANS

SAXS and SANS address the questions of the association states of macromolecules in solution, their shape or conformation. General reviews on the method and applications are, e.g., Jacrot and Zaccai (1981), Koch, Vachette, and Svergun (2003), Mertens and Svergun (2010), Serdyuk, Zaccai, and Zaccai (2007), and Zaccai and Jacrot (1983). Approaches developed in the last years investigate the information given from the whole scattering curves (Petoukhov & Svergun, 2007; Svergun, 2010), in terms of low-resolution structure (i.e., maximum 1 nm resolution). It allows ascertaining in solution a structure determined by crystallography, or, in the absence of model structure, determining an *ab initio* envelope. Importantly, contrast variation in SANS allows to focus on one component within a complex structure, which is of high interest for membrane proteins. AUC is complementary for checking the requested sample homogeneity and determining the protein absolute concentration and association state, and the amount of bound detergent. For a review concerning specifically membrane proteins, see Breyton, Gabel, et al. (2013). After a short presentation of

Figure 3—Cont'd All extrusions were performed at room temperature. The sample was diluted 20 times for AUC experiments. DAF was stored in the dark at 4 °C as a 5 m*M* stock solution in DMSO. Before AUC experiment, it was diluted in DMSO to 50 μ*M* and 1 μL added to the 500 μL liposome sample, and DLS measurements were done using a DynaPro NanoStar (Wyatt). SV experiments were performed at 20 °C and 18,000 rpm, using 12 mm-path length double-sector cells (each of the sector can be filled with a different solution, given the probe level is similar). Analysis in Sedfit was done, in view of the large size of the liposomes, with the least square $g^*(s)$ analysis (Schuck, Perugini, Gonzales, Howlett, & Schubert, 2002), considering typically 200 particles in the appropriate *s*-range, fitting the meniscus, and radial and time-independent noises, and with a confidence level (*F*-ratio) of 0.68. The numerical values used for analyzing the results are: for the buffer, the density of 1.006 g mL^{-1} and viscosity of 1.00 cp calculated with the program Sednterp (http://sednterp.unh.edu/), neglecting DMSO contribution; for DPhPC: $\bar{v}_l = 1.0234\,\text{mL}\,\text{g}^{-1}$ from chemical composition and tabulated volumes (Durchschlag & Zipper, 1997).

the method, we will review the more recent strategies allowing membrane proteins' structural investigation.

3.1 General Information on SAS Devices and Analysis

The practical requirements for an SAS experiment are: ≈100 μL of solution with the protein in the mg mL^{-1} range (possibly labeled, or in different H_2O/D_2O solvents, and/or at different concentrations); an accurate measurement of the macromolecular concentration (in mg mL^{-1}); and the access to an X-ray or neutron small-angle instrument and competent help to plan the experiment.

Instrumentation is schematically described in Fig. 4. The collimated beam, with a wavelength, λ, of typically 0.5–1.5 Å, irradiates the sample and is scattered by the solution. Considering a dilute solution, the macromolecules are distributed randomly in both position and orientation, and the scattering intensity depends only on the scattering vector $q=4\pi\sin\theta/\lambda$, with 2θ the scattering angle. Table 1 shows some specificities of SAXS and SANS.

In the forward direction ($q=0$), global information on the macromolecule composition and molar mass is obtained (see below). The radius of gyration, R_g, determined from the data at small angle (qR_g below ≈1), informs on the global shape of the macromolecule. For diluted samples, the scattering curve is determined by the sum of the scattering of the individual particles in solution. The internal distances within the particles represent the most information that can be derived. It is possible to calculate the scattering curve from a structure defined by the atomic coordinates of its atoms. Determining a three-dimensional structure from the scattering curve is referred as the inverse scattering problem. It has no unique solution. However, for high-quality samples (only one type of macromolecule in solution), given the compact shape of proteins (defining connectivity constraints in the reconstituted structure), it is possible to determine low-resolution *ab initio*

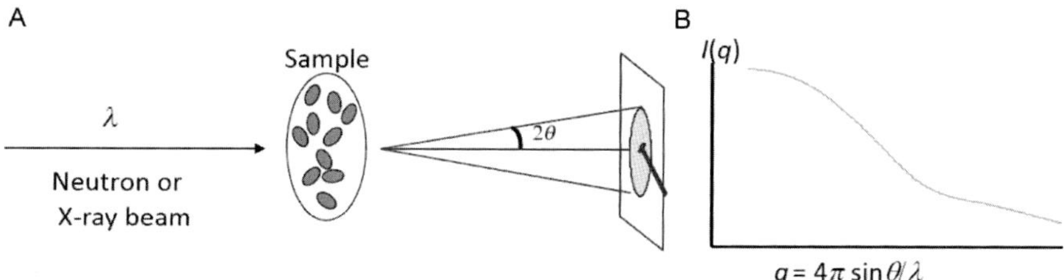

Figure 4 Schematic representation of a small-angle scattering experiment (A) and scattering curve (B).

Table 1 SAXS and SANS Specificities and Differences

	SANS	SAXS
Beam lines example	D22 at ILL FR	BM29 at ESRF FR
Beam	Neutron	X-Rays
Scattering by	Electrons	Nuclei
Typical acquisition time	15–30 min	0.1–1 s
Radiation damage	Nondestructive	Possible
Specificity	Contrast variation	Time-resolved scattering

structures with confidence from the two-dimensional scattering curves (see references above).

3.2 Contrast and Forward Intensity in SAS

X-Rays are scattered by the electrons of the atoms, and neutrons by nuclei. All atoms are characterized by scattering lengths determining the scattering amplitude. In SAXS, the scattering length is proportional to the number of the atom electrons, which changes considerably with the atomic number (Table 2). In SANS, it varies in a limited range of magnitude depending on the atoms, but, importantly, depends on the isotopic type, with, in particular, a negative value for the H atom, while the other atoms that are usually represented in biology have positive values (Table 2). In SANS and SAXS, which are low-resolution techniques, we consider the components (protein, detergent, or part of it—see below—solvent) as homogeneous in composition. Scattering densities (formally scattering length densities) calculated from the atom composition and partial specific volume are thus considered. The difference between the scattering density of the molecule and of the solvent determines the scattering intensity and is referred to the contrast. When the contrast is null, the component does not contribute to scattering: it is masked.

The forward intensity $I(0)$ is extrapolated from experimental data, after subtraction of the signal of the buffer and normalization. It is the sum of the scattering contributions of the different types, i, of molecules (pd complex, detergent micelle, and monomer):

$$I(0) = \sum_i \left(\frac{1}{N_A}\right) c_i M_i \left(\frac{\partial \rho_X}{\partial c_i}\right)^2_{\mu}, \tag{10}$$

Table 2 Scattering Amplitudes of Atoms

	SAXS	SANS
Atom	**Number of Electrons**	**Coherent Scattering Length (10^{-12} cm)**
H	1	−0.37409
D	1	0.6674
C	6	0.66484
N	7	0.936
O	8	0.5805
F	9	0.5654
P	15	0.513
S	16	0.2847

In SAXS, the scattering amplitude is nf_{el}, with n the number of electrons and $f_{el} = 2.8 \times 10^{-14}$ cm the classical radius of an electron. In SANS, it is the coherent scattering length.

with M_i the molar mass, c_i (mg mL^{-1}) the concentration, and $(\partial\rho_X/\partial c_i)_\mu$ the electron or neutron density increment, index X being for el, in SAXS or N, in SANS. The contribution for the pd complex can be expressed very similarly to the buoyant factor $(\partial\rho/\partial c_p)_\mu$ in AUC (Section 2.2), with ρ_X and b_X scattering length densities per volume and gram unit, respectively, with label ° for the solvent:

$$c_p M_p \left(\frac{\partial \rho_X}{\partial c_p}\right)_\mu^2 = c_p M_p \left[\left(b_{Xp} - \rho^\circ{}_X \bar{v}_p\right) + \delta_d \left(b_{Xd} - \rho^\circ{}_X \bar{v}_d\right)\right]^2. \quad (11)$$

As mentioned above (Section 2.2), a more complex equation can be written with explicit lipid, hydration, or any other bound component contributions, here integrated in the "detergent" contribution.

In SANS, specific deuteration of macromolecules and/or solvent offers the possibility to modulate the scattering amplitude of the solvent and/or of specific components. It allows determining the particle composition and low-resolution structures of specific components within multicomponent particles. Figure 5 illustrates the principle of a contrast variation series, using a solvent containing different H_2O/D_2O, for a pd complex plus detergent micelle sample. The contrast match point (CMP) is the D_2O % in the solvent allowing to mask a given component. It can be calculated as shown in Fig. 6. The CMP is the intercept between the scattering density

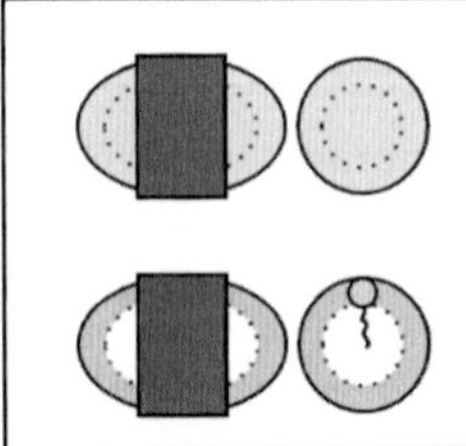 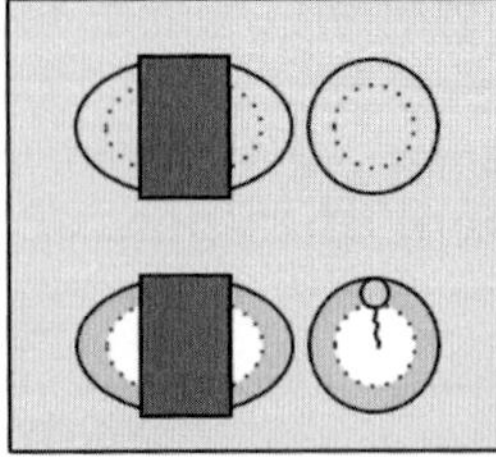 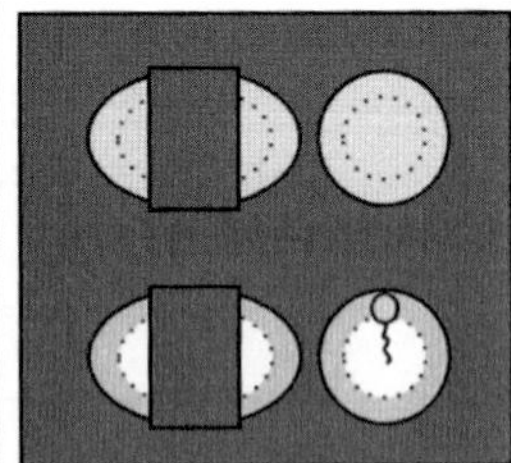

Figure 5 Principle of contrast variation for membrane proteins. The white, light gray, or dark gray backgrounds, from right to left panels, represent the scattering density of the solvent which is larger with increased D_2O content. The dark gray rectangles represent the protein, which would be masked in the right panel. On the top part of each panel, the light gray discs and semi-discs represent the detergent micelle and bound detergent, respectively, which would be masked at intermediate D_2O content. The bottom parts illustrate the case of a chemically heterogeneous detergent that will not be homogeneously masked over the whole scattering curve.

curves plotted for different D_2O %s, for the solvent and for the component. Figure 6 shows contrast variation plots for water, protein, and variously deuterated *n*-octyl glucoside (OG) detergent. Hydrogenated OG and protein have rather similar CMPs (19% and 42%, respectively), which would not allow to obtain data with a high signal/noise at the match point of the detergent. Perdeuterated OG is impossible to match (calculated CMP of 119%). Tail-deuterated OG (d17-OG) is the best in this series, with a CMP of 90%. This molecule, however, exemplifies the limit of applying the contrast variation approach to match detergent. Detergent is indeed structured with chemical heterogeneity correlated to spatial redistribution: the hydrophobic tail of detergents is rich in CH_2 or CD_2 groups and colocalize to avoid contact with the solvent; their hydrophilic head group has (in general) a different content in H-atoms and is exposed to the solvent (Fig. 5). The hydrophilic head of OG has scattering properties that are close to that of proteins, the deuterated tail behaving dramatically differently (Fig. 6, right panel). The detergent will be only globally matched at 90% D_2O—in the forward direction, its contribution vanishes—but there will be nonnegligible residual scattering contributions out of the forward direction. For some other detergents, such as Foscholine12, Brij35, and F_6-DigluM (see below), the calculated SANS contrast variation curves of the head and tail are close enough, suggesting that at the CMP of the molecule, no residual signal will occur (Breyton, Gabel, et al., 2013; Breyton, Flayhan, et al., 2013).

In SAXS, there are no possibilities to match the detergent. Imperfect strategies adding sucrose in the solvent were however used (Le Maire,

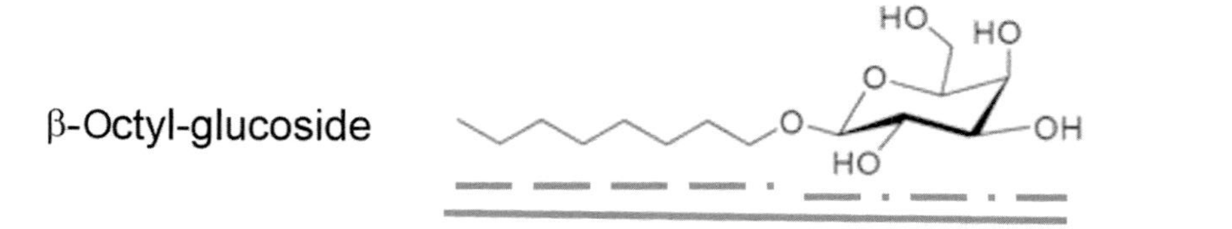

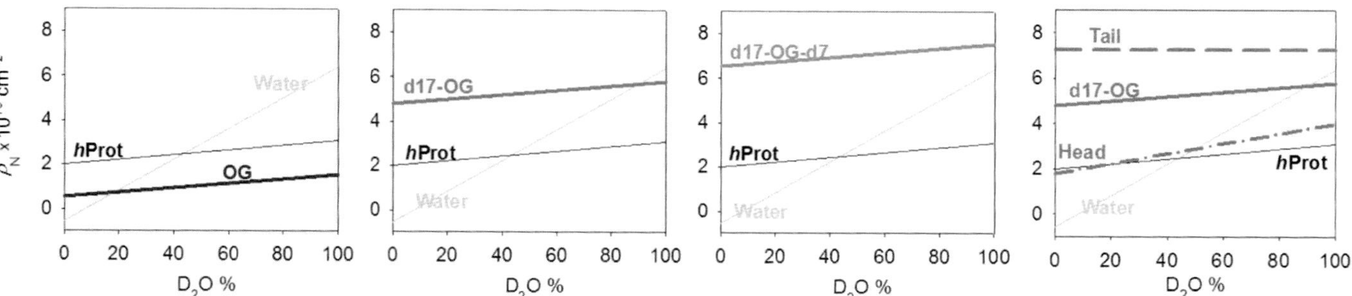

Figure 6 Contrast curves for OG with different deuterations and protein. Neutron scattering length densities of water (mixture of H_2O and D_2O) (light blue (light gray in the print version)) and of hydrogenated protein (gray) are plotted as a function of D_2O content. Curves for OG with different deuteration are drawn, from right to left panels, for hydrogenated OG (dark blue (dark gray in the print version), bold), tail-deuterated OG (d17-OG, dark green (dark gray in the print version), bold), and perdeuterated OG (d17-OGd7, dark khaki (gray in the print version), bold). The right panel shows the contribution of the tail (dash) and head (dash-dot) for d17-OG.

Moller, & Tardieu, 1981), but adding a solvent component generates solvent redistribution that may contribute to scattering.

3.3 Review on Recent Strategies to Investigate Membrane Proteins

SAS has been used to investigate membrane protein samples only in the past decade. This is due to the difficulties of preparing homogeneous pd samples, on one hand, and to the presence of the bound detergent and free detergent micelles, on the other hand. Strategies developed to focus on the membrane protein are either deconvoluting or matching the detergent contribution. For recent reviews, see, e.g., Clifton, Neylon, and Lakey (2013) and Heller (2010). We have reviewed the membrane protein-SAS literature until 2012 in Breyton, Gabel, et al. (2013). Since then, strategies have refined, allowing more sophisticated results to be obtained, in particular regarding the arrangement of the detergent belt around the protein.

In SAXS, the problem of the precise subtraction of the free detergent micelles contribution is elegantly solved by mounting on-line SEC. This allows one to separate the excess micelles from the pd complex, and to remove the scattering contribution of the solvent detergent micelle: from the scattering of the mobile phase containing the pd complex is subtracted that of the chromatography buffer, which contains exactly the same amount of free detergent micelles. What remains to be modeled is the detergent belt around the protein. Perez and collaborators have used the Aquaporin-0 (AQP0)–DDM complex as a study case, as the atomic structure of AQP0 is available. They propose a "model-free" *ab initio* coarse-grained fitting algorithm that takes into account the different electron densities (detergent head and tail, protein) and general physical properties of the system. Application of this algorithm provides high-quality fits of the SAXS data and a more objective estimation of the detergent belt shape. The data, combined with molecular dynamics simulation, then provide a detailed all-atom model consistent with the experimental results. The authors derive some rules about which strategy to adopt in further studies with different proteins, and conclude that, as intuitively expected, the detergent belt around the tetrameric AQP0 follows the overall shape of the transmembrane surface and tends to adopt the square-like shape of the AQP0 tetramer (Koutsioubas, Berthaud, Mangenot, & Perez, 2013; Perez & Koutsioubas, 2015). A very similar conclusion is obtained by Le et al., who studied the arrangement of the DDM belt around the trimeric Photosystem I (PSI) by SANS. *Ab initio* models from PSI–DDM complexes measured at the CMP of DDM, where only the protein should contribute, and 100% D_2O, where the whole protein–detergent complex does, are compared. The authors conclude to a clover-like detergent belt around the trimeric protein, rather than a torus (Le et al., 2014). Whereas the conclusion is reasonable, it should be noted that at its CMP, DDM is not homogeneously masked, and the residual signal will interfere with the protein signal (see below and, e.g., Zimmer, Doyle, & Grossmann, 2006).

In Gabel et al., we investigated by SANS the solution structure of FhaC, the outer-membrane, β-barrel transporter of the *Bordetella pertussis* filamentous hemagglutinin adhesin. In the crystal structure, helix H1 folds into the β-barrel of the protein, obstruding the pore. Could H1 fold outside of the barrel in solution? FhaC solubilized in d17-OG was measured at the CMP of the detergent and of the protein. As stated above, the detergent cannot be matched over the whole scattering curve. Molecular dynamics simulations generated an ensemble of protein–detergents models that were compared to the experimental data. Good fits were obtained for relatively compact, connected detergent belts and allowed to conclude that helix H1 clearly sits inside the pore as in the crystal structure (Gabel et al., 2014).

Fluorinated surfactant F_6-DigluM is one of the detergents that can be masked over the whole scattering curve. It has the additional particularity of having its CMP very close to that of proteins, so that hydrogenated proteins and F_6-DigluM are masked together, and proteins have to be deuterated—75% is enough—to be visible at the match point of the detergent (Fig. 7). These features give the opportunity to selectively investigate the structure of proteic partners within a membrane protein complex. We used this strategy to investigate the conformational changes induced upon binding of phage T5 Receptor Binding Protein pb5 to its *E. coli* outer-membrane receptor FhuA, which *in vivo* triggers phage infection. We produced hydrogenated and 75% deuterated forms of the two proteins and all the combinations of the complexes. At 46% D_2O, where both proteins and F_6-DigluM are matched, we checked that only the deuterated protein partner contributes to the signal. The structure of the individual partners alone and within the complex, and that of the whole complex determined at low resolution (Fig. 7), showed that the binding of pb5 to FhuA does not induce large conformational changes within each of the proteins (Breyton, Flayhan, et al., 2013).

An interesting alternative to manipulate membrane proteins is the use of the so-called nanodiscs. These are patches of lipids solubilized by a membrane-scaffold protein MSP1D1. For a recent work providing a

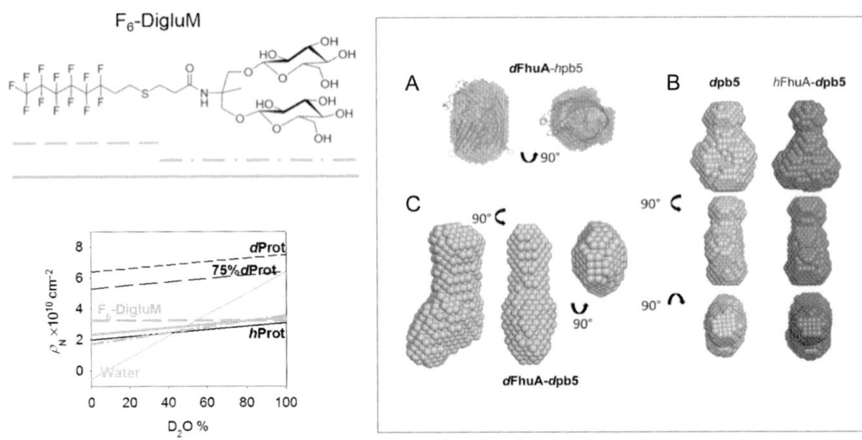

Figure 7 Low-resolution structures determined in SANS with the surfactant F_6-DigluM. Left top: chemical structure of F_6-DigluM. Left bottom: neutron scattering densities of water (mixture of H_2O and D_2O) (light blue); of protein (black lines) either hydrogenated (*h*Prot, continuous), 75% deuterated (75% *d*Prot, long dash), or deuterated (*d*Prot, small dash); and of F_6-DigluM (green lines, bold), with tail (green long dash) and head (green dash-dot) contributions, are plotted as a function of D_2O content. This figure was originally published in Breyton, Gabel, et al. (2013), with kind permission of *The European Physical Journal*. Right: *ab initio* envelopes from SANS at 46% D_2O, where F_6-DigluM and *h*proteins are masked, of *d*FhuA-*h*pb5 (red mesh), superimposed with the crystal structure of FhuA (2FCP, green) (A); *d*pb5 (orange) and *h*FhuA-*d*pb5 (blue) (B); and *d*FhuA-*d*pb5 (C). Index *d* and *h* are for hydrogenated and deuterated proteins. This research was originally published in Breyton, Flayhan, et al. (2013). © The American Society for Biochemistry and Molecular Biology. (See the color plate.)

comprehensive biophysical characterization on nanodisc in the AUC, see Inagaki, Ghirlando, and Grisshammer (2013). Nanodisc-solubilized proteins are thus inserted into a lipid bilayer, closer to their native environment than when solubilized in detergent. To allow SANS investigation of nanodisc-solubilized membrane proteins, Maric et al. have investigated the possibility of contrast matching the whole lipid-MSP1D1 particle (Maric et al., 2014). This required different deuteration levels of phosphatidylcholines head groups and tails, which could be separately controlled via their respective biosynthetic pathways, and the partial deuteration of the MSP1D1. The resulting deuterated analogues of nanodisc and liposomes give a minimal contribution to the neutron scattering data when used in 100% D_2O. This allows investigating the low-resolution structure of large membrane proteins in a lipidic environment.

ACKNOWLEDGMENTS

This work used the platforms of the Grenoble Instruct Center (ISBG: UMS 3518 CNRS-CEA-UJF-EMBL) with support from FRISBI (ANR-10-INSB-05-02) and GRAL (ANR-10-LABX-49-01) within the Grenoble Partnership for Structural Biology (PSB). This work was supported in part by the CEA, the CNRS, Université Grenoble Alpes, and grants from the ANR (Programme Blanc DYN FHAC ANR-10-BLAN-1306), and by the Intramural Research Program of the National Institute of Biomedical Imaging and Bioengineering, National Institutes of Health. We thank Pierre Moscatello for participation to AUC experiments, and Chad Brautigan for having implemented calculation facilities in the program Gussi.

REFERENCES

Balbo, A., Minor, K. H., Velikovsky, C. A., Mariuzza, R. A., Peterson, C. B., & Schuck, P. (2005). Studying multiprotein complexes by multisignal sedimentation velocity analytical ultracentrifugation. *Proceedings of the National Academy of Sciences of the United States of America*, *102*(1), 81–86.

Breyton, C., Flayhan, A., Gabel, F., Lethier, M., Durand, G., Boulanger, P., et al. (2013). Assessing the conformation changes of pb5, the receptor binding protein of phage T5, upon binding to its E. coli receptor FhuA. *The Journal of Biological Chemistry*, *288*, 20763–20772.

Breyton, C., Gabel, F., Lethier, M., Flayhan, A., Durand, G., Jault, J. M., et al. (2013). Small angle neutron scattering for the study of solubilised membrane proteins. *The European Physical Journal. E, Soft Matter*, *36*(7), 71.

Brown, P. H., Balbo, A., Zhao, H., Ebel, C., & Schuck, P. (2011). Density contrast sedimentation velocity for the determination of protein partial-specific volumes. *PLoS One*, *6*(10), e26221.

Burgess, N. K., Stanley, A. M., & Fleming, K. G. (2008). Determination of membrane protein molecular weights and association equilibrium constants using sedimentation equilibrium and sedimentation velocity. *Methods in Cell Biology*, *84*, 181–211.

Butler, P. J., Ubarretxena-Belandia, I., Warne, T., & Tate, C. G. (2004). The Escherichia coli multidrug transporter EmrE is a dimer in the detergent-solubilised state. *Journal of Molecular Biology*, *340*(4), 797–808.

Casassa, E. F., & Eisenberg, H. (1964). Thermodynamic analysis of multicomponent solutions. *Advances in Protein Chemistry*, *19*, 287–395.

Clifton, L. A., Neylon, C., & Lakey, J. H. (2013). Examining protein-lipid complexes using neutron scattering. *Methods in Molecular Biology*, *974*, 119–150.

Dach, I., Olesen, C., Signor, L., Nissen, P., le Maire, M., Moller, J. V., et al. (2012). Active detergent-solubilized H+, K+-ATPase is a monomer. *Journal of Biological Chemistry*, *287*(50), 41963–41978.

Durchschlag, H., & Zipper, P. (1997). Calculation of partial specific volumes and other volumetric properties of small molecules and polymers. *Journal of Applied Crystallography*, *30*(2), 803–807.

Ebel, C. (2011). Sedimentation velocity to characterize surfactants and solubilized membrane proteins. *Methods*, *54*, 56–66.

Ebel, C., Møller, J. V., & le Maire, M. (2007). Analytical ultracentrifugation: Membrane protein assemblies in the presence of detergent. In E. Pebay-Peyroula (Ed.), *Biophysical analysis of membrane proteins. Investigating structure and function* (pp. 91–120). Wiley: Weinheim (chapter 4).

Eisenberg, H. (1976). *Biological macromolecules and polyelectrolytes in solution*. Oxford: Clarendon Press.

Fagerberg, L., Jonasson, K., von Heijne, G., Uhlen, M., & Berglund, L. (2010). Prediction of the human membrane proteome. *Proteomics*, *10*(6), 1141–1149.

Flayhan, A., Wien, F., Paternostre, M., Boulanger, P., & Breyton, C. (2012). New insights into pb5, the receptor binding protein of bacteriophage T5, and its interaction with its Escherichia coli receptor FhuA. *Biochimie*, *94*(9), 1982–1989.

Fleming, K. G., Ren, C. C., Doura, A. K., Eisley, M. E., Kobus, F. J., & Stanley, A. M. (2004). Thermodynamics of glycophorin A transmembrane helix dimerization in C14 betaine micelles. *Biophysical Chemistry*, *108*(1–3), 43–49.

Gabel, F., Lensink, M. F., Clantin, B., Jacob-Dubuisson, F., Villeret, V., & Ebel, C. (2014). Probing the conformation of FhaC with small-angle neutron scattering and molecular modeling. *Biophysical Journal*, *107*(1), 185–196.

Ghirlando, R., Balbo, A., Piszczek, G., Brown, P. H., Lewis, M. S., Brautigam, C. A., et al. (2013). Improving the thermal, radial, and temporal accuracy of the analytical ultracentrifuge through external references. *Analytical Biochemistry*, *440*(1), 81–95.

Heller, W. T. (2010). Small-angle neutron scattering and contrast variation: A powerful combination for studying biological structures. *Acta Crystallographica. Section D, Biological Crystallography*, *66*(Pt. 11), 1213–1217.

Hellman, L. M., Zhao, C., Melikishvili, M., Tao, X., Hopper, J. E., Whiteheart, S. W., et al. (2011). Histidine-tag-directed chromophores for tracer analyses in the analytical ultracentrifuge. *Methods*, *54*(1), 31–38.

Inagaki, S., Ghirlando, R., & Grisshammer, R. (2013). Biophysical characterization of membrane proteins in nanodiscs. *Methods*, *59*(3), 287–300.

Jacrot, B., & Zaccai, G. (1981). Determination of molecular weight by neutron scattering. *Biopolymers*, *20*, 2413–2426.

Josse, D., Ebel, C., Stroebel, D., Fontaine, A., Borges, F., Echalier, A., et al. (2002). Oligomeric states of the detergent-solubilized human serum paraoxonase (PON1). *The Journal of Biological Chemistry*, *277*(36), 33386–33397.

Koch, M. H., Vachette, P., & Svergun, D. I. (2003). Small-angle scattering: A view on the properties, structures and structural changes of biological macromolecules in solution. *Quarterly Reviews of Biophysics*, *36*(2), 147–227.

Koutsioubas, A., Berthaud, A., Mangenot, S., & Perez, J. (2013). Ab initio and all-atom modeling of detergent organization around Aquaporin-0 based on SAXS data. *The Journal of Physical Chemistry. B*, *117*(43), 13588–13594.

Le, R. K., Harris, B. J., Iwuchukwu, I. J., Bruce, B. D., Cheng, X., Qian, S., et al. (2014). Analysis of the solution structure of Thermosynechococcus elongatus photosystem I in n-dodecyl-beta-D-maltoside using small-angle neutron scattering and molecular dynamics simulation. *Archives of Biochemistry and Biophysics*, *550–551*, 50–57.

le Maire, M., Arnou, B., Olesen, C., Georgin, D., Ebel, C., & Møller, J. V. (2008). Gel chromatography and analytical ultracentrifugation to determine the extent of detergent binding and aggregation, and Stokes radius of membrane proteins using sarcoplasmic reticulum Ca2+-ATPase as an example. *Nature Protocols*, *3*(11), 1782–1795.

le Maire, M., Champeil, P., & Møller, J. V. (2000). Interaction of membrane proteins and lipids with solubilizing detergents. *Biochimica et Biophysica Acta*, *1508*(1–2), 86–111.

Le Maire, M., Moller, J. V., & Tardieu, A. (1981). Shape and thermodynamic parameters of a CA-2+-dependent ATPASE—A solution X-ray-scattering and sedimentation equilibrium study. *Journal of Molecular Biology*, *150*(2), 273–296.

Le Roy, A., Breyton, C., & Ebel, C. (2014). Analytical ultracentrifugation and size-exclusion chromatography coupled with light scattering for the characterization of membrane proteins in solution. In I. Mus-Veteau (Ed.), *Membrane proteins production for structural analysis* (pp. 267–287). New York: Springer.

Le Roy, A., Nury, H., Wiseman, B., Sarwan, J., Jault, J. M., & Ebel, C. (2013). Sedimentation velocity analytical ultracentrifugation in hydrogenated and deuterated solvents for the characterization of membrane proteins. *Methods in Molecular Biology, 1033*, 219–251.

Lustig, A., Engel, A., Tsiotis, G., Landau, E. M., & Baschong, W. (2000). Molecular weight determination of membrane proteins by sedimentation equilibrium at the sucrose or nycodenz-adjusted density of the hydrated detergent micelle. *Biochimica et Biophysica Acta, 1464*(2), 199–206.

MacGregor, I. K., Anderson, A. L., & Laue, T. M. (2004). Fluorescence detection for the XLI analytical ultracentrifuge. *Biophysical Chemistry, 108*(1–3), 165–185.

Maric, S., Skar-Gislinge, N., Midtgaard, S., Thygesen, M. B., Schiller, J., Frielinghaus, H., et al. (2014). Stealth carriers for low-resolution structure determination of membrane proteins in solution. *Acta Crystallographica. Section D, Biological Crystallography, 70*(Pt. 2), 317–328.

Mertens, H. D. T., & Svergun, D. I. (2010). Structural characterization of proteins and complexes using small-angle X-ray solution scattering. *Journal of Structural Biology, 172*(1), 128–141.

Nury, H., Manon, F., Arnou, B., le Maire, M., Pebay-Peyroula, E., & Ebel, C. (2008). Mitochondrial bovine ADP/ATP carrier in detergent is predominantly monomeric but also forms multimeric species. *Biochemistry, 47*(47), 12319–12331.

Overington, J. P., Al-Lazikani, B., & Hopkins, A. L. (2006). How many drug targets are there? *Nature Reviews. Drug Discovery, 5*(12), 993–996.

Perez, J., & Koutsioubas, A. (2015). Memprot: A program to model the detergent corona around a membrane protein based on SEC-SAXS data. *Acta Crystallographica. Section D, Biological Crystallography, 71*(Pt. 1), 86–93.

Petoukhov, M. V., & Svergun, D. I. (2007). Analysis of X-ray and neutron scattering from biomacromolecular solutions. *Current Opinion in Structural Biology, 17*(5), 562–571.

Plancon, L., Janmot, C., le Maire, M., Desmadril, M., Bonhivers, M., Letellier, L., et al. (2002). Characterization of a high-affinity complex between the bacterial outer membrane protein FhuA and the phage T5 protein pb5. *Journal of Molecular Biology, 318*(2), 557–569.

Pownall, H. J., Pao, Q., Brockman, H. L., & Massey, J. B. (1987). Inhibition of lecithin-cholesterol acyltransferase by diphytanoyl phosphatidylcholine. *The Journal of Biological Chemistry, 262*(19), 9033–9036.

Ravaud, S., Do Cao, M. A., Jidenko, M., Ebel, C., Le Maire, M., Jault, J. M., et al. (2006). The ABC transporter BmrA from Bacillus subtilis is a functional dimer when in a detergent-solubilized state. *The Biochemical Journal, 395*(2), 345–353.

Reynolds, J. A., & Tanford, C. (1976). Determination of molecular weight of the protein moiety in protein-detergent complexes without direct knowledge of detergent binding. *Proceedings of the National Academy of Sciences of the United States of America, 73*(12), 4467–4470.

Rosenbusch, J. P., Lustig, A., Grabo, M., Zulauf, M., & Regenass, M. (2001). Approaches to determining membrane protein structures to high resolution: Do selections of subpopulations occur? *Micron, 32*(1), 75–90.

Salvay, A. G., & Ebel, C. (2006). Analytical ultracentrifuge for the characterization of detergent in solution. *Progress in Colloid and Polymer Science, 131*, 74–82.

Salvay, A. G., Santamaria, M., le Maire, M., & Ebel, C. (2007). Analytical ultracentrifugation sedimentation velocity for the characterization of detergent-solubilized membrane proteins Ca++-ATPase and ExbB. *Journal of Biological Physics, 33*, 399–419.

Schuck, P. (2000). Size-distribution analysis of macromolecules by sedimentation velocity ultracentrifugation and Lamm equation modeling. *Biophysical Journal, 78*(3), 1606–1619.

Schuck, P., Perugini, M. A., Gonzales, N. R., Howlett, G. J., & Schubert, D. (2002). Size-distribution analysis of proteins by analytical ultracentrifugation: Strategies and application to model systems. *Biophysical Journal, 82*(2), 1096–1111.

Serdyuk, I. N., Zaccai, N. R., & Zaccai, G. (2007). X-ray and neutron diffraction. In I. N. Serdyuk, N. R. Zaccai, & G. Zaccai (Eds.), *Methods in molecular biophysics: Structure, dynamics, function* (1 ed, pp. 765–881). Cambridge, England: Cambridge University Press.

Svergun, D. I. (2010). Small-angle X-ray and neutron scattering as a tool for structural systems biology. *Biological Chemistry, 391*(7), 737–743.

Tanford, C., Nozaki, Y., Reynolds, J. A., & Makino, S. (1974). Molecular characterization of proteins in detergent solutions. *Biochemistry, 13*(11), 2369–2376.

Tanford, C., & Reynolds, J. A. (1976). Characterization of membrane proteins in detergent solutions. *Biochimica et Biophysica Acta, 457*(2), 133–170.

Terstappen, G. C., & Reggiani, A. (2001). In silico research in drug discovery. *Trends in Pharmacological Sciences, 22*(1), 23–26.

Zaccai, G., Blasie, J. K., & Schoenborn, B. P. (1975). Neutron-diffraction studies on location of water in lecithin bilayer model membranes. *Proceedings of the National Academy of Sciences of the United States of America, 72*(1), 376–380.

Zaccai, G., & Jacrot, B. (1983). Small angle neutron scattering. *Annual Review of Biophysics and Bioengineering, 12*, 139–157.

Zhao, H., Casillas, E., Jr., Shroff, H., Patterson, G. H., & Schuck, P. (2013). Tools for the quantitative analysis of sedimentation boundaries detected by fluorescence optical analytical ultracentrifugation. *PLoS One, 8*(10), e77245.

Zhao, H., Lomash, S., Glasser, C., Mayer, M. L., & Schuck, P. (2013). Analysis of high affinity self-association by fluorescence optical sedimentation velocity analytical ultracentrifugation of labeled proteins: Opportunities and limitations. *PLoS One, 8*(12), e83439.

Zhao, H., Ma, J., Ingaramo, M., Andrade, E., MacDonald, J., Ramsay, G., et al. (2014). Accounting for photophysical processes and specific signal intensity changes in fluorescence-detected sedimentation velocity. *Analytical Chemistry, 86*(18), 9286–9292.

Zhao, H. Y., Mayer, M. L., & Schuck, P. (2014). Analysis of protein interactions with picomolar binding affinity by fluorescence-detected sedimentation velocity. *Analytical Chemistry, 86*(6), 3181–3187.

Zimmer, J., Doyle, D. A., & Grossmann, J. G. (2006). Structural characterization and pH-induced conformational transition of full-length KcsA. *Biophysical Journal, 90*(5), 1752–1766.

CHAPTER THIRTEEN

Hydrodynamic Models of G-Quadruplex Structures

Jonathan B. Chaires[1], William L. Dean, Huy T. Le, John O. Trent[1]
James Graham Brown Cancer Center, University of Louisville, Louisville, Kentucky, USA
[1]Corresponding authors: e-mail address: j.chaires@louisville.edu; john.trent@louisville.edu

Contents

Abstract

G-quadruplexes are noncannonical four-stranded DNA or RNA structures formed by guanine-rich repeating sequences. Guanine nucleotides can hydrogen bond to form a planar tetrad structure. Such tetrads can stack to form quadruplexes of various molecularities with a variety of types of single-stranded loops joining the tetrads. High-resolution structures may be obtained by X-ray crystallography or NMR spectroscopy for quadruplexes formed by short ($\approx$25 nt) sequences but these methods have yet to succeed in characterizing higher order quadruplex structures formed by longer sequences. An integrated computational and experimental approach was implemented in our laboratory to obtain structural models for higher order quadruplexes that might form in longer telomeric or promoter sequences. In our approach, atomic-level models are built using folding principles gleaned from available high-resolution structures and then optimized by molecular dynamics. The program HYDROPRO is then used to construct bead models of these structures to predict experimentally testable hydrodynamic

Methods in Enzymology, Volume 562
ISSN 0076-6879
http://dx.doi.org/10.1016/bs.mie.2015.04.011

properties. Models are validated by comparison of these properties with measured experimental values obtained by analytical ultracentrifugation or other biophysical tools. This chapter describes our approach and practical procedures.

1. INTRODUCTION

The human genome has about 3 billion base pairs in its 23 pairs of chromosomes. Each chromosome has one long piece of DNA. Most of this DNA is in cannonical Watson–Crick duplex form, but certain sequences can adopt alternate forms such as three- or four-stranded structures called triplexes and quadruplexes, respectively. Quadruplex structures can form in guanine-rich repetitive sequences with the motif -G_xN_y G_xN_y G_xN_y G_xN_y-, with $x>2$ and $y=1$–7 (Lane, Chaires, Gray, & Trent, 2008). Strand segments in folded or assembled quadruplexes may be oriented in either parallel or antiparallel directions. Uni- and bimolecular quadruplexes feature stacked planar G-tetrads linked by single-stranded loops of various types. Quadruplex-forming sequences are highly conserved and are distributed nonrandomly in the genome (Huppert, 2008; Huppert & Balasubramanian, 2005, 2007). Telomeric and gene promoter sequences are particularly enriched in quadruplex-forming sequences. The structure and function of genomic quadruplexes are of great current interest and are under active investigation. X-ray crystallography and NMR spectroscopy have been used to study quadruplex structures formed by folding of short oligonucleotides sequences, with some 190 high-resolution structures deposited in the structural databases. These are, with few exceptions, single quadruplex structures (with or without bound ligands) that are either uni-, bi-, or tetramolecular and typically containing roughly 20–25 nt. The choice of these sequences for crystallographic or NMR studies is driven by expediency. Short pieces of DNA or RNA are chosen because they will crystallize or they contain few enough atoms to be tractable for structure determination by NMR studies. Such sequences are often arbitrarily clipped from their longer genomic contexts, and are then often manipulated or modified from their actual wild-type sequence in order to provide a more homogenous conformational distribution. Longer, unmodified genomic sequences might well form more complex higher order quadruplex structures and methods are needed to explore that possibility.

Hydrodynamic bead modeling offers an approach for the structural characterization of biological macromolecules refractory to high-resolution methods (Byron, 2008; also see Chapter Hydrodynamic modeling and its application in AUC by Rocco and Byron, of this volume). There are many possible approaches for constructing models, but Garcia de la Torre has provided the particularly useful HYDROPRO (Garcia de la Torre & Harding, 2013; Garcia De La Torre, Huertas, & Carrasco, 2000; Ortega, Amoros, & Garcia de la Torre, 2011) suite of programs to facilitate hydrodynamic bead modeling. HYDROPRO has been validated by studies of both short DNA oligonucleotides (Fernandes, Ortega, Lopez Martinez, & Garcia de la Torre, 2002) and of longer DNA polymers (Amorós, Ortega, & García de la Torre, 2011). Our laboratory has used HYDROPRO to study a variety of G-quadruplex structures. In an initial application, we showed that a parallel propeller-shaped quadruplex structure determined in crystals by X-ray diffraction did not predict hydrodynamic properties consistent with experimental values and was therefore not the predominant form in solution (Li, Correia, Wang, Trent, & Chaires, 2005). In contrast, an antiparallel basket quadruplex structure in Na^+ solution determined by NMR spectroscopy predicted hydrodynamic properties fully consistent with experimentally determined values. Recently, we reported a more extensive validation of HYDROPRO predictions for a variety of quadruplex structures obtained by NMR (Le, Buscaglia, Dean, Chaires, & Trent, 2013; Le, Dean, Buscaglia, Chaires, & Trent, 2014; Le et al., 2012; Miller et al., 2011). We have also used this approach to access higher order quadruplex structures (Petraccone, Garbett, Chaires, & Trent, 2010; Petraccone, Trent, & Chaires, 2008; Petraccone et al., 2011). hTERT promoter sequences were explored by building plausible models, predicting their hydrodynamic properties, and testing the predictions by comparison to measured biophysical properties (Chaires et al., 2014). These higher order structures could not otherwise be determined by NMR or crystallography because of current limitations in methodology.

Figure 1 shows a schematic of our approach. The process begins on the computational side with either available structures taken from databases or constructed using principles obtained from relevant available structures.

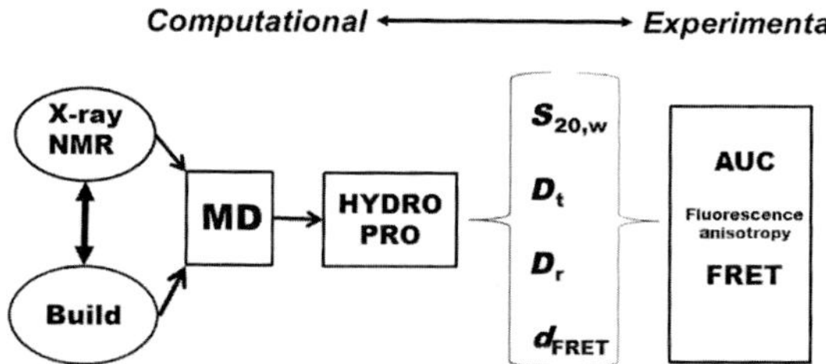

Figure 1 Work flow for hydrodynamic modeling of quadruplex structures.

Structures are then optimized by molecular dynamics simulations. HYDROPRO software is then employed to construct bead models from which hydrodynamic properties may be calculated. On the experimental side, measurements are made using sequences identical to those used in the calculations to obtain actual values for hydrodynamic properties for comparison and validation. The sedimentation coefficient ($s_{20,w}$) is the most convenient and useful parameter for comparisons, along with translational diffusion coefficients (D_t). Both of these may be readily obtained by analytical ultracentrifugation studies. We have on occasion used steady-state fluorescence anisotropy methods to estimate rotational diffusion coefficients (D_r) to provide an additional parameter for comparison. Fluorescence resonance energy transfer (FRET) methods could in principle be used to test models by measuring end-to-end distances or other distances (d_{FRET}) between strategically placed FRET pairs. This chapter will describe the methods used at the various steps shown in Fig. 1, with comments about key parameters needed for analysis and possible pitfalls and problems encountered all the way.

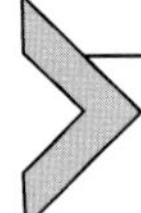

2. CORRELATING MOLECULAR STRUCTURE WITH EXPERIMENTAL SOLUTION HYDRODYNAMIC MEASUREMENTS—GENERAL COMMENTS

HYDROPRO is a program that calculates many hydrodynamic properties based on three-dimensional structure and hydrodynamic bead modeling. These properties include sedimentation and translational coefficients, intrinsic viscosity, rotational relaxation times, and the radius of gyration. The sedimentation coefficient ($s_{20,w}$) is the most accessible to compare directly with the experimentally derived analytical ultracentrifugation measurement. Thus, we correlate an assumed structural model to actual experimental solution measurements. This is a powerful strategy to infer what structures are possible and, importantly, what structures are *not* possible in solution. It is important to note that this strategy cannot determine the detailed three-dimensional structure, but can only show what structures are consistent with the experimentally observed data, i.e., $s_{20,w}$. We note that structures derived from X-ray crystallography and NMR spectroscopy are also only models, albeit models that are derived from a more extensive set of experimental constraints (reflection intensities and angles or distance constraints). In the case of molecular models, it is logical to build with known structural topologies and work with combinations of such. The limitation of this approach is that we may not have three-dimensional structural examples of all possible quadruplex topologies. This in itself can be useful in identifying that a sequence forms a yet unidentified topology. This was the case in our initial report (Li et al., 2005) on this approach investigating the human telomere sequence. We had shown that the crystal structure of d[AGGG (TTAGGG)$_3$] was not the predominant form in solution, but also that the other known topologies were not the best fit either. Subsequently, the hybrid-1 (Ambrus et al., 2006) and -2 (Luu, Phan, Kuryavyi, Lacroix, & Patel, 2006) topologies were reported and are well correlated with the experimental $s_{20,w}$. It can be seen in Fig. 2 that the parallel form of the human telomere and hybrid-1 have dramatically different shape, which leads to different hydrodynamic properties and hydrodynamic bead models.

3. HYDROPRO SOFTWARE

The software package HYDROPRO is available for download at http://leonardo.inf.um.es/macromol/programs/hydropro/hydropro.htm.

It is available as a Windows or Unix binary executable file and the two required files to run it are the binary executable and a text input file "hydropro.dat."

4. GETTING STRUCTURE FILES AND BUILDING MODELS

The structure files for known structures are available from the Protein Databank (www.pdb.org) or the Nucleic Acids Database (ndbserver.rutgers.edu) in PDB format.

5. SOFTWARE FOR MODEL BUILDING

The initial models can be constructed in any modeling software package that has the following features: (1) it is able to read a PDB-formatted file correctly, (2) it can create standard DNA/RNA bases and strands, (3) it can perform minimization/implicitly solvated molecular dynamics of the structures, (4) it can freeze regions of the structure during minimization/implicitly solvated molecular dynamics (for example, the quartet guanines constrained/restrained when minimizing new loop configurations), and (5) it can write a correctly formatted PDB file. Our model building software preference is the Schrödinger Maestro package and Macromodel, although we use our in-house program *DNAsar* to take any DNA-containing PDB file and correctly format the bases in order with correct atom names and

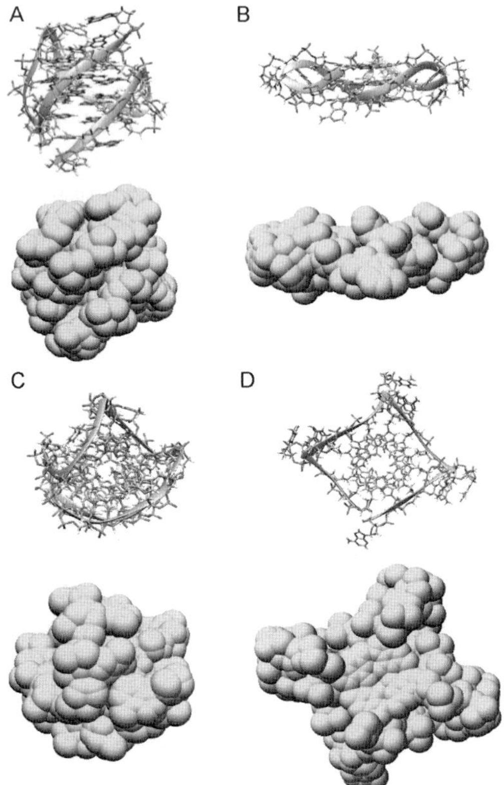

Figure 2 Hydrodynamic bead modeling of the human quadruplex hybrid-1 and parallel forms. Side view of hybrid-1 form (2HY9.pdb) (A) and parallel form (1KF1.pdb) (B) in structural representations with the hydrodynamic bead model below. Top view of hybrid-1 form (2HY9.pdb) (C) and parallel form (1KF1.pdb) (D) in structural representations with the hydrodynamic bead model below.

connectivity. Maestro does not automatically do this. We subsequently optimize the models by running fully solvated molecular dynamics using the AMBER suite (Case et al., 2005, 2014) of programs (leap, sander, and pmemd, current version 14). Numerous errors will occur in leap if the PDB file does not have the correct atom names and connectivity.

6. PROTOCOL FOR RUNNING HYDROPRO ON QUADRUPLEX NUCLEIC ACID STRUCTURE FILES

1. Obtain a structure file in PDB format. This can either be from X-ray crystallography, NMR, or from a constructed molecular model.

2. Strip the PDB file of non-nucleic acid components unless they are part of what you are actually observing in solution. If it is a NMR-derived file, separate out individual entries and run individually. When dealing with PDB files, it is important to check the files carefully, including the addition of hydrogen atoms to X-ray crystal structures, missing bases, or dual occupancy. Do not automatically assume that the PDB file is correct and complete.
3. Calculate the molecular weight of the nucleotide in the PDB file. For quadruplex DNA, add the mass for the central quartet coordinated ions, i.e., K^+ or Na^+ to the molecular weight.
4. Change the **Name** and molecular weight variables in the hydropro.dat file (in bold in Fig. 3).
5. Run HYDROPRO with the following default hydropro.dat file (Fig. 3) for an atomic-level primary model shell calculation. This has been shown (Le et al., 2014) to reproduce the experimental values within 2% error.
6. The summary of input data and the results, including the sedimentation coefficient value, is in the **Name**-res.txt output file. The primary hydrodynamic model is in the PDB-formatted file **Name**-pri.bea and VRML-formatted file **Name**-pri.vrml file.
7. f/f_0 can be calculated by dividing the translational equivalent radii by the volume, although we find that this is less useful than the $s_{20,w}$.

```
Name                !Name of molecule
Name                !Name for output file
Name.pdb            !Structural (PDB) file
1                   !Type of calculation
2.53,               !AER, radius of primary elements
-1,                 !NSIG
1.0,                !Minimum radius of beads in the shell (SIGMIN)
2.0,                !Maximum radius of beads in the shell (SIGMAX)
20.,                !T (temperature, centigrade)
0.01002,            !ETA (Viscosity of the solvent in poises
7069.8,             !RM (Molecular weight)
0.55,               !Partial specific volume, cm3/g
0.99823,            !Solvent density, g/cm3
21                  !Number of values of Q
2.e+7,              !QMAX
30,                 !Number of intervals for the distance distribution
-1.0,               !RMAX
1000,               !Number of trials for MC calculation of covolume
1                   !IDIF=1 (yes) for full diffusion tensors
*                   !End of file
```

Figure 3 Example of a "hydropro.dat" file with typical values for modeling quadruplex structures.

An example of an actual quadruplex HYDROPRO data file is shown in Fig. 3, with comments.

The standard parameters for an aqueous solvated system are given in Fig. 3. We have successfully applied this strategy to quadruplex-forming sequences from the human telomere to several promoter regions. The hydrodynamic radius of the elements in the atomic-level primary shell model (Type of calculation = 1) of AER = 2.53 was parameterized using extensive simulations (Le et al., 2014) of the different morphologies of the human telomere quadruplex. The default value of AER = 2.9 was recommended for standard protein analysis.

7. EXTENDING THE STATIC HYDRODYNAMIC BEAD MODEL BY ANALYZING MOLECULAR DYNAMICS TRAJECTORIES

Molecular dynamics trajectories can be analyzed to obtain a representative distribution of sedimentation coefficient values by extracting the frames into individual PDB files and consecutively running HYDROPRO calculations. This can be easily automated by scripting a loop around the trajectory coordinate extraction and the HYDROPRO calculation as only the **Name** variable changes in the required hydropro.dat file for trajectory analysis. For example, an AMBER pmemd molecular dynamics trajectory can be analyzed by looping the extraction of the nucleic acid residues into a PDB file from the mdcrd file using cpptraj with the desirable start, stop, and offset values to sample the trajectory, and subsequently running HYDROPRO on the individual PDB files. Care should be taken over the looped **Name** variable so that it is consecutive under Unix commands if the trajectory is to be analyzed with respect to time. Please note that while there is no limit to the number of atoms used in HYDROPRO, operating system and memory limitations may become an issue on larger systems. Another useful analysis is to cluster the trajectory and obtain representative sedimentation coefficient values for the different resident substates and compare with experimental data (Le et al., 2014). This can be performed in two ways to generate useful information. The first is to cluster structurally by RMSD and determine the range of sedimentation coefficient values associated with a substate cluster. The second is to cluster by sedimentation coefficient values to obtain sets of conformations that are consistent with the range of values.

We have used these methods to examine the reported structural forms of the human telomere sequence, Fig. 4. The distributions of the calculated (red (light gray in the print version) and blue (dark gray in the print version)) and measured $s_{20,w}$ are in close agreement. There are two major outliers, 143D and 1KF1. The 143D structure was obtained under sodium buffer conditions, and in fact the sequence sediments with an experimental $s_{20,w}$ of 1.97 in potassium buffer, in close agreement with the predicted value. The distributions for the human parallel form 1KF1, Fig. 4B, are not in agreement and this reproduces the original observation that the human telomere parallel form is not the predominant form in solution.

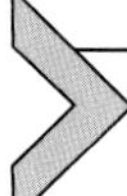

8. ACCELERATED MOLECULAR DYNAMICS CAN EXPLORE MORE CONFORMATIONAL SPACE

To explore conformational states on a longer timescale, accelerated molecular dynamics can be used (Chaires et al., 2014). In this study, we examined the hTERT promoter as there are several recently reported mutations in the quadruplex-forming sequence of the promoter that are overexpressed in some forms of cancer. There was an existing proposed (Palumbo, Ebbinghaus, & Hurley, 2009) topology for the hTERT promoter region, but no three-dimensional structure or model reported. This topology did not explain how the mutants would affect quadruplex stability. To explore possible structures consistent with the observed circular dichroism and AUC-determined $s_{20,w}$, several molecular models were built, including that of the previous topology. The experimental $s_{20,w}=4.0$ for this sized sequence was unusually high indicating that it is a highly compact structure. The calculated $s_{20,w}=3.3$ for the modeled existing topological form was inconsistent with the $s_{20,w}$ as well as the circular dichroism spectra. A new folded form of the hTERT promoter was modeled that combined the parallel forms, supported by the circular dichroism spectra, Fig. 5. Explicitly solvated molecular dynamics was employed, followed by accelerated molecular dynamics to cover more conformational space, particularly of the loop regions, as starting with molecular models can lead to bias for preconceived conformations or can lock into local minima. The analysis of the molecular dynamics trajectory using HYDROPRO showed that the newly proposed molecular model was fully consistent with all experimental observations.

One major advantage of accelerated molecular dynamics sampling arises when investigating new topologies. This is particularly relevant when there

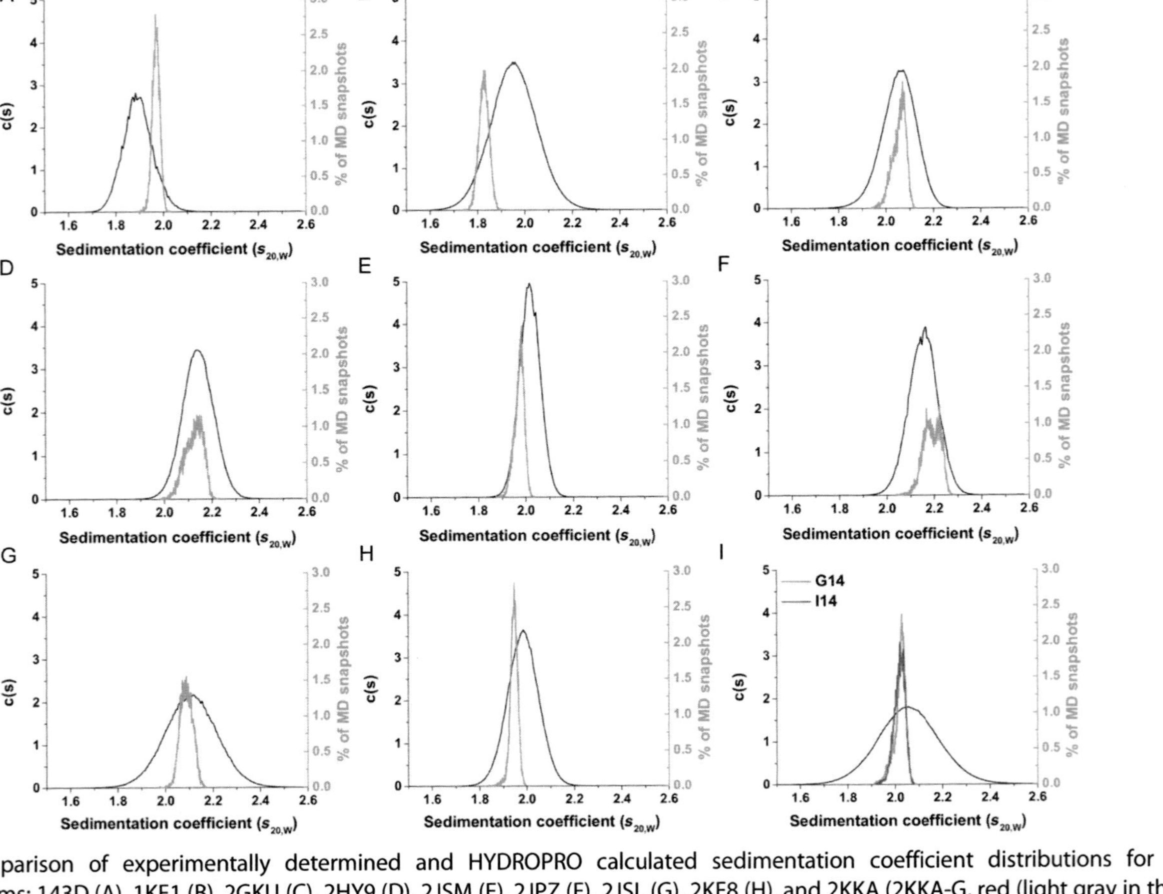

Figure 4 Comparison of experimentally determined and HYDROPRO calculated sedimentation coefficient distributions for hybrid quadruplex forms: 143D (A), 1KF1 (B), 2GKU (C), 2HY9 (D), 2JSM (E), 2JPZ (F), 2JSL (G), 2KF8 (H), and 2KKA (2KKA-G, red (light gray in the print version); 2KKA-I, blue (dark gray in the print version)) (I). For each G-quadruplex structure, sedimentation coefficients ($s_{20,w}$) were determined experimentally by AUC (black) and calculated from MD "snapshots" using HYDROPRO (red (light gray in the print version) and blue (dark gray

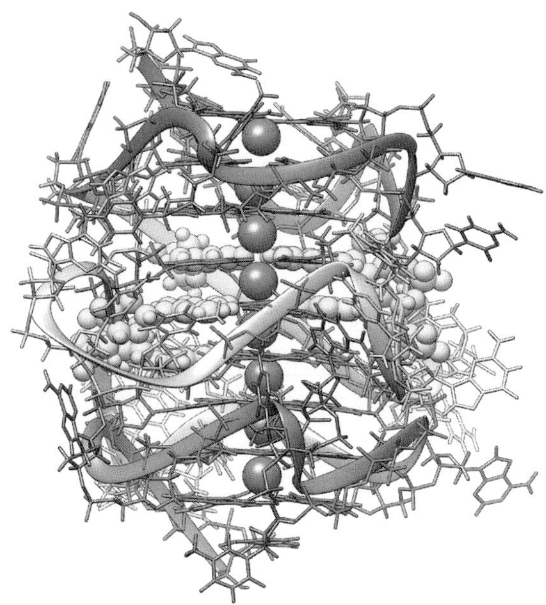

Figure 5 Improved hTERT structure. Molecular model of a structure consistent with all biophysical data. Quadruplex guanine bases in red, potassium ions in purple, and guanine mutations in yellow ball and stick representation. (See the color plate.)

are long quadruplex loop regions that have no known structural reference, or higher order quadruplex-component systems. Longer loops will have their own conformational substates, so efficient sampling will greatly aid in answering the fundamental question of "Is the calculated $s_{20,w}$ of the structural model consistent with the experimental $s_{20,w}$?"

9. EXAMINING HIGHER ORDER QUADRUPLEXES FROM THE HUMAN TELOMERE SEQUENCE

There has been considerable interest in the human telomere single-strand overhang comprising up to ~400 nt (Cimino-Reale et al., 2001). However, the detailed structural study of single-strand multimeric quadruplex forms is intractable by current X-ray and NMR methods. Therefore, molecular dynamics and hydrodynamic measurement may provide some insight into these structural motifs (Petraccone et al., 2008, 2010, 2011). Three questions that have been examined are "what topological forms are present?" "do successive quadruplexes form from the single

Table 1 Sedimentation Values of Higher Order Telomere Sequences

Sequence	MW_{calc}[a]	ε_{260}	$s_{20,w}$	MW_{Expt}	f/f_0	HYDROPRO Calculated $s_{20,w}$
$(TTAGGG)_4$	7574	244,600	2.21	7740	1	2.11
$(TTAGGG)_8$	15,211	489,000	2.80	13,600	1.25	Stacked 2.87, unstacked 2.85
$(TTAGGG)_{12}$	22,848	733,400	3.56	20,800	1.31	Stacked 3.49, unstacked 2.98
$(TTAGGG)_{16}$	30,485	977,800	3.87	29,200	1.42	Stacked 4.20, unstacked 3.68,
$(TTAGGG)_{32}$	61,033	1,955,400	5.43	59,600	1.65	Stacked 5.45, unstacked 4.83

[a]Nucleotide bases, central potassium ions are included in the HYDROPRO calculated $s_{20,w}$.

strand?" and "what interactions are there between successive quadruplexes?" The last question is summed up by determining if the successive quadruplexes are "beads on a string" or dependent on stacking interactions. To examine this we extended the human telomere sequence of interest from 24 nt to 48 nt, 72 nt, 96 nt, and 192 nt capable of forming 2, 3, 4, and 8 successive quadruplexes, respectively, Table 1. Molecular models of various stacked and unstacked conformations were created based on our previous studies (Petraccone et al., 2008, 2010, 2011) and allowed to evolve using molecular dynamics simulations. Overall, there appears to be some interaction between successive quadruplexes since the measured $s_{20,w}$ is between those calculated for stacked and unstacked models. Representative molecular models that are consistent (and inconsistent) with the experimental values are shown in Fig. 6. The successive quadruplexes are not independent as in "beads on a string," but they are also not rigidly stacked.

10. EXPERIMENTAL DETERMINATION BY ANALYTICAL ULTRACENTRIFUGATION

Sedimentation and diffusion coefficients were determined by standard sedimentation velocity methods and analysis. Oligonucleotides of defined sequence are obtained from commercial sources (Integrated DNA Technologies, Inc., Coralville, Iowa 52241). For long sequences, it is essential to purchase gel-purified material to minimize sample heterogeneity. In sample preparation, the annealing protocol used can play an important role in

Figure 6 Simulated, experimentally validated higher order telomeric quadruplex structures. The experimentally consistent structures are shown with a green (light gray in the print version) ribbon for the single, dimer, trimer, tetramer, and octamer quadruplexes. The inconsistent trimer and tetramer are shown with a red (dark gray in the print version) backbone.

G-quadruplex formation (Le et al., 2012). Samples at working concentrations of 2–10 μ*M* (strand) are annealed by immersion into boiling water contained in a 1 L beaker for 5 min, after which the beaker is removed from the heat source and allowed to slow cool to room temperature overnight (12–24 h).

AUC is carried out in a Beckman Coulter Proteome Lab XL-A analytical ultracentrifuge using standard 2 sector cells with quartz windows in 4- or 8-hole rotors. All sedimentation velocity experiments are at 50,000 rpm with a preset scan number of 100 per cell. The reference sector is filled with 450 μL of buffer and the sample sector is filled with 430 μL of sample with an absorbance at 260 nm (1 cm path length) between 0.2 and 1.0. Centrifugation is started after the rotor has equilibrated to 20 °C for approximately 1 h. Buffer density is measured at 20.0 °C in a Mettler/Paar Calculating Density Meter DMA 55A. Buffer viscosity is measured in an Anton Parr AMVn Automated Microviscometer. Data are analyzed using the program Sedfit (www.sedfit.com). Hydrodynamic parameters are determined using the continuous c(s) model for $s_{20,w}$ and the non-interacting discrete species model for $D_{20,w}$ in the Sedfit program. The instrument has been calibrated for elapsed time, scan velocity, temperature, and radial magnification (Zhao et al., 2015). $s_{20,w}$ for several compact quadruplex structures was determined in the range of 1–4 μ*M* (absorbance at 260 nm from 0.2 to 1.0) and essentially no concentration dependence was observed.

Pitfalls: (1) There is pronounced dependency of the sedimentation coefficient on monovalent salt concentration for nucleic acids due to their charge. For quadruplex structures, salt concentrations of 0.2–0.4 *M* are needed to obtain $s_{20,w}$ values that are independent of salt concentration for comparison to calculated hydrodynamic parameters. (2) For most quadruplex-forming sequences that are not heavily modified to reduce polymorphism, there are likely to be more than one species in solution. Such polymorphism may be manifested in multiple peaks or by broad, possibly asymmetric, single peaks in c(s) profiles. It may be difficult in these cases to compare experimental values to computed ones. Additional purification steps are then needed to prepare homogeneous samples. Finally for longer sequences that may have extended structures, extrapolation to zero concentration is necessary to obtain correct $s_{20,w}$ values.

11. THE PARTIAL SPECIFIC VOLUME PROBLEM

There is no convenient way to compute partial specific volume (psv) estimates for DNA oligonucleotides from their composition, unlike for proteins in which psv may usually be reliably estimated from amino acid composition. For most nucleic acids, a value of 0.55 g/L appears to be applicable. Use of this value for quadruplexes was validated by fitting sedimentation equilibrium data of samples of known molecular weight to obtain psv estimates (Li et al., 2005). More recently, Hellman, Rodgers, and Fried (2010) determined psv values by varying solution density using either NaCl or KCl in the range of 0–2 *M* and obtained values near 0.55 g/L. Since Na^+ and K^+ bind to quadruplex structures as a function of salt concentration and also affect solution ideality, the reported results may contain subtle errors due to uncertainties in the contribution of ion binding to the molecular weight. We further investigated the psv problem by additional studies using ^{18}O water to vary solution densities.

For partial specific volume determinations, buffers are prepared in 0–80% $^{18}O\text{-}H_2O$ and resulting densities and viscosities are measured directly. $M(1-\nu\rho)$ is calculated using Sedfit from both sedimentation velocity measurements at 50,000 rpm and sedimentation equilibrium measurements at several rotor speeds. Using the method of Edelstein and Schachman (1967), $M(1-\nu\rho)$ is plotted versus solution density to allow for simultaneous determination of apparent partial specific volume (ν) and molecular weight (M), as shown in Fig. 7. The resultant values for ν are 2JSM $= 0.58 \pm 0.07$; 2JSL $= 0.56 \pm 0.07$; Tel22 $= 0.58 \pm 0.07$. These values

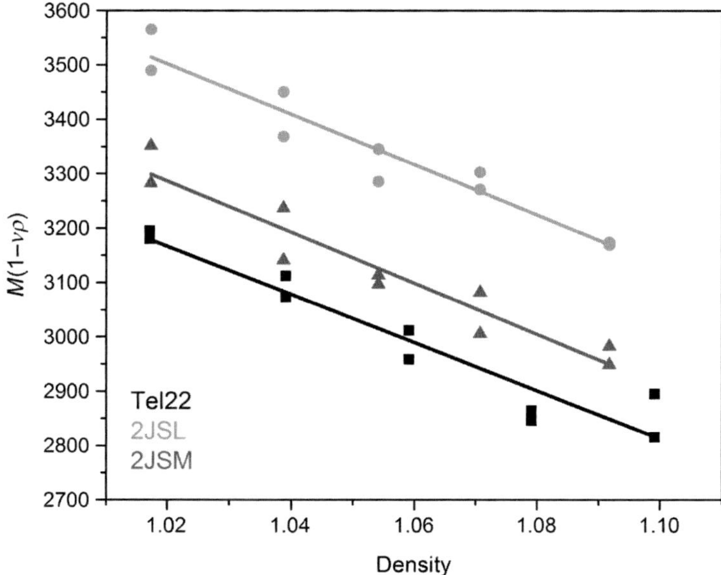

Figure 7 Partial specific volume data obtained in 0.4 *M* KCl.

for the partial specific volume are slightly larger for all three structures than the value of 0.55 mL/g used previously but within the standard error of that value. The standard error is similar to that reported by Hellman, Rodgers, and Fried using this method. However the Fried laboratory used KCl and NaCl to manipulate solution density in the range of 0–2 *M*, and we have shown that KCl binding to quadruplex structures such as hTel22 and 2GKU is considerably greater than the value of $2K^+$ assumed by Fried over the range of KCl concentrations used (Gray & Chaires, 2011). The increased partial specific volume in 0.4 *M* KCl indicates that significant amounts of K^+ are bound thus increasing the apparent partial specific volume and the molecular weight of hTel22, 2JSL, and 2JSM.

12. CORRELATION OF CALCULATED AND MEASURED HYDRODYNAMIC PROPERTIES

HYDROPRO has been proven to be able to accurately predict the hydrodynamic properties of short DNA duplex structures (Fernandes et al., 2002), longer worm-like DNA duplexes (Amorós et al., 2011), and monomeric folded quadruplex structures (Le et al., 2014). These validations lend confidence to efforts to use HYDROPRO to predict more complex

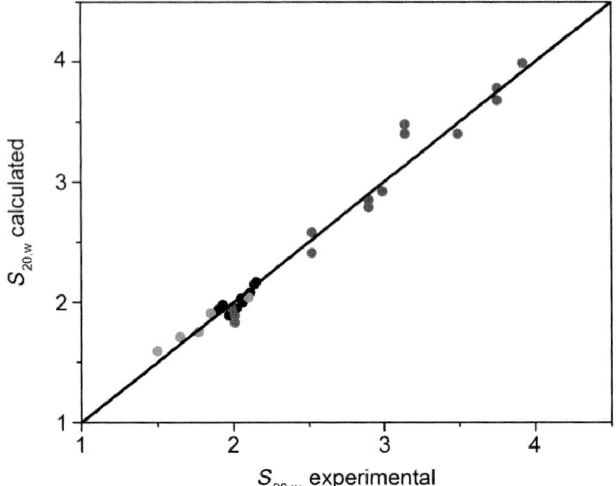

Figure 8 Correlation of predicted and measured sedimentation coefficients. The points in red are for short DNA duplexes taken from Fernandes et al. (2002). The black points are for known human telomere structures taken from Le et al. (2014). The blue points are for a variety of higher order quadruplex structures determined in our laboratory. (See the color plate.)

higher order quadruplex structures and to use the predictions to select plausible models for structures that would otherwise be intractable by current crystallographic and NMR methods. Figure 8 shows the correlation between predicted and measured sedimentation coefficients for a wide range of structures. The correlation is excellent (Pearson's $r = 0.991$; slope $= 0.999 \pm 0.008$), validating our approach for the characterization of complex quadruplex structures.

REFERENCES

Ambrus, A., Chen, D., Dai, J., Bialis, T., Jones, R. A., & Yang, D. (2006). Human telomeric sequence forms a hybrid-type intramolecular G-quadruplex structure with mixed parallel/antiparallel strands in potassium solution. *Nucleic Acids Research*, *34*(9), 2723–2735.

Amorós, D., Ortega, A., & García de la Torre, J. (2011). Hydrodynamic properties of wormlike macromolecules: Monte Carlo simulation and global analysis of experimental data. *Macromolecules*, *44*(14), 5788–5797. http://dx.doi.org/10.1021/ma102697q.

Byron, O. (2008). Hydrodynamic modeling: The solution conformation of macromolecules and their complexes. *Methods in Cell Biology*, *84*, 327–373.

Case, D. A., Babin, V., Berryman, J. T., Betz, R. M., Cai, Q., Cerutti, D. S., et al. (2014). *AMBER 14*. San Francisco: University of California.

Case, D. A., Cheatham, T. E., 3rd, Darden, T., Gohlke, H., Luo, R., Merz, K. M., Jr., et al. (2005). The Amber biomolecular simulation programs. *Journal of Computational Chemistry*, *26*(16), 1668–1688.

Chaires, J. B., Trent, J. O., Gray, R. D., Dean, W. L., Buscaglia, R., Thomas, S. D., et al. (2014). An improved model for the hTERT promoter quadruplex. *PLoS One*, *9*(12), e115580. http://dx.doi.org/10.1371/journal.pone.0115580.

Cimino-Reale, G., Pascale, E., Battiloro, E., Starace, G., Verna, R., & D'Ambrosio, E. (2001). The length of telomeric G-rich strand 3'-overhang measured by oligonucleotide ligation assay. *Nucleic Acids Research*, *29*(7), E35 (Research Support, Non-U.S. Gov't).

Edelstein, S. J., & Schachman, H. K. (1967). The simultaneous determination of partial specific volumes and molecular weights with microgram quantities. *The Journal of Biological Chemistry*, *242*(2), 306–311.

Fernandes, M. X., Ortega, A., Lopez Martinez, M. C., & Garcia de la Torre, J. (2002). Calculation of hydrodynamic properties of small nucleic acids from their atomic structure. *Nucleic Acids Research*, *30*(8), 1782–1788.

Garcia de la Torre, J., & Harding, S. E. (2013). Hydrodynamic modelling of protein conformation in solution: ELLIPS and HYDRO. *Biophysical Reviews*, *5*(2), 195–206. http://dx.doi.org/10.1007/s12551-013-0102-6.

Garcia De La Torre, J., Huertas, M. L., & Carrasco, B. (2000). Calculation of hydrodynamic properties of globular proteins from their atomic-level structure. *Biophysical Journal*, *78*(2), 719–730.

Gray, R. D., & Chaires, J. B. (2011). Linkage of cation binding and folding in human telomeric quadruplex DNA. *Biophysical Chemistry*, *159*(1), 205–209.

Hellman, L. M., Rodgers, D. W., & Fried, M. G. (2010). Phenomenological partial-specific volumes for G-quadruplex DNAs. *European Biophysics Journal*, *39*(3), 389–396.

Huppert, J. L. (2008). Hunting G-quadruplexes. *Biochimie*, *90*(8), 1140–1148.

Huppert, J. L., & Balasubramanian, S. (2005). Prevalence of quadruplexes in the human genome. *Nucleic Acids Research*, *33*(9), 2908–2916.

Huppert, J. L., & Balasubramanian, S. (2007). G-quadruplexes in promoters throughout the human genome. *Nucleic Acids Research*, *35*(2), 406–413. http://dx.doi.org/10.1093/nar/gkl1057.

Lane, A. N., Chaires, J. B., Gray, R. D., & Trent, J. O. (2008). Stability and kinetics of G-quadruplex structures. *Nucleic Acids Research*, *36*(17), 5482–5515.

Le, H. T., Buscaglia, R., Dean, W. L., Chaires, J. B., & Trent, J. O. (2013). Calculation of hydrodynamic properties for G-quadruplex nucleic acid structures from in silico bead models. *Topics in Current Chemistry*, *330*, 179–210.

Le, H. T., Dean, W. L., Buscaglia, R., Chaires, J. B., & Trent, J. O. (2014). An investigation of G-quadruplex structural polymorphism in the human telomere using a combined approach of hydrodynamic bead modeling and molecular dynamics simulation. *The Journal of Physical Chemistry B*, *118*(20), 5390–5405.

Le, H. T., Miller, M. C., Buscaglia, R., Dean, W. L., Holt, P. A., Chaires, J. B., et al. (2012). Not all G-quadruplexes are created equally: An investigation of the structural polymorphism of the c-Myc G-quadruplex-forming sequence and its interaction with the porphyrin TMPyP4. *Organic & Biomolecular Chemistry*, *10*(47), 9393–9404.

Li, J., Correia, J. J., Wang, L., Trent, J. O., & Chaires, J. B. (2005). Not so crystal clear: The structure of the human telomere G-quadruplex in solution differs from that present in a crystal. *Nucleic Acids Research*, *33*(14), 4649–4659.

Luu, K. N., Phan, A. T., Kuryavyi, V., Lacroix, L., & Patel, D. J. (2006). Structure of the human telomere in K+ solution: An intramolecular (3+1) G-quadruplex scaffold. *Journal of the American Chemical Society*, *128*(30), 9963–9970.

Miller, M. C., Le, H. T., Dean, W. L., Holt, P. A., Chaires, J. B., & Trent, J. O. (2011). Polymorphism and resolution of oncogene promoter quadruplex-forming sequences. *Organic & Biomolecular Chemistry*, *9*(22), 7633–7637.

Ortega, A., Amoros, D., & Garcia de la Torre, J. (2011). Prediction of hydrodynamic and other solution properties of rigid proteins from atomic- and residue-level models. *Biophysical Journal*, *101*(4), 892–898.

Palumbo, S. L., Ebbinghaus, S. W., & Hurley, L. H. (2009). Formation of a unique end-to-end stacked pair of G-quadruplexes in the hTERT core promoter with implications for inhibition of telomerase by G-quadruplex-interactive ligands. *Journal of the American Chemical Society*, *131*(31), 10878–10891. http://dx.doi.org/10.1021/ja902281d.

Petraccone, L., Garbett, N. C., Chaires, J. B., & Trent, J. O. (2010). An integrated molecular dynamics (MD) and experimental study of higher order human telomeric quadruplexes. *Biopolymers*, *93*(6), 533–548. http://dx.doi.org/10.1002/bip.21392.

Petraccone, L., Spink, C., Trent, J. O., Garbett, N. C., Mekmaysy, C. S., Giancola, C., et al. (2011). Structure and stability of higher-order human telomeric quadruplexes. *Journal of the American Chemical Society*, *133*(51), 20951–20961.

Petraccone, L., Trent, J. O., & Chaires, J. B. (2008). The tail of the telomere. *Journal of the American Chemical Society*, *130*(49), 16530–16532.

Zhao, H. et al. (2015). A multilaboratory comparison of calibration accuracy and the performance of external references in analytical ultracentrifugation. *PLoS One*, in press.

CHAPTER FOURTEEN

Analytical Ultracentrifugation as a Tool to Study Nonspecific Protein–DNA Interactions

Teng-Chieh Yang*, Carlos Enrique Catalano†, Nasib Karl Maluf‡,1

*Pfizer, Pharmaceutical Research and Development, Chesterfield, Missouri, USA
†Skaggs School of Pharmacy and Pharmaceutical Sciences, University of Colorado, Anschutz Medical Campus, Aurora, Colorado, USA
‡Alliance Protein Laboratories, San Diego, California, USA
[1]Corresponding author: e-mail address: karl.maluf@ap-lab.com

Contents

Abstract

Analytical ultracentrifugation (AUC) is a powerful tool that can provide thermodynamic information on associating systems. Here, we discuss how to use the two fundamental AUC applications, sedimentation velocity (SV), and sedimentation equilibrium (SE), to study nonspecific protein–nucleic acid interactions, with a special emphasis on how to analyze the experimental data to extract thermodynamic information. We discuss three specific applications of this approach: (i) determination of nonspecific binding stoichiometry of *E. coli* integration host factor protein to dsDNA, (ii) characterization of nonspecific binding properties of Adenoviral IVa2 protein to dsDNA using SE-AUC, and (iii) analysis of the competition between specific and nonspecific DNA-binding

Methods in Enzymology, Volume 562
ISSN 0076-6879
http://dx.doi.org/10.1016/bs.mie.2015.04.009

interactions observed for *E. coli* integration host factor protein assembly on dsDNA. These approaches provide powerful tools that allow thermodynamic interrogation and thus a mechanistic understanding of how proteins bind nucleic acids by both specific and nonspecific interactions.

1. INTRODUCTION

Fundamental biological processes such as DNA replication, repair, recombination, transcription, and translation are regulated via protein–nucleic acid interactions. At any time during the cell cycle, a nucleic acid will be covered with a variety of proteins that compete for binding sites, and this competition dictates how the cell responds to its environment. These proteins bind the nucleic acid both specifically—where a protein recognizes a certain sequence of bases with much higher affinity than it does with other sequences, and nonspecifically—where a protein binds indiscriminately to any sequence. Any attempt to model this competition—and therefore develop a predictive understanding of how a cell responds appropriately to its environment—requires a mechanistic understanding of how proteins bind nucleic acids by both specific and nonspecific interactions.

Proteins that bind specifically to a site on a nucleic acid (a unique base sequence) must be able to select this sequence from what is essentially a vast excess of random base sequences (i.e., nonspecific sites). Thus, while the specific protein–nucleic acid interaction is important for controlling cellular functions, the nonspecific interaction is equally important in consideration of the biological outcome. So how does the protein molecule find a specific site buried within the vast nonspecific background, especially when this site is obscured with nonspecifically bound proteins? Do nonspecific sequences serve as a reservoir for sequestering specific binding proteins close to their sites to minimize their search space and thus optimize specific binding rates? How does one evaluate a protein whose biological role requires both specific and nonspecific DNA-binding interactions? To answer these fundamental questions, we need rigorous experimental approaches to measure these interactions and sophisticated computational tools to appropriately analyze the data within the context of meaningful molecular models.

Analytical ultracentrifugation (AUC) has proved to be a reliable approach for studying macromolecular interactions. With modern computational power and the newly developed fluorescence detection system,

AUC allows one to conduct experiments over a wide concentration range and to directly apply first-order physical principles to analyze the data. More importantly, different AUC experimental approaches—sedimentation velocity (SV-AUC) and sedimentation equilibrium (SE-AUC)—provide rigorous hydrodynamic (molecular shape distributions) and thermodynamic (strengths of interactions at equilibrium) information (Demeler & van Holde, 2004; Laue & Stafford, 1999; Lebowitz, Lewis, & Schuck, 2002; Philo, 2011; Ucci & Cole, 2004; Yphantis, 1964). Here, we discuss how to study nonspecific protein–DNA interactions using AUC approaches and emphasize in particular how to model AUC data via nonlinear least squares (NLLS) approaches using models that account for nonspecific nucleic acid binding. We have organized this review as follows: First we discuss the elegant theory, originally developed by McGhee and von Hippel (McGhee & von Hippel, 1974), which describes the binding of a ligand to a one-dimensional lattice and application of the model to interrogate nonspecific protein–nucleic acid interactions. Next we discuss a more complex scenario, where specific and nonspecific sites compete for protein binding. Then we review four different case studies originally developed by Record and coworkers (Tsodikov, Holbrook, Shkel, & Record, 2001) that deal with the competition between specific and nonspecific binding and discuss how we have applied these models to the analysis of AUC data collected in our laboratories.

2. NONSPECIFIC INTERACTION MODEL

2.1 Nonspecific Binding of Proteins to Linear Nucleic Acids (Non-Cooperative Versus Cooperative)

A DNA-binding protein that interacts with a nonspecific DNA sequence is characterized by three thermodynamic parameters: (1) the intrinsic binding affinity, K, (2) the binding (occluded) site size, n, and (3) the cooperativity parameter, ω. In this case, the intrinsic binding affinity quantifies the strength with which a protein binds to a nonspecific site on the nucleic acid lattice and the binding site size represents the space that single protein occupies, or occludes, on the lattice. For a nucleic acid molecule that can accommodate more than one protein, the binding of the first protein may influence binding of the second protein adjacent to it, and this tendency is quantified by the cooperativity parameter. Positive cooperativity refers to the case where binding of the first protein increases the probability of binding of the second, while negative cooperativity implies that the first binding event decreases the *probability* of the second event.

A crucial aspect of the nonspecific interaction is that, if a given space on the nucleic acid is occupied by one protein, it is assumed to be physically impossible for another protein to interact with any of the occluded site. Moreover, if two proteins are bound close enough to each other such that the number of bases between them is less than the occluded site size, then this segment of the nucleic acid is unavailable for binding by additional proteins (the so-called overlapping effect). This creates a conundrum for developing a model to quantify the number of proteins bound to the nucleic acid since the concentration of available nucleic acid binding sites depends not only on the number of proteins bound to the lattice, but also on the configuration with which the proteins are bound—the same number of proteins can be bound to the nucleic acid in multiple configurations, and these configurations in general have differing numbers of remaining sites for the next protein-binding event (Fig. 1).

To resolve this complication, McGhee and von Hippel (MvH) used an approach based on conditional probabilities to account for all the possible configurations for protein bound nonspecifically to an infinite, one-dimensional lattice, all in terms of the intrinsic binding affinity and

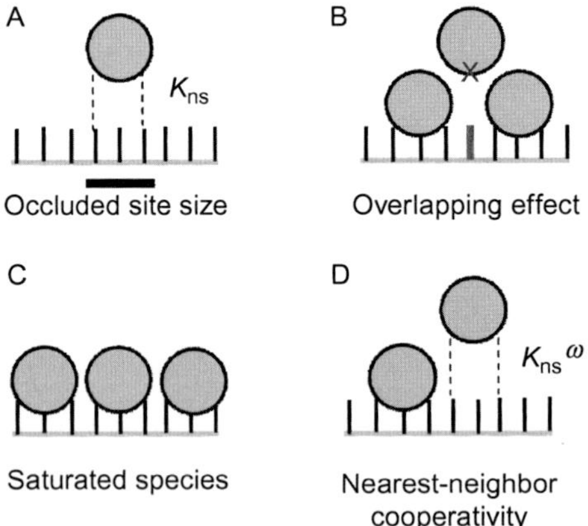

Figure 1 The McGhee and von Hippel approach is used to resolve the nonspecific binding properties of protein–DNA interactions. This approach assumes that the protein interacts with a one-dimensional DNA lattice and uses conditional probabilities to analyze three nonspecific binding properties: (A) the occluded site size (number of bases occluded by the bound protein), (B) the overlapping effect, and (D) the nearest-neighbor cooperativity. Because of the overlapping effect, under this nonspecific binding scenario, it is difficult to fully saturate the duplex (C).

the occluded binding site size. They also derived expressions that incorporate "nearest-neighbor" cooperativity. This concept assumes that the only possibility for cooperative interactions occurs when two ligands are bound immediately next to each other; if any two proteins have at least one free (unbound) nucleic acid base between them, there can be no cooperativity.

The MvH model was later extended by Epstein, who developed exact expressions that accounted for a protein binding to a nucleic acid of finite length using a combinatorial approach (Epstein, 1978). While this approach takes into account all possible nonspecifically bound protein states, it does not produce a simple closed-form expression convenient for NLLS modeling of binding data. Since many studies have been carried out using rather short nucleic acids, and since the MvH formalism provides a straightforward method for NLLS data analysis, Record and coworkers modified the MvH model, using a weighted conditional probability approach, to provide approximate (although highly accurate), closed-form expressions suitable for analysis of data collected on short nucleic acid lattices of finite length (Tsodikov et al., 2001). In addition to the MvH and Epstein models, which consider only one mode of nonspecific interaction, more complex binding models including multiple nonspecific binding modes have been considered (Bujalowski, 2006; Bujalowski, Lohman, & Anderson, 1989; Bujalowski, Overman, & Lohman, 1988). Several groups have adapted these models to analyze data from electrophoretic mobility shift (EMS) (Aeling et al., 2006; Sanyal, Yang, & Catalano, 2014), calorimetry (Holbrook, Tsodikov, Saecker, & Record, 2001) and filter binding assays (Bujalowski, 2006; Lohman & Bujalowski, 1991) with success. In this review, we will discuss how to analyze AUC data to interrogate nonspecific binding of proteins to nucleic acid lattices of finite length, by adapting the modified MvH model presented by Record and coworkers.

2.2 MvH Model

In the following discussion, we will assume that the nucleic acid lattice is dsDNA. The MvH model assumes a dynamic equilibrium exists between the free protein and the free protein-binding site on a one-dimensional duplex, described by the following reaction scheme:

$$\text{Free protein-binding site on DNA} + \text{free protein} \overset{K_{ns}}{\Leftrightarrow} \text{bound protein on DNA}$$

where K_{ns} is the intrinsic, nonspecific association binding constant. Applying mass conservation to this reaction, we obtain:

$$\frac{\overline{P_b}}{P_f} = K_{\text{ns}}.\overline{D_f} \quad (1)$$

where $\overline{P_b}$ is the average number of proteins bound per DNA, P_f is the concentration of free protein, and $\overline{D_f}$ is the average number of free protein-binding sites per DNA. This is essentially a Scatchard formulation that takes into account the nonspecific binding of multiple copies of a ligand (e.g., protein) to a one-dimensional lattice of finite length (e.g., DNA).

Consider a DNA molecule that is N-bp long and where each bound protein occupies (occludes) n bp. According to the MvH formalism, the general equation for the nonspecific binding isotherm is given by:

$$\frac{\upsilon}{K_{\text{ns}}.P_f} = (1 - n\upsilon)(ff)^{n-1}\left(\frac{N - n + 1}{N}\right) \quad (2)$$

where υ represents the protein binding density, i.e., moles of protein bound per mole of *base pair* of DNA. Logically, $\overline{P_b}$, from Eq. (1), is a function of υ, N, and the concentration of DNA molecules ([DNA_t]), and can be expressed as:

$$\overline{P_b} = \upsilon N[\text{DNA}_\text{t}] \quad (3)$$

According to mass conservation, $P_f = P_t - \overline{P_b} = P_t - \upsilon N[\text{DNA}_\text{t}]$, where P_t is the total protein concentration. The term $(ff)^{n-1}$ in Eq. (2) is the conditional probability of randomly selecting $(n - 1)$ consecutive bp, that are adjacent to any given free bp, but that are not bound by protein. Since the probability of finding one free bp is $(1 - n\upsilon)$, the product $(1 - n\upsilon)(ff)^{n-1}$ represents the probability of finding n consecutive, unoccupied base pairs (i.e., a stretch of n free base pairs). The term $\left(\frac{N-n+1}{N}\right)$ is an approximation that has been shown by Record and coworkers to adequately account for the finite length of the DNA lattice (i.e., end effects) (Tsodikov et al., 2001).

The term $(ff)^{n-1}$ depends upon different binding scenarios. In the case of non-cooperative binding, $(ff)^{n-1}$ is given by $\left(\frac{1-n\upsilon}{1-(n-1)\upsilon}\right)^{n-1}$. For proteins that display nearest-neighbor cooperativity (either favorable or unfavorable), we have to consider different scenarios for how different "protein clusters" are found on the DNA. A cooperativity parameter, ω, is introduced to describe this likelihood of having one free protein bound immediately adjacent to another protein. Thus, $(ff)^{n-1}$ must depend on ω and is expressed as

$$(ff)^{n-1} = \left(\frac{(2\omega-1)(1-nv)+v-R}{2(\omega-1)(1-nv)}\right)^{n-1} \cdot \left(\frac{1-(n+1)v+R}{2(1-nv)}\right)^2 \quad (4a)$$

where $R = \sqrt{[1-(n+1)v]^2 + 4\omega v(1-nv)}$. Note that the above equation as written is undefined for the non-cooperative case, i.e., when $\omega = 1$. In this case, Bujalowski et al. derived a single, closed-form expression of the equation,

$$(ff)^{n-1} = \left(\frac{2\omega(1-nv)}{(2\omega-1)(1-nv)+v+R}\right)^{n-1} \cdot \left(\frac{1-(n+1)v+R}{2(1-nv)}\right)^2 \quad (4b)$$

which is applicable for either non-cooperative or cooperative binding. This expression is easier to incorporate into NLLS fitting algorithms since a single equation can be used in the fitting algorithm for analysis of the experimental data over the range $0 < \omega < \infty$ (Bujalowski et al., 1989). It is important to keep in mind that as written, the binding density v is an implicit variable (i.e., it is not possible to solve Eq. 2 for v). For any NLLS data analysis algorithm, v must be determined at each value of P_f, and this is accomplished by an iterative procedure that examines successive values of v within the range $0 < v < \frac{1}{n}$, where $\frac{1}{n}$ represents the maximum value of v (when the DNA is completely saturated with ligand). This is done until the right- and left-hand sides of Eq. (2) are equal to each other within a defined tolerance. In all the cases we consider here, we have used the built-in implicit solver available in the commercially available Scientist software (MicroMath, St. Louis MO).

2.3 Specific and Nonspecific Competitive Binding of Proteins to Linear Nucleic Acids

In general, DNA-binding proteins can interact with DNA both specifically and nonspecifically. In some cases, specific binding is so much stronger than nonspecific binding that the latter can be ignored during data analysis. However, in many cases nonspecific interactions cannot be ignored. This can occur in cases where discrimination between specific and nonspecific binding sites is modest (i.e., KNS<KSP<=~10 × KNS). In other cases, the concentration of nonspecific sites is large enough such that nonspecific binding can compete significantly with specific binding. This can occur experimentally (i.e., *in vitro*) if the duplex used in the experiment is long relative to the site-specific site size; indeed, this is commonly encountered *in vivo*, where the number of specific binding sites are always vastly outnumbered by the total number of nonspecific sites in the genome. For example, in the case of the *Escherichia coli lac* operon, it has been estimated that there are 10–30 lac repressor proteins that must find one to four copies of the specific *lac* operon out of a total of $\sim 10^7$ nonspecific DNA-binding sites in the *E. coli* genome (von Hippel & Berg, 1986). In either case, competitive nonspecific binding must be accounted for to properly analyze the binding isotherms. Record and coworkers have developed specific models for this purpose, and we discuss two particularly useful cases below (Tsodikov et al., 2001).

In the first case (case 1), they consider the binding of a protein to a relatively short DNA molecule that contains a single specific DNA-binding sequence. For this model, it is important to keep in mind that it is common for specific DNA-binding proteins to occlude more DNA upon binding to their specific site than they do when they bind nonspecifically, i.e., it is usually the case that the specific binding site size is larger than the nonspecific site size. Thus, in this model they consider the case where a DNA molecule is bound to its protein either entirely specifically or entirely nonspecifically—there are no DNA molecules that have both specific and nonspecifically bound proteins since there is not enough space on the short duplex to simultaneously accommodate both protein binding modes. Thus, if a protein is bound to the specific DNA site, the binding stoichiometry is 1:1. However, if the protein is bound nonspecifically, then it is possible to have multiple proteins bound, up to the saturation limit (Fig. 1C).

The second case (case 2) stipulates that the DNA contains a single specific site with flanking regions that are much longer than the length of the specific site so that proteins can simultaneously interact with both the specific and nonspecific DNA-binding sites. Note that if a protein is bound nonspecifically to a site that overlaps with the specific site, it is assumed that this physically prevents another protein from binding to the site. Thus, the entire specific site must be available for a protein to bind in a specific binding mode.

In these two cases, we must consider three states of the protein ligand: protein that is free in solution (P_f), protein that is bound to DNA in a nonspecific manner ($P_{b,\mathrm{ns}}$), and protein that is bound to a specific site in the duplex ($P_{b,\mathrm{sp}}$). According to Eq. (2), P_f can be expressed as:

$$P_f = \frac{N\upsilon}{K_{\mathrm{ns}}(N-n+1)(ff)^{n-1}(1-n\upsilon)} \tag{5}$$

For case 1, $P_{b,\mathrm{ns}}$ can be expressed as (Tsodikov et al., 2001):

$$P_{b,\mathrm{ns}} = \upsilon N[\mathrm{DNA_t}]\left(\frac{K_{\mathrm{ns}}(N-n+1)}{K_{\mathrm{ns}}(N-n+1)+NK_{\mathrm{sp}}\upsilon(ff)^{N-n}}\right) \tag{6}$$

where K_{sp} is the association constant for the specific binding of the protein to the DNA and the expression in parenthesis accounts for the competition of protein binding to nonspecific sites by the specific site. Similarly, the expression for $P_{b,\mathrm{sp}}$ is given by:

$$P_{b,\mathrm{sp}} = \upsilon N[\mathrm{DNA_t}]\left(\frac{K_{\mathrm{sp}}(ff)^{N-n}}{K_{\mathrm{ns}}(N-n+1)+NK_{\mathrm{sp}}\upsilon(ff)^{N-n}}\right) \tag{7}$$

The total protein concentration is then obtained by combining Eqs. (5)–(7), as shown in Eq. (8):

$$P_t = \frac{N\upsilon}{K_{\mathrm{ns}}(N-n+1)(ff)^{n-1}(1-n\upsilon)} + \upsilon N[\mathrm{DNA_t}]\left(\frac{K_{\mathrm{ns}}(N-n+1)+K_{\mathrm{sp}}(ff)^{N-n}}{K_{\mathrm{ns}}(N-n+1)+NK_{\mathrm{sp}}\upsilon(ff)^{N-n}}\right) \tag{8}$$

For case 2, since there are many more nonspecific binding sites than the single, specific site, the expression for $P_{b,\mathrm{ns}}$ is simply Eq. (3). However, $P_{b,\mathrm{sp}}$ is more complicated since nonspecifically bound protein can overlap with the specific site, preventing specific binding. Therefore, $P_{b,\mathrm{sp}}$ also depends on the location of the specific binding sequence within the DNA duplex. With these considerations, $P_{b,\mathrm{sp}}$ can be expressed as:

$$P_{b,\mathrm{sp}} = \frac{K_{\mathrm{sp}}\lambda P_f[\mathrm{DNA_t}]}{1+K_{\mathrm{sp}}\lambda P_f}, \text{ with } \lambda = \frac{1}{N}(1-n\upsilon)\frac{2-(ff)^{N_{f1}}-(ff)^{N_{f2}}}{1-(ff)} \tag{9}$$

Where N_{f1} and N_{f2} are the known lengths of the flanking regions (in bp) on either side of the specific DNA site. Combining Eqs. (3), (5), and (8) affords an expression for the total protein concentration as shown in Eq. (10):

$$P_t \cong \frac{N\upsilon}{K_{\mathrm{ns}}(N-n+1)(ff)^{n-1}(1-n\upsilon)} + [\mathrm{DNA_t}]\left(\frac{N\upsilon+(N\upsilon+1)\lambda K_{\mathrm{sp}}P_f}{1+\lambda K_{\mathrm{sp}}P_f}\right) \tag{10}$$

In many experimental situations, what is actually measured is not the moles of ligand bound per bp, but rather the fraction of DNA that is bound (e.g., EMS data). For these cases, it is necessary to obtain expressions for the fraction of DNA that is free and the fraction of DNA that has at least one bound protein (i.e., the fraction of DNA that is bound). These expressions are given below in Eqs. (11) and (12), originally derived in the following references (Tsodikov et al., 2001; Yang & Maluf, 2014):

$$F_f = (1 - nv)(ff)^{N-1} \tag{11}$$

$$F_b = 1 - (1 - nv)(ff)^{N-1} \tag{12}$$

where F_f is the fraction of free DNA and F_b is the fraction of DNA that has at least one protein bound. Note that Eq. (11) represents the probability of finding a stretch of $(N-1)$ free bp next to a free DNA bp; i.e., it represents the probability of finding a DNA with N free bp, i.e., a free DNA molecule.

For the data-fitting procedures, one must solve for two implicit variables: P_f and v. We obtain P_f over the range $0 < P_f < P_t$, and v over the range $0 < v < \frac{1}{n}$. The independent variables are $[P_t]$ and $[DNA_t]$. For case 1, we use Eqs. (2), (4a), (4b), and (8) to calculate v and P_f at any $[P_t]$ and $[DNA_t]$. For case 2, we use Eqs. (2), (4a), (4b), and (10). Then, depending on the type of data to be analyzed, we use Eqs. (11) and (12) to calculate the fraction of free or bound DNA. This procedure allows us to determine the best-fit parameters K_{ns}, K_{sp}, ω, and n.

3. CASE STUDIES

3.1 Determination of Binding Stoichiometry Using SE-AUC

A key measurement in the study of any protein–DNA binding reaction is the binding stoichiometry, which is defined as the number of protein ligands bound per DNA molecule at saturation. SE-AUC is an ideal technique for this measurement since it reports directly the average molar mass of the protein–DNA complex. In our laboratories, we have successfully carried out these experiments using DNA molecules that have been conjugated with a Cy3 chromophore. This allows us to selectively monitor the absorbance of the chromophore at 522 nm, a wavelength where protein molecules do not typically absorb light. At this wavelength, we detect all the DNA species—the free DNA along with the ensemble of protein-bound DNA species—without pollution of the signal with any free protein species. This experimental design simplifies the data analysis.

E. coli integration host factor (IHF) is a basic, heterodimeric DNA-binding protein that belongs to a class of histone-like proteins capable of bending and wrapping DNA. As such, it plays an important role in condensation of the bacterial nucleoid, binding nonspecifically to duplex DNA. In addition, it serves as a regulatory protein associated with a number of cellular processes, including transcription, DNA replication, and site-specific recombination. In this role, IHF binds to a specific sequence with high affinity to introduce a strong bend (>140°) into the duplex. This provides a duplex architecture conducive to the cooperative assembly of additional regulatory proteins at the site. Thus, IHF must interact with DNA in both specific and nonspecific binding modes to exert distinct biological functions.

In Fig. 2, we show an application of the approach described above to determine the binding stoichiometry of IHF to a 10 bp nonspecific DNA duplex using sedimentation equilibrium approaches. Each experiment (i.e., data point) was carried out at three rotor speeds (22, 27, and 33 K RPM) and then globally fit to an ideal single-species model,

$$A(r) = A_{\mathrm{ref}}.\exp^{\sigma.\left(\frac{r^2 - r_{\mathrm{ref}}^2}{2}\right)} + b, \qquad \sigma = \frac{M_{b,\mathrm{app}}}{RT}\omega^2 \tag{13}$$

where $A(r)$ and A_{ref} are the total absorbance of all DNA species (free and protein bound) at radial position, r, and reference radial position, r_{ref}, respectively; b is the baseline offset, R is the gas constant, T is the absolute

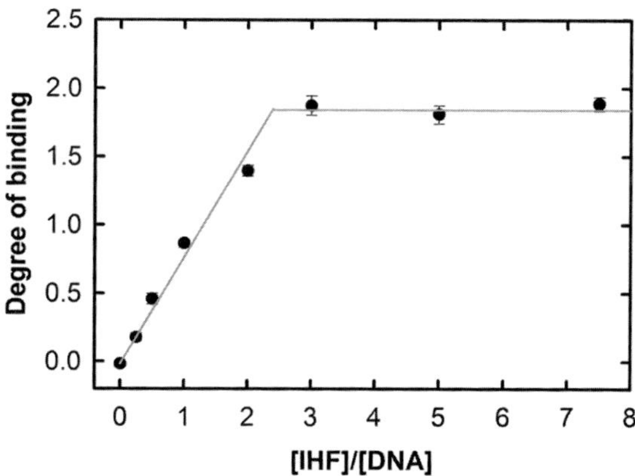

Figure 2 Stoichiometry of IHF binding to a 10 bp nonspecific duplex determined by SE-AUC. Samples were prepared by mixing various concentrations of IHF with 400 n*M* DNA; SE-AUC experiments were performed at three different rotor speeds (22, 27, and 33 K RPM) at 4 °C. The data were analyzed by an ideal single-species model (Eq. 13) to approximate the average molecular weight of all species, from which the degree of binding was calculated using Eq. (15). The error bar indicates 68.3% confidence intervals. *Data provided by Dr. Saurarshi J. Sanyal.*

temperature, ω is the angular speed of the rotor, and $M_{b,\text{app}}$ is the apparent buoyant molecular weight, given by:

$$M_{b,\text{app}} = M_{\text{app}}\left(1 - \overline{\nu}_{\text{app}}\rho\right) \tag{14}$$

where M_{app} is the apparent weight-average molecular weight of all species (Yphantis, 1964) and $\overline{\nu}_{\text{app}}$ is the apparent weight-average partial specific volume of all species, and ρ is the buffer density. The fitted $M_{b,\text{app}}$ is then used to calculate the approximate average number of IHF, L, bound per DNA duplex (i.e., the degree of binding) according to Eq. (15):

$$M_{b,\text{app}} = M_{b,\text{DNA}} + L.M_{b,\text{IHF}} \tag{15}$$

Note that this approach assumes there is no significant volume change associated with formation of the protein–DNA complexes. $M_{b,\text{DNA}}$ and $M_{b,\text{IHF}}$ can either be calculated from their respective sequences or can be measured experimentally. In this case, we calculate $M_{b,\text{IHF}} = 5.6\text{kDa}$ using the program SEDNTERP (Laue, Shah, Ridgeway, & Pelletier, 1992), and utilized SE-AUC to measure $M_{b,\text{DNA}} = 2.7(\pm 0.1)\text{kDa}$.

As shown in Fig. 2, a plot of the degree of binding as a function of [IHF]/[DNA] increases linearly and then reaches a plateau at about 1.9. This feature indicates that the experiment is carried out under stoichiometric binding conditions (i.e., the total DNA concentration is at least 10 times greater than the apparent dissociation constant) and that this 10 bp duplex DNA can accommodate two IHF proteins at saturation. This suggests that each IHF occupies about 5 bp of the DNA ($n=5$), which is in agreement with previous ITC studies (carried out under similar buffer conditions) (Holbrook et al., 2001).

A common problem with the interpretation of a stoichiometric binding curve is that the linear-plateau breakpoint is dependent upon not only the stoichiometry of the interaction but also the *activity* of the purified protein. For instance, a breakpoint of two can result from (i) a binding stoichiometry of 2:1 if the protein is 100% active or (ii) a stoichiometry of 1:1 if the protein is only 50% active. A significant advantage when using SE-AUC to measure protein–DNA binding stoichiometries is that the plateau provides a *direct* measure of the buoyant mass of the complex. This allows the researcher to make a direct comparison between the binding stoichiometry estimated from the plateau value with that implied by the breakpoint. If 100% of the protein molecules are active for DNA binding, then these two measurements should provide identical values for the binding stoichiometry. If, however, the breakpoint yields an estimate of the binding stoichiometry that is larger than that measured from the plateau, this most likely indicates that the protein preparation is not 100% active. In Fig. 2, comparison of the break point with the plateau indicates that this preparation of IHF is 83% active for nonspecific DNA binding.

3.2 Evaluation of Protein Nonspecific Binding Parameters Using SE-AUC

In the above model-independent analysis, we have essentially ignored the rich information content of the SE-AUC data to provide a simple method to estimate binding stoichiometries. It is also possible to analyze the primary data directly from first principles to obtain information on the nonspecific binding parameters. This requires the construction of a fitting model to directly analyze the concentration distributions. Here, we discuss how we analyzed SE-AUC data collected for the nonspecific binding of the Adenoviral IVa2 protein to a Cy3-labeled, duplex DNA, and compare these results with orthogonal data collected using the filter binding technique. Other examples are available that rigorously analyze nonspecific protein–DNA interactions using SE-AUC using the Epstein model (Ucci & Cole, 2004). In the approach presented here, using the MvH model, we can treat n as a continuous fitting parameter, which significantly simplifies the data analysis procedure.

To directly analyze the SE concentration distributions using the MvH nonspecific binding model, we substituted the expression for an ideal sedimenting free protein into Eq. (2), which yields Eq. (16):

$$\upsilon(r) = K_{\text{ns}} \cdot \left[P_{f,m} . \exp\left(\frac{M_p\left(1-\overline{\upsilon}_p\rho\right).\omega^2}{RT} . \frac{r^2 - r_m^2}{2} \right) \right] .(1 - n\upsilon(r)) . (ff(r))^{(n-1)} \cdot \left(\frac{N-n+1}{N} \right) \tag{16}$$

where $P_{f,m}$ denotes the concentration of free protein at meniscus position, r_m, M_p is the molar mass of free protein, and $\overline{\upsilon}_p$ is the partial specific volume of free protein. The binding density, $\upsilon(r)$ and $(ff(r))$, which are functions of the radial position, is calculated implicitly at each radial position, over the range $0 < \upsilon(r) < \frac{1}{n}$.

Next, in order to express the total DNA concentration in terms of the radial position, we multiplied Eq. (11) by $[\mathrm{DNA_t}]$, which provides the free DNA concentration, $[\mathrm{DNA_f}]$. Solving for $[\mathrm{DNA_t}]$ yields:

$$[\mathrm{DNA_t}] = \frac{[\mathrm{DNA_f}]}{(1 - nv(r))(ff(r))^{N-1}} \tag{17}$$

Note that Eq. (17) is expressed in terms of the independent variables $[\mathrm{DNA_f}]$ and $[P_f]$. To obtain an expression for the total DNA concentration $[\mathrm{DNA_t}](r)$, as a function of radial position, r, we substituted Eq. (18), which is the expression for an ideal free DNA species at sedimentation equilibrium into Eq. (16), which yields Eq. (19):

$$[\mathrm{DNA_f}](r) = [\mathrm{DNA_{f,ref}}].\exp^{\left(\frac{M_D(1-\bar{v}_D\rho)\omega^2}{RT}.\frac{r^2-r_{\mathrm{ref}}^2}{2}\right)} \tag{18}$$

$$[\mathrm{DNA_t}](r) = \frac{[\mathrm{DNA_{f,ref}}].\exp^{\left(\frac{M_D(1-\bar{v}_D\rho)\bar{\omega}^2}{RT}.\frac{r^2-r_{\mathrm{ref}}^2}{2}\right)}}{(1 - nv(r))(ff(r))^{N-1}} \tag{19}$$

where $[\mathrm{DNA_{f,ref}}]$ denotes the concentration of free DNA at the reference radial position (usually taken as the meniscus position), M_D is the molecular weight of the free DNA, and $\bar{v}_D$ is the partial specific volume of the free DNA. To directly analyze the absorbance data, we substituted Eq. (15) into Beer's law, $\mathrm{Abs}(r) = [\mathrm{DNA_t}](r).\varepsilon_\lambda.l$, to obtain Eq. (20):

$$\mathrm{Abs}(r) = \frac{[\mathrm{DNA_{f,ref}}].\exp^{\left(\frac{M_D(1-\bar{v}_D\rho)\bar{\omega}^2}{RT}.\frac{r^2-r_{\mathrm{ref}}^2}{2}\right)}}{(1 - nv(r))(ff(r))^{N-1}}.\varepsilon_\lambda.l + b \tag{20}$$

where b is the baseline offset, l is the path length; and ε_λ is the extinction coefficient of the Cy3-labeled DNA at 522 nm. During the data-fitting procedure, Eqs. (16), (4a), and (4b) are used to implicitly solve for $v(r)$ at each radial position. Then each pair of $v(r)$ and r is substituted into Eq. (20) to compare with the experimental data using NLLS.

For this system, the total DNA concentration was high enough to ensure the binding reaction occurred under stoichiometric conditions. However, the distribution of bound states is affected by the parameters n and ω. In general, proteins that bind DNA with positive cooperativity ($\omega > 1$) tend to produce longer clusters of bound protein, and therefore more of the DNA sediments with a higher average molar mass than for systems that display lower cooperativity. On the other hand, larger values of n tend to disfavor formation of longer clusters, since it becomes more difficult to load the next protein onto the DNA. For this reason, even under stoichiometric conditions, the shapes of the isotherms will provide information of the relative values of ω and n. To test for consistency, we fitted the SE-AUC data using the values determined for K_{ns}, ω, and n from an independent filter binding study (Yang & Maluf, 2014), while allowing only the free protein and DNA concentrations at reference radial positions and baseline offsets to float. The excellent agreement between the model and the data indicate the values obtained for K_{ns}, ω, and n from filter binding are consistent with the SE-AUC data (Fig. 3), and thus, the two orthogonal approaches are in agreement.

3.3 Examination of Specific and Nonspecific Competitive Binding Using SV-AUC

Sedimentation equilibrium approaches were used in the above two examples to determine the stoichiometry, affinity, and cooperativity of proteins binding to nonspecific duplex DNA. While powerful, SE approaches require that the nucleoprotein complexes under investigation are stable during the time course of the experiment, which can be up to 10 days in length. In contrast, a sedimentation velocity experiment can be completed in a matter of hours. Here, we demonstrate that SV-AUC, by resolving different interacting species, can be used to dissect the competition between specific and nonspecific DNA-binding interactions. Again, *E. coli* IHF serves as an excellent example because this protein binds to DNA in distinct specific (bent) and nonspecific (condensed) complexes (*vide supra*, Section 3.1).

IHF binds to its specific recognition element with nanomolar affinity (Aeling et al., 2006; Holbrook et al., 2001; Ortega & Catalano, 2006; Sanyal et al., 2014). By labeling the DNA molecule with fluorescein, we can use the AU-FDS (Fluorescence Detection System) to monitor the sedimentation of the IHF-DNA complexes at sufficiently low DNA concentrations (4 n*M*) such that equilibrium binding conditions are ensured. In this case, the data can be analyzed to afford equilibrium association constants. SV-AUC data were collected for a series of samples containing labeled DNA in the presence of the indicated concentrations of IHF. The raw data were analyzed by Sedfit (Schuck, 2003) to obtain c(s) sedimentation coefficient distributions and the weight-average sedimentation coefficient, $\langle s \rangle$, for each mixture of IHF and DNA. This analysis yields a sedimentation coefficient for 1.96 S for both the specific and nonspecific 27 bp duplexes. When these DNA molecules are titrated with IHF, $\langle s \rangle$ increases, indicating the

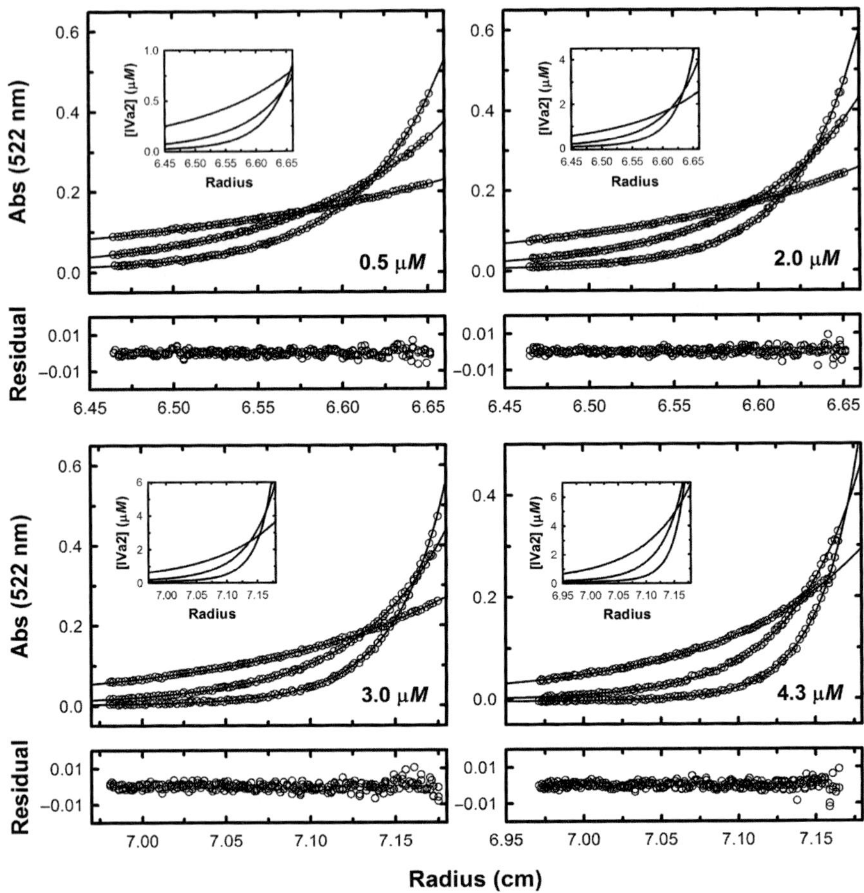

Figure 3 NLLS analysis of the primary sedimentation equilibrium data using the MvH model. The data in this figure were analyzed by NLLS according to the MvH model. The smooth curves are the result of this analysis fixing K_{ns}, ω, and n at the values determined from other approaches: $K_{ns} = 3.9 \times 10^4 M^{-1}$, $\omega = 125$, and $n = 18.8$bp (Yang & Maluf, 2014). In this analysis, the concentration of free IVa2 protein and free DNA at the meniscus position were allowed to float. The insets show the predicted concentration of the total IVa2 protein calculated at 5, 7.5, and 10 K RPM, using the best-fit parameters from the NLLS analysis (Yang & Maluf, 2014).

formation of IHF–DNA complexes (Fig. 4A and B). Each value of $\langle s \rangle$ reflects the ensemble distribution of IHF–DNA complexes at a given total IHF concentration.

A plot of $\langle s \rangle$ as a function of IHF concentration is presented in Fig. 4C; close inspection of the data reveals that while IHF binds to the nonspecific DNA in a monotonic manner, it appears that the binding curve for

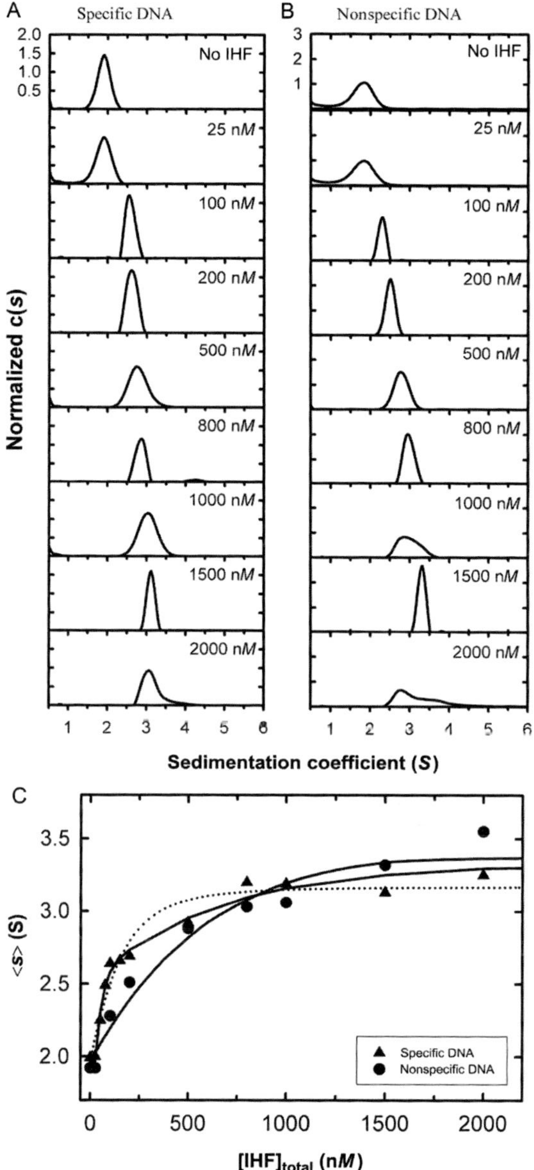

Figure 4 Interrogation of binding of IHF to minimal DNA duplexes using sedimentation velocity analytical ultracentrifugation (SV-AUC). Increasing concentrations of IHF were added to 27 bp specific and nonspecific DNA model duplexes, and their sedimentation behavior was monitored by SV-AUC. The c(*s*) distribution for each binding experiment was calculated using Sedfit. (A) Normalized c(*s*) profiles for the specific DNA. (B) Normalized c(*s*) profiles for the nonspecific DNA. (C) Weight-average sedimentation coefficients for each of the c(*s*) distributions shown in panels A (triangles) and B (triangles) were calculated using Sedfit and are plotted as a function of IHF concentration. The dotted line represents the best fit of the specific binding data to the nonspecific finite lattice DNA binding model, which does not adequately describe the data. The solid lines represent the best fit of global analysis of the nonspecific and specific binding data to the (i) nonspecific finite lattice DNA binding and (ii) case 1 models, respectively.

the specific DNA is biphasic. This strongly suggests that a high affinity (likely specific) binding interaction has been detected by SV-AUC. Ultimately, the specific and nonspecific binding curves reach similar plateau values. We interpret this to indicate that at high enough IHF concentration, nonspecific binding out-competes the specific binding interaction and dominates the sedimentation behavior of the complex. To address this feature, we can express $\langle s \rangle$ for either the specific (Eq. 21) or the nonspecific (Eq. 22) DNA as follows:

$$\langle s \rangle = s_{\text{free}} F_{\text{free}} + \langle s_{b,\text{ns}} \rangle F_{b,\text{ns}} + s_{b,\text{sp}} F_{b,\text{sp}} \tag{21}$$

$$\langle s \rangle = s_{\text{free}} F_{\text{free}} + \langle s_{b,\text{ns}} \rangle F_{b,\text{ns}} \tag{22}$$

where s_{free} is the sedimentation coefficient of free DNA, $\langle s_{b,\text{ns}} \rangle$ is the weight-average sedimentation coefficient for the ensemble of nonspecific protein–DNA complexes, $s_{b,\text{sp}}$ is the sedimentation coefficient for the specific protein–DNA complex, F_{free} is the fraction of free DNA (given by Eq. 11), $F_{b,\text{ns}}$ is the fraction of protein nonspecifically bound DNA, and $F_{b,\text{sp}}$ is the fraction of protein specifically bound DNA, at a given protein concentration. Note $F_{b,\text{ns}}$ represents the DNA with at least one protein bound in a nonspecific manner. Thus, when fitting data collected for the DNA with a specific site, one uses Eq. (21), and when fitting data collected for a fully nonspecific DNA, one uses Eq. (22).

The specific DNA duplex in our experiment is short (27 bp) and only contains a single specific binding site, such that it can accommodate either one specifically bound IHF protein or multiple IHF nonspecific interactions. Hence, the most suitable model to analyze the specific binding isotherm is the specific/nonspecific competition model (case 1, Section 2.2), which requires that only one IHF can bind in a specific mode, while multiple IHF molecules can bind in a nonspecific mode, *but not both* (i.e., there is competition between specific and nonspecific binding). Therefore, in this case the concentration of specifically bound IHF is the same as the concentration of DNA that is specifically bound (since the *specific* binding stoichiometry is 1:1, and is usually different from the *nonspecific* binding stoichiometry; see Section 3.1). Next, we can rearrange Eq. (7) to get an expression for $F_{b,\text{sp}}$:

$$F_{b,\text{sp}} = \upsilon N \left(\frac{K_{\text{sp}} (ff)^{N-n}}{K_{\text{ns}} (N - n + 1) + N K_{\text{sp}} \upsilon (ff)^{N-n}} \right) \tag{23}$$

and $F_{b,\text{ns}}$ can be calculated according to:

$$F_{b,\,\text{ns}} = 1 - \left(F_{\text{free}} + F_{b,\text{sp}}\right) \tag{24}$$

where F_f is given by Eq. (11). To fit data collected for titration of the *specific* DNA, Eqs. (23) and (24) were substituted into Eq. (21) to yield an expression that describes $\langle s \rangle$ as a function of IHF concentration. To fit data collected for titration of the *nonspecific* DNA, Eq. (24) (with $F_{b,\text{sp}} = 0$) is substituted into Eq. (22). Importantly, it is not possible to extract both the specific and nonspecific parameters by carrying out experiments using only the specific DNA—there is insufficient information to constrain the NLLS analysis. However, when titrations are carried out using both specific and nonspecific DNA, as was done in this case, the ensemble of datasets can be analyzed globally (i.e., simultaneously); the nonspecific DNA titration provides information exclusively on the nonspecific DNA-binding parameters (Eq. 23) while the specific DNA titration provides information on both the specific- and nonspecific binding parameters (Eq. 21). The global fitting approach provides adequate constraints in the data analysis to accurately resolve all of the fitting parameters. In this fitting routine, the nonspecific binding site size (n), DNA length (N), and experimentally determined s for free DNA were fixed as global constants; $s_{b,\text{sp}}$ and $\langle s_{b,\text{ns}} \rangle$ were local variables that were allowed to float for the specific and nonspecific datasets, respectively, and K_{ns}, K_{sp}, and ω were global floating parameters for the combined datasets. The implicit variables P_f and υ were calculated as described in Section 2.2 and the results of this analysis are shown as the solid curves in Fig. 4C. This analysis returns the local parameters: $s_{b,\,\text{sp}} = 2.59(\pm 0.12)\text{S}$ and $\langle s_{b,\,\text{ns}} \rangle = 3.37(\pm 0.04)\text{S}$, and the global parameters: $K_{\text{ns}} = 1.7(\pm 0.6) \times 10^6 M^{-1}$, $K_{\text{sp}} = 2.0(\pm 0.6) \times 10^8 M^{-1}$, and $\omega = 10(\pm 4)$. Moreover, according to our analysis, we calculate a binding specificity $(\frac{K_{\text{sp}}}{K_{\text{ns}}})$ for IHF of 117. Notably, these parameters agree well with those determined previously using ITC to interrogate IHF–DNA binding interactions (under slightly different solution conditions) (Holbrook et al., 2001).

3.4 Identification of a Conformational Change Upon Nonspecific Protein–DNA Binding Using SV-AUC

SV-AUC is a hydrodynamic technique and as such is not only useful to measure the mass change upon protein–DNA binding, but it is also sensitive to the overall shape of the complexes. In this way, it is able to detect conformational differences in the sedimenting species (Correia & Stafford, 2015; Stafford et al., 2001). IHF provides an exceptional model system with which to study protein-induced conformational changes in DNA molecules, as follows. IHF binds to its cognate element and introduces a severe bend into the duplex (Khrapunov, Brenowitz, Rice, & Catalano, 2006). In contrast, the protein weakly bends and wraps the duplex when bound in a nonspecific mode. While these distinct binding modes are not evident using the minimal length duplexes described above, a different picture emerges when a longer DNA is employed.

In this study, two longer 274 bp substrates were examined; (i) a specific DNA (cos-274) that contains a single, centrally positioned 27 bp specific IHF recognition element and (ii) a 274 bp DNA of random sequence (Sanyal et al., 2014). EMS assays have demonstrated that IHF binds to the specific site in cos-274 with a low nanomolar affinity to afford a specific, strongly bent nucleoprotein complex (Ortega & Catalano, 2006; Sanyal et al., 2014). In contrast, the protein binds weakly to ns-274 to afford a smeared complex on the gel. Here, we demonstrate by AUC that addition of IHF to the specific cos-274 substrate results in a progressive, concentration-dependent increase in $\langle s \rangle$ from 4.2 S for free DNA to a maximal value of ~10 S at apparent saturation (Fig. 5A). A plot of $\langle s \rangle$ as a function of IHF concentration is presented in Fig. 5C. The concentration dependence and the magnitude of the shift suggest that in contrast to what is observed in the EMS studies, multiple IHF proteins weakly assemble on the longer duplex. Further, we note that $\langle s \rangle$ increases up to 1.6 μM IHF, but subsequently decreases slightly as the concentration of IHF is further increased. This is discussed further below.

The concentration of nonspecific flanking sequence (247 bp) is much greater than the unique 27 bp specific site in the cos-274 DNA, therefore, we hypothesized that nonspecific binding would dominate the binding reaction observed in the AUC. We therefore examined IHF binding to the nonspecific 274 bp DNA. In contrast to the relatively simple binding curve observed with the specific DNA, binding of IHF to ns-274 is remarkable. An increase in $\langle s \rangle$ is initially observed similar to that for the cos-274 DNA (Fig. 5B); however, a 2.5 S species emerges as the protein concentration is further increased, at the expense of the larger s species. We note that the appearance of two distinct peaks in the c(s) distributions collected for the nonspecific DNA suggests that the kinetics of interconversion between these two populations is slow relative to the timescale of SV-AUC experiment (a few hours). We further note that this new species has a sedimentation coefficient *smaller* than that of the free DNA (4.2 S). This remarkable

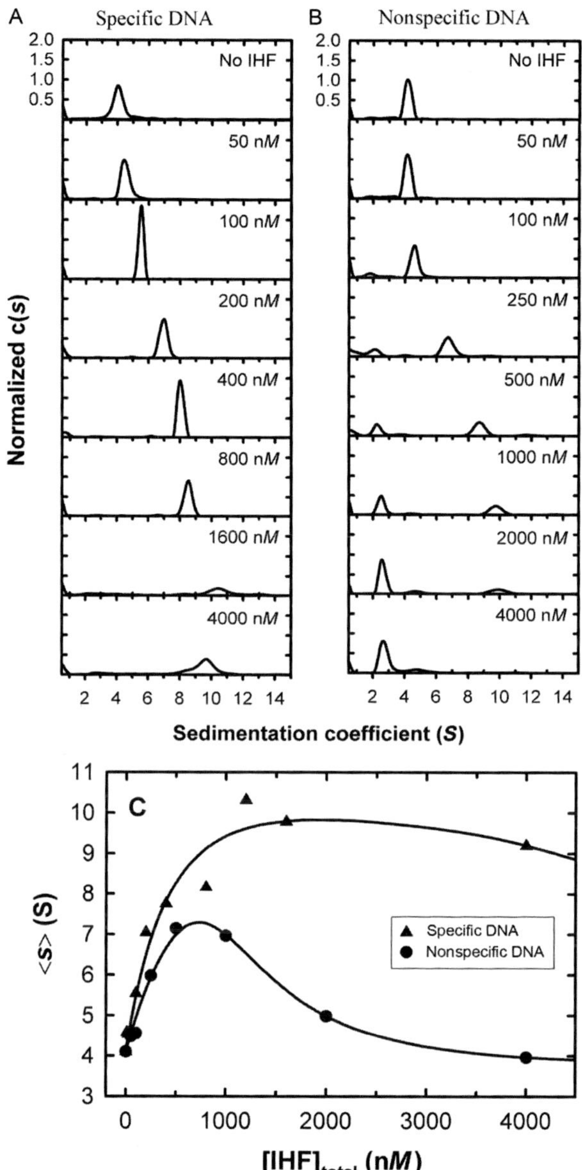

Figure 5 Interrogation of IHF assembly on specific and nonspecific 274 bp DNAs using SV-AUC. The specific duplex has a 27 bp IHF recognition element embedded at the center of the duplex while the nonspecific duplex is of random sequence. Increasing concentrations of IHF were added to each duplex, and their sedimentation behavior was monitored by SV-AUC. The c(*s*) distribution for each binding experiment were calculated using Sedfit. (A) Normalized c(*s*) profiles for the specific 274 bp DNA. (B) Normalized c(*s*) profiles for the nonspecific 274 bp DNA. (C) Weight-average sedimentation coefficients for each of the c(*s*) distributions shown in panel A (circles, specific DNA) and panel B (circles, nonspecific DNA) were calculated using Sedfit and are plotted as a function of IHF concentration. The solid lines represent the best fits of global analysis of the ensemble of binding data to the DNA unbending model (Fig. 6).

transition is quite apparent in the plot of $\langle s \rangle$ as a function of IHF concentration presented in Fig. 5C, which further shows that a similar, though less striking trend is observed with the specific cos-274 DNA. How can this observation be rationalized?

EMS studies confirm that IHF binds to both duplexes in this concentration range to afford complexes of increasing mass (Sanyal et al., 2014). This is consistent with the initial increase in $\langle s \rangle$; however, s is also a function of particle shape and *decreases* as the complex becomes more elongated in solution (Correia & Stafford, 2015; Stafford et al., 2001). Thus, the most parsimonious explanation is that while both species represent IHF–DNA complexes of increasing buoyant mass, they ultimately afford complexes that differ significantly in shape. Specifically, the 2.5 S species must be elongated, even relative to the free, unbound duplex (4.2 S). Within this context, the contour length of a 274 bp DNA is about 93 nm, which is larger than the persistence length of DNA (35–50 nm depending upon solution conditions) (Bustamante, Bryant, & Smith, 2003; Kratky & Porod, 1949; Wang, Yin, Landick, Gelles, & Block, 1997). This means that the free 274 bp DNA molecule is flexible during sedimentation and it will behave like a floppy polymer in solution. Hence, we propose that the transition from the ~10 to the 2.5 S "species" is due to a dramatic conformational change in the DNA resulting from a substantial increase in the stiffness of nucleoprotein complex as the duplex becomes saturated with IHF. Thus, even though the molar mass of a nearly saturated DNA species is approximately four to fivefold higher than the molar mass of the free DNA, the sedimentation coefficient is significantly smaller, due to a dramatic increase in the frictional coefficient of the "rigid rod" relative to that of free DNA.

We developed a quantitative model to analyze IHF binding to both specific and nonspecific 274 bp duplex substrates and to determine the energetics associated with such a conformational transition (Fig. 6; Sanyal et al., 2014). Briefly, the model takes into account cooperative, nonspecific assembly of IHF on DNA such that depending on the number of IHF molecules bound, the DNA will eventually transition from a flexible strand to an extended rigid, rod-like nucleoprotein complex. The model incorporates three DNA species: free DNA, flexible IHF–DNA complexes, and rigid, extended IHF–DNA complexes. At low IHF concentrations, IHF binds weakly and cooperatively to bend, wrap, and condense the DNA (i.e., at least one IHF bound); the increase in the buoyant mass and the decrease in the frictional ratio results in a significant increase in the sedimentation coefficient of the complex. Importantly, the occluded site size for IHF

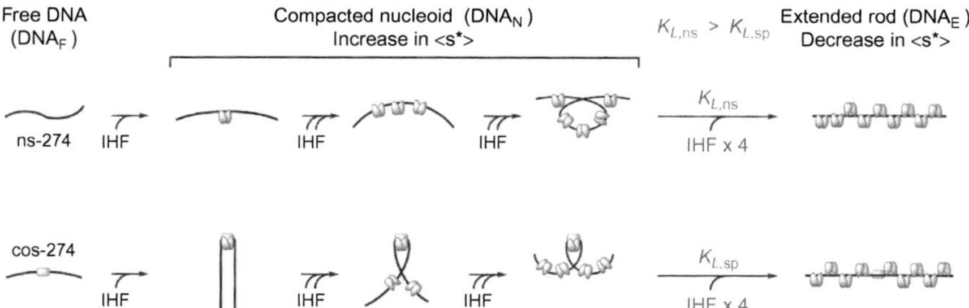

Figure 6 Model for IHF–DNA complexes. Details are provided in the text and in Sanyal et al. (2014).

Table 1 Analysis of the ns-274 and cos-274 AUC Data to Approximate DNA Unbending Energy

	m^a	$\langle s \rangle$ (S)	K_L (M^{-4})	Apparent $K_{L,app}{}^b$ (M^{-1})	ΔG^c (kcal/mol)
ns-274	4 ± 2	(10.0 ± 0.4)	$K_{L,ns} = (1.2 \pm 0.8) \times 10^{36}$	$K_{L,app\text{-}ns} = (1.0 \pm 0.2) \times 10^9$	-11.5
cos-274	4 (2, N/A[d])	(3.7 ± 0.4)	$K_{L,sp} \leq 3.7 \times 10^{31}$	$K_{L,app\text{-}sp} \leq 7.8 \times 10^7$	≥ -10.1

[a] m is the number of IHF required to unbent the DNA.
[b] $K_{L,app}$ values were calculated from $(K_L)^{1/m}$.
[c] The free energy change was calculated from $\Delta G = -RT \ln(K_{L,app})$.
[d] The upper limit of m for cos-274 DNA is not available due to the insufficient unbending of specific DNA.

binding in a bent complex (~35 bp) (Holbrook et al., 2001) is significantly larger than that of a nonspecific, linear complex (~8 bp) (Holbrook et al., 2001) and the DNA must therefore be "unbent" in order to fully saturate the duplex. Thus, above a threshold value, the ensemble of IHF–DNA complexes is "linearized" or "unbent" so that all of the available nonspecific sites can be bound by IHF. This increases the stiffness and extends the shape of the nucleoprotein complex, even relative to free DNA. The model assumes this "unbending" is driven by the binding of additional IHF molecules, and this reaction is described by a single equilibrium constant (essentially a Hill-binding model). Details of the mathematical model and fitting procedures are presented in Sanyal et al. (2014).

The best-fit parameters obtained from this analysis are shown in Table 1 and the results of the fit are shown as the smooth curves in Fig. 5C. The model well describes the experimental data and implies that the unbending (linearization) energetic barrier is at least 1.4 kcal/mol. In other words, it is at least 13 times easier to linearize a nonspecific duplex with IHF bound than a specific duplex with IHF bound. This reflects the stability of the severe bend introduced by IHF bound to its cognate site. In addition, the model suggests that four IHF dimers are required to trigger the unbending transition. Importantly, the AUC studies discussed here are consistent with electron microscopy images (Sarkar et al., 2009) that show rigid, extended IHF–DNA complexes and we have provided a solution-based, thermodynamic analysis of the transition from condensed to extended nucleoprotein complexes.

4. SUMMARY

In this review we have described the power of sedimentation equilibrium and sedimentation velocity approaches to interrogate protein–DNA binding interactions. We have paid special attention to how these data can be analyzed using appropriate mathematical models to obtain biologically relevant, solution-based thermodynamic information on the interactions. These approaches are particularly well suited to these studies because they are founded on well-understood physical principles and consequently provide rigorous quantitative information on the interactions, largely free from simplifying assumptions. In addition, the physical responses to protein–DNA interactions—the increase in the molar mass (detected by sedimentation equilibrium and velocity) and any associated changes in nucleoprotein complex conformation (detected by sedimentation velocity), can be probed and quantified in a rigorous manner. The mechanism of specific recognition of a DNA-binding element within the context of a vast excess of nonspecific DNA can be addressed, which is essential since this is the situation *in vivo*; regulatory proteins must not only compete with nonspecific binding proteins for their cognate element, but they must also overcome their own nonspecific binding interactions to avoid sequestration by the cellular genome. Importantly, the tools described here are general and can be applied to any experimental system to provide mechanistic insight into biological function of proteins within a cell.

ACKNOWLEDGMENTS

We thank Dr. Saurarshi J. Sanyal for generating the data in Fig. 2 and Dr. John Sumida for guidance on the AUC fluorescence detection system.

REFERENCES

Aeling, K. A., Opel, M. L., Steffen, N. R., Tretyachenko-Ladokhina, V., Hatfield, G. W., Lathrop, R. H., et al. (2006). Indirect recognition in sequence-specific DNA binding by Escherichia coli integration host factor: The role of DNA deformation energy. *The Journal of Biological Chemistry*, *281*(51), 39236–39248.

Bujalowski, W. (2006). Thermodynamic and kinetic methods of analyses of protein-nucleic acid interactions. From simpler to more complex systems. *Chemical Reviews*, *106*(2), 556–606.

Bujalowski, W., Lohman, T. M., & Anderson, C. F. (1989). On the cooperative binding of large ligands to a one-dimensional homogeneous lattice: The generalized three-state lattice model. *Biopolymers*, *28*(9), 1637–1643.

Bujalowski, W., Overman, L. B., & Lohman, T. M. (1988). Binding mode transitions of Escherichia coli single strand binding protein-single-stranded DNA complexes. Cation, anion, pH, and binding density effects. *The Journal of Biological Chemistry*, *263*(10), 4629–4640.

Bustamante, C., Bryant, Z., & Smith, S. B. (2003). Ten years of tension: Single-molecule DNA mechanics. *Nature*, *421*(6921), 423–427.

Correia, J. J., & Stafford, W. (2015). Sedimentation velocity: A classical perspective. Methods in Enzymology, 562, 49–80.

Demeler, B., & van Holde, K. E. (2004). Sedimentation velocity analysis of highly heterogeneous systems. *Analytical Biochemistry*, *335*(2), 279–288.

Epstein, I. R. (1978). Cooperative and non-cooperative binding of large ligands to a finite one-dimensional lattice. A model for ligand-oligonucleotide interactions. *Biophysical Chemistry*, *8*(4), 327–339.

Holbrook, J. A., Tsodikov, O. V., Saecker, R. M., & Record, M. T., Jr. (2001). Specific and non-specific interactions of integration host factor with DNA: Thermodynamic evidence for disruption of multiple IHF surface salt-bridges coupled to DNA binding. *Journal of Molecular Biology*, *310*(2), 379–401.

Khrapunov, S., Brenowitz, M., Rice, P. A., & Catalano, C. E. (2006). Binding then bending: A mechanism for wrapping DNA. *Proceedings of the National Academy of Sciences of the United States of America*, *103*(51), 19217–19218.

Kratky, O., & Porod, G. (1949). Diffuse small-angle scattering of X-rays in colloid systems. *Journal of Colloid Science*, *4*(1), 35–70.

Laue, T. M., Shah, B. D., Ridgeway, T. M., & Pelletier, S. L. (1992). *Computer-aided interpretation of analytical sedimentation data for proteins*. Cambridge: Royal Society of Chemistry.

Laue, T. M., & Stafford, W. F., 3rd. (1999). Modern applications of analytical ultracentrifugation. *Annual Review of Biophysics and Biomolecular Structure*, *28*, 75–100.

Lebowitz, J., Lewis, M. S., & Schuck, P. (2002). Modern analytical ultracentrifugation in protein science: A tutorial review. *Protein Science*, *11*(9), 2067–2079.

Lohman, T. M., & Bujalowski, W. (1991). Thermodynamic methods for model-independent determination of equilibrium binding isotherms for protein-DNA interactions: Spectroscopic approaches to monitor binding. *Methods in Enzymology*, *208*, 258–290.

McGhee, J. D., & von Hippel, P. H. (1974). Theoretical aspects of DNA-protein interactions: Co-operative and non-co-operative binding of large ligands to a one-dimensional homogeneous lattice. *Journal of Molecular Biology*, *86*(2), 469–489.

Ortega, M. E., & Catalano, C. E. (2006). Bacteriophage lambda gpNu1 and Escherichia coli IHF proteins cooperatively bind and bend viral DNA: Implications for the assembly of a genome-packaging motor. *Biochemistry*, *45*(16), 5180–5189.

Philo, J. S. (2011). Limiting the sedimentation coefficient for sedimentation velocity data analysis: Partial boundary modeling and g(s) approaches revisited. *Analytical Biochemistry*, *412*(2), 189–202.

Sanyal, S. J., Yang, T. C., & Catalano, C. E. (2014). Integration host factor assembly at the cohesive end site of the bacteriophage lambda genome: Implications for viral DNA packaging and bacterial gene regulation. *Biochemistry, 53*(48), 7459–7470.

Sarkar, T., Petrov, A. S., Vitko, J. R., Santai, C. T., Harvey, S. C., Mukerji, I., et al. (2009). Integration host factor (IHF) dictates the structure of polyamine-DNA condensates: Implications for the role of IHF in the compaction of bacterial chromatin. *Biochemistry, 48*(4), 667–675.

Schuck, P. (2003). On the analysis of protein self-association by sedimentation velocity analytical ultracentrifugation. *Analytical Biochemistry, 320*(1), 104–124.

Stafford, W. F., Jacobsen, M. P., Woodhead, J., Craig, R., O'Neall-Hennessey, E., & Szent-Gyorgyi, A. G. (2001). Calcium-dependent structural changes in scallop heavy meromyosin. *Journal of Molecular Biology, 307*(1), 137–147.

Tsodikov, O. V., Holbrook, J. A., Shkel, I. A., & Record, M. T., Jr. (2001). Analytic binding isotherms describing competitive interactions of a protein ligand with specific and non-specific sites on the same DNA oligomer. *Biophysical Journal, 81*(4), 1960–1969.

Ucci, J. W., & Cole, J. L. (2004). Global analysis of non-specific protein-nucleic interactions by sedimentation equilibrium. *Biophysical Chemistry, 108*(1–3), 127–140.

von Hippel, P. H., & Berg, O. G. (1986). On the specificity of DNA-protein interactions. *Proceedings of the National Academy of Sciences of the United States of America, 83*(6), 1608–1612.

Wang, M. D., Yin, H., Landick, R., Gelles, J., & Block, S. M. (1997). Stretching DNA with optical tweezers. *Biophysical Journal, 72*(3), 1335–1346.

Yang, T. C., & Maluf, N. K. (2014). Characterization of the non-specific DNA binding properties of the adenoviral IVa2 protein. *Biophysical Chemistry, 193–194*, 1–8.

Yphantis, D. A. (1964). Equilibrium ultracentrifugation of dilute solutions. *Biochemistry, 3*, 297–317.

CHAPTER FIFTEEN

Characterization of Homogeneous, Cooperative Protein–DNA Clusters by Sedimentation Equilibrium Analytical Ultracentrifugation and Atomic Force Microscopy

Ingrid Tessmer*[,1], Michael G. Fried[†,1]
*Rudolf Virchow Center for Experimental Biomedicine, University of Würzburg, Würzburg, Germany
[†]Center for Structural Biology, Department of Molecular and Cellular Biochemistry, University of Kentucky, Lexington, Kentucky, USA
[1]Corresponding authors: e-mail address: Ingrid.tessmer@virchow.uni-wuerzburg.de; michael.fried@uky.edu

Contents

Abstract

Strong, positively cooperative binding can lead to the clustering of proteins on DNA. Here, we describe one approach to the analysis of such clusters. Our example is based on recent studies of the interactions of O^6-alkylguanine DNA alkyltransferase (AGT) with high-molecular-weight DNAs (Adams et al., 2009; Tessmer, Melikishvili, & Fried, 2012). Cooperative cluster size distributions are predicted using the simplest homogeneous binding and cooperativity (HBC) model, together with data obtained by sedimentation equilibrium analysis. These predictions are tested using atomic force microscopy

Methods in Enzymology, Volume 562
ISSN 0076-6879
http://dx.doi.org/10.1016/bs.mie.2015.06.036

imaging; for AGT, measured cluster sizes are found to be significantly smaller than those predicted by the HBC model. A mechanism that may account for cluster size limitation is briefly discussed.

1. INTRODUCTION

1.1 Homogeneous Cooperative Assemblies

Many DNA–protein complexes are formed in reactions in which the binding of one protein is modulated by interactions with others. Such cooperative assemblies play important roles in cellular functions, including DNA packaging (Azzaz et al., 2014; Mamoon, Song, & Wellman, 2005), DNA replication and DNA repair (Kozlov et al., 2015; Melikishvili, Rasimas, Pegg, & Fried, 2008; Tessmer et al., 2005), and transcription (Hieb, Gansen, Böhm, & Langowski, 2014; Newman, Cooper, Aitkenhead, & Gileadi, 2015). Cooperative protein–DNA complexes can be homomeric (containing only one kind of protein) or heteromeric (containing more than one kind of protein). Our focus here will be on homomeric complexes. Examples of homomeric protein–DNA complexes include some formed by RecA (Takahashi, 1989) and single-strand binding proteins (Karpel, 2002; Kowalczykowski et al., 1986), the *Escherichia coli* CAP protein (Saxe & Revzin, 1979), human interferon-inducible protein IFI16 (Morrone et al., 2014), and the human DNA repair protein, AGT (Fried, Kanugula, Bromberg, & Pegg, 1996). In the population of homomeric complexes, we distinguish ones in which all proteins make identical contacts with DNA (these are accordingly sequence-nonspecific) and ones that make identical contacts with other protein molecules in the assembly. Deviation from homogeneity in protein–protein interactions may occur when two or more distinct protein–protein contacts are possible, and at the ends of cooperative clusters, where the terminal protein in a linear array has fewer neighbors than one that is inside the array. Deviation from homogeneity in protein–DNA contacts usually results in preferential binding to some DNA sequences and/or structures (binding specificity).

The most widely used model of cooperative DNA binding is that of McGhee and von Hippel (1974). This posits a linear lattice of DNA sites with identical protein affinities (K) and identical protein–protein interaction equilibrium constants (ω). We will refer to this as the homogeneous binding and cooperativity (HBC) model. This simplicity makes possible a closed-form expression relating free protein concentration $[P]$ and binding density (ν) to K and ω (Eq. 1)

$$\frac{\nu}{[P]} = K(1-s\nu)\left(\frac{(2\omega-1)(1-s\nu)+\nu-R}{2(\omega-1)(1-s\nu)}\right)^{n-1}\left(\frac{1-(s+1)\nu+R}{2(1-s\nu)}\right)^{2}$$
$$R = \left((1-(s+1)\nu)^{2}+4\omega\nu(1-s\nu)\right)^{1/2}. \quad (1)$$

Here, s is the size of the site (in nucleotides or base pairs) that a protein molecule occupies to the exclusion of others and other terms are defined above. This model assumes that the lattice of binding sites on DNA is long enough that end effects are not significant (McGhee & von Hippel, 1974). Other models have been developed to deal with inhomogeneous binding (Tsodikov, Holbrook, Shkel, & Record, 2001), short DNA lattices (McGhee & von Hippel, 1974; Tsodikov et al., 2001), overlapping binding sites (Saroff, 1995; Wolfe & Meehan, 1992), DNA allosterism (Dattagupta, Hogan, & Crothers, 1980), and heterogeneous cooperative interactions (Bujalowski, Lohman, & Anderson, 1989; Wolfe & Meehan, 1992). That said, the HBC model provides a good starting point for characterizing new interactions and a valuable basis for comparison when more specialized models are used.

On long DNA substrates, positive cooperativity results in clusters of contiguously bound proteins (Epstein, 1978; Kowalczykowski et al., 1986; Schwarz & Watanabe, 1983). The distribution of the number of proteins in a population of clusters depends on the binding density ν and the values of s and ω at which complexes are formed. For systems with HBC, the average number of protein monomers per cluster is

$$\bar{C} = \frac{2\nu(\omega-1)}{(s-1)\nu-1+R} \quad (2)$$

with ν, ω, s, and R defined as in Eq. (1) (Kowalczykowski et al., 1986). The product $\bar{C}\cdot s$ gives the mean length of DNA occupied by one cluster to the exclusion of others.[1] It was recognized early that values of $\bar{C}$ and $\bar{C}\cdot s$ are testable predictions of binding models. Measurements of the length of cooperative clusters of RecA and gene 32 proteins on DNA and rho protein on RNA were made by negative-stain electron microscopy (Bear et al., 1988; Dunn, Chrysogelos, & Griffith, 1982; Ruyechan & Wetmer, 1975). More

[1] When protein binding sites on the DNA do not overlap, $\bar{C}\cdot s$ approximates the average contour length of clusters. When sites overlap, $\bar{C}\cdot s$ underestimates the contour length.

recently, high-resolution cryo-electron microscopy, atomic force microscopy (AFM), and other single-molecule methods have expanded the technical options for characterizing protein–DNA clusters (Galletto, Amitani, Baskin, & Kowalczykowski, 2006; Joo et al., 2006; Morrone et al., 2014; Tessmer et al., 2012, 2005; van der Heijden et al., 2007). Here, we describe the use of sedimentation equilibrium (SE) data to derive a simple, realistic cooperativity model for the DNA binding of human O^6-alkylguanine DNA alkyltransferase (AGT), the use of that model to predict $\bar{C}$ and the use of AFM to test that prediction.[2] Finally, we discuss a mechanism by which predicted and measured values of $\bar{C}$ may differ and how values of $\bar{C}$ can provide clues to possible functions of the protein cluster of interest.

2. MEASUREMENT OF DNA BINDING

2.1 Samples

AFM imaging reveals all macromolecules and other particles present in the sample. The purity of protein, DNA, and buffer solutions is important, as is the absence of inert particles and "dust" that are often undetectable by other methods. AFM deposition buffers were filtered through a syringe filter (0.02 μm pore, Anotop, Whatman); protein and DNA samples were purified as described below.

Human AGT, with wild-type sequence modified by a C-terminal $(His)_6$-tag replacing residues 202–207, was encoded on plasmid pQE-hAGT (Daniels et al., 2004). Protein was expressed in *E. coli* XL1-Blue (Stratagene) and purified by immobilized metal ion affinity chromatography (IMAC) chromatography as described (Daniels et al., 2004). The protein was transferred into 20 m*M* Tris (pH 8.0 at 20 °C), 250 m*M* NaCl, 1 m*M* DTT buffer by Sephadex G-50 chromatography. Aliquots were stored at −80 °C. AGT concentrations were measured spectrophotometrically using $\varepsilon_{280} = 2.64 \times 10^4$ *M*/cm (Rasimas, Pegg, & Fried, 2003). Samples of this protein were >95% active in DNA binding and in repair of short DNAs containing O^6-methylguanine lesions (Melikishvili et al., 2008).

Plasmid pUC19 DNA was obtained from New England Biolabs and linearized by digestion with EcoR1 endonuclease. A 1000-bp fragment of pUC19 DNA (spanning residues 1414–2414) was obtained by polymerase chain reaction amplification (Innis & Gelfand, 1990) using pUC19 DNA as

[2] A related analysis of nonspecific protein–DNA interactions using sedimentation velocity analytical ultracentrifugation is described by K. Maluf and C. Catalano in Chapter 23.

template. The 1000-bp fragment was purified by gel electrophoresis and recovered using a PCR cleanup kit from Qiagen. Stock DNA concentrations were measured spectrophotometrically, using $\varepsilon_{260} = 1.31 \times 10^4\ /M\,\mathrm{cm}$ (per base pair).

2.2 Sedimentation Equilibrium Analyses

AGT protein and DNA samples were dialyzed into 10 mM Tris (pH 7.6 at 20 °C), 1 mM EDTA, 1 mM DTT, and 100 mM NaCl buffer. Analytical ultracentrifugation was performed in a Beckman XL-A centrifuge using an AN60Ti rotor, operated at 20 °C. Radial scans were obtained at 260 nm. Equilibration was considered complete when scans taken 6 h apart were indistinguishable. Typically, this required equilibration times ≥24 h. Five equilibrium scans were averaged for each sample at each rotor speed. With DNAs large enough to accommodate several cooperative protein clusters, the concentration of protein-free DNA molecules becomes negligible well before all available binding sites are saturated. Accordingly, these solutions contain mixtures of free protein and protein–DNA complex, with the weight-average molecular weight of the complex increasing smoothly with protein concentration. At SE, this system has a radial absorbance distribution given by Eq. (3):

$$A(r) = \alpha_{\mathrm{P}} \exp\left[\sigma_{\mathrm{P}}\left(r^2 - r_0^2\right)\right] + \alpha_{\mathrm{P_nD}} \exp\left[\sigma_{\mathrm{P_nD}}\left(r^2 - r_0^2\right)\right] + \zeta \tag{3}$$

Here, $A(r)$ is the absorbance at radial position r, α_{P}, and $\alpha_{\mathrm{P_nD}}$ are absorbances of protein and protein–DNA complex at the reference position r_0, and ζ is a baseline offset that accounts for position-independent differences in the absorbances of different cell assemblies. The reduced molecular weights of AGT protein and protein–DNA complexes are given by $\sigma_{\mathrm{P}} = M_{\mathrm{P}}(1 - \bar{v}_{\mathrm{P}}\rho)\omega^2/(2RT)$ and $\sigma_{\mathrm{P_nD}} = (nM_{\mathrm{P}} + M_{\mathrm{D}})(1 - \bar{v}_{\mathrm{P_nD}}\rho)\omega^2/(2RT)$, respectively. Here, M_{P} and M_{D} are molecular weights of protein and DNA, n is the protein:DNA ratio of the complex, ρ is the solvent density, ω is the rotor angular velocity, R is the gas constant, and T is the temperature (Kelvin). The density of sample buffer and the partial specific volume of AGT were calculated using the public-domain program SEDNTERP (available from http://www.rasmb.bbri.org/) (Laue, Shah, Ridgeway, & Pelletier, 1992). The partial specific volume of duplex NaDNA at 0.1 M NaCl ($\bar{v}_{\mathrm{D}} = 0.55\,\mathrm{ml/g}$) was estimated by interpolation of the data of Cohen and Eisenberg (1968). Partial specific volumes of protein–DNA complexes were calculated using Eq. (4):

$$\bar{v}_{\mathrm{P_nD}} = \frac{(nM_{\mathrm{P}}\bar{v}_{\mathrm{P}} + M_{\mathrm{D}}\bar{v}_{\mathrm{D}})}{M_{\mathrm{P}} + M_{\mathrm{D}}}. \tag{4}$$

Shown in Fig. 1A is a subset of the SE data used for determining weight-average stoichiometries (n). Values of n allow calculation of free protein concentrations from $[P] = [P]_{\mathrm{input}} - n[\mathrm{DNA}]_{\mathrm{input}}$. Figure 1B gives graphs of stoichiometry as functions of free [AGT] for the 1000 bp and pUC19 (2686 bp) substrates. Large values of n found at high [AGT] (~150 for the 1000-bp fragment and ~350 for pUC19) are striking, although they correspond to smaller binding densities than those found with short duplexes (Melikishvili et al., 2008). Scatchard plots are shown in the figure insets. The solid curves are fits of the McGhee–von Hippel equation (Eq. 1) to the data, using $\nu = n/$(number of base pairs per DNA) and other terms as defined above. These analyses gave $K = 9667 \pm 1499\ M^{-1}$, $\omega = 35.9 \pm 6.8$, and $s = 6.32 \pm 0.12$ for the 1000-bp DNA and $K = 7960 \pm 916\ M^{-1}$, $\omega = 44.2 \pm 3.8$, and $s = 6.81 \pm 0.14$ for pUC19. These values of K and ω are somewhat smaller than those found for short double-stranded oligonucleotides (Melikishvili et al., 2008) and the limiting binding site sizes are larger (compare $s = 6.8$ bp/protein monomer for linear pUC19 with $s = 4$ bp/protein monomer for binding duplex 16-mers (Rasimas et al., 2003)). We speculate that these differences reflect packing inhomogeneities that are expected with large DNAs and are largely absent on short DNAs near binding saturation.

2.3 Prediction of Cluster Sizes

On long DNA substrates, positive binding cooperativity results in clusters of bound proteins. The distribution of the number of proteins in a population of clusters depends on the binding density ν and the values of ω, and s at which complexes are formed. For systems that conform to the HBC model, the average number of protein monomers per cluster, $\bar{C}$, is given by Eq. (2). Shown in Fig. 2 are predicted values of $\bar{C}$ calculated using the ν, ω, and s parameters obtained from the binding analyses shown in Fig. 1. These calculations predict a mean cluster size of <2 molecules for [AGT] ~ 1.5 μM. At such low protein concentrations, singly bound AGT molecules dominate binding distributions. In contrast, at [AGT] = 10 μM, these calculations predict $\bar{C} \sim 8.5$ molecules/cluster for the 1000-bp fragment and $\bar{C} \sim 11$ molecules/cluster of AGT for pUC19. Even larger clusters are predicted at higher [AGT]. These predictions can be tested by AFM as described below.

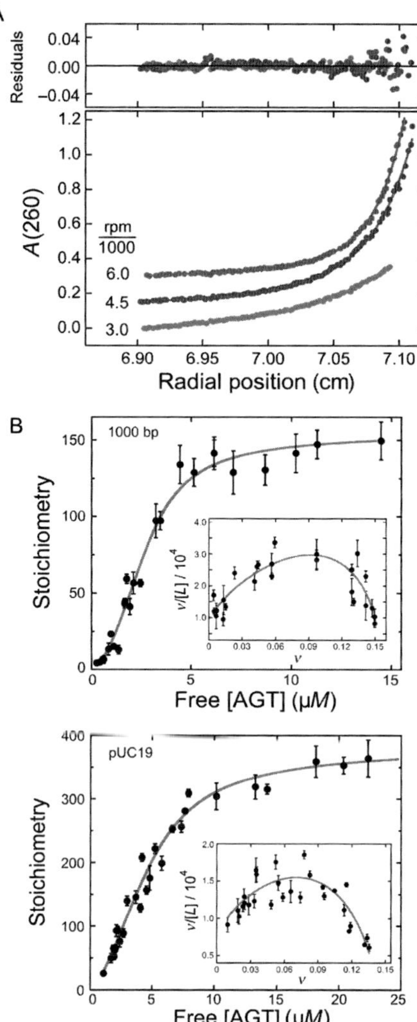

Figure 1 Characterization of AGT–DNA interactions at sedimentation equilibrium. (A) Data for an AGT–DNA mixture obtained at 20 ± 0.1 °C. This sample contained the 1000-bp DNA fragment (0.015 μ*M*) and AGT protein (3.5 μ*M*) in 10 m*M* Tris (pH 7.6 at 20 °C), 1 m*M* EDTA, 100 m*M* NaCl, and 1 m*M* DTT. Radial scans were taken at 3000 rpm, 4500 rpm, and 6000 rpm as indicated here, they are offset vertically for clarity. The smooth curves are fits of Eq. (1) to these data. Small, symmetrically distributed residuals (upper panel) indicate that the two-species model of Eq. (1) is consistent with the distributions of DNA in these samples. (B) Relation of binding stoichiometry to free [AGT] for the 1000-bp fragment (upper panel) and linear pUC19 DNA (lower panel). Stoichiometries were inferred from weight-average molecular weights determined at sedimentation equilibrium. The error bars give 95% confidence limits. The smooth curves were calculated using parameters determined from the Scatchard plots shown in the insets. Insets: Scatchard plots for the data sets shown in the main panels. The solid curves are fits of Eq. (1), returning $K = 9667 \pm 1499$, $\omega = 35.9 \pm 6.8$, and $s = 6.32 \pm 0.12$ for the 1000-bp fragment and $K = 7960 \pm 916$, $\omega = 44.2 \pm 3.8$, and $s = 6.81 \pm 0.14$ for the linear pUC19 DNA.

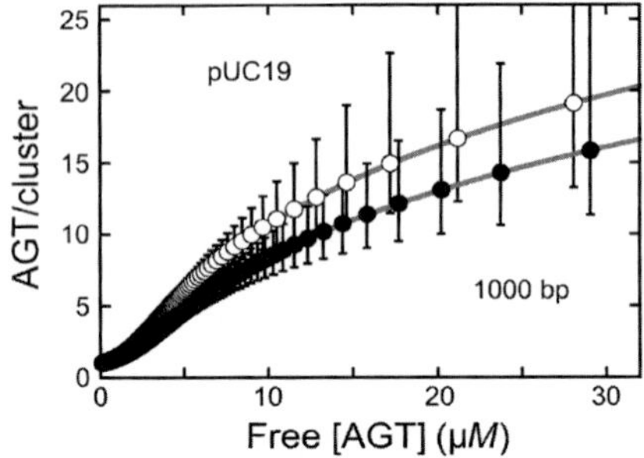

Figure 2 Theoretical cluster sizes for 1000 bp and linear pCU19 DNAs. Values of $\bar{C}$ were calculated with Eq. (2) using experimental values of ν and values of ω, and s obtained from the analyses shown in Fig. 1. Ninety-five percent confidence intervals (bars) were calculated by propagating corresponding intervals for measured ω and s parameters through Eq. (2).

3. AFM TESTS OF THE HBC MODEL

The AFM method has features that make it well suited for studies of cooperative protein–DNA complexes. These include rapid sample preparation that requires only small sample volumes and macromolecular concentrations, gentle sample handling, and the ability to image unmodified biomolecules under native-solution conditions (Buechner & Tessmer, 2013). On the other hand, the use of a mechanical probe in AFM results in the convolution of sample and probe geometries. This limits the resolution of the technique (Winzer, Kraft, Bhushan, Stepanenko, & Tessmer, 2012). However, realistic corrections for probe contributions to apparent sample dimensions are available (Vesenka et al., 1992; Wang & Chen, 2007; Winzer et al., 2012) and one will be discussed below. Important to the analyses described here, AFM provides the ability to inspect large-scale features of protein–DNA assemblies (Fig. 3) and thus, to test predictions of the HBC model. These predictions include homogeneous (sequence and structure independent) DNA binding (tested by analysis of protein distribution along the DNA contour), homogeneous cooperative interactions (tested by analysis of the contour length distribution of complexes), and cluster size distributions dependent solely on ν, ω, and s as described in Eq. (2) (tested by measurement of mean contour length of complexes). Details of these tests for the AGT–DNA system are described below.

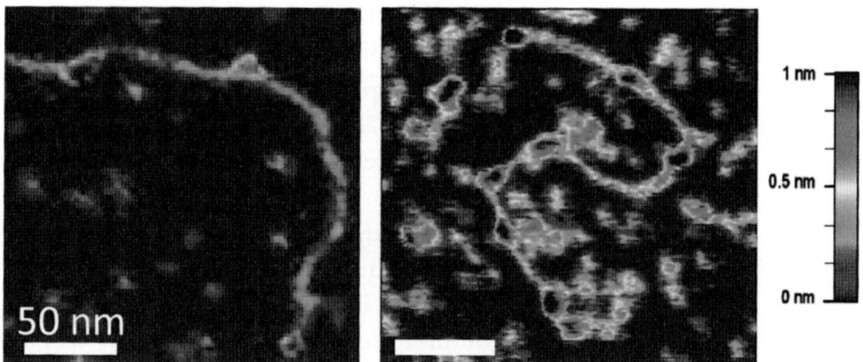

Figure 3 Short cooperative clusters formed by AGT on the 1000-bp DNA fragment. With increasing protein concentration (left 6 μ*M*, right 12 μ*M*), the frequency of DNA-bound clusters increases but their lengths appear to remain unchanged. Relative deflection of the AFM tip provides AFM topographical heights and is encoded in color with dark purple representing the smallest values and red-to-black the greatest, as shown in the color bar on the right. Scale bars in the images indicate 50 nm. (See the color plate.)

3.1 Sample Handling and Imaging of Protein–DNA Complexes

AFM uses a fine probe to directly scan the sample surface and produce topographical maps of the sample based on the detection of interaction forces between sample particles and the probe tip (for an introduction of the technique see for example, Tessmer, Kaur, Lin, & Wang, 2013). For AFM imaging, protein and DNA are incubated in an appropriate buffer solution and deposited onto a smooth substrate surface. Commonly used surfaces are mica or silicon owing to their high degree of smoothness (<0.1 nm RMSD). At physiological pH, both mica and silicon surfaces are negatively charged and negatively charged DNA polymers are bound to the substrate surface by divalent cations provided in the sample deposition buffer (Buechner & Tessmer, 2013; Tessmer et al., 2013). For AGT–DNA mixtures, samples were incubated in 10 m*M* Tris (pH 7.6 at 20 °C), 1 m*M* EDTA, 1 m*M* DTT, and 100 m*M* NaCl for 30 min at ambient temperature; diluted 40- to 300-fold for deposition in AFM deposition buffer (25 m*M* HEPES (pH 7.5 at 20 °C), 25 m*M* sodium acetate and 10 m*M* magnesium acetate) to achieve optimum surface coverage; and deposited on the surface of freshly cleaved mica. The mica surfaces were then rinsed with nanopure water and dried in a stream of nitrogen. This renders the sample particles stably fixed on the substrate surface so they can, in principle, be imaged for days (Buechner & Tessmer, 2013).

As contaminants can complicate data analysis, sample purity is important. Buffers, especially those used for substrate washing and sample deposition should be filtered through a 0.02-μm pore syringe filter (for example, Anotop, Whatman). Proteins should be purified by gel filtration chromatography and DNA substrates by agarose gel electrophoresis followed by extraction. Alternatively, stable protein–DNA complexes can be isolated by size exclusion chromatography or spin column filtering (Evrin et al., 2009; Shlyakhtenko et al., 2011; Shlyakhtenko, Lushnikov, Miyagi, & Lyubchenko, 2012; Verhoeven, Wyman, Moolenaar, & Goosen, 2002; Verhoeven, Wyman, Moolenaar, Hoeijmakers, & Goosen, 2001) and directly deposited on the AFM substrate. While purification of protein–DNA complexes can reduce the background of unbound protein molecules in samples to be imaged, it also has the potential to bias the binding equilibria under study. For this reason, the strategy of purifying preformed protein–DNA complexes should be followed with caution.

The AFM "tapping" imaging mode is often employed for studies of soft samples like protein–DNA complexes. In this mode, the AFM probe is oscillated closely above the sample surface, so that at the bottom of the oscillation cycle the probe can interact with molecules on the surface. Deflection of the probe is detected by a laser beam reflected onto a position sensitive photodetector. Interaction forces between probe and sample surface are detected through modification of the oscillation amplitude. The brief, vertical interaction of probe with molecules minimizes lateral force components that could displace molecules during the scanning process.

3.2 AFM Probe Calibration

AFM images of AGT–DNA complexes are shown in Fig. 3. Wider views of these and other images show particles with elongated structures and contour lengths similar to those of the naked DNAs (Tessmer et al., 2012). Distributed along the contours of these structures are distinct segments with diameters and heights greater than that of naked DNA. Their increasing frequency with increasing [AGT] (Fig. 3) supports the interpretation that these are AGT complexes. However, to accurately determine the dimensions of these complexes, particle and probe dimensions must be deconvoluted.

Scans perpendicular to DNA fragments (diameter ~2.0 nm (Mandelkern, Elias, Eden, & Crothers, 1981; Watson & Crick, 1953)) provide a calibration standard.[3] The encounter of an AFM tip with the DNA cylinder can be modeled with the tip of the AFM probe represented by a hemisphere and the DNA by a rectangular box with height h_{DNA} as measured from the AFM images. The radius of the AFM probe, r_{probe}, can then

[3] Measurements of molecular dimensions were made with the public-domain program *ImageJ* obtained from http://rsbweb.nih.gov/ij/.

be estimated from the deviation of the measured widths of the DNA cross-sections, W, from its theoretical width of 2 nm (Winzer et al., 2012):

$$r_{\text{probe}} \approx \frac{(W - 2\text{nm})^2}{8h_{\text{DNA}}} \tag{5}$$

Electron microscopy imaging shows that this simple model very slightly overestimates the AFM probe size. This may be because DNA molecules prepared for AFM imaging in air retain a hydration layer that results in apparent diameters slightly larger than the theoretical 2 nm value (Winzer et al., 2012). With knowledge of the approximate AFM tip dimension, lower limit values of molecular dimensions L_{LL} can then be obtained from the measured dimensions L_{Apparent} and measured heights, h_s using Eq. (6):

$$L_{\text{LL}} = L_{\text{Apparent}} - \sqrt{8h_S r_{\text{probe}} - 4h_S^2}. \tag{6}$$

On the other hand, L_{Apparent} values represent the dimensions that would be obtained with a probe of zero radius and hence represent an upper limit to possible molecular dimensions.

3.3 Characterization of AGT Complexes

Measured by AFM, the dimensions of AGT–DNA complexes are significantly larger than those of free AGT molecules in parallel samples (Fig. 4). This supports the notion that the observed AGT–DNA complexes are formed of clusters containing several protein molecules, as predicted by

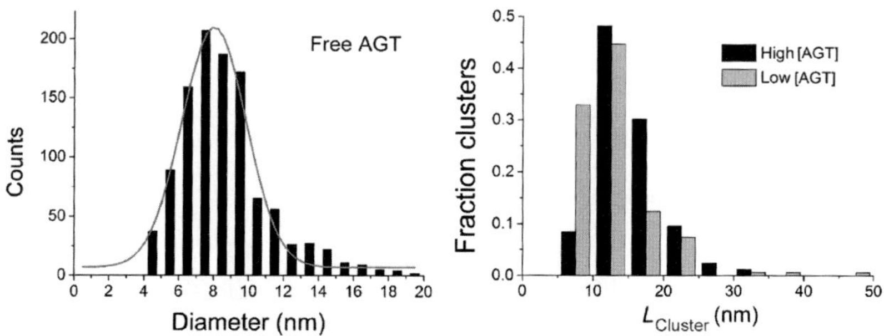

Figure 4 Dimensions of free AGT and AGT–DNA complexes measured by AFM. Values shown are not corrected for tip dimensions. Left: Histogram of particle diameters for free AGT. A Gaussian fit to the ensemble of measurements returned a mean of 8.0 ± 1.9 nm ($R^2 > 0.97$, $n = 1078$). Right: Histogram of cluster lengths measured parallel to the DNA axis. High AGT values were obtained at [AGT] = 12 μ*M*; low AGT values were obtained at [AGT] = 6 μ*M*.

the HBC model. The homogeneous binding model requires the absence of sequence- and structure-specific interactions. This was tested by measuring contour lengths between cluster centers and DNA ends. A graph of occupation frequency as a function of position on the DNA showed no obvious preference for any internal position but a modest preference for DNA ends (Fig. 5). This indicates that any sequence specificity is below the detection threshold but also shows a structural specificity at odds with the HBC model.

The size distribution of cooperative clusters provides another test for the HBC model. Cluster lengths were measured along the local DNA axis for several AGT concentrations and corrected for AFM tip diameter as described. Values of $\bar{C}$ were calculated using a binding density of 1 protein/4 bp, as seen with short DNAs at binding saturation (Melikishvili et al., 2008; Rasimas et al., 2003). This gave lower limit estimates of ~3 AGT molecules/cluster at [AGT] = 2 μ*M* and ~6 AGT molecules/cluster for [AGT] > 10 μ*M* (Fig. 6). Upper limit estimates (based on lengths uncorrected for AFM tip diameter as described) ranged from ~6 molecules/cluster at [AGT] ~ 2 μ*M* to ~8 molecules/cluster for [AGT] > 10 μ*M*. These values of $\bar{C}$ accord with the predictions of the HBC model for [AGT] ≤ 10 μ*M*, but they are significantly below the predicted range at higher AGT concentrations. A mechanism that has the potential to reconcile the differences between model predictions and experimental results is considered briefly below.

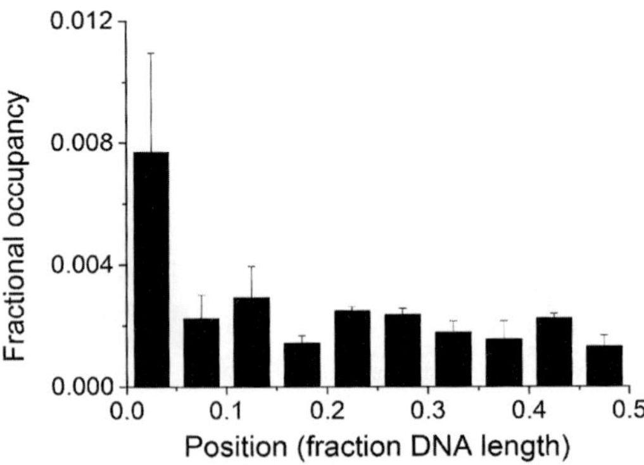

Figure 5 Distribution of AGT clusters along the DNA contour. Fractional occupancies for 50-bp-long sections of the 1000-bp DNA fragment, demonstrating nearly uniform protein coverage of internal DNA segments and preferential binding to DNA ends. Because unmodified DNA ends could not be distinguished, locations are reported in units of fractional DNA length ranging from 0% (at a DNA end) to 50% (at the DNA center).

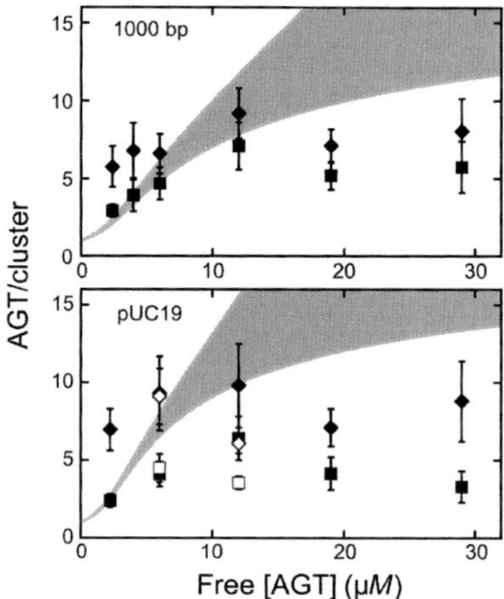

Figure 6 Comparison of measured cluster sizes with predictions from the HBC model. Two values are given for each set of measurements; r_{tip}-corrected values (filled squares) as lower limit estimates of the number of AGT monomers per cluster and uncorrected values (filled diamonds) as upper limit estimates. Open symbols (open square, open diamond) denote a few measurements made after glutaraldehyde crosslinking (Tessmer et al., 2012), indicating that cluster length limitations are not an artifact due to dilution of the protein–DNA samples for AFM sample deposition. Error bars give standard deviations for each sample population. Ranges for $\bar{C}$ predicted for the HBC binding model (gray zones) were calculated with Eq. (2), using experimental values of ν, ω, and s and their corresponding error ranges as described for Fig. 2.

3.4 A Mechanism That Accounts for Differences Between Model and Experiment

The disparity between predicted and observed cluster sizes shows that the HBC model fails to capture an essential feature of the AGT–DNA interaction. Factors contributing to $\bar{C}$ include binding density ν, occluded site size s, and cooperativity factor ω (Eq. 2). The results in Fig. 6 show that $\bar{C}$ is nearly independent of binding density over the experimental concentration range. Occluded site size is weakly dependent on DNA length and binding density (Adams et al., 2009; Melikishvili et al., 2008); however, the data in Fig. 6 show that dependence of s on ν cannot be large enough to influence observed values of $\bar{C}$. These considerations focus attention on ω. AGT binding widens the DNA minor groove by ~3 Å and bends the DNA toward the major groove by ~15° (Daniels et al., 2004). In the process, it also unwinds the DNA by ~7°/protein molecule (Adams et al., 2009). Stress from these distortions should contribute positively to the cooperative free energy. As each protein is anchored to its neighbors and to the DNA, stress introduced by each added protein monomer will accumulate as clusters grow. Ultimately, binding should become noncooperative (i.e., $\omega = 1$) when ΔG(deformation) $= -\Delta G$(protein–protein interaction). At that point, a free protein would be just as likely to bind an isolated site as it would to add to the end of an existing cluster.

A simple calculation shows that the torsional free energy is comparable to that of protein–protein interaction for clusters similar in size to those seen with AGT. The dependence of ΔG(DNA twist) on the twist angle ϕ (in radians) is given by Eq. (7), in which H is the torsional Hooke's constant for duplex DNA ($\sim 3 \times 10^{-19}$ erg/cm (Crothers, Drak, Kahn, & Levene, 1992; Delrow, Heath, & Schurr, 1997)) and L is the DNA length over which the twist acts (Bloomfield, Crothers, & Tinoco, 2000).

$$\Delta G_{\text{twist}} = \frac{1}{2} H \left(\frac{\phi}{L}\right)^2 L. \tag{7}$$

Graphs of this function are shown in Fig. 7 for several unwinding angles. Also graphed are horizontal lines indicating the cooperative free energies measured for pUC19 and 1000 bp DNAs. At the intersections of these functions, ΔG(twist) $= -\Delta G$(cooperative). For a cluster size of ~ 6 (typical of

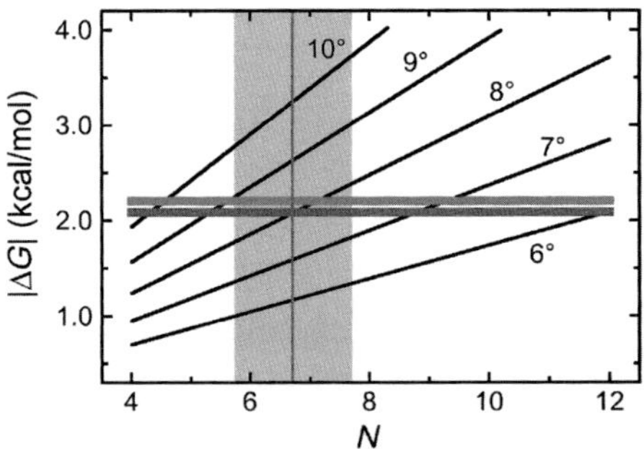

Figure 7 Comparison of predicted cooperative free energies (ΔG(cooperative)) and torsional free energies (ΔG(twist)) for clusters of AGT proteins. ΔG is given as its absolute value. Equation (7) was used to calculate the cumulative ΔG(twist) as functions of number of proteins/cluster (N) for torsional displacements of 6–10°/protein. The red (light gray in the print version) and blue (dark gray in the print version) horizontal lines give ΔG(cooperative) for the addition of a single protein molecule to complexes formed on linear pUC19 DNA and the 1000-bp fragment, respectively. The intersections of these functions occur where ΔG(twist) $= -\Delta G$(cooperative). A vertical gray line is plotted for a cluster length of 6.7 proteins, the mean of length estimates for [AGT] $\geq 6\ \mu M$, for the 1000-bp DNA. The vertical light gray zone spans mean values of minimum and maximum estimates of cluster length.

values at high [AGT], Fig. 6), this equivalence point is reached at a net unwinding of ~8.5°/protein. This value is only slightly greater than one previously measured by topoisomerase assay (Adams et al., 2009).

Although this coincidence does not prove that torsional stress limits the size of AGT clusters, it shows that such a mechanism is capable of doing so and might account for the observation that AGT clusters are smaller than those predicted by the HBC model. A consideration of the possible roles of cooperative clusters in AGT function is beyond the scope of the current chapter. However, it is worth pointing out that small cooperative clusters are likely to form and decay at rates quite different from larger ones and that this difference may have functional significance. The combination of model building using SE data and model testing by AFM (or other appropriate imaging method) is a uniquely powerful approach for the analysis of cooperative clusters.

ACKNOWLEDGMENTS

These studies were supported by the Deutsche Forschungsgemeinschaft (DFG, Forschungszentrum FZ82 to I.T.) and National Institutes of Health (NIH, GM-070662 to M.G.F.).

REFERENCES

Adams, C. A., Melikishvili, M., Rodgers, D. W., Rasimas, J. J., Pegg, A. E., & Fried, M. G. (2009). Topologies of complexes containing O^6-alkylguanine-DNA alkyltransferase and DNA. *Journal of Molecular Biology*, *389*(2), 248–263.

Azzaz, A. M., Vitalini, M. W., Thomas, A. S., Price, J. P., Blacketer, M. J., Cryderman, D. E., et al. (2014). Human heterochromatin protein 1α promotes nucleosome associations that drive chromatin condensation. *The Journal of Biological Chemistry*, *289*, 6850–6861.

Bear, D. G., Hicks, P. S., Escudero, K. W., Andrews, C. L., McSwiggen, J. A., & von Hippel, P. H. (1988). Interactions of Escherichia coli transcription termination factor rho with RNA. II. Electron microscopy and nuclease protection experiments. *Journal of Molecular Biology*, *199*, 623–635.

Bloomfield, V. A., Crothers, D. M., & Tinoco, I., Jr. (2000). *Nucleic acids: Structures, properties, and functions*. Sausalito: University Science Books, 176–181.

Buechner, C. N., & Tessmer, I. (2013). DNA substrate preparation for atomic force microscopy studies of protein-DNA interactions. *Journal of Molecular Recognition*, *26*(12), 605–617.

Bujalowski, W., Lohman, T. M., & Anderson, C. F. (1989). On the cooperative binding of large ligands to a one-dimensional homogeneous lattice: The generalized three-state lattice model. *Biopolymers*, *28*, 1637–1643.

Cohen, G., & Eisenberg, H. (1968). Deoxyribonucleate solutions: Sedimentation in a density gradient, partial specific volumes, density and refractive density increments and preferential interactions. *Biopolymers*, *6*, 1077–1100.

Crothers, D. M., Drak, J., Kahn, J., & Levene, S. D. (1992). DNA bending, flexibility and helical repeat by cyclization kinetics. *Methods in Enzymology*, *212*, 3–29.

Daniels, D. S., Woo, T. T., Luu, K. X., Noll, D. M., Clarke, N. D., Pegg, A. E., et al. (2004). DNA binding and nucleotide flipping by the human DNA repair protein AGT. *Nature Structural & Molecular Biology*, *11*, 714–720.

Dattagupta, N., Hogan, M., & Crothers, D. M. (1980). Interaction of netropsin and distamycin with deoxyribonucleic acid: Electric dichroism study. *Biochemistry*, *19*, 5998–6005.

Delrow, J. J., Heath, P. J., & Schurr, J. M. (1997). On the origin of the temperature dependence of the supercoiling free energy. *Biophysical Journal*, *73*(5), 2688–2701.

Dunn, K., Chrysogelos, S., & Griffith, J. (1982). Electron microscopic visualization of RecA-DNA filaments: Evidence for a cyclic extension of duplex DNA. *Cell*, *28*, 757–765.

Epstein, I. R. (1978). Cooperative and non-cooperative binding of large ligands to a finite one-dimensional lattice. A model for ligand-oligonucleotide interactions. *Biophysical Chemistry*, *8*, 327–339.

Evrin, C., Clarke, P., Zech, J., Lurz, R., Sun, J., Uhle, S., et al. (2009). A double-hexameric Mcm2-7 complex is loaded onto origin DNA during licensing of eukaryotic DNA replication. *Proceedings of the National Academy of Sciences of the United States of America*, *106*, 20240–20245.

Fried, M. G., Kanugula, S., Bromberg, J. L., & Pegg, A. E. (1996). DNA binding mechanisms of O^6-alkylguanine-DNA alkyltransferase: Stoichiometry and effects of DNA base composition and secondary structures on complex stability. *Biochemistry*, *35*, 15295–15301.

Galletto, R., Amitani, I., Baskin, R. J., & Kowalczykowski, S. C. (2006). Direct observation of individual RecA filaments assembling on single DNA molecules. *Nature*, *443*, 875–878.

Hieb, A. R., Gansen, A., Böhm, V., & Langowski, J. (2014). The conformational state of the nucleosome entry-exit site modulates TATA box-specific TBP binding. *Nucleic Acids Research*, *42*, 7561–7576.

Innis, M. A., & Gelfand, D. H. (1990). *PCR protocols: A guide to methods and applications*. New York: Academic Press.

Joo, C., McKinney, S. A., Nakamura, M., Rasnik, I., Myong, S., & Ha, T. (2006). Real-time observation of RecA filament dynamics with single monomer resolution. *Cell*, *126*, 515–527.

Karpel, R. L. (2002). LAST motifs and SMART domains in gene 32 protein: An unfolding story of autoregulation? *IUBMB Life*, *53*, 161–166.

Kowalczykowski, S. C., Paul, L. S., Lonberg, N., Newport, J. W., McSwiggen, J. A., & von Hippel, P. H. (1986). Cooperative and noncooperative binding of protein ligands to nucleic acid lattices: Experimental approaches to the determination of thermodynamic parameters. *Biochemistry*, *25*, 1226–1240.

Kozlov, A. G., Weiland, E., Mittal, A., Waldman, V., Antony, E., Fazio, N., et al. (2015). Intrinsically disordered C-terminal tails of E. coli single-stranded DNA binding protein regulate cooperative binding to single-stranded DNA. *Journal of Molecular Biology*, *427*, 763–774.

Laue, T. M., Shah, B. D., Ridgeway, T. M., & Pelletier, S. L. (1992). Computer-aided interpretation of analytical sedimentation data for proteins. In S. E. Harding, A. J. Rowe, & J. C. Horton (Eds.), *Analytical ultracentrifugation in biochemistry and polymer science* (pp. 90–125). Cambridge, England: The Royal Society of Chemistry.

Mamoon, N. M., Song, Y., & Wellman, S. E. (2005). Binding of histone H1 to DNA is described by an allosteric model. *Biopolymers*, *77*, 9–17.

Mandelkern, M., Elias, J. G., Eden, D., & Crothers, D. M. (1981). The dimensions of DNA in solution. *Journal of Molecular Biology*, *152*, 153–161.

McGhee, J., & von Hippel, P. H. (1974). Theoretical aspects of DNA-protein interactions: Co-operative and non-co-operative binding of large ligands to a one-dimensional homogeneous lattice. *Journal of Molecular Biology*, *86*, 469–489.

Melikishvili, M., Rasimas, J. J., Pegg, A. E., & Fried, M. G. (2008). Interactions of human O^6-alkylguanine-DNA alkyltransferase (AGT) with short double-stranded DNAs. *Biochemistry*, *47*, 13754–13763.

Morrone, S. R., Wang, T., Constantoulakis, L. M., Hooy, R., Delannoy, M. J., & Sohn, J. (2014). Cooperative assembly of IFI16 filaments on dsDNA provides insights into host defense strategy. *Proceedings of the National Academy of Sciences of the United States of America*, *111*, E62–E71.

Newman, J. A., Cooper, C. D., Aitkenhead, H., & Gileadi, O. (2015). Structural insights into the autoregulation and cooperativity of the human transcription factor Ets-2. *The Journal of Biological Chemistry*, *290*, 8539–8549.

Rasimas, J. J., Pegg, A. E., & Fried, M. G. (2003). DNA-binding mechanism of O^6-alkylguanine-DNA alkyltransferase. Effects of protein and DNA alkylation on complex stability. *The Journal of Biological Chemistry*, *278*, 7973–7980.

Ruyechan, W. T., & Wetmer, J. G. (1975). Studies on the cooperative binding of the Escherichia coli DNA unwinding protein to single-stranded DNA. *Biochemistry*, *14*, 5529–5534.

Saroff, H. A. (1995). Energetics of protein-DNA interactions: An exact calculation for binding of ligands to a lattice of overlapping sites. *Biopolymers*, *36*, 121–134.

Saxe, S. A., & Revzin, A. (1979). Cooperative binding to DNA of catabolite activator protein of *Escherichia coli*. *Biochemistry*, *18*, 255–263.

Schwarz, G., & Watanabe, F. (1983). Thermodynamics and kinetics of cooperative protein-nucleic acid binding. I. General aspects of analysis of data. *Journal of Molecular Biology*, *163*, 467–484.

Shlyakhtenko, L. S., Lushnikov, A. Y., Li, M., Lackey, L., Harris, R. S., & Lyubchenko, Y. L. (2011). Atomic force microscopy studies provide direct evidence for dimerization of the HIV restriction factor APOBEC3G. *The Journal of Biological Chemistry*, *286*, 3387–3395.

Shlyakhtenko, L. S., Lushnikov, A. Y., Miyagi, A., & Lyubchenko, Y. L. (2012). Specificity of binding of single-stranded DNA-binding protein to its target. *Biochemistry*, *51*, 1500–1509.

Takahashi, M. (1989). Analysis of DNA-RecA protein interactions involving the protein self-association reaction. *The Journal of Biological Chemistry*, *264*, 288–295.

Tessmer, I., Kaur, P., Lin, J., & Wang, H. (2013). Investigating bioconjugation by atomic force microscopy. *Journal of Nanobiotechnology*, *11*, 25.

Tessmer, I., Melikishvili, M., & Fried, M. G. (2012). Cooperative cluster formation, DNA bending and base-flipping by O^6-alkylguanine-DNA alkyltransferase. *Nucleic Acids Research*, *40*, 8296–8308.

Tessmer, I., Moore, T., Lloyd, R. G., Wilson, A., Erie, D. A., Allen, S., et al. (2005). AFM studies on the role of the protein RdgC in bacterial DNA recombination. *Journal of Molecular Biology*, *350*(2), 254–262.

Tsodikov, O. V., Holbrook, J. A., Shkel, I. A., & Record, M. T., Jr. (2001). Analytic binding isotherms describing competitive interactions of a protein ligand with specific and nonspecific sites on the same DNA oligomer. *Biophysical Journal*, *81*, 1960–1969.

van der Heijden, T., Seidel, R., Modesti, M., Kanaar, R., Wyman, C., & Dekker, C. (2007). Real-time assembly and disassembly of human RAD51 filaments on individual DNA molecules. *Nucleic Acids Research*, *35*, 5646–5657.

Verhoeven, E. E., Wyman, C., Moolenaar, G. F., & Goosen, N. (2002). The presence of two UvrB subunits in the UvrAB complex ensures damage detection in both DNA strands. *The EMBO Journal*, *21*, 196–205.

Verhoeven, E. E., Wyman, C., Moolenaar, G. F., Hoeijmakers, J. H., & Goosen, N. (2001). Architecture of nucleotide excision repair complexes: DNA is wrapped by UvrB before and after damage recognition. *The EMBO Journal*, *20*, 601–611.

Vesenka, J., Guthold, M., Tang, C. L., Keller, D., Delaine, E., & Bustamante, C. (1992). Substrate preparation for reliable imaging of DNA molecules with the scanning force microscope. *Ultramicroscopy*, *42–44*(Pt. B), 1243–1249.

Wang, Y., & Chen, X. (2007). Carbon nanotubes: A promising standard for quantitative evaluation of AFM tip apex geometry. *Ultramicroscopy*, *107*(4–5), 293–298.

Watson, J. D., & Crick, F. H. (1953). The structure of DNA. *Cold Spring Harbor Symposia on Quantitative Biology*, *18*, 123–131.

Winzer, A. T., Kraft, C., Bhushan, S., Stepanenko, V., & Tessmer, I. (2012). Correcting for AFM tip induced topography convolutions in protein-DNA samples. *Ultramicroscopy*, *121*, 8–15.

Wolfe, A. R., & Meehan, T. (1992). Use of binding site neighbor-effect parameters to evaluate the interactions between adjacent ligands on a linear lattice. Effects on ligand-lattice association. *Journal of Molecular Biology*, *223*, 1063–1087.

CHAPTER SIXTEEN

Sedimentation Velocity Analysis of Large Oligomeric Chromatin Complexes Using Interference Detection

Ryan A. Rogge, Jeffrey C. Hansen[1]

Department of Biochemistry and Molecular Biology, Colorado State University, Fort Collins, Colorado, USA

[1]Corresponding author: e-mail address: jeffrey.c.hansen@colostate.edu

Contents

Abstract

Sedimentation velocity experiments measure the transport of molecules in solution under centrifugal force. Here, we describe a method for monitoring the sedimentation of very large biological molecular assemblies using the interference optical systems of the analytical ultracentrifuge. The mass, partial-specific volume, and shape of macromolecules in solution affect their sedimentation rates as reflected in the sedimentation coefficient. The sedimentation coefficient is obtained by measuring the solute concentration as a function of radial distance during centrifugation. Monitoring the concentration can be accomplished using interference optics, absorbance optics, or the

Methods in Enzymology, Volume 562
ISSN 0076-6879
http://dx.doi.org/10.1016/bs.mie.2015.05.007

fluorescence detection system, each with inherent advantages. The interference optical system captures data much faster than these other optical systems, allowing for sedimentation velocity analysis of extremely large macromolecular complexes that sediment rapidly at very low rotor speeds. Supramolecular oligomeric complexes produced by self-association of 12-mer chromatin fibers are used to illustrate the advantages of the interference optics. Using interference optics, we show that chromatin fibers self-associate at physiological divalent salt concentrations to form structures that sediment between 10,000 and 350,000S. The method for characterizing chromatin oligomers described in this chapter will be generally useful for characterization of any biological structures that are too large to be studied by the absorbance optical system.

1. INTRODUCTION

Analytical ultracentrifugation has been used extensively for the characterization of macromolecular assemblies like chromatin fibers (Correll, Schubert, & Grigoryev, 2012; Fan, Gordon, Luger, Hansen, & Tremethick, 2002; Lu & Hansen, 2004; Muthurajan, McBryant, Lu, Hansen, & Luger, 2011). Sedimentation velocity experiments yield information regarding the mass and shape of molecules in solution allowing researchers to investigate molecular interactions and conformational changes. These experiments rely on determining the concentration in solution as a function of radial distance under centrifugal force. This is accomplished most frequently using the absorbance optics, which takes about 1 min to scan a cell. The time to collect a scan is a limiting factor in monitoring massive macromolecular complexes that sediment very rapidly. The use of the interference optics, which collects data much faster, allows for monitoring molecules with sedimentation coefficients order of magnitude larger than those that can be seen with the absorbance optics. The following protocol details the use of the interference system to characterize chromosome-sized oligomeric complexes of chromatin fibers using sedimentation velocity, but the principles will allow for expanded sedimentation velocity analyses of a variety of much larger biological assemblies.

2. THEORY

The signal from the interference system is a pattern of light and dark horizontal bands called a fringe pattern (Fig. 1). The vertical displacement (fringe shift, ΔY) of these fringes is due to differences in the optical path between the sample and reference channel. The interference system

Figure 1 A typical fringe pattern for a sedimentation velocity experiment of self-associated chromatin fibers. The sample here is 601(207 bp) 12-nucleosome arrays with endogenous chicken octamer self-associated in 8 m*M* $MgCl_2$. The vertical arrow marks the meniscus of the sample sector and the double-headed horizontal arrow marks the boundary region. Note the increased light scattering in the region to the right of the boundary area due to the presence of large particles.

measures concentration by tracking changes in refractive index in solution. The refractive index of a solution is altered by its component solutes. The fringe shift magnitude is given by $\Delta Y = lc(\mathrm{d}n/\mathrm{d}c)/\lambda$, where l is the path length, c is the concentration, λ is the wavelength, and $\mathrm{d}n/\mathrm{d}c$ is a property of the unique individual solutes. For DNA $\mathrm{d}n/\mathrm{d}c$ is ~0.17 mL/g (Chincholi, Havlik, & Vold, 1974) and for proteins it averages ~0.189 mL/g; however, differences in protein composition can change this considerably (Zhao, Brown, & Schuck, 2011). Polysaccharides and phospholipids are difficult to use with the absorbance optics but have an average $\mathrm{d}n/\mathrm{d}c$ of ~0.15 and ~0.16, respectively, which makes interference measurements ideal for these molecules (Tumolo, Angnes, & Baptista, 2004). Importantly, the fringe pattern is projected onto a camera which makes data capture along the radius of the analytical ultracentrifuge (AUC) cell simultaneously. A Fourier transform of the fringe pattern results in a concentration profile that is obtained in a matter of seconds compared to the minute necessary for the absorbance optics.

The fast data acquisition of the interference system makes it ideal for monitoring the sedimentation of large molecules that sediment quickly. For example, self-associated chromatin oligomers have been examined using absorbance-based sedimentation velocity experiments previously, but the experiments have failed to capture more than a single complete boundary. The interference system has some unique properties to consider when compared to the absorbance system. The first is that interference generally requires a significantly more concentrated sample than absorbance. However, this is dependent on the extinction coefficient of the molecule being studied. For instance, interference measurement of a nucleic acid sample will require around a 10-fold higher concentration than monitoring the samples absorbance at 260 nm. A second consideration is the increased presence of systematic noise. Because all changes in the optical path will affect the fringe pattern, the concentration profiles contain a large amount of time invariant noise. Vibrations as well as instability of the Fourier transform will lead to radially invariant noise. Both radially invariant and time invariant noise can be minimized by good maintenance of the system, and can be further accounted for during data analysis.

3. EQUIPMENTS

Beckman XL-I AUC
An50-Ti or An60Ti rotor
AUC counterbalance and weights
Assembled charcoal filled Epon 2-channel centerpiece AUC cell with sapphire windows

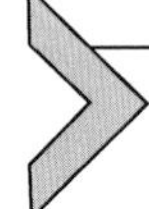

4. SETTING UP THE INTERFEROMETER LASER: LASER DELAY AND DURATION

In ProteomeLab, choose Interference > Laser Setup in the dropdown menu in order to access the laser delay and duration settings for each cell. These options allow you to fine-tune the laser to generate an optimal fringe pattern. The laser delay controls where the laser turns on in the course of a 360° rotation of the rotor. The laser duration controls how long the laser is on after the laser delay triggers. The laser delay defaults to values of 180°, 90°, and 0° for cell positions 1, 2, and 3, respectively. The laser duration default is set to 0.6°, which should be close to appropriate. These values should be adjusted slightly to produce an optimal fringe pattern with good contrast. Under these conditions, the laser turns on as the cell crosses the condenser lens, and turns off just as it has passed. For identical cells and rotors, these values should not need to be changed.

The XL-I AUC must be spinning at a minimum of 3000 rpm in order to generate interference data and to set the laser delay and duration. For very high-molecular-weight samples, the slowest possible speed is chosen to maximize the amount of data that can be collected during the experiment. Setting up the laser and the radial calibration should be performed on a double sector cell with both channels containing reference buffer. A cell with sapphire windows should be used to limit distortion of the windows. Set the laser conditions using this cell. Autodetection can be used to allow the software to select a laser delay and is a good option for those unfamiliar with the process. The software autoselect option will adjust the delay to optimize the fringe pattern at the radial position selected by clicking in the fringe display (red (gray in the print version) text in top left of display). The entire fringe pattern can then be optimized by making small changes manually, with high contrast, unbroken bands of light and dark at all radial positions being the goal. To set the laser delay manually, find the upper and lower limits at which the fringe pattern begins to deteriorate, and then select the midpoint. The laser duration should be adjusted to maximize fringe intensity without saturating the image. Values between 0.5° and 1° are typical for laser duration. The process of setting up the laser should also be done for the counterbalance reference holes. Clear fringe patterns should be visible through both reference holes.

5. RADIAL CALIBRATION OF INTERFERENCE DETECTOR

The counterbalance contains two reference holes that are used to determine radial position. When examining the fringe display of the counterbalance, this results in two regions of fringe pattern separated by a lack of signal. The inside edges of these holes, and thus the edges of their fringe patterns mark the radial positions of 5.85 and 7.15 cm. Opening the Interference > Radial Calibration menu allows you to define these positions in the fringe display. To set the radial points, first click within the fringe display at the inside edge of the fringe pattern (vertical position does not matter). Then click the inside/outside option and set radius to set the location to 5.85 (inside) or 7.15 cm (outside).

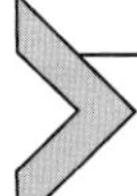

6. FINAL CONSIDERATIONS FOR INTERFERENCE SEDIMENTATION VELOCITY RUN

There are a few parameters to adjust before starting an experimental run. The first is to set the number of pixels per fringe. This setting is in the Details menu of each cell. Selecting automatic will have the software calculate this value. This value can be estimated manually by dividing 96 (the number of pixels in a column) by the number of visible fringes. If the automatic value is significantly different from the manual estimation, the optical system is likely misaligned.

The next two options will affect the data files generated by ProteomeLab and are used to compensate for systematic noise. Note that the systematic noise can be removed using specific types of data analysis (see below), making these options unnecessary. However, if you wish to analyze interference data using a method does not compensate for systematic noise, it would be wise to consider the following options. Misalignment of the interference detector can lead to fringe patterns which are not horizontal. This can be corrected by manipulating the detector, or by rotating the fringe data in the Interference > Fringe Rotation menu by a set number of degrees. The other option for removing systematic noise is to subtract a blank scan from every scan generated. Blank scans should be taken from empty rotor positions so that subtracting them will negate noise from defects inherent to the optical system. If subtracting a blank scan, the blank data should be collected using identical laser settings to those used for sample data collection. These options can reduce systematic noise in the data, but the subtraction of a blank scan will increase the amount of stochastic noise in the data.

7. EXPERIMENTAL

7.1 Reconstitution of Chromatin Fibers and Assembly of Chromatin Oligomers

Oligomeric chromatin fibers provide a good example of very large biological complexes that can be studied using sedimentation velocity only in conjunction with interference optics. Model chromatin fibers are obtained by reconstituting purified histone octamers onto tandem repeats of nucleosome positioning DNA using a salt dialysis method as described (Rogge et al., 2013). This protocol yields equally spaced arrays of nucleosomes with a defined length and composition (see below).

The structure of the chromatin fibers in solution and within the cell is highly dependent on ionic strength, particularly the concentration of divalent cations (Korolev et al., 2010; Strick, Strissel, Gavrilov, & Levi-Setti, 2001). In low salt, the model chromatin fibers are monomeric and adopt an extended, beads-on-a-string structure. As divalent salts (e.g., $MgCl_2$) are first titrated into solution, the chromatin fibers fold into helical 30 nm structures (Hansen, Ausio, Stanik, & van Holde, 1989). At physiologically $MgCl_2$ concentrations (i.e., >3 m*M*), the chromatin fibers self-associate to form large oligomeric complexes (Schwarz, Felthauser, Fletcher, & Hansen, 1996; Fig. 2). Sedimentation velocity analysis of the chromatin oligomers is not ideal using absorbance optics because the oligomers pellet before a set of scans with complete boundaries is collected, even at 3000 rpm.

In our experiments, chromatin fibers consisting of 12 spaced nucleosomes were reconstituted from tandemly repeated "601" nucleosome

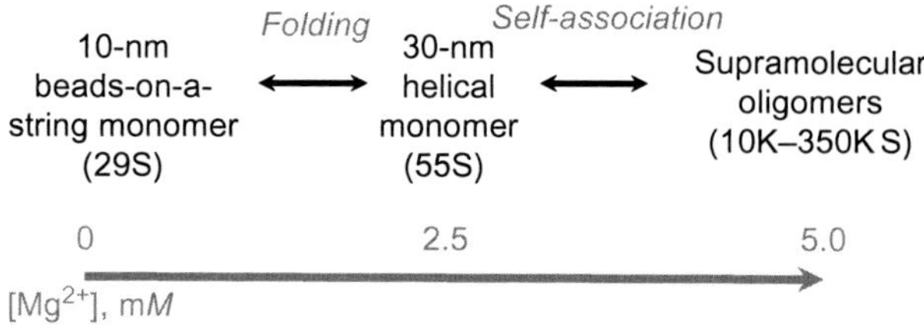

Figure 2 Dynamics of a 12-nucleosome chromatin fiber in solution across a range of Mg^{2+} concentrations. Sedimentation coefficient values for monomeric extended and folded 12-nucleosome fibers are prevalent in the literature (Hansen & Lohr, 1993). Sedimentation coefficients for the oligomeric self-associated chromatin are demonstrated in Fig. 4.

positioning DNA and purified chicken histone octamers and stored in a buffer of 10 m*M* Tris, pH 7.8, 0.25 m*M* EDTA, and 2.5 m*M* NaCl. The chromatin fibers were then diluted to a concentration of 0.215 mg/mL, and $MgCl_2$ added to a final concentration of 8 m*M* to assemble the oligomeric complexes. The rotor, optics, and AUC chamber were precooled to one degree below the final desired temperature. The speed was 3000 rpm (see below). Data were collected using the interference optics as described in Sections 4–6. The oligomeric samples formed broad but discrete boundaries (Fig. 4A), which were analyzed using the time derivative method as described in Section 8.1. Previous sedimentation velocity experiments of these oligomers have only captured either single or incomplete boundaries of chromatin oligomers (Blacketer, Feely, & Shogren-Knaak, 2010; Schwarz et al., 1996).

7.2 Choosing the Rotor Speed

The rotor speed, number of cells per run, and number of scans per run should be chosen based on how rapidly the samples sediment. The interference optical system captures data from the fringe display about every 8 s. In our analysis of the chromatin oligomers, the rotor speed was set at 3000 rpm and only one cell was sedimented at a time. These conditions were necessary because the oligomers sediment in the range of 10,000–350,000S. For samples that sediment in the range of hundreds to thousands of S, it may be preferable to use faster rotor speeds or analyze more than one cell per run. Scanning multiple cells increases the amount of time it takes to scan each individual cell, which decreases the maximum sedimentation coefficient that can be monitored (Fig. 3).

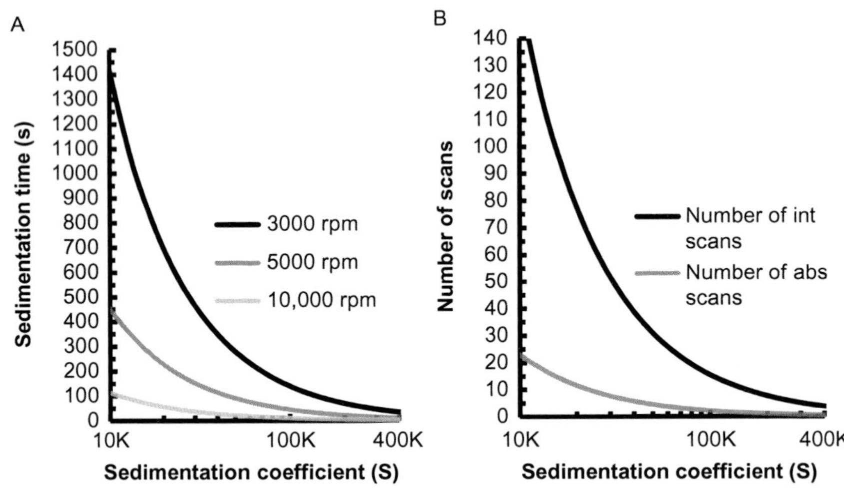

Figure 3 Calculations of amount of time spinning and scanning until an object has sedimented in the centrifuge. (A) Time until sedimentation to bottom of cell. This calculation assumes the use of a two-sector cell and a fluid column of 1 cm. Rotor acceleration time is not considered so actual time will be shorter. (B) Maximum number of possible scans collected for samples spun at the minimum rotor speed of 3000 rpm. Scan times are approximated at 9 s for the interference system and 60 s for the absorbance system.

The amount of time necessary to sediment the entire sample can be estimated if the sedimentation coefficient is known, according to $\times 10^{-13} S = \frac{\frac{0.01m}{t}}{730 \times 9.8\frac{m}{s^2}}$, where S is the sedimentation coefficient in Svedbergs and t is the number of seconds for the object to sediment. The value of 0.01 is the length of the fluid column in meters, and 730 is the g force generated by the rotor at its minimum operational speed. Enter the desired number of scans in the Method menu. If the sample is being characterized for the first time, enter the maximum number of scans and stop the run after the boundary has reached the bottom of the cell. Under the Options menu, select stop XL-I after last scan, so that the centrifuge will stop after the last scan is taken. Once the sample has been characterized in this manner, the rotor speed, number of cells sedimented, and number of scans collected can be modified as appropriate.

If the sample is very heterogeneous, it is possible that smaller solution components will remain at the meniscus after the larger components have sedimented. For example, at intermediary Mg^{2+} concentrations, chromatin samples will contain populations of both small monomeric and large oligomeric chromatin fibers. To characterize the smaller components, a second run using higher speeds can be started after the first is completed. Smaller solutes are likely to take longer to sediment, so to avoid very large data sets use a scan delay time that can be found in the Method menu. The scan delay time adds time between captures of the fringe pattern, so that over the course of a longer run a smaller data set is generated.

8. DATA ANALYSIS

8.1 Systematic Noise Considerations

As mentioned above, interference data have relatively high amounts of systematic noise. The systematic noise can be further categorized as radially invariant and time invariant noise. Radially invariant noise arises due to vibrations in the detector as well and the nature of the fast Fourier transform performed in the ProteomeLab software. Radially invariant noise appears visually as an offset in the baseline of the scan. The radially invariant component of the noise can be removed by aligning the scans in a consistent region, such as the air-to-air region, during data editing. Time invariant noise arises due to differences in the optical path between the reference and the sample sector. The differences due to the interference optical system and not the AUC cell can be removed by subtracting a blank scan, but this will lead to an increase in the stochastic noise. Alternatively, the use of the time derivative method of data analysis, or the use of modeling software will allow for the removal of time invariant noise.

8.2 Time Derivative Method

The time derivative method for determining distributions works by subtracting the subsequent scan in pairs of scans to determine how much the concentration has changed over time. A comprehensive explanation of the method can be found here (Stafford, 1992). Subtracting the subsequent scan from a pair removes time invariant noise, much in the same way subtracting a blank scan would; however, the subsequent scan contains time invariant noise components which arise due to both the cell components and the interference optical system. Because this time invariant noise removal is more comprehensive than the subtraction of blank scan, it is an ideal method for the analysis of interference sedimentation velocity data. The primary limitation of this method for smaller solutes is that the sedimentation coefficient distribution will not be corrected for the effects of diffusion on the boundary. However, since diffusion is inversely proportional to solute size, the sedimentation coefficient distributions obtained using the time derivative method of very large macromolecular assemblies will be an accurate reflection of the actual composition of the sample.

8.3 Other Analysis Methods

Other data analysis can be used with the interference optics as well. Methods of modeling the data using software such as UltraScanIII and Sedfit contain methods for decomposing and removing systematic noise contributions to the data (Brown & Schuck, 2006; Demeler, 2005; Schuck & Demeler, 1999). These programs generate a solution with a finite number of elements with sedimentation and diffusion coefficients, the model solutions concentration profile can then be compared against the experimental data. With prior knowledge or assumptions regarding the frictional coefficient or partial-specific volume, these models allow for the determination of molecular weight. The noise contributions determined during modeling are useful for other analysis methods such as the second moment and van Holde–Weischet (Holde & Weischet, 1978) methods, although with noise files subtracted these are no longer model-independent analyses.

9. DISCUSSION

9.1 Chromatin Oligomers

The core of a eukaryotic chromosome is a single long chromatin fiber composed of 10^4–10^5 nucleosomes. An important outstanding question in the chromatin field is how the conformational dynamics of short array of nucleosomes *in vitro* relates to the structure and assembly of a chromosome in the cell. Most previous attention has focused on the local folding of the chromatin fiber into helical 30 nm structures (Allahverdi et al., 2011; Dorigo, Schalch, Bystricky, & Richmond, 2003; Fletcher, Serwer, & Hansen, 1994). Sedimentation velocity experiments using absorbance optics have proven very useful for analysis of this intramolecular conformational change (Ausio, 2000; Fierz et al., 2011; Huynh, Robinson, & Rhodes, 2005). However, physicochemical studies of chromatin fiber oligomerization have lagged behind due to difficulties in characterizing the extremely large size of the oligomeric complexes. To overcome this hurdle, we have developed sedimentation velocity together with the interference optical system as a quantitative assay for the structural features of chromatin oligomers. Our results demonstrate that the oligomers sediment in the range of 10,000–350,000S (Fig. 4B), spanning the size of range of eukaryotic chromosomes. This result indicates the self-association of a large number of the monomeric arrays, which sediment in the range of 30S (Fig. 4B, inset). By comparison, bacteriophage T7 (Dubin, Benedek, Bancroft, & Freifelder, 1970) and the largest amyloid measured (MacRaild, Hatters, Lawrence, & Howlett, 2003) sediment at 875S and 3000S, respectively. This makes the chromatin oligomers the largest biological assemblies yet to be characterized by sedimentation velocity AUC.

It is always preferable to combine sedimentation velocity data with information obtained using complementary techniques. Toward this end, we have studied the chromatin oligomers using fluorescence microscopy, transmission electron microscopy, and small angle X-ray scattering (Maeshima et al., 2015, manuscript in preparation). Collectively, our studies have revealed that the oligomers are globular, assembled from smaller globular

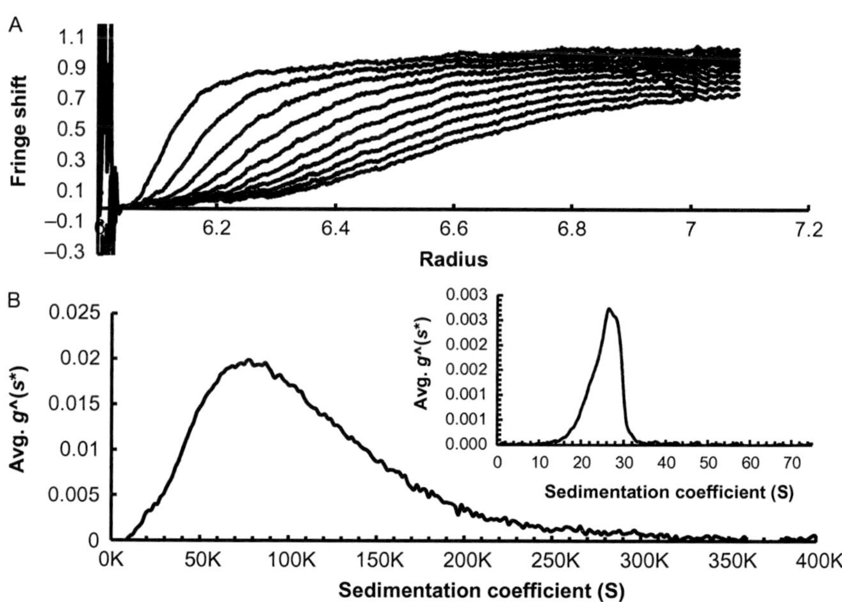

Figure 4 Sedimentation velocity experiments of self-associated chromatin fibers. This experiment used 601(207 bp) 12-mer DNA reconstituted with endogenous chicken octamers. (A) A typical scan set used for the analysis of chromatin oligomer sedimentation. (B) The $g^\wedge(s^*)$ distribution of sedimentation coefficients for a sample of chromatin oligomers in 8 mM Mg^{2+}. The inset shows the $g^\wedge(s^*)$ distribution for monomeric arrays in the absence of Mg^{2+}.

intermediates, and packaged as interdigitated 10 nm fibers. These physicochemical properties of the chromatin oligomers mimic the hierarchical organization of chromatin within an interphase chromosome (Bolzer et al., 2005; Joti et al., 2012; Lieberman-Aiden et al., 2009), indicating that future studies of chromatin fiber oligomerization using sedimentation velocity and the interference optics will yield much new molecular-based insight into how eukaryotic chromosomes are structured, assembled, and maintained.

9.2 Large Complexes in General

One of the major challenges facing biochemists and biophysicists is to develop ways to deal with sample complexity. One common type of complexity is compositional heterogeneity. Sedimentation velocity AUC allows analysis of complex mixtures due to the availability of data analysis methods that determine sedimentation coefficient distributions (Demeler & van Holde, 2004). Another type of complexity is sample size. The line between biochemistry and cell biology is increasingly becoming blurred as it becomes possible to isolate or reconstitute supramolecular biological assemblies. Successful characterization of these large macromolecular complexes requires a technical means to quantitatively characterize the structure and assembly of complex samples that sediment in excess of 5000S. Our studies of chromatin oligomers pave the way for general application of sedimentation velocity together with interference optics to the analysis of very large biological samples in heterogeneous mixtures.

REFERENCES

Allahverdi, A., Yang, R., Korolev, N., Fan, Y., Davey, C. A., Liu, C. F., et al. (2011). The effects of histone H4 tail acetylations on cation-induced chromatin folding and self-association. *Nucleic Acids Research*, *39*(5), 1680–1691. http://dx.doi.org/10.1093/nar/gkq900.

Ausio, J. (2000). Analytical ultracentrifugation for the analysis of chromatin structure. *Biophysical Chemistry*, *86*(2-3), 141–153.

Blacketer, M. J., Feely, S. J., & Shogren-Knaak, M. A. (2010). Nucleosome interactions and stability in an ordered nucleosome array model system. *The Journal of Biological Chemistry*, *285*(45), 34597–34607. http://dx.doi.org/10.1074/jbc.M110.140061.

Bolzer, A., Kreth, G., Solovei, I., Koehler, D., Saracoglu, K., Fauth, C., et al. (2005). Three-dimensional maps of all chromosomes in human male fibroblast nuclei and prometaphase rosettes. *PLoS Biology*, *3*(5), 0826–0842. http://dx.doi.org/10.1371/journal.pbio.0030157.

Brown, P. H., & Schuck, P. (2006). Macromolecular size-and-shape distributions by sedimentation velocity analytical ultracentrifugation. *Biophysical Journal*, *90*(12), 4651–4661. http://dx.doi.org/10.1529/biophysj.106.081372.

Chincholi, B., Havlik, A., & Vold, R. (1974). Specific refractive index increments of polymer systems at four wavelengths. *Journal of Chemical and Engineering Data*, *19*(2), 4–8. http://dx.doi.org/10.1021/je60061a006.

Correll, S. J., Schubert, M. H., & Grigoryev, S. A. (2012). Short nucleosome repeats impose rotational modulations on chromatin fibre folding. *The EMBO Journal*, *31*(10), 2416–2426. http://dx.doi.org/10.1038/emboj.2012.80.

Demeler, B. (2005). UltraScan: A comprehensive data analysis software package for analytical ultracentrifugation experiments. In D. J. Scott, S. E. Harding, & A. J. Rowe (Eds.), *Modern analytical ultracentrifugation: Techniques and methods* (pp. 210–230). UK: Royal Society of Chemistry

Demeler, B., & van Holde, K. E. (2004). Sedimentation velocity analysis of highly heterogeneous systems. *Analytical Biochemistry*, *335*(2), 279–288. http://dx.doi.org/10.1016/j.ab.2004.08.039.

Dorigo, B., Schalch, T., Bystricky, K., & Richmond, T. J. (2003). Chromatin fiber folding: Requirement for the histone H4 N-terminal tail. *Journal of Molecular Biology*, *327*, 85–96. http://dx.doi.org/10.1016/S0022-2836(03)00025-1.

Dubin, S. B., Benedek, G. B., Bancroft, F. C., & Freifelder, D. (1970). Molecular weights of coliphages and coliphage DNA. *Journal of Molecular Biology*, *54*, 547–556.

Fan, J. Y., Gordon, F., Luger, K., Hansen, J. C., & Tremethick, D. J. (2002). The essential histone variant H2A.Z regulates the equilibrium between different chromatin conformational states. *Nature Structural Biology*, *9*(3), 172–176. http://dx.doi.org/10.1038/nsb767.

Fierz, B., Chatterjee, C., McGinty, R. K., Bar-Dagan, M., Raleigh, D. P., Muir, T. W., et al. (2011). Histone H2B ubiquitylation disrupts local and higher-order chromatin compaction. *Nature Chemical Biology*, 7(2), 113–119. http://dx.doi.org/10.1038/nchembio.501.

Fletcher, T., Serwer, P., & Hansen, J. (1994). Quantitative analysis of macromolecular conformational changes using agarose gel electrophoresis: Application to chromatin folding. *Biochemistry*, *33*(36), 10859–10863. Retrieved from: http://pubs.acs.org/doi/abs/10.1021/bi00202a002.

Hansen, J. C., Ausio, J., Stanik, V. H., & van Holde, K. E. (1989). Homogeneous reconstituted oligonucleosomes, evidence for salt-dependent folding in the absence of histone H1. *Biochemistry*, *28*(23), 9129–9136. http://dx.doi.org/10.1021/bi00449a026.

Hansen, J. C., & Lohr, D. (1993). Assembly and structural properties of subsaturated chromatin arrays. *The Journal of Biological Chemistry*, *268*(8), 5840–5848. Retrieved from: http://www.ncbi.nlm.nih.gov/pubmed/8449950.

Huynh, V. A. T., Robinson, P. J. J., & Rhodes, D. (2005). A method for the in vitro reconstitution of a defined "30 nm" chromatin fibre containing stoichiometric amounts of the linker histone. *Journal of Molecular Biology*, *345*(5), 957–968. http://dx.doi.org/10.1016/j.jmb.2004.10.075.

Joti, Y., Hikima, T., Nishino, Y., Kamada, F., Hihara, S., Takata, H., et al. (2012). Chromosomes without a 30-nm chromatin fiber. *Nucleus (Austin, Tex.)*, *3*(5), 404–410. http://dx.doi.org/10.4161/nucl.21222.

Korolev, N., Allahverdi, A., Yang, Y., Fan, Y., Lyubartsev, A. P., & Nordenskiöld, L. (2010). Electrostatic origin of salt-induced nucleosome array compaction. *Biophysical Journal*, *99*(6), 1896–1905. http://dx.doi.org/10.1016/j.bpj.2010.07.017.

Lieberman-Aiden, E., van Berkum, N. L., Williams, L., Imakaev, M., Ragoczy, T., Telling, A., et al. (2009). Comprehensive mapping of long-range interactions reveals folding principles of the human genome. *Science (New York, N.Y.)*, *326*(5950), 289–293. http://dx.doi.org/10.1126/science.1181369.

Lu, X., & Hansen, J. C. (2004). Identification of specific functional subdomains within the linker histone H10 C-terminal domain. *The Journal of Biological Chemistry*, *279*(10), 8701–8707. http://dx.doi.org/10.1074/jbc.M311348200.

MacRaild, C. A., Hatters, D. M., Lawrence, L. J., & Howlett, G. J. (2003). Sedimentation velocity analysis of flexible macromolecules: Self-association and tangling of amyloid fibrils. *Biophysical Journal*, *84*(4), 2562–2569. http://dx.doi.org/10.1016/S0006-3495(03)75061-9.

Maeshima, K., Rogge, R., Joti, Y., Hikima, T., Tamura, S., & Szerlong, H., et al. (2015). Nucleosomal arrays self-assemble into hierarchical globular structures lacking 30 nm fibers. Manuscript in Preparation.

Muthurajan, U. M., McBryant, S. J., Lu, X., Hansen, J. C., & Luger, K. (2011). The linker region of macroH2A promotes self-association of nucleosomal arrays. *The Journal of Biological Chemistry*, *286*(27), 23852–23864. http://dx.doi.org/10.1074/jbc.M111.244871.

Rogge, R. A., Kalashnikova, A. A., Muthurajan, U. M., Porter-Goff, M. E., Luger, K., & Hansen, J. C. (2013). Assembly of nucleosomal arrays from recombinant core histones and nucleosome positioning DNA. *Journal of Visualized Experiments*, (79), e50354. http://dx.doi.org/10.3791/50354.

Schuck, P., & Demeler, B. (1999). Direct sedimentation analysis of interference optical data in analytical ultracentrifugation. *Biophysical Journal*, *76*(4), 2288–2296. http://dx.doi.org/10.1016/S0006-3495(99)77384-4.

Schwarz, P. M., Felthauser, A., Fletcher, T. M., & Hansen, J. C. (1996). Reversible oligonucleosome self-association: Dependence on divalent cations and core histone tail domains. *Biochemistry*, *35*(13), 4009–4015. http://dx.doi.org/10.1021/bi9525684.

Stafford, W. F. (1992). Boundary analysis in sedimentation transport experiments: A procedure for obtaining sedimentation coefficient distributions using the time derivative of the concentration profile. *Analytical Biochemistry*, *203*, 295–301. http://dx.doi.org/10.1016/0003-2697(92)90316-Y.

Strick, R., Strissel, P. L., Gavrilov, K., & Levi-Setti, R. (2001). Cation-chromatin binding as shown by ion microscopy is essential for the structural integrity of chromosomes. *The Journal of Cell Biology*, *155*(6), 899–910. http://dx.doi.org/10.1083/jcb.200105026.

Tumolo, T., Angnes, L., & Baptista, M. S. (2004). Determination of the refractive index increment (dn/dc) of molecule and macromolecule solutions by surface plasmon resonance. *Analytical Biochemistry*, *333*, 273–279. http://dx.doi.org/10.1016/j.ab.2004.06.010.

van Holde, K., & Weischet, W. (1978). Boundary analysis of sedimentation velocity experiments with monodisperse and paucidisperse solutes. *Biopolymers*, *17*(6), 1387–1403. http://dx.doi.org/10.1002/bip.1978.360170602.

Zhao, H., Brown, P. H., & Schuck, P. (2011). On the distribution of protein refractive index increments. *Biophysical Journal*, *100*(9), 2309–2317. http://dx.doi.org/10.1016/j.bpj.2011.03.004.

CHAPTER SEVENTEEN

Dissecting Steroid Receptor Function by Analytical Ultracentrifugation

David L. Bain*[,1], Rolando W. De Angelis*, Keith D. Connaghan*, Qin Yang*, Gregory D. Degala[†], James R. Lambert[†]

*Department of Pharmaceutical Sciences, University of Colorado Anschutz Medical Campus, Aurora, Colorado, USA

[†]Department of Pathology, University of Colorado Anschutz Medical Campus, Aurora, Colorado, USA

[1]Corresponding author: e-mail address: david.bain@ucdenver.edu

Contents

Abstract

Steroid receptors comprise a family of ligand-activated transcription factors. The members include the androgen receptor (AR), estrogen receptor (ER), glucocorticoid receptor (GR), mineralocorticoid receptor (MR), and progesterone receptor (PR). Each receptor controls distinct sets of genes associated with development, metabolism, and homeostasis. Although a qualitative understanding of how individual receptors mediate gene expression has come into focus, quantitative insight remains less clear. As a step toward delineating the physical mechanisms by which individual receptors activate their target genes, we have carried out a systematic dissection of receptor interaction energetics with their multisite regulatory elements. Analytical ultracentrifugation (AUC) has proved indispensable in these studies, in part by revealing the energetics of receptor self-association and its thermodynamic coupling to DNA binding. Here, we discuss these findings in the context of understanding specificity of receptor-mediated gene control. We first highlight the role of sedimentation velocity and sedimentation equilibrium in addressing receptor assembly state, and present a comparative analysis across the

Methods in Enzymology, Volume 562
ISSN 0076-6879
http://dx.doi.org/10.1016/bs.mie.2015.04.005

receptor family. We then use these results for understanding how receptors assemble at multisite regulatory elements, and hypothesize how these findings might play a role in receptor-specific gene regulation. Finally, we examine receptor behavior in a cellular context, with a view toward linking our *in vitro* studies with *in vivo* function.

1. INTRODUCTION

Steroid receptors comprise an evolutionarily conserved family of ligand-activated transcription factors (Tsai & O'Malley, 1994). The members include the androgen receptor (AR), estrogen receptor (ER), glucocorticoid receptor (GR), mineralocorticoid receptor (MR), and progesterone receptor (PR). As shown in Fig. 1A, phylogenetic studies demonstrate that

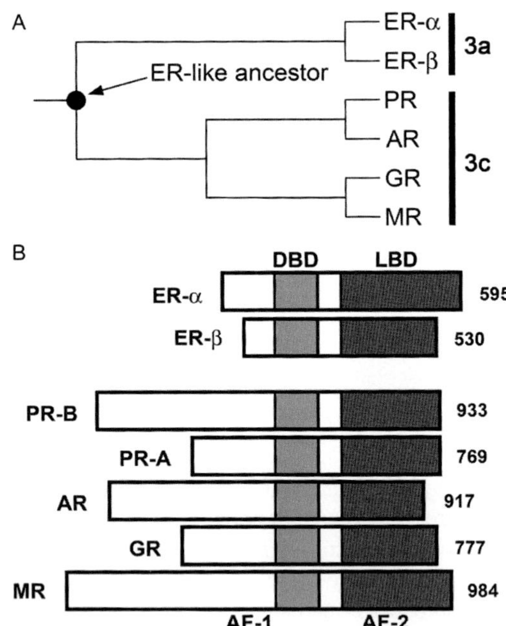

Figure 1 Phylogenetics and modular structure of the steroid receptor family. (A) Phylogenetic tree displaying steroid receptor evolution. The filled circle represents the node of an ER-like common ancestor for subfamilies 3A (ER-α and ER-β) and 3C (PR, AR, GR, and MR). Because the two PR isoforms (PR-A and PR-B) are transcribed from the same gene, they are not shown. (B) Schematic representing the modular structure and number of amino acids of the steroid receptor family members. Functional domains are labeled as DBD, DNA-binding domain; LBD, ligand-binding domain; AF, activation functions are encoded in both the N-terminal region and the LBD.

all receptors descend from a common ER-like ancestor, with AR, GR, PR, and MR forming subgroup 3C, and ER forming the more distantly related subgroup 3A (Nuclear Receptors Nomenclature Committee, 1999; Thornton, 2001). ER exists naturally as two functionally distinct isoforms (ER-α and ER-β), as does PR (PR-A and PR-B). All receptors share a common modular structure, containing a centrally located and highly conserved DNA-binding domain, a C-terminal hormone-binding domain (HBD), and an N-terminal intrinsically disordered region (Fig. 1B). Activation functions are located within both the N- and C-terminal sequences.

The traditional model of receptor function is that upon binding their steroidal ligand, receptor dimers assemble at hormone-response elements (HREs) located within target promoters and enhancers. Receptor–DNA assembly is coupled to chromatin remodeling, recruitment of coactivating proteins, and initiation of transcription via RNA polymerase II. Due to their homology, receptors bind largely identical response elements, yet activate largely distinct gene networks (Monroe et al., 2003; Richer et al., 2002; Wan & Nordeen, 2002). Although we have a qualitative understanding for how receptors generate this specificity of gene control, a quantitative framework—one capable of predictive insight—is still lacking.

Lack of insight is due in part to the great complexity associated with receptor function. As already noted, transcriptional activation occurs in multiple steps, with at least a subset of reactions being allosterically coupled (Kumar & McEwan, 2012; Métivier et al., 2003; Stavreva, Varticovski, & Hager, 2012; Vicent et al., 2010). Moreover, activation is associated with scores of coregulatory proteins and with interactions occurring on the second to minute timescale (Hager, McNally, & Misteli, 2009). Such complexity raises the question of whether reductionist approaches, proven successful in describing simpler regulatory switches (Segal & Widom, 2009; Shea & Ackers, 1985), have a place in deciphering higher-order gene control. Here, we discuss efforts in using reductionist thinking to understand complex systems, with a focus on identifying and characterizing the primary forces responsible for receptor-specific transcriptional activation.

As a step toward identifying mechanisms of receptor-specific gene control, we have carried out a systematic dissection of receptor–promoter interactions under a single "standard-state" condition (Connaghan-Jones, Heneghan, Miura, & Bain, 2006, 2007; De Angelis, Yang, Miura, & Bain, 2013; Heneghan, Berton, Miura, & Bain, 2005; Heneghan, Connaghan-Jones, Miura, & Bain, 2006; Moody, Miura, Connaghan, & Bain, 2012; Robblee, Miura, & Bain, 2012). Here, we present a summary of these studies, focusing on the role of analytical ultracentrifugation (AUC). We first highlight the role of sedimentation velocity and sedimentation equilibrium in addressing receptor assembly state and then present a comparative analysis across the receptor family. We next use these results for understanding how receptors assemble at multisite regulatory elements, and hypothesize how these findings might play a role in receptor-specific gene regulation. Finally, we examine receptor behavior in a cellular context, with a view toward linking our *in vitro* studies with *in vivo* function.

2. AUC ANALYSIS OF STEROID RECEPTOR SELF-ASSEMBLY

Rigorous characterization of receptor–promoter interactions first requires highly purified protein. Toward this end, we developed methods for generating milligram amounts of highly pure, full-length human steroid receptors (as His-tagged, FLAG-tagged, or untagged receptors). We then subjected them to detailed sedimentation velocity and sedimentation equilibrium studies to determine receptor assembly energetics. These studies have been carried out for nearly all of the steroid receptors; we present the highlights of our more recent work below.

2.1 Homologous Steroid Receptors Display a Vast Range of Self-Assembly Energetics

Our early work on receptor self-assembly energetics was carried out on the two PR isoforms, PR-A and PR-B (Connaghan-Jones et al., 2006; Heneghan et al., 2005). This work is discussed in detail in an earlier volume of this series (Connaghan-Jones & Bain, 2009), and so we will only summarize it here. Briefly, we found that PR isoforms indeed underwent reversible dimerization, consistent with the traditional model noted in Section 1. However, the energetics of dimerization were significantly weaker than those estimated by semiquantitative studies. For example, early analyses of PR dimerization suggested nanomolar or stronger dimerization affinities (DeMarzo, Beck, Onate, & Edwards, 1991; Rodriguez, Weigel, O'Malley, & Schrader, 1990; Skafar, 1991), whereas our sedimentation velocity and sedimentation equilibrium results independently confirmed that receptor dimerization takes place only with micromolar affinity ($\Delta G_{\text{dim}} = -7$ kcal/mol). A key prediction of such weak dimerization is that at the low nanomolar concentrations of receptor thought to exist in cells, essentially all PRs must be in a monomeric state. Consistent with this, we also demonstrated that PR monomers assemble at isolated half-sites at natural promoters, with stabilization provided by longer-range cooperative interactions between sites (Connaghan-Jones, Heneghan, Miura, & Bain, 2008). Furthermore, simulations suggested that weak dimerization and cooperativity might play a role in specificity of receptor–promoter interactions (Connaghan-Jones et al., 2007; Robblee et al., 2012).

2.2 GR and AR Show No Evidence of Reversible Dimerization

Our unexpected results for the PR isoforms prompted us to examine the self-assembly energetics of the remaining steroid receptors. We initially carried out this work on GR and AR (De Angelis et al., 2013; Robblee et al., 2012), two receptors closely related to the PR isoforms. Shown in Fig. 2A are representative sedimentation velocity data of full-length human GR, with the solid lines representing the best fit by *c*(*s*) analysis. Like our work on PR, these studies were carried out over a range of GR concentrations to detect the presence of reversible assembly. As shown in Fig. 2B, the majority of GR sediments as a single peak with a sedimentation coefficient ($s_{20,w}$) of 4.1 *S* regardless of concentration. This result is strongly indicative of a single stoichiometric state. Surprisingly, however, a *c*(*M*) analysis returned a molecular mass of 80.4 kDa, suggesting that the 4.1 *S* peak represents a GR monomer rather than the expected dimer. (The calculated mass of the GR monomer is 90,925 Da.) Although we do observe a minor peak at 5.7 *S* consistent with a GR dimer (returning a mass of 159 kDa), the percentage of this species is invariant with receptor concentration, indicating

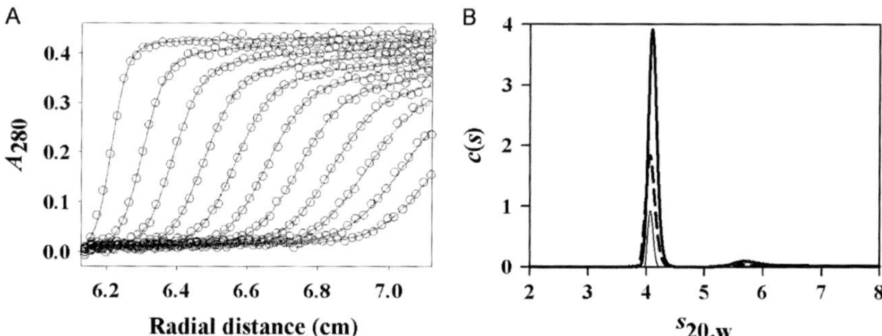

Figure 2 Sedimentation velocity analysis of GR. (A) Sedimentation velocity data for 5 μ*M* GR. Open circles represent absorbance data collected at 230 nm as a function of radial distance and time. For clarity, only every seventh data point is displayed. Continuous lines represent the best fit from *c*(*s*) analysis as implemented in the program Sedfit. (B) *c*(*s*) distributions for three concentrations of GR: 10.0 (thick line), 5.0 (broken line), and 2.0 μ*M* (thin line).

that it is irreversibly formed—it is not in equilibrium with a monomer population. Thus, contrary to early studies indicating that GR reversibly dimerizes with nanomolar affinity (Perlmann, Eriksson, & Wrange, 1990; Segard-Maurel et al., 1996), we find that the liganded receptor is effectively monomeric up to and above micromolar concentrations, showing no evidence of reversible dimerization.

Noting that our findings on GR conflicted with earlier reports, we used sedimentation equilibrium to directly measure the masses of the putative GR monomer and (irreversible) dimer. We carried out sedimentation equilibrium studies at multiple GR concentrations and rotor speeds, and under buffer conditions identical to those of the velocity studies. Because we observed a noninteracting, dimer species in the velocity studies, we globally fit the equilibrium data to a model allowing for two noninteracting species. Shown in Fig. 3 are the results of that analysis. The noninteracting, two-species model describes well all the data by both visual inspection and fitting statistics (Schuster & Laue, 1994). Moreover, the resolved mass of the first species is 88.0 ± 2.9 kDa, within 3% of the calculated mass of the GR monomer, and the stoichiometry of the second species is 2.0 ± 0.4, indicating that it indeed represents a GR dimer. By contrast, global fitting of the data to either a single-species model or a monomer–dimer equilibrium model resulted in statistically poorer fits and nonrandom residuals (not shown). Thus, both the sedimentation velocity and equilibrium studies demonstrate that GR is overwhelmingly monomeric with only a small population of irreversibly formed dimer. Moreover, the concordance between the sedimentation velocity and sedimentation equilibrium results indicates that GR is structurally homogenous, lending credence to our DNA-binding studies presented in Section 3.

We next examined full-length human AR. Shown in Fig. 4A are representative sedimentation velocity scans for wild-type (wt) AR. Shown in Fig. 4B are the corresponding sedimentation coefficient distributions, determined over a range of protein concentrations, for wt AR and a point mutant implicated in prostate cancer, T877A. Comparable to our findings for GR, both AR proteins sediment predominantly as single species regardless of concentration, with a buffer and temperature-corrected sedimentation coefficient ($s_{20,w}$) of 4.8 *S*. Similarly, the absence of a concentration-dependent change in the sedimentation coefficient indicates that neither receptor undergoes reversible self-association over this concentration range. With regard to assembly state, $c(M)$ analysis of the 4.8 *S* peak yields an average molecular mass of 94 kDa for wt AR and 84 kDa for T877A, suggesting that the proteins are monomeric up to micromolar concentrations. (Human AR

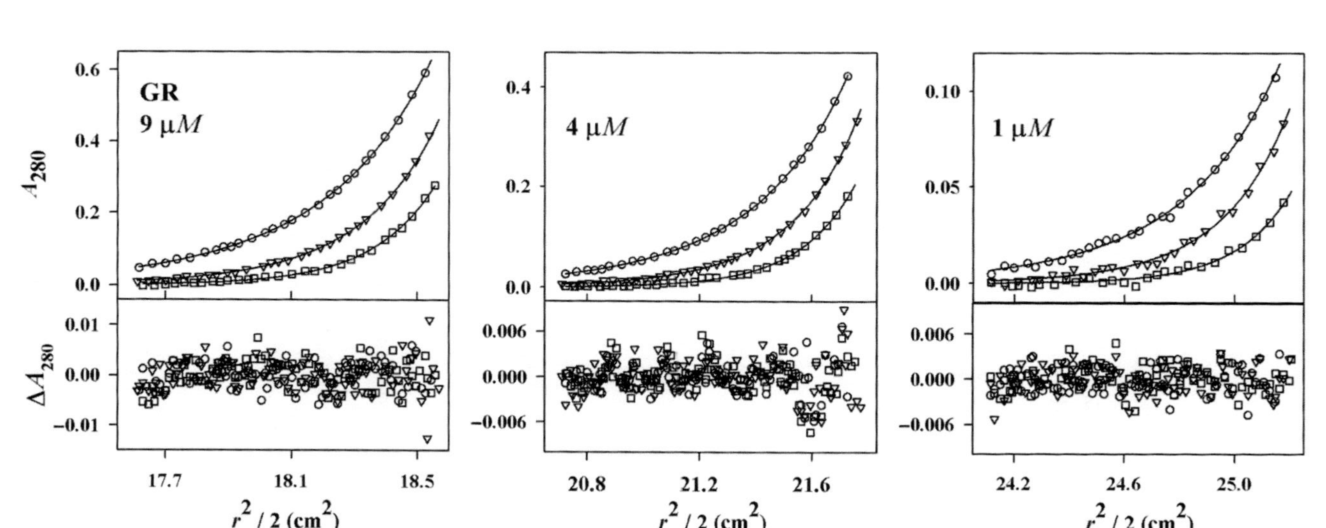

Figure 3 Sedimentation equilibrium (SE) analysis of GR plotted as absorbance at 280 nm versus $r^2/2$ for three different GR loading concentrations. From left to right: 9.3, 4.2, and 0.9 μ*M*. Symbols represent GR absorbance at three rotor speeds: 15,000 (open circles), 18,000 (inverted triangles), and 21,000 rpm (open squares). For clarity, only every third data point is displayed. Solid lines represent best global fit to a two-species noninteracting model. Standard deviation of global fit was 0.0024 absorbance units. Residuals are plotted below the data and best-fit lines (all data points are displayed).

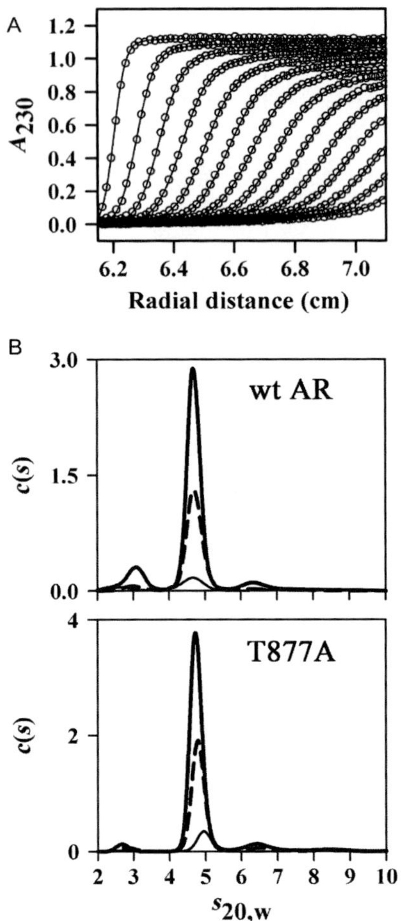

Figure 4 Sedimentation velocity analysis of wt AR and T877A. (A) Sedimentation velocity data for 1.4 μ*M* wt AR. Open circles represent absorbance data collected at 230 nm as a function of radial distance and time. Continuous lines represent the best fit from *c*(*s*) analysis as implemented in the program Sedfit. For clarity, only every seventh data point is displayed. (B) *c*(*s*) distributions for three concentrations of wt AR and T877A: 1.4 (thick line), 0.7 (broken line), and 0.14 μ*M* (thin line).

has a calculated molecular mass of 104 kDa.) However, we also observed two minor peaks, one at ~2.8 *S* and another at 6.5 *S*. Each peak represents ~5% of the total *c*(*s*) distribution regardless of receptor concentration, indicating that they represent irreversibly formed conformers or aggregates. The 2.8 *S* species undoubtedly reflects a misfolded or partially unfolded monomer, since simple hydrodynamic calculations show that a fully unfolded, random coil AR dimer could sediment no slower than 3.4 *S*. The 6.5 *S* species must represent a higher-order assembly state since it sediments too quickly to represent an AR monomer, regardless of conformation. We note that the presence of these species, even at such a low amount, causes the underestimate in molecular mass for the putative 4.8 *S* monomer (Philo, 2000; Schuck, 2000).

As we did for PR and GR, we also carried out sedimentation equilibrium studies of AR under conditions identical with those of sedimentation velocity. Shown in Fig. 5A are sedimentation equilibrium scans for wt AR carried out at three concentrations and three rotor speeds. Analogous scans for T877A are shown in Fig. 5B. Based on the sedimentation velocity results indicating the presence of an irreversible 6.5 *S* aggregate, we globally fit the data to a noninteracting, two-species model as we did for GR. (It is unnecessary to account for the 2.8 *S* partially unfolded monomer since, being the same mass as the 4.8 *S* monomer, it would be invisible by sedimentation equilibrium.) As seen by the lines through the data and random residuals, this model readily described all data for both receptors. Moreover, for wt AR, the fit resolved a species 1 molecular mass of 104 ± 1 kDa indicative of an AR monomer, and a stoichiometry for species 2 of 3.0 ± 0.1, indicative of AR trimer. An essentially identical result was seen for T877A, resolving a species 1 molecular mass of 104 ± 1 kDa and a stoichiometry of 2.6 ± 0.2. Attempts to fit the data to either a simple monomer model or monomer–N-mer equilibrium models resulted in poorer fits, akin to our analyses of GR. Thus like GR, AR exists only as a monomer with a small amount of irreversibly formed higher-order species—in this case, trimer. Contrary to dogma, we find no evidence of AR dimerization. These results were confirmed over a range of buffer conditions.

Our results are at odds with earlier studies indicating that AR and GR reversibly dimerize with high affinity (Liao, Zhou, & Wilson, 1999; Perlmann et al., 1990; Segard-Maurel et al., 1996). Why this should be the case is unclear. However, we note that these studies were necessarily constrained by the use of partially purified receptor preparations and/or semiquantitative analytical approaches. In the case of crystallographic work showing a dimeric GR HBD (Bledsoe et al., 2002), dimerization may have been induced by the high concentration of protein, crystal packing, or nonphysiological buffer conditions. Finally, we note that for both AR and GR, we find that insufficient reducing conditions lead to enhanced formation of polydisperse aggregates—including irreversibly formed dimers and trimers. The observation that dimers and trimer appear in conjunction with widespread receptor aggregation suggests that such species may be functionally irrelevant.

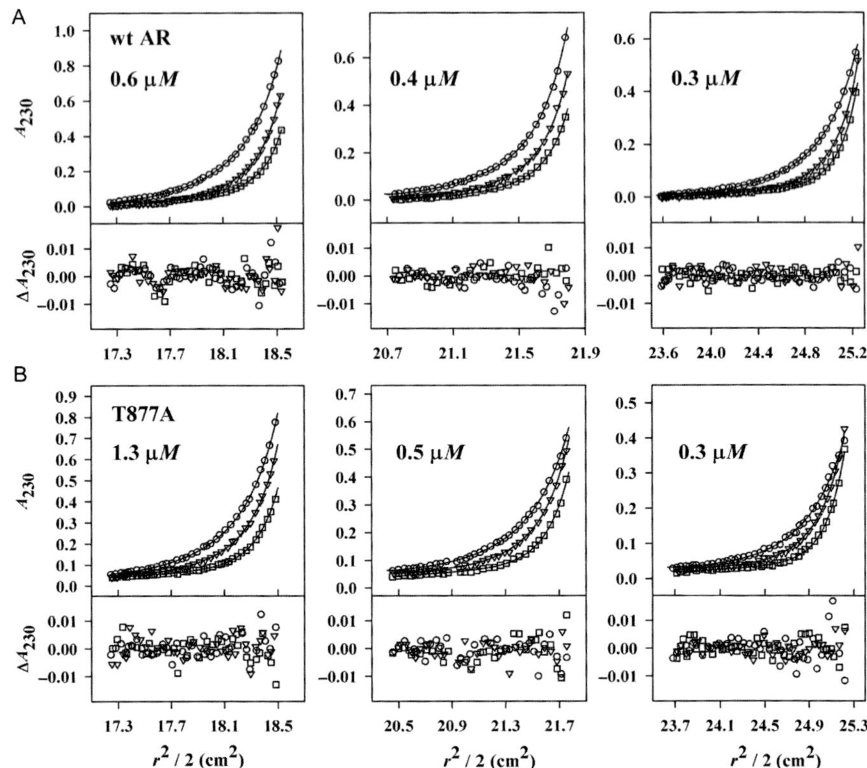

Figure 5 Sedimentation equilibrium analysis of wt AR and T877A. (A) wt AR sedimentation equilibrium data plotted as absorbance (230 nm) versus $r^2/2$ for three different wt AR loading concentrations. From left to right: 0.6, 0.4, and 0.3 μ*M*. Symbols represent wt AR absorbance at three rotor speeds: 14,000 (open circles), 17,000 (inverted triangles), and 21,000 rpm (open squares). Continuous lines represent best global fit to a two-species noninteracting model. Standard deviation of this global fit was 0.0039 absorbance units. Residuals are plotted below the data and best-fit lines. For clarity, only every third data point is displayed. (B) T877A sedimentation equilibrium data plotted as absorbance (230 nm) versus $r^2/2$ for three different T877A loading concentrations. From left to right: 1.3, 0.5, and 0.3 μ*M*. Symbols corresponding rotor speeds and model are identical to those described above. The standard deviation of the global fit was 0.0043 absorbance units. Residuals are plotted below the data and best-fit lines. For clarity, only every third data point is displayed.

2.3 ER-α Is the Only Receptor to Date That Exhibits Strong Self-Association

Our work on AR, GR, and the PR isoforms revealed that the receptors show little evidence of the strong dimerization predicted by the traditional model of receptor function. However, our work on ER-α proved to be the

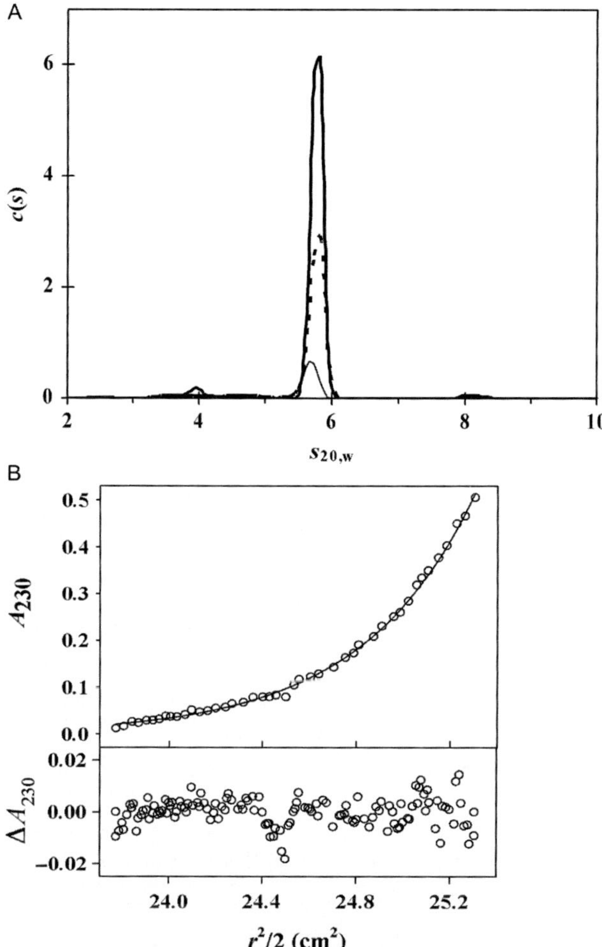

Figure 6 Sedimentation velocity and equilibrium analyses of ER-α. (A) *c*(*s*) distributions for three concentrations of ER-α: 1.7 (thick line), 1.0 (broken line), and 0.3 μ*M* (thin line). (B) Sedimentation equilibrium data plotted as absorbance (230 nm) versus $r^2/2$. ER-α loading concentration was 0.5 μ*M*. Open circles represent ER-α absorbance at a 10,800 rpm rotor speed. For clarity, only every third data point is displayed. Continuous lines represent best fit to a single-species model. Standard deviation of the fit was 0.0054 absorbance units. Residuals are plotted below the data and best-fit line (every data point is displayed).

exception to this rule. Shown in Fig. 6A are sedimentation coefficient distributions for full-length ER-α again determined at multiple concentrations. Like AR and GR, ER-α sediments as a single species regardless of protein concentration. However, in sharp contrast to these receptors, a *c*(*M*) analysis of the 5.7 s peak resolved a molecular weight of 124 kDa, close to the predicted molecular weight of a ER-α dimer (141,079 Da). These results indicate that ER-α exists as single, nondissociating species with a molecular weight consistent with that of dimer.

In order to rigorously confirm the assembly state of His-ER-α, we carried out sedimentation equilibrium studies, again as a function of protein concentration. However, empirical studies revealed that after 20 h, ER-α undergoes irreversible aggregation at concentrations greater than 1 μ*M*. We were therefore only able to collect useful data at 0.5 μ*M* His-ER-α. As shown in Fig. 6B, the data were fit to a single-species model, resolving a molecular mass of 139.0 ± 9.8 kDa. This estimate is in excellent agreement with the calculated mass of the dimer. Noting that ER-α sediments as a dimer even at the lowest protein concentration analyzed (0.2 μ*M*, or the limit of detection by absorbance optics), and that sedimentation techniques are sensitive enough to detect 10% monomer, a putative dimerization dissociation constant must be 20 n*M* (−9.9 kcal/mol) or tighter. In fact, our DNA-binding studies allowed an indirect estimate of ER-α dimerization, placing it at 0.4 n*M* (−12 kcal/mol). To put this in perspective, if GR is able to reversibly dimerize, simple calculations place the dimerization constant as no stronger than 100 μ*M*, or over five orders of magnitude weaker than ER-α. The potential functional implications of such large differences for receptor–promoter interactions are explored in more detail below.

3. INTEGRATING AUC RESULTS INTO UNDERSTANDING RECEPTOR–PROMOTER INTERACTIONS

The above sedimentation studies revealed that closely related steroid receptors maintain large differences in dimerization energetics (ΔG_{dim} in Fig. 7A). In parallel with these studies, we had also dissected the energetics of receptor–promoter interactions, finding that steroid receptors also exhibit large difference in cooperative binding energetics (ΔG_c in Fig. 7B) (Connaghan-Jones et al., 2007; De Angelis et al., 2013; Moody et al., 2012; Robblee et al., 2012). For example, ER-α displays negligible cooperativity ($\Delta G_c = -0.2$ kcal/mol) on the two-site promoter in Fig. 7 (HRE_2), whereas as PR, GR, and AR exhibit cooperativity up to −4 kcal/mol, corresponding to a 1000-fold enhancement in receptor–promoter stability. Intriguingly, the differences in dimerization and cooperativity among receptors parallel the phylogenetic divergence of the receptor family: Class 3C receptors such as AR, GR, and PR exhibit weak or

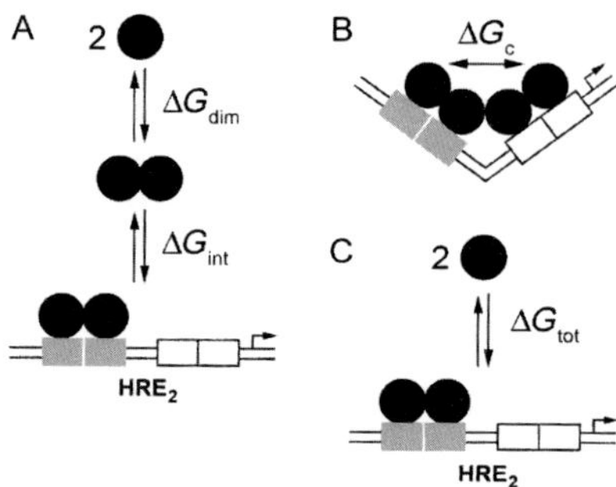

Figure 7 Schematic representing steroid receptor–DNA interactions at the HRE_2 promoter. (A) Representative dimer-binding pathway for receptor HRE_2 assembly. Monomers form dimers (black circles) with free energy ΔG_{dim}. Preformed dimers bind response elements with free energy ΔG_{int}. (B) Potential intersite cooperative free energy between two DNA-bound receptor dimers is represented by ΔG_c. (C) Schematic representing the overall reaction for two receptor monomers binding at a response element, regardless of pathway. Total free energy is represented by ΔG_{tot}. Free energy changes are related to their interaction constants through the standard relationship, $\Delta G_i = -RT \ln k_i$.

nonexistent dimerization but strong cooperativity, whereas class 3A receptors such as ER-α exhibit the inverse—strong dimerization and little cooperativity. (Our unpublished work on MR supports this premise, finding that this receptor also shows little evidence for dimer assembly.)

These observations suggested to us that such large differences in receptor–promoter interaction energetics could play a role in triggering receptor-specific transcriptional regulation. We have specifically hypothesized that receptors view their promoter (and enhancer) binding sites across the genome (i.e., the cistrome) as a collection of affinity landscapes, with receptors preferentially selecting from this landscape via their unique energetic signatures. One prediction of this thinking is that receptors should differentially regulate their cooperative energetics as a function of different promoter architectures. To test this, we analyzed the cooperative binding energetics of ER-α, GR, and PR-B on an array of promoter layouts (Connaghan, Yang, Miura, Moody, & Bain, 2014).

Shown in Fig. 8 is a series of two-site promoters (HRE_2) with center-to-center distances ranging from 25 to 65 bp between response elements, or 85–221 Å of B-form DNA. These distances encompass up to 40 bp or almost four turns of the DNA helix. Spacing between sites was varied in 5-bp increments to examine the role of in-phase or out-of-phase receptor binding sites. Thermodynamic cooperativity, defined as the excess free

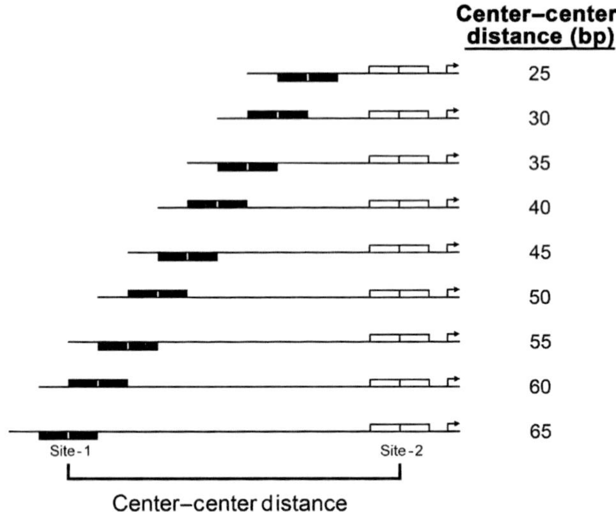

Figure 8 Schematic representing the architecture of the nine HRE_2 promoters used to determine steroid receptor intersite cooperativity. From top to bottom, the spacing between the two response elements increases by 5 bp between the center of site 1 (filled rectangle) and the center of site 2 (open rectangle). The center-to-center distance (in bp) between sites is listed on the right. Relative phasing of the two sites is indicated by the top-bottom (out-of-phase) or top-top (in-phase) orientation of sites 1 and 2. The arrow indicates the direction of transcription.

energy beyond the sum of the free energies for binding at the isolated sites (ΔG_c), was determined experimentally by carrying out quantitative footprint titrations of each HRE_2 promoter in conjunction with a reduced-valency template (HRE_{1-}) containing only a single binding site. The resultant individual site isotherms were globally analyzed to resolve ΔG_c as follows.

Because PR-B and ER-α undergo dimerization, the free receptor monomer concentration (x) is calculated using the experimentally determined dimerization constants (k_{dim}) and the appropriate conservation of mass equation. (Since GR does not dimerize in the absence of DNA, the free monomer concentration constitutes the total receptor concentration.) The fractional saturation ($\bar{Y}$) of PR-B and ER-α dimer binding to a single-site promoter is then:

$$\bar{Y}_{HRE_{1-}} = \frac{k_{dim} k_{int} x^2}{1 + k_{dim} k_{int} x^2} \tag{1}$$

where k_{dim} and x are as defined previously, and k_{int} is the intrinsic association constant for a preformed dimer binding to a HRE (depicted as ΔG_{int} in Fig. 7A).

Using the same theoretical approach, the equation describing the fractional saturation of an individual site at an HRE_2 promoter is:

$$\bar{Y}_{\mathrm{HRE}_2} = \frac{k_{\mathrm{dim}} k_{\mathrm{int}} x^2 + k_{\mathrm{dim}}^2 k_{\mathrm{int}}^2 k_c x^4}{1 + 2k_{\mathrm{dim}} k_{\mathrm{int}} x^2 + k_{\mathrm{dim}}^2 k_{\mathrm{int}}^2 k_c x^4} \quad (2)$$

where k_c corresponds to the intersite cooperativity. Because each HRE is identical in sequence, Eq. (2) also describes binding to site 2 of the promoter.

Because human GR does not reversibly form dimers, Eqs. (1) and (2) could not be used to analyze the data. Instead, the reduced-valency GR binding isotherms were first to the Hill equation to determine a Hill coefficient. This value was statistically identical to 2 for all GREs (2.2 ± 0.2), indicating that GR binding to the palindrome is highly cooperative and the singly ligated monomer GR–DNA species is not significantly populated (Robblee et al., 2012). Thus, the isotherms were fit to a simplified Adair equation to resolve K_{tot} the total affinity for two GR monomers to fully ligate a palindromic response element regardless of pathway (depicted as ΔG_{tot} in Fig. 7C):

$$\bar{Y}_{\mathrm{HRE}_{1-}} = \frac{K_{\mathrm{tot}} x^2}{1 + K_{\mathrm{tot}} x^2} \quad (3)$$

Using a similar approach, the HRE_2 promoter data were fit to:

$$\bar{Y}_{\mathrm{HRE}_2} = \frac{K_{\mathrm{tot}} x^2 + K_{\mathrm{tot}}^2 k_c x^4}{1 + 2K_{\mathrm{tot}} x^2 + K_{\mathrm{tot}}^2 k_c x^4} \quad (4)$$

where k_c again corresponds to GR-mediated intersite cooperativity (ΔG_c in Fig. 7B).

Shown in Fig. 9A is a footprint titration of GR assembly at an HRE_2 promoter containing a 35 bp center-to-center distance. Briefly, increasing concentrations of GR were equilibrated with the promoter and exposed to DNase under thermodynamically valid conditions (Brenowitz, Senear, Shea, & Ackers, 1986). Shown in Fig. 9B are the individual site-binding isotherms for this promoter and the HRE_{1-} template. The analogous results for PR-B and ER-α are shown in Fig. 9C and D. Visual inspection indicates large differences in intersite cooperativity among the receptors, as evidenced by an increase in apparent binding affinity and a steeper binding transition at

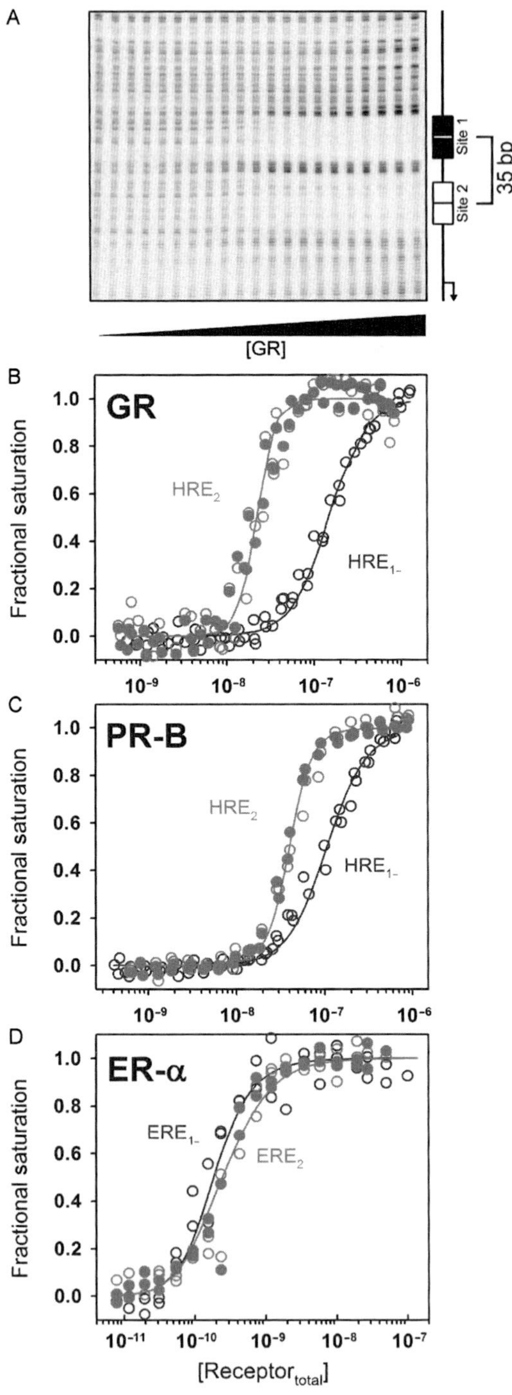
A
Site 1
Site 2
35 bp
[GR]
B
GR
HRE2
HRE1-
Fractional saturation
C
PR-B
HRE2
HRE1-
Fractional saturation
D
ER-α
ERE1-
ERE2
Fractional saturation
[Receptor_total]

the HRE_2 promoter relative to HRE_{1-}. For example, GR binding to the HRE_2 promoter shows an approximately ninefold increase in apparent binding affinity relative to HRE_{1-} (20 n*M* vs. 180 n*M*), and PR-B shows an approximately threefold increase (35 n*M* vs. 100 n*M*). Likewise, for both receptors, binding to the HRE_2 promoter generates steeper binding transitions. ER-α shows neither an increase in apparent binding affinity nor an increase in isotherm steepness, indicative of negligible cooperativity.

In actuality, the differences in cooperativity are much more dramatic than they appear. Fundamental theory demonstrates that cooperativity between identical binding sites will be visualized only as the square root of k_c (Ackers, Shea, & Smith, 1983). In addition, any linked reaction (e.g., PR dimerization in the absence of DNA) will also skew interpretation. Thus, although GR appears to exhibit a 10-fold cooperative enhancement, the actual value as determined by global analysis of the isotherms using fundamental theory is 1091-fold or −3.8 kcal/mol. This degree of cooperativity meets or exceeds that observed for highly cooperative classical systems such as cI repressor (Ackers, Johnson, & Shea, 1982). Using the same approach, PR-B cooperativity is found to be 41-fold or −2.0 kcal/mol. ER-α exhibits no cooperativity with a k_c of 0.5 or +0.3 kcal/mol. Thus, closely related steroid receptors are capable of enormous differences in cooperativity on identical promoter architectures.

Finally, we note that receptor–DNA-binding affinities are also misrepresented by visual inspection. For example, although the apparent affinity

Figure 9 Quantitative DNase footprint titrations of GR, PR-B, and ER-α at the 35 bp center-to-center HRE_2 promoter. (A) Representative autoradiogram of GR binding at the 35 bp center-to-center HRE_2 promoter. GR concentration increases from left to right as depicted by the solid triangle. Positions of site 1 (filled rectangle) and site 2 (open rectangle) are depicted on the right. (B) Individual site-binding isotherms constructed from analysis of GR footprint titrations at the HRE_{1-} and HRE_2 promoters. Filled and open circles represent binding to sites 1 and 2 of the HRE_2 promoter, respectively. Open squares represent binding to site 2 of the HRE_{1-} promoter. Continuous lines represent best global fit to all isotherms using Eqs. (3) and (4). Fit lines for both HRE_2 sites overlay because the sequences of sites 1 and 2 are identical. (C) Individual site-binding isotherms constructed from analysis of PR-B footprint titrations at the HRE_{1-} and HRE_2 promoters under the same buffer conditions as GR. Symbols corresponding to binding site and promoter are identical to those described above. Data were fit to Eqs. (1) and (2). (D) Analysis of ER-α binding under identical conditions, except that the two response elements now correspond to AGATCAcagTGACCT rather than TGTACAggaTGTTCT. Data were fit using Eqs. (1) and (2). Symbols are the same as above. (See the color plate.)

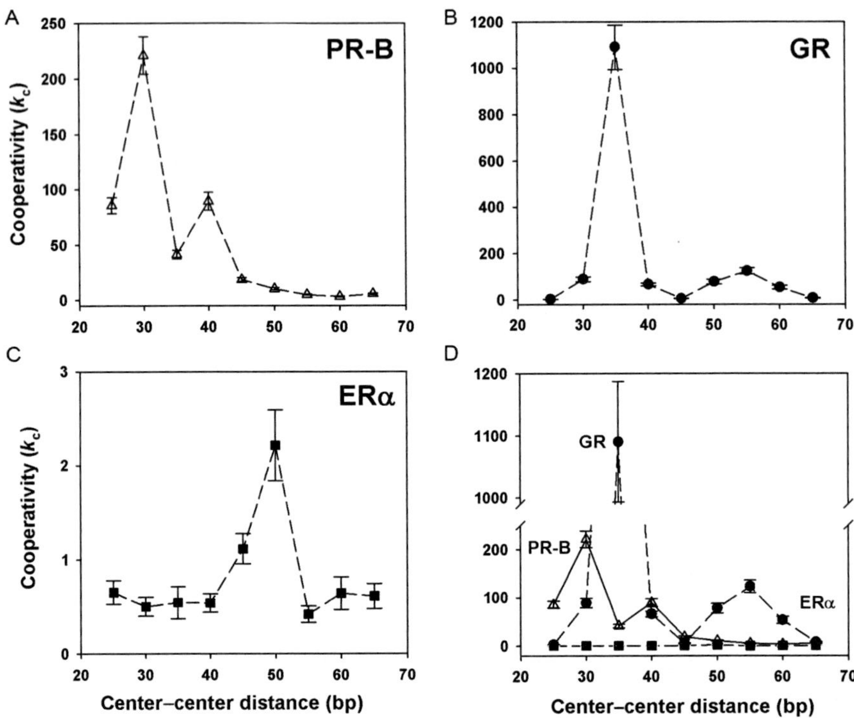

Figure 10 Intersite cooperativity as a function of promoter architecture for GR, PR-B, and ER-α. Resolved intersite cooperativity (k_c) as a function of center–center response element distance for (A) PR-B, (B) GR, and (C) ER-α. (D) Overlay plot of PR-B, GR, and ER-α. Error bars represent one standard deviation as determined by the program Scientist (Micromath). (See the color plate.)

of PR-B binding to a single response element is only of modest affinity (100 n*M*), the intrinsic affinity (ΔG_{int} in Fig. 7) is 2.3 n*M*, once the dimerization reaction is accounted for. This simple example speaks to the importance of explicitly dissecting each of the reactions linked to DNA binding.

Noting the large and receptor-specific differences in cooperativity we observe in Fig. 9, we repeated our studies on all the promoters in Fig. 8. The resolved free energy of cooperativity for PR-B, GR, and ER-α as a function of promoter layout is shown in Fig. 10A–C. It is evident that PR-B shows maximal cooperativity at promoters with closely spaced and in-phase binding sites (e.g., promoters with center–center distances of 30 and 40 bp). By contrast, GR cooperativity is maintained over greater distances between sites, is larger energetically, and shows a markedly different phase dependence, with cooperativity maximized only for two helical turns of DNA and when binding sites are out of phase. Finally, ER-α appears incapable of cooperativity regardless of promoter architecture. An overlay is shown in Fig. 10D.

Noting the observed cyclical patterns for GR and PR-B cooperativity that decay as the distance between sites increased, our results suggest that the structural basis of cooperativity involves direct protein–protein contacts. Although this interpretation does not exclude other contributions such as DNA bending, we note the persistence length of DNA under these conditions is ~150 bp or ~500 Å. Regardless, GR and PR-B clearly distribute their cooperative energetics differently on a range of different promoters. Structural work of intact receptors will be necessary to better understand the basis of these differences, particularly the unexpected two-turn dependency of GR, but the results could suggest that PR-B and GR cooperativity occur via different mechanisms.

What might be a functional role for receptor-specific differences in cooperativity? Shown in Fig. 11 are simulations of GR and PR-B binding to each of the promoters in Fig. 8 but under conditions in which both receptors are competing for promoter binding. The microstate energetics determined from our AUC and footprinting studies (k_{dim}, k_{int}, k_c) were used to calculate the probability of the fully ligated promoter for each receptor (i.e., the presumed transcriptionally active species). Thus, even though the receptors bind with similar apparent affinities to isolated response elements and have structurally identical DNA-binding domains, differences in cooperativity between the receptors predict receptor-specific and promoter-specific occupancy. For example, promoters with spacing of 35 or 55 bp between response elements afford preferential occupancy by GR, whereas promoters with spacing of 25 or 30 bp result in PR-B dominance. Promoters with 50 or 60 bp spacing between sites allow comparable occupancy. For reference, the average estimate of intracellular receptor concentration is shown as a shaded bar.

The simulations thus suggest that by simply manipulating promoter architecture, it might be possible to drive either preferential or dual receptor–promoter occupancy in the cell. Because receptors such as GR and PR regulate distinct but overlapping gene networks, differences in cooperative energetics may serve as a discriminator for generating receptor-specific transcriptional control. For cases in which cooperativity is minimal (e.g., a promoter with 65 bp spacing), neither receptor is capable of significant occupancy at the intracellular receptor concentration, again suggestive of a critical role for cooperative interactions. Experimental

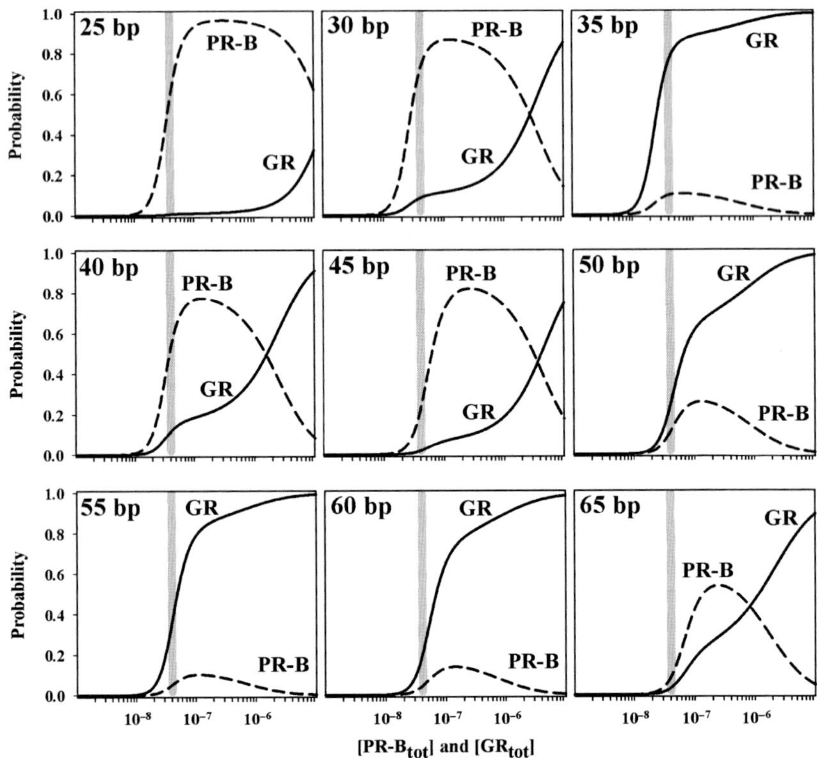

Figure 11 Simulated probabilities of GR and PR-B occupancy at the nine promoters under competitive binding conditions. Plots represent probability of the fully ligated state for GR (continuous line) and PR-B (broken line) assembly in the presence of equimolar concentrations of both receptors ranging from 1 n*M* to 10 μ*M*. Simulations were carried out with the energetics listed in Connaghan et al. (2014) and assuming no heterodimerization or heterocooperativity. Gray bar represents estimate of intracellular receptor concentration (40 n*M*) (Theofan & Notides, 1984). (See the color plate.)

studies are currently under way to test whether receptors indeed compete for promoter assembly *in situ*, both at natural and synthetic promoters.

In summary, we have found that subgroup 3C receptors (e.g., GR and PR-B) are capable of strong cooperativity, but weak or nonexistent dimerization. By contrast, the more distantly related ER-α exhibits no evidence of cooperativity regardless of promoter layout, and exceptionally strong dimerization. Why subgroup 3C receptors should not tightly dimerize would initially seem unclear, since self-association provides a significant advantage in DNA-binding strength and specificity. Noting, however, that these receptors derive from an ER-like ancestor, it may be the case that dimerization was dispensed with in order to gain cooperativity. As seen by the simulations in Fig. 11, the addition of cooperativity to the subgroup 3C repertoire suggests a powerful mechanism for triggering receptor- and promoter-specific occupancy.

If cooperativity and dimerization are somehow interconnected evolutionarily, then this could suggest that the C-terminal HBD, long known to be the primary mediator of receptor dimerization, mediates cooperativity. This possibility can be better appreciated by examining the overlaid distributions for ER-α, PR-B, and GR cooperativity (Fig. 10D). Not only is cooperativity different for each receptor as already discussed, but so too is the plasticity. For example, GR cooperativity shows the greatest absolute amplitude, the largest dynamic range, and covers the greatest range of distances between binding sites. PR-B is intermediate in all of the above, followed by ER-α. Interestingly, this rank order follows the rank ordering for receptor self-association: Under conditions identical to those used for measuring cooperativity, GR dimerization is undetectable, PR-B dimerization is weak (micromolar), and ER-α dimerization is strong (nanomolar). This inverse correlation between dimerization and cooperative energetics may indicate that self-association by the HBD serves as a quaternary constraint on cooperativity; release of this constraint by reducing dimerization energetics enhances the plasticity of cooperative interactions. Consistent with such thinking, we find that AR, which like GR shows no evidence for dimerization, exhibits strong GR-like cooperativity (De Angelis et al., 2013). Likewise, a receptor chimera with dimerization energetics similar to ER-α displays no cooperativity (Connaghan, Miura, Maluf, Lambert, & Bain, 2013). In summary, our observations suggest a mechanistic link between dimerization and cooperativity, possibly as means to generate more sophisticated gene control by subgroup 3C receptors.

4. IMPLICATIONS FOR RECEPTOR-MEDIATED TRANSCRIPTIONAL REGULATION

Although it is not yet possible to directly test whether dimerization or cooperativity energetics contribute directly to transcriptional activity, the broader question of whether DNA-binding energetics contribute has been examined by several groups (Geserick, Meyer, & Haendler, 2005; Klinge, 2001; Meijsing et al., 2009). Interestingly, these studies found only weak correlation between energetics (more specifically, receptor–DNA-binding affinity) and the extent of transcriptional activity. Although this type of analysis would seem to be straightforward, closer inspection reveals it to be problematic. This can best be seen by recognizing that receptor–DNA-binding affinity is determined by analyzing binding over a *range* of receptor concentrations, whereas transcriptional activity is determined at a *single* receptor concentration. A more appropriate analysis would therefore be to determine the extent of receptor-response element binding over a range of receptor concentrations and then compare the results to transcriptional activity measured over a range of receptor concentrations in the cell. In the case of the free energy measurements we have presented here, the "extent of receptor-response element binding" corresponds to the probability of receptor-response element occupancy, and the "transcriptional activity measured over a range of receptor concentrations in the cell" corresponds to a traditional dose–response curve.

Shown in Fig. 12 is a scatter plot representation of such an approach. Plotted on the *x*-axis is GR-mediated transcriptional activity determined *in vivo* (as a fold increase in activity), using 11 unique response elements and covering a 500-fold range of cellular receptor levels. Plotted on the *y*-axis is the expected result, based on the probability of *in vitro*

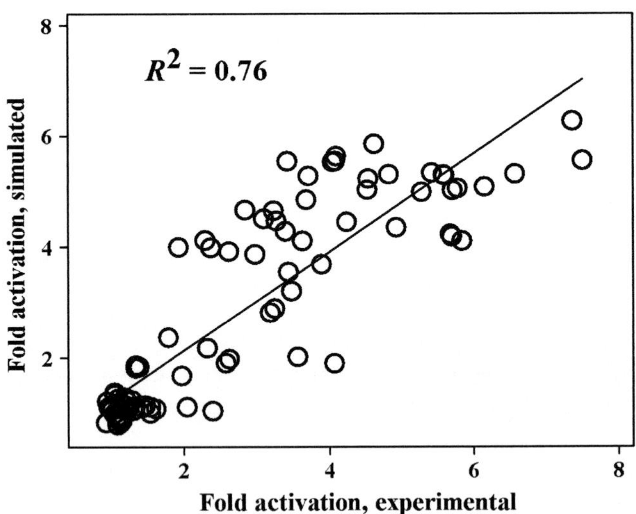

Figure 12 Scatter plot of experimental GR-mediated transcriptional activity versus simulated activity using GR–DNA interaction energetics. *x*-Axis represents experimentally measured activity (via transient transfection assay) using 11 unique hormone-response elements and a 500-fold range of GR expression levels. *y*-Axis represents simulated transcriptional activity based on the *in vitro* probability of GR–DNA binding and an affinity-based interaction model. Probabilities were normalized to experimental transcriptional activity using a linear scaling factor.

GR-response element binding (Bain et al., 2012). Although GR binding affinity toward the 11 response elements spans a nearly 700-fold range, binding probability is highly correlated with cellular transcriptional activity ($R^2=0.76$). We emphasize that if these data (or any subset) were analyzed using a traditional affinity-activity plot, little or no correlation would be observed. The results in Fig. 12 thus directly connect GR–DNA energetics measured in the test-tube with transcriptional activity in the cell and therefore suggest that binding energetics are a dominant contributor to *in vivo* GR function. Since steroid receptors share a highly conserved DNA-binding domain and bind largely identical response elements, we anticipate that energetics-based gene control will be common to the entire receptor family.

The strong correlation we observe between GR–DNA-binding energetics and transcriptional activity implies a thermodynamic control mechanism. That is, receptor–DNA interactions in the cell must be significantly faster than other linked reactions such as coactivator recruitment, nucleosome rearrangement, or RNA polymerase isomerization. Under this condition, the ratio of receptor-bound DNA to free receptor and free DNA will closely approximate the equilibrium ratio (i.e., the equilibrium constant for binding) at all times. This will be the case even though the levels of receptor and coactivating proteins in the cell are not at equilibrium, and that transcription itself is an inherently irreversible process. Recent live-cell imaging studies provide some qualitative support for this interpretation. For example, receptor interactions with chromatin have average residence times of only a few seconds, whereas coactivator exchange and nucleosomal turnover occur over seconds to minutes (Becker et al., 2002; Coulon, Chow, Singer, & Larson, 2013; Voss & Hager, 2014). Finally, receptor–DNA exchange on the second timescale is also observed in the test-tube, again consistent with our interpretation (De Angelis & Bain, unpublished data).

Nonetheless, it should be pointed out that our study in Fig. 13 used transient transfection assays and synthetic reporter constructs. A more direct and physiologically meaningful measurement would be to directly measure *in vivo* receptor occupancy on endogenous promoters and correlate the extent of occupancy with the extent of mRNA output from the cognate gene. Shown in Fig. 13 are the results of such a study using chromatin immunoprecipitation assays to measure promoter occupancy in the cell and quantitative RT-PCR to measure mRNA output. As indicated by the R^2 values, *in vivo* promoter occupancy and mRNA output are well correlated for three endogenous promoters, FKBP5, GILZ, and SGK. These

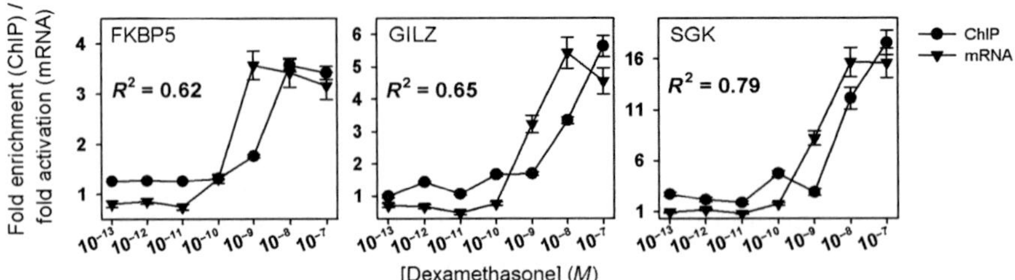

Figure 13 *In vivo* analyses of receptor–DNA occupancy and mRNA production at three endogenous promoters. Promoter occupancy and corresponding mRNA levels as a function of ligand concentration (dexamethasone) at three endogenous promoters (FKBP5, GILZ, and SGK). Promoter occupancy (filled circles) was determined by chromatin immunoprecipitation (ChIP). mRNA production (filled inverted triangles) was determined by quantitative reverse transcription PCR (RT-PCR). Error bars represent the standard error of the mean of three independent experiments. The coefficient of determination (R^2) represents the correlation between promoter occupancy (ChIP) and mRNA production (RT-PCR). (See the color plate.)

studies thus provide independent support for receptor–promoter energetics, playing a dominant role in transcriptional regulation.

5. CONCLUSIONS AND FUTURE DIRECTIONS

The results presented here highlight the importance of reductionist approaches in understanding steroid receptor-mediated transcriptional control. We note that several advances have made this possible, including new developments in high-yield receptor expression and purification; application of rigorous characterization methods and fundamental theory; and novel approaches for linking *in vitro* measurements to *in vivo* function. For the AUC studies in particular, analysis over a wide range of receptor concentrations and testing of multiple models proved essential in accurately characterizing receptor self-association. This latter approach has obvious implications for examining therapeutic protein formulations, for which accurate identification of high-order states is of direct clinical relevance. Future challenges will be to continue to link thermodynamic and kinetic studies of receptor function with transcriptional control *in vivo*, both at the single-cell and genome-wide level.

ACKNOWLEDGMENTS

We are grateful to Drs Karl Maluf, Amie Moody, and James Robblee for critical input.

Funding. This work was supported by grants from the National Institutes of Health (DK-88843) and the ALSAM Foundation Skaggs Scholars Program.

REFERENCES

Ackers, G. K., Johnson, A. D., & Shea, M. A. (1982). Quantitative model for gene regulation by lambda phage repressor. *Proceedings of the National Academy of Sciences of the United States of America*, *79*, 1129–1133.

Ackers, G. K., Shea, M. A., & Smith, F. R. (1983). Free energy coupling within macromolecules. The chemical work of ligand binding at the individual sites in co-operative systems. *Journal of Molecular Biology*, *170*(1), 223–242.

Bain, D. L., Yang, Q., Connaghan, K. D., Robblee, J. P., Miura, M. T., Degala, G. D., et al. (2012). Glucocorticoid receptor-DNA interactions: Binding energetics are the primary determinant of sequence-specific transcriptional activity. *Journal of Molecular Biology*, *422*(1), 18–32. http://dx.doi.org/10.1016/j.jmb.2012.06.005.

Becker, M., Baumann, C., John, S., Walker, D. A., Vigneron, M., McNally, J. G., et al. (2002). Dynamic behavior of transcription factors on a natural promoter in living cells. *EMBO Reports*, *3*(12), 1188–1194. http://dx.doi.org/10.1093/embo-reports/kvf244.

Bledsoe, R. K., Montana, V. G., Stanley, T. B., Delves, C. J., Apolito, C. J., McKee, D. D., et al. (2002). Crystal structure of the glucocorticoid receptor ligand binding domain reveals a novel mode of receptor dimerization and coactivator recognition. *Cell*, *110*(1), 93–105.

Brenowitz, M., Senear, D. F., Shea, M. A., & Ackers, G. K. (1986). "Footprint" titrations yield valid thermodynamic isotherms. *Proceedings of the National Academy of Sciences of the United States of America*, *83*(22), 8462–8466. http://dx.doi.org/10.1073/pnas.83.22.8462.

Connaghan, K. D., Miura, M. T., Maluf, N. K., Lambert, J. R., & Bain, D. L. (2013). Analysis of a glucocorticoid-estrogen receptor chimera reveals that dimerization energetics are under ionic control. *Biophysical Chemistry*, *172*, 8–17. http://dx.doi.org/10.1016/j.bpc.2012.12.005.

Connaghan, K. D., Yang, Q., Miura, M. T., Moody, A. D., & Bain, D. L. (2014). Homologous steroid receptors assemble at identical promoter architectures with unique energetics of cooperativity. *Proteins*, *82*(9), 2078–2087. http://dx.doi.org/10.1002/prot.24563.

Connaghan-Jones, K. D., & Bain, D. L. (2009). Using thermodynamics to understand progesterone receptor function: Method and theory. *Methods in Enzymology*, *455*, 41–70. http://dx.doi.org/10.1016/S0076-6879(08)04202-X.

Connaghan-Jones, K. D., Heneghan, A. F., Miura, M. T., & Bain, D. L. (2006). Hydrodynamic analysis of the human progesterone receptor A-isoform reveals that self-association occurs in the micromolar range. *Biochemistry*, *45*(39), 12090–12099. http://dx.doi.org/10.1021/bi0612317.

Connaghan-Jones, K. D., Heneghan, A. F., Miura, M. T., & Bain, D. L. (2007). Thermodynamic analysis of progesterone receptor–promoter interactions reveals a molecular model for isoform-specific function. *Proceedings of the National Academy of Sciences of the United States of America*, *104*(7), 2187–2192. http://dx.doi.org/10.1073/pnas.0608848104.

Connaghan-Jones, K. D., Heneghan, A. F., Miura, M. T., & Bain, D. L. (2008). Thermodynamic dissection of progesterone receptor interactions at the mouse mammary tumor virus promoter: Monomer binding and strong cooperativity dominate the assembly reaction. *Journal of Molecular Biology*, *377*(4), 1144–1160. http://dx.doi.org/10.1016/j.jmb.2008.01.052.

Coulon, A., Chow, C. C., Singer, R. H., & Larson, D. R. (2013). Eukaryotic transcriptional dynamics: From single molecules to cell populations. *Nature Reviews. Genetics*, *14*(8), 572–584. http://dx.doi.org/10.1038/nrg3484.

De Angelis, R. W., Yang, Q., Miura, M. T., & Bain, D. L. (2013). Dissection of androgen receptor-promoter interactions: Steroid receptors partition their interaction energetics in parallel with their phylogenetic divergence. *Journal of Molecular Biology*, *425*(22), 4223–4235. http://dx.doi.org/10.1016/j.jmb.2013.07.033.

DeMarzo, A. M., Beck, C. A., Onate, S. A., & Edwards, D. P. (1991). Dimerization of mammalian progesterone receptors occurs in the absence of DNA and is related to the release of the 90-kDa heat shock protein. *Proceedings of the National Academy of Sciences of the United States of America, 88*(1), 72–76.

Geserick, C., Meyer, H.-A., & Haendler, B. (2005). The role of DNA response elements as allosteric modulators of steroid receptor function. *Molecular and Cellular Endocrinology, 236*(1–2), 1–7. http://dx.doi.org/10.1016/j.mce.2005.03.007.

Hager, G. L., McNally, J. G., & Misteli, T. (2009). Transcription dynamics. *Molecular Cell, 35*(6), 741–753. http://dx.doi.org/10.1016/j.molcel.2009.09.005.

Heneghan, A. F., Berton, N., Miura, M. T., & Bain, D. L. (2005). Self-association energetics of an intact, full-length nuclear receptor: The B-isoform of human progesterone receptor dimerizes in the micromolar range. *Biochemistry, 44*(27), 9528–9537. http://dx.doi.org/10.1021/bi050609i.

Heneghan, A. F., Connaghan-Jones, K. D., Miura, M. T., & Bain, D. L. (2006). Cooperative DNA binding by the B-isoform of human progesterone receptor: Thermodynamic analysis reveals strongly favorable and unfavorable contributions to assembly. *Biochemistry, 45*(10), 3285–3296. http://dx.doi.org/10.1021/bi052046g.

Klinge, C. M. (2001). Estrogen receptor interaction with estrogen response elements. *Nucleic Acids Research, 29*(14), 2905–2919.

Kumar, R., & McEwan, I. J. (2012). Allosteric modulators of steroid hormone receptors: Structural dynamics and gene regulation. *Endocrine Reviews, 33*(2), 271–299. http://dx.doi.org/10.1210/er.2011-1033.

Liao, M., Zhou, Z. x, & Wilson, E. M. (1999). Redox-dependent DNA binding of the purified androgen receptor: Evidence for disulfide-linked androgen receptor dimers. *Biochemistry, 38*(30), 9718–9727. http://dx.doi.org/10.1021/bi990589i.

Meijsing, S. H., Pufall, M. A., So, A. Y., Bates, D. L., Chen, L., & Yamamoto, K. R. (2009). DNA binding site sequence directs glucocorticoid receptor structure and activity. *Science, 324*, 407–410. http://dx.doi.org/10.1126/science.1164265.

Métivier, R., Penot, G., Hübner, M. R., Reid, G., Brand, H., Kos, M., et al. (2003). Estrogen receptor-alpha directs ordered, cyclical, and combinatorial recruitment of cofactors on a natural target promoter. *Cell, 115*(6), 751–763.

Monroe, D. G., Getz, B. J., Johnsen, S. A., Riggs, B. L., Khosla, S., & Spelsberg, T. C. (2003). Estrogen receptor isoform-specific regulation of endogenous gene expression in human osteoblastic cell lines expressing either ERalpha or ERbeta. *Journal of Cellular Biochemistry, 90*(2), 315–326. http://dx.doi.org/10.1002/jcb.10633.

Moody, A. D., Miura, M. T., Connaghan, K. D., & Bain, D. L. (2012). Thermodynamic dissection of estrogen receptor-promoter interactions reveals that steroid receptors differentially partition their self-association and promoter binding energetics. *Biochemistry, 51*(3), 739–749. http://dx.doi.org/10.1021/bi2017156.

Nuclear Receptors Nomenclature Committee. (1999). A unified nomenclature system for the nuclear receptor superfamily. *Cell, 97*(100), 161–163. http://dx.doi.org/10.1016/S0092-8674(00)80726-6.

Perlmann, T., Eriksson, P., & Wrange, O. (1990). Quantitative analysis of the glucocorticoid receptor-DNA interaction at the mouse mammary tumor virus glucocorticoid response element. *The Journal of Biological Chemistry, 265*(28), 17222–17229.

Philo, J. S. (2000). A method for directly fitting the time derivative of sedimentation velocity data and an alternative algorithm for calculating sedimentation coefficient distribution functions. *Analytical Biochemistry, 279*(2), 151–163. http://dx.doi.org/10.1006/abio.2000.4480.

Richer, J. K., Jacobsen, B. M., Manning, N. G., Abel, M. G., Wolf, D. M., & Horwitz, K. B. (2002). Differential gene regulation by the two progesterone receptor isoforms in human breast cancer cells. *The Journal of Biological Chemistry, 277*(7), 5209–5218. http://dx.doi.org/10.1074/jbc.M110090200.

Robblee, J., Miura, M., & Bain, D. (2012). Glucocorticoid receptor–promoter interactions: Energetic dissection suggests a framework for the specificity of steroid receptor-mediated gene regulation. *Biochemistry*, *51*(22), 4463–4472. http://dx.doi.org/10.1021/bi3003956.

Rodriguez, R., Weigel, N. L., O'Malley, B. W., & Schrader, W. T. (1990). Dimerization of the chicken progesterone receptor in vitro can occur in the absence of hormone and DNA. *Molecular Endocrinology (Baltimore, Md.)*, *4*(12), 1782–1790. http://dx.doi.org/10.1210/mend-4-12-1782.

Schuck, P. (2000). Size-distribution analysis of macromolecules by sedimentation velocity ultracentrifugation and lamm equation modeling. *Biophysical Journal*, *78*, 1606–1619.

Schuster, T. M., & Laue, T. M. (Eds.), (1994). *Modern analytical ultracentrifugation*. Boston, MA: Birkhäuser. http://dx.doi.org/10.1007/978-1-4684-6828-1.

Segal, E., & Widom, J. (2009). From DNA sequence to transcriptional behaviour: A quantitative approach. *Nature Reviews. Genetics*, *10*(7), 443–456. http://dx.doi.org/10.1038/nrg2591.

Segard-Maurel, I., Rajkowski, K., Jibard, N., Schweizer-Groyer, G., Baulieu, E. E., & Cadepond, F. (1996). Glucocorticosteroid receptor dimerization investigated by analysis of receptor binding to glucocorticosteroid responsive elements using a monomer-dimer equilibrium model. *Biochemistry*, *35*(5), 1634–1642. http://dx.doi.org/10.1021/bi951369h.

Shea, M. A., & Ackers, G. K. (1985). The OR control system of bacteriophage lambda. A physical-chemical model for gene regulation. *Journal of Molecular Biology*, *181*(2), 211–230.

Skafar, D. F. (1991). Differential DNA binding by calf uterine estrogen and progesterone receptors results from differences in oligomeric states. *Biochemistry*, *30*(25), 6148–6154.

Stavreva, D. A., Varticovski, L., & Hager, G. L. (2012). Complex dynamics of transcription regulation. *Biochimica et Biophysica Acta*, *1819*(7), 657–666. http://dx.doi.org/10.1016/j.bbagrm.2012.03.004.

Theofan, G., & Notides, A. C. (1984). Characterization of the calf uterine progesterone receptor and its stabilization by nucleic acids. *Endocrinology*, *114*(4), 1173–1179. http://dx.doi.org/10.1210/endo-114-4-1173.

Thornton, J. W. (2001). Evolution of vertebrate steroid receptors from an ancestral estrogen receptor by ligand exploitation and serial genome expansions. *Proceedings of the National Academy of Sciences of the United States of America*, *98*(10), 5671–5676. http://dx.doi.org/10.1073/pnas.091553298.

Tsai, M. J., & O'Malley, B. W. (1994). Molecular mechanisms of action of steroid/thyroid receptor superfamily members. *Annual Review of Biochemistry*, *63*, 451–486. http://dx.doi.org/10.1146/annurev.bi.63.070194.002315.

Vicent, G. P., Nacht, A. S., Zaurín, R., Ballaré, C., Clausell, J., & Beato, M. (2010). Minireview: Role of kinases and chromatin remodeling in progesterone signaling to chromatin. *Molecular Endocrinology (Baltimore, Md.)*, *24*(11), 2088–2098. http://dx.doi.org/10.1210/me.2010-0027.

Voss, T. C., & Hager, G. L. (2014). Dynamic regulation of transcriptional states by chromatin and transcription factors. *Nature Reviews. Genetics*, *15*(2), 69–81. http://dx.doi.org/10.1038/nrg3623.

Wan, Y., & Nordeen, S. K. (2002). Overlapping but distinct gene regulation profiles by glucocorticoids and progestins in human breast cancer cells. *Molecular Endocrinology (Baltimore, Md.)*, *16*(6), 1204–1214. http://dx.doi.org/10.1210/mend.16.6.0848.

CHAPTER EIGHTEEN

Ultracentrifuge Methods for the Analysis of Polysaccharides, Glycoconjugates, and Lignins

Stephen E. Harding*[,1], Gary G. Adams*[,†], Fahad Almutairi*, Qushmua Alzahrani*, Tayyibe Erten*, M. Samil Kök*[,‡], Richard B. Gillis*[,†]

*National Centre for Macromolecular Hydrodynamics, University of Nottingham, Nottingham, United Kingdom

[†]School of Health Sciences, University of Nottingham, Queen's Medical Centre, Nottingham, United Kingdom

[‡]Department of Food Engineering, Abant Izzet Baysal University, Bolu, Turkey

[1]Corresponding author: e-mail address: steve.harding@nottingham.ac.uk

Contents

Methods in Enzymology, Volume 562
ISSN 0076-6879
http://dx.doi.org/10.1016/bs.mie.2015.06.043

Abstract

Although like proteins, polysaccharides are synthesized by enzymes, unlike proteins there is no template. This means that they are polydisperse, do not generally have compact folded structures, and are often very large with greater nonideality behavior in solution. This chapter considers the relevant analytical ultracentrifuge methodology available for characterizing these and related carbohydrate-based systems and information this methodology supplies, in terms of sizes, shapes, and interactions using a comprehensive range of examples, including glycoconjugates and lignins. The relevance and potential of recent software developments such as SEDFIT-MSTAR, the Extended Fujita algorithm, and HYDFIT are considered.

1. INTRODUCTION

Analytical ultracentrifugation (AUC) provides a powerful, matrix-free method for the characterization of the heterogeneity, molecular weight (molar mass) distribution, and interactions of "carbohydrate polymers," namely polysaccharides and glycoconjugates, and also related materials such as lignins (Harding, 2005a, 2005b, 2005c). In addition, when used in conjunction with viscometry and light scattering—particularly the powerful probe of size exclusion chromatography coupled to multiangle light scattering (Wyatt, 1992, 2013)—AUC provides a method for assessing the conformation and conformational flexibility (via the persistence length L_p and related parameters) of these macromolecules (García de la Torre & Harding, 2013; Harding, 1995a, 1995b, 1997a, 1997b, 2012a, 2012b, 2013a).

Compared with proteins, carbohydrate polymers provide a different challenge to characterization methodologies. Although they are synthesized by enzymes, there is no template to direct this synthesis (except for the protein or peptide components of glycoconjugates), and as a result they are polydisperse and do not usually have highly defined folded structures. Secondary structure may exist in the carbohydrate, e.g., as helices or double (or even triple) helices but very few are compact (glycogen and possibly gum Arabic are notable exceptions). Carbohydrate polymers are usually of higher molecular weight and are highly swollen in solution through solvation effects and in the case of polysaccharides can be very highly charged (Harding, 1994a, 1994b, 1994c). This means that besides being polydisperse and poorly defined, they can be very nonideal in the thermodynamic sense due to high exclusion volume effects and, if the concentration of supporting electrolyte is not sufficient, also have large polyelectrolyte behavior (Winzor, Carrington, Deszczynski, & Harding, 2004).

Like with protein research, for carbohydrate polymers, both sedimentation velocity and sedimentation equilibrium have strong complementary roles to play:

Sedimentation velocity

Simple matrix-free (no columns or membranes) and highly resolving assay for heterogeneity (sedimentation coefficient distributions), interactions (self-associations and ligand binding through co-sedimentation assays), and sedimentation coefficient determinations (Dam & Schuck, 2004; Harding, 1994b, 1994c, 2000; Stafford & Correia, 2015), in seconds, s or Svedberg units ($1S = 10^{-13}$ s).

Sedimentation equilibrium

Absolute molar mass or molecular weight (primarily the weight average) in g/mol or Da and molecular weight distribution determination (Harding, 1988, 1992, 1994a, 1994d).

Following the early pioneering work in the 1930s and 1940s on the AUC of carbohydrates by Svedberg and colleagues (see Jullander, 1987 for a review), work on carbohydrates became eclipsed by the great focus of biochemical science on proteins and nucleic acids. This led to carbohydrates, and polysaccharides in particular, acquiring a sort of "Cinderella" status (see Harding, 1993). This has dramatically changed in the last two decades, particularly with the increasing recognition of carbohydrates in molecular recognition phenomena (see Flint et al., 2004) and the increasing awareness of the importance of carbohydrate polymers or "glycopolymers" (Harding, 1994e) in food, pharmaceuticals, and biotechnology (Tombs & Harding, 1998). There has been a concomitant increase in awareness of the usefulness of AUC for the characterization of these materials—and its complementarity to other methodologies for characterizing the solution properties of polysaccharides such as light scattering and viscosity (Harding, 1994f; Harding, Vårum, Stokke, & Smidsrød, 1991, Jumel, Fiebrig, & Harding, 1996, Jumel, Harding, & Mitchell, 1996; Jumel, Harding, Mitchell, To, et al., 1996; Morris & Harding, 2013). It has become particularly useful for the study of the stability of polysaccharide/glycoconjugate-based biopharmaceutical formulations (Harding, 2010a), particularly glycoconjugate vaccines (Harding, Abdelhameed, & Morris, 2010).

We consider the advances in analytical ultracentrifuge methodology for characterizing molecular weight, molecular weight distribution, conformation, and flexibility of carbohydrate (or carbohydrate derived or related) polymers, and for characterizing molecular interactions (self-associations and ligand interactions). First, we define the types of polymer we are considering (Fig. 1).

2. POLYSACCHARIDES

Polysaccharides (see Tombs & Harding, 1998) are polymers of hexose or pentose sugar residues, linked together by glycosidic bonds, resulting in expulsion of one water molecule per bond. A two-residue substance is known as a disaccharide (e.g., maltose, lactose, sucrose, cellobiose), 2–10 residues is an oligosaccharide (e.g., raffinose), and >10 residues a polysaccharide (Fig. 1A). Since the molecular weight of a residue is usually 160–250 g/mol, this means polysaccharides have (average) molecular weights usually in excess of ~2000 g/mol. Heparin is a low molecular

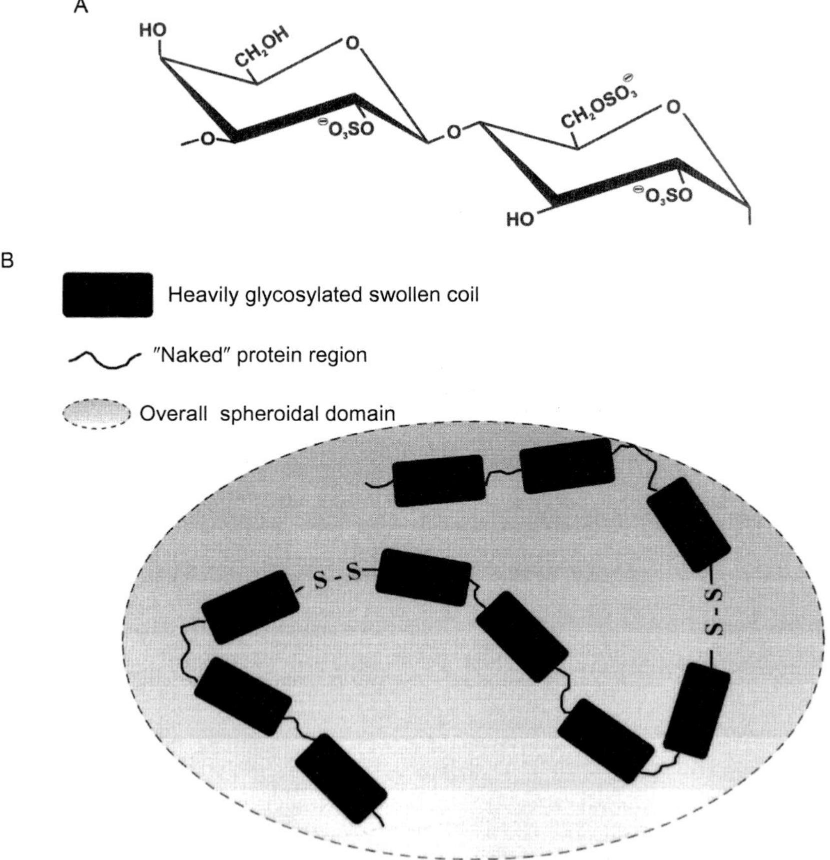

Figure 1 (A) A polysaccharide repeat structure: lambda-carrageenan. The degree of polymerization can be ~3500 with considerable polydispersity. (B) Linear random coil model for pig colonic mucin. The shadowed area shows the "effective" overall spheroidal volume of influence of this glycoprotein.

(*Continued*)

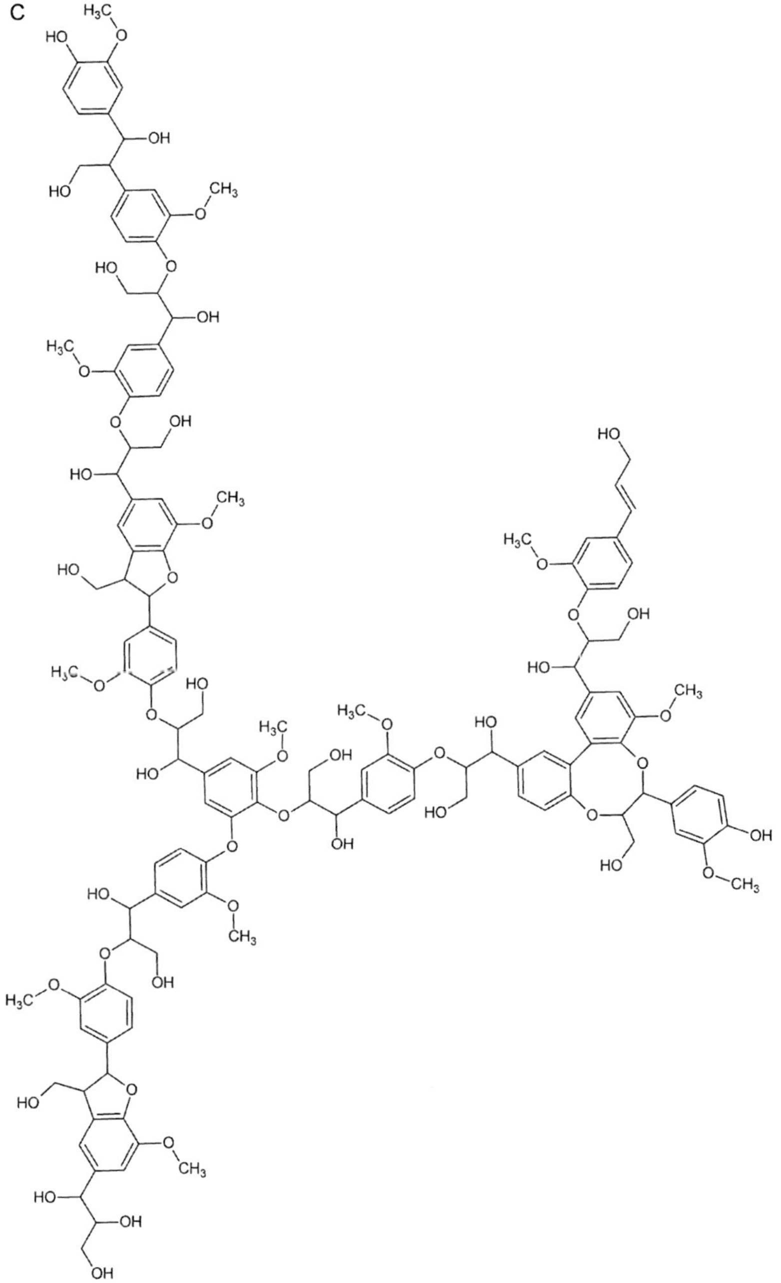

Figure 1—Cont'd (C) Complex aromatic structural model for part of a softwood lignin. *(A) Reprinted from Almutairi et al. (2013) with permission from Elsevier. (B) Reprinted with permission of Springer and Jumel et al. (1997). (C) Redrawn based on Brunow (2001). Copyright Wiley-VCH Verlag GmbH & Co. KGaA. Reproduced with permission.*

weight polysaccharide ~8000 g/mol. The largest can be in excess of 50 million g/mol (e.g., amylopectin). They can be neutral (e.g., the galactomannans, pullulans, dextrans), polyanionic (alginate, pectin, xanthan), or polycationic (chitosans, aminocelluloses).

3. GLYCOCONJUGATES

The most well-known glycoconjugates are deoxyribose nucleic acid (DNA) and ribonucleic acid (RNA) involving conjugates of sugar residues with nucleic acid bases, with the sugar backbone held together not by glycosidic bonds but by phosphodiester bonds. Historically, ultracentrifuge methods have been used, for example, to show that replication of DNA was semiconservative (Meselson & Stahl, 1958), and earlier to confirm the purity of calf-thymus DNA used by Creeth and colleagues in the first demonstration of hydrogen bonds in DNA using viscometry (Creeth, 1947; Creeth, Gulland, & Jordan, 1947; see also Harding, 2010b; Harding & Winzor, 2010; Watson, 2012). Modern application of ultracentrifuge methods to DNA has been covered by, for example, Clay, Carels, Douady, and Bernadi (2005) and for RNA by Mitra (2014) and will not be considered further in this chapter other than in terms of condensation interactions with polysaccharides.

Glycoproteins consist of a protein backbone with sugar residues attached. Ovalbumin and antibodies are examples of glycoproteins with low degrees of glycosylation: since their properties are dictated by the protein component, again they will not be considered in this review. Mucin glycoproteins (Harding, 1989; Fig. 1B) also have a protein backbone by which sugar chains are attached (through O-linkages via serine or threonine residues on the backbone). By contrast, these, however, are heavily glycosylated (generally >80%)—and their properties are close to those of polysaccharides—so they will be considered here.

4. LIGNINS

Lignins are not carbohydrates but are closely associated with carbohydrate polymers in plant cell wall structures and are cross-linked racemic macromolecules (Fig. 1C). They are a class of natural, highly branched phenylpropanoid macromolecules, which have a random and amorphous 3D structure, in which the partly cross-linked chains are hydrophobic, heterogeneous, and polydisperse. The carbon content of the aromatic lignins is around 60–63%, i.e., higher than that of the accompanying polysaccharides (see Alzahrani et al., 2015; Daly, Maluk, Zwirek, & Halpin, 2012). There is now a major push in the field of the enzymatic breakdown and manipulation of lignins and cellulosic type of materials to facilitate biofuel production from plant biomass (Daly, Maluk, Zwirek, & Halpin, 2012). There is also considerable interest now in the importance of these substances in wood and the mechanisms responsible for the decay, particularly in archaeological wood (Harding, 2012a). One outstanding example is the alum induced decay of cellulose in archaeological wood structures from the Oseberg ship find of 1904 in Norway—involving the search for alternative carbohydrate-based polymer consolidants to replace the cellulose and which will interact with the lignin structures and give long-term strength.

5. RELEVANT RECENT ADVANCES IN AUC AND RELATED PROCEDURES

5.1 Sedimentation Velocity: SEDFIT

The SEDFIT procedure (Dam & Schuck, 2004) for obtaining distributions of sedimentation coefficient is well known and has been widely applied mainly to protein-based systems. The procedure solves the Lamm equation describing the change of concentration distribution with radial position with time in terms of a distribution of sedimentation coefficients, $g(s)$ versus s, where s is the sedimentation coefficient (Fig. 2A). The (differential) distribution of sedimentation coefficients $g(s)$ can be defined as the population—weight fraction—of species with a sedimentation coefficient between $s + ds$ (Dam & Schuck, 2004). Corrections can be made for diffusion broadening of peaks of monodisperse and paucidisperse systems to yield a modified distribution known as a $c(s)$ versus s plot, and this can be further adjusted, based on assumptions involving the conformation/friction coefficient to a molar mass distribution plot $c(M)$ versus M. For polydisperse polymers—such as polysaccharides, which have a continuous rather than discrete distribution of sizes, the $c(s)$ versus s method is not so suitable (except for discrete components of narrow distribution); however, since many polysaccharide/glycoconjugate systems are quite large ($M > 100{,}000$ g/mol), diffusive effects are much slower and the $g(s)$ versus s is more representative of the true distribution of sizes. In addition, there is a method for transforming $g(s)$ versus s to a molecular weight distribution $f(M)$ versus M for polysaccharides and other polymers with a quasicontinuous polydispersity—the *Extended Fujita Method*—which we consider below.

To eliminate the effects of nonideality (co-exclusion and any residual polyelectrolyte effects), distributions should be obtained at the lowest possible

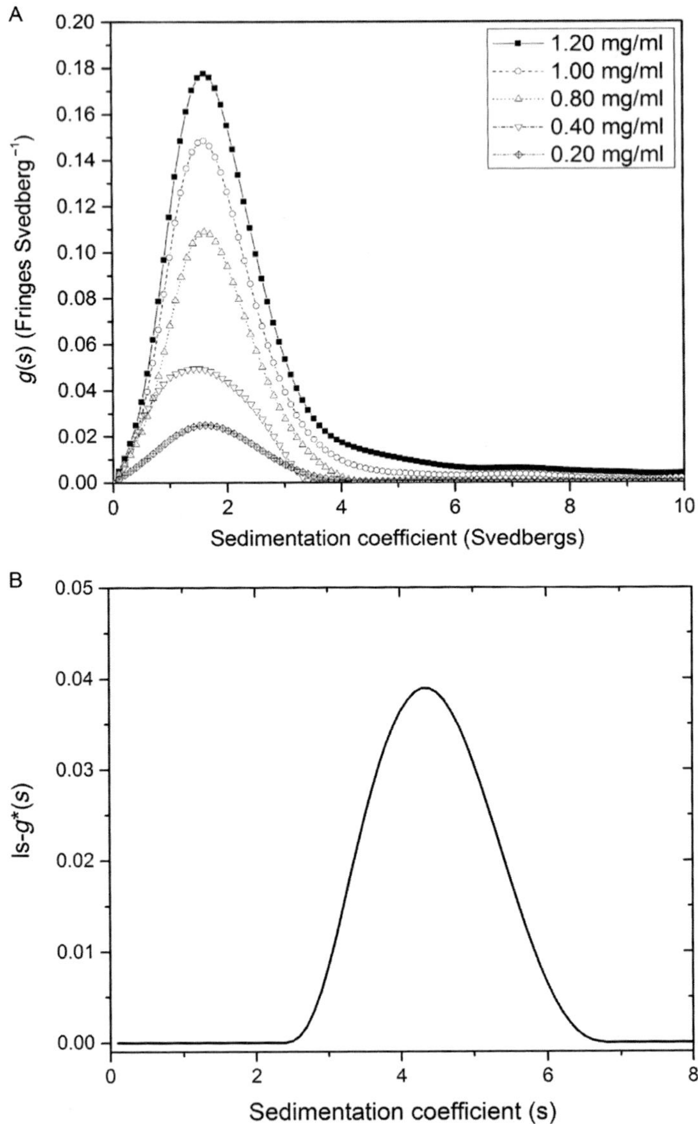

Figure 2 Sedimentation coefficient $g(s)$ versus s distribution profiles (A) chitosan in 0.2 M acetate buffer, pH 4.3, for a range of concentrations; (B) for an alginate in 0.3 M NaCl at a loading concentration of 0.03 mg/ml.

concentration (concentrations as low as 0.03 mg/ml are possible—Fig. 2). The weight average sedimentation coefficient s (corrected to standard conditions of the density and viscosity of water at 20.0 °C, to give $s_{20,w}$) can then be used to obtain other information (such as conformation or flexibility). Alternatively, an extrapolation can be made of $(1/s_{20,w})$ against concentration, c to infinite dilution $c=0$, to yield $s^{\circ}_{20,w}$ (Fig. 3A) using the (Gralen, 1944) relation

$$\left(\frac{1}{s_{20,w}}\right) = \left(\frac{1}{s^{\circ}_{20,w}}\right)(1 + k_s c) \quad (1)$$

where k_s is the "Gralen" coefficient (ml/g), or equations with higher order in c in cases of more severe nonideality, such as for alginate (Fig. 3B).

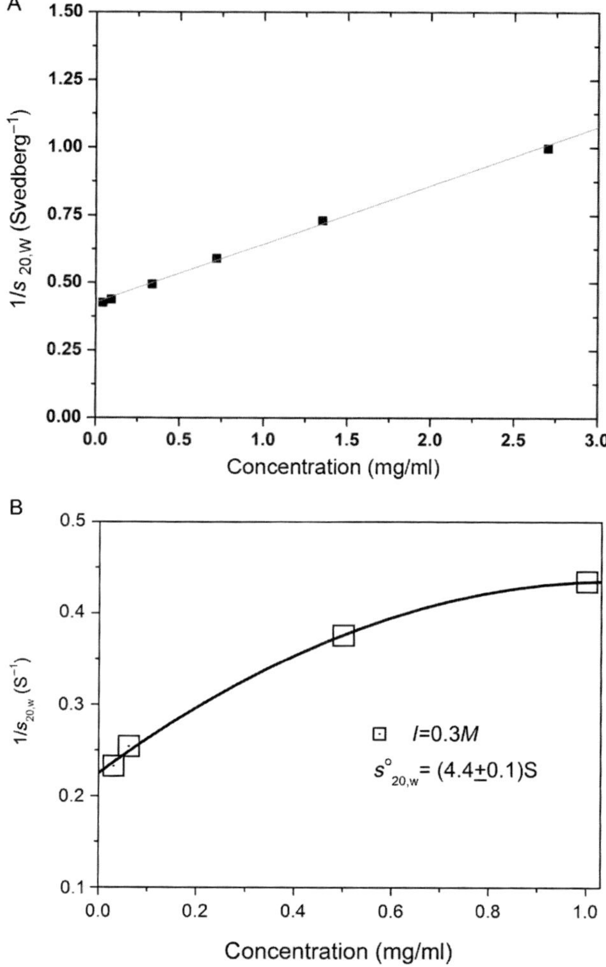

Figure 3 Concentration dependence (reciprocal) of sedimentation coefficient plots for (A) chitosan in 0.2 *M* acetate buffer (Almutairi et al., 2015)—and a linear fit (B) alginate in 0.3 *M* NaCl. *Reprinted from Almutairi et al. (2015) with permission from Elsevier.*

5.1.1 Hypersharpening

Nonideality, besides lowering the apparent (weight average) sedimentation coefficient, can also lead to skewing of the distribution due to hypersharpening of the sedimenting boundary, namely faster moving components being slowed down by sedimenting through a solution of the slower moving components—see Harding (1989, 1992), Harding (1994a, 1994b, 1994c, 1994d, 1994e, 1994f), Dhami, Cölfen, and Harding (1995), Dhami, Harding, Jones, and Hughes (1995), and Harding et al. (1996). Figure 4A shows a classical hypersharpening of a sedimenting boundary from an experiment on a xanthan using the Schlieren optical system (Dhami, Cölfen, et al., 1995; Dhami, Harding, Elizabeth, et al., 1995; Dhami, Harding, Jones, et al., 1995). Figure 4B shows a modern equivalent—$g(s)$ versus s profile of a xanthan at a similar concentration (0.23 mg/ml). For comparison, Fig. 4B also shows how the hypersharpening disappears at very low concentration, 0.075 mg/ml (Erten, Adams, Foster, & Harding, 2014). Another excellent example of hypersharpening using the Schlieren system is for the polysaccharide xylinan in a study by Harding et al. (1996).

The situation is in some ways analogous to the existence of a steady state between the effects of diffusion broadening countered by nonideality, explored in detail by Creeth (1964) and more recently by Scott, Harding, and Winzor (2015). The observation of hypersharpening of boundaries reinforces the importance of either working at very low concentrations or performing an extrapolation to zero concentration to obtain a reliable sedimentation coefficient for further use in hydrodynamic modeling or molecular weight analysis.

5.2 Sedimentation Equilibrium: SEDFIT-MSTAR

The long-established MSTAR method (Cölfen & Harding, 1997; Harding, Horton, & Morgan, 1992), in which the M^* function is computed over the whole length of the solution column has recently been incorporated into the SEDFIT suite of algorithms (Schuck et al., 2014). The M^* function is defined (Creeth & Harding, 1982a, 1982b) as a function of radial position r from the center of rotation and concentration $c(r)$ by

$$M^*(r) = (c(r) - c_{\mathrm{m}}) \Big/ \left\{ kc_{\mathrm{m}}\left(r^2 - r_{\mathrm{m}}^2\right) + 2k\int_{r_{\mathrm{m}}}^{r} (c(r) - c_{\mathrm{m}})r\mathrm{d}r \right\} \qquad (2)$$

where k is an experimental constant depending on the rotor speed, partial specific volume of the solute, and density of the solvent and the subscript

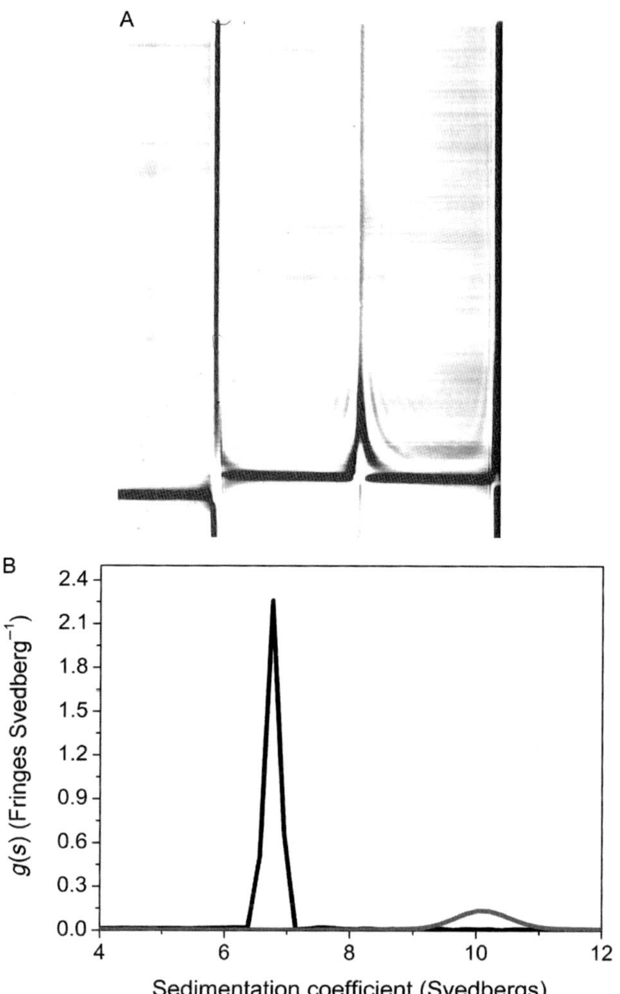

Figure 4 Hypersharpening of sedimentation profiles in polysaccharides (A) classical Schlieren pattern (Dhami, Harding, Jones, et al., 1995) for keltrol xanthan. Rotor speed 30,000 rev/min, temperature 20.0 °C; solvent $I=0.01$ *M*. The single hypersharp boundary is indicative of sample purity and also high thermodynamic nonideality, even at the low loading concentration used (0.2 mg/ml) (B) corresponding plots for TSF xanthan using $g(s)$ versus s (Erten et al., 2014). Note the strong hypersharpening again at low loading concentration of 0.23 mg/ml and high ionic strength $I=0.3$ *M* (black trace) and how it disappears at very low concentration (0.075 mg/ml) when non-ideality ~ negligible (blue trace; dark gray in the print version). *(A) Reprinted from Dhami, Harding, Jones, et al. (1995), with permission from Elsevier. (B) Reprinted from Erten et al. (2014), with permission from Elsevier.*

"m" denotes the air-solution meniscus. The original Creeth and Harding (1982a, 1982b) paper lists several useful properties of $M^*(r)$, the most important of which is that extrapolation of $M^*(r)$ to the radial position at the cell base ($r=r_b$) yields an estimate of $M_{w,app}$ for the whole solution as loaded into the cell: a "whole cell weight average molecular weight." As described in Schuck et al. (2014), SEDFIT-MSTAR marks an advance on previous versions of MSTAR inasmuch as the (necessary) extrapolations of data values to the meniscus and cell base radial positions are now based upon the properties of the whole data set, rather than upon simple local polynomial "pieces," facilitating evaluation of M^* $(r=r_b)=M_{w,app}$ for the whole distribution.

The routine also evaluates $M_{w,app}(r)$ as a function of radial position in the cell, r, or the equivalent local concentration $c(r)$. The value of $M_{w,app}(r)$ at the "hinge point" of the cell, the radial position where $c(r)=c^o$, the original cell-loading concentration prior to redistribution, provides an alternative measure for the whole distribution $M_{w,app}$ (Schuck et al., 2014). SEDFIT-MSTAR employs a smart-smooth method for obtaining baselines using Savitsky–Golay filters, and also provides a (low-resolution) estimate of the molecular weight distribution $c(M)$ versus M. Figure 5 gives an example for lambda-carrageenan. Another recent example is the application to the absolute molecular weight characterization of wood and nonwood lignins (Alzahrani et al., 2015).

Molecular weights obtained from SEDFIT-MSTAR are apparent ones—that is not corrected for thermodynamic nonideality (through co-exclusion and polyelectrolyte behavior). Unfortunately, whereas for sedimentation velocity it is possible to run experiments at concentrations as low as 0.01 mg/ml, the minimum concentration for a sedimentation equilibrium experiment in conventional 12 mm optical path length cells (Beckman XL-I ultracentrifuge) is ~0.4–0.5 mg/ml. Thirty-millimeter path length cells were commercially available until the 1980s with the Beckman Model E analytical ultracentrifuge. Encouragingly, 20 mm path length cells have recently become available from Nanolytics Ltd. (Potsdam, Germany), allowing a minimum concentration of 0.2–0.3 mg/ml. These concentrations may still though be too high to be able to rule out the effects of nonideality for some materials. Harding (1992, 2005a) gives comparative tables assessing the underestimate of measured molecular weights for a range of polysaccharide materials, measured at these minimum concentrations without correction. For some polysaccharides $M_w \sim M_{w,app}$ measured at these

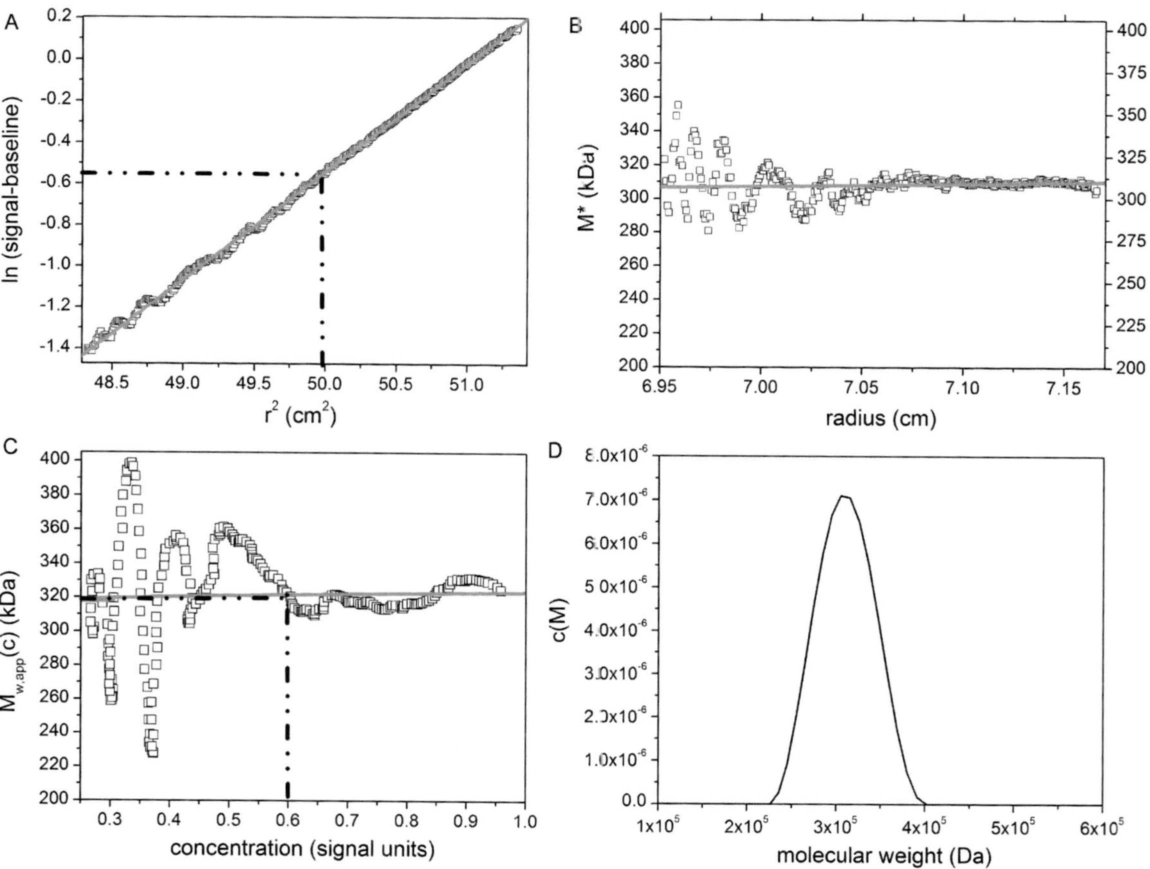

Figure 5 See legend on next page.

concentrations are a reasonable approximation; for others, like xanthan and alginate, it is not, even at 0.2 mg/ml. In these cases, an extrapolation of $1/M_{\mathrm{w,app}}$ versus c, the loading concentration to $c=0$ is necessary, with the hinge point estimates for $M_{\mathrm{w,app}}$ giving the optimal extrapolation, as explained in Schuck et al. (2014). This extrapolation is normally linear:

$$\left(\frac{1}{M_{\mathrm{w,app}}}\right)=\left(\frac{1}{M_{\mathrm{w}}}\right)(1+2BM_{\mathrm{w}}c) \tag{3}$$

where B is the second thermodynamic virial coefficient. Although useful for many systems, Eq. (3) does not universally apply, particularly for strongly polyanionic or polycationic polysaccharides. For example, even in high ionic strength (0.3 M or higher) solvents, an extra virial term in c^2 has been shown necessary for alginates (Harding, 1992; Horton, Harding, & Mitchell, 1991).

This requirement for an extrapolation is a disadvantage for sedimentation equilibrium compared with SEC-MALS as with the latter, due to dilution on the SEC-columns, nonideality effects can be very small and such extrapolations are unnecessary.

5.3 Higher Resolution Molecular Weight Distribution Evaluation—the *Extended Fujita* Approach

For a higher resolution of the molecular weight distribution, two other approaches are possible. First—and assuming thermodynamic non-ideality is negligible—the best fit to the concentration distribution to a mul-

Figure 5—Cont'd SEDFIT-MSTAR output for analysis of lambda-carrageenan (at a loading concentration of 0.3 mg/ml) by sedimentation equilibrium (A) log concentration ln $c(r)$ versus r^2 plot, where r is the radial distance from the center of rotation; (B) M^* versus r plot (open squares) and fit: the value of M^* extrapolated to the cell base $=M_{\mathrm{w,app}}$ the apparent weight average molecular weight for the whole distribution. (C) Point or local apparent weight average molecular weight at radial position r (open square) plotted against the local concentration $c(r)$ for different radial position (D) molecular weight distribution $c(M)$ versus M plot. The dash-dot line gives the estimate for $M_{\mathrm{w,app}}$ at the hinge point. The two estimates for $M_{\mathrm{w,app}}$ are in good agreement, (310,000 $\pm$ 5000) g/mol from the M^* extrapolation and (320,000 $\pm$ 20,000) g/mol from the hinge point. *Reproduced from Schuck et al. (2014) with permission from the Royal Society of Chemistry.*

ticomponent system can be applied to the data in the routine MULTISIG, and a quasicontinuous distribution can be generated. A description of this method is given in Gillis, Adams, Heinze, et al. (2013).

Second, the *Extended Fujita* method (Harding, Schuck, et al., 2011) is recommended. This involves transformation of the sedimentation coefficient $g(s)$ versus s profile into a molecular weight distribution $f(M)$ versus M. The method was originally provided by Fujita for randomly coiled types of polymer and then recently extended by Harding, Schuck, et al. (2011) to cover any polymer conformational type. To transform the $g(s)$ versus s profile requires three extra pieces of information. First, the weighted average molecular weight M_w from either sedimentation equilibrium (SEDFIT-MSTAR) or SEC-MALS, and second, the corresponding weighted average sedimentation coefficient from sedimentation velocity. Third, some knowledge of the approximate conformation of the polymer is required (sphere, rod, coil, or something in between). The method, like SEDFIT-MSTAR, has been built into the SEDFIT routine (Harding, Schuck, et al., 2011).

The transformation is as follows:

$$f(M) = \left(\frac{ds}{dM}\right)g(s) \tag{4}$$

with

$$M = \left(\frac{s}{\kappa_s}\right)^{1/b} \tag{5}$$

and

$$\frac{ds}{dM} = b\kappa_s^{1/b}s^{(b-1)/b} \tag{6}$$

b is the conformation/scaling parameter, with limits of ~0.15 for a rod, 0.4–0.5 for a random coil, and 0.67 for a compact sphere (see Harding, Vårum, et al., 1991) and κ_s can be found from Eq. (5) provided that at least one value of M (e.g., M_w from sedimentation equilibrium or SEC-MALS) is known for one value of s (e.g., the weight average s value).

Figure 6A shows an example of a determination for sodium alginate in $I=0.3$ M solvent and at very low loading concentration (0.03 mg/ml) to

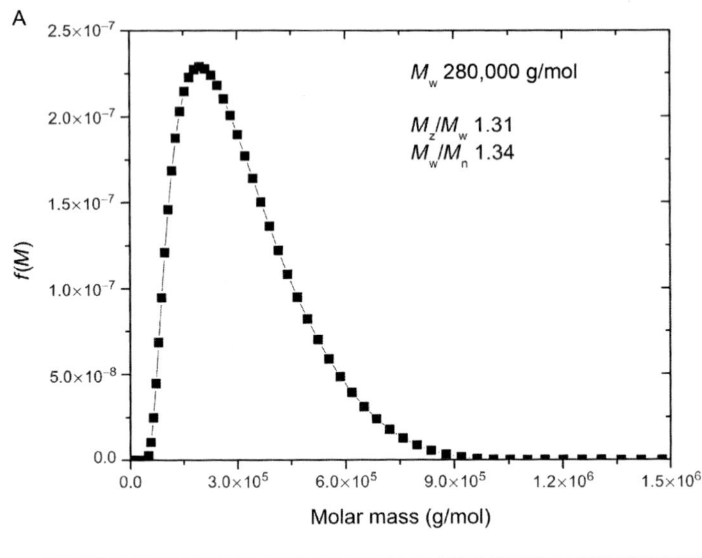

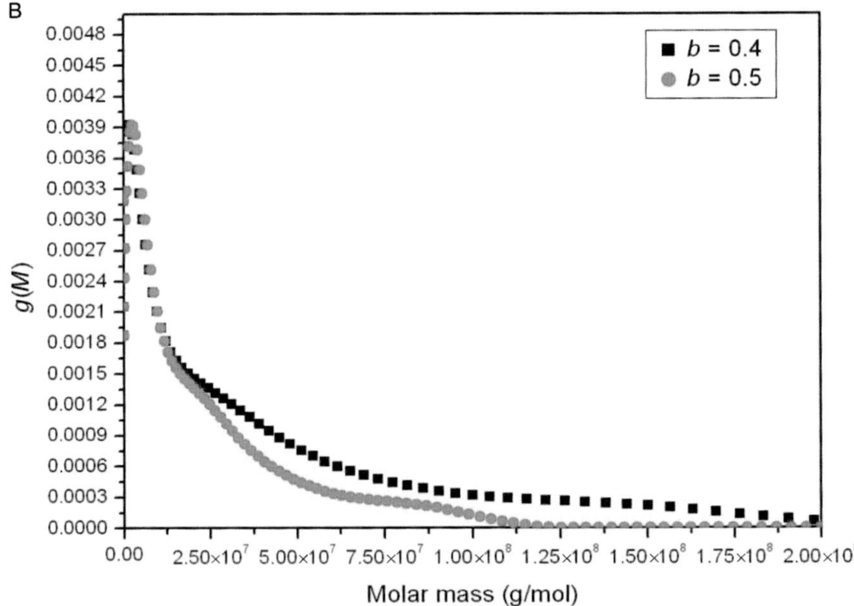

Figure 6 Normalized molecular weight (molar mass) distribution for (A) alginate in 0.3 *M* NaCl obtained from transformation of the *g*(*s*) versus *s* distribution at concentration (0.03 mg/ml), and (B) a large glycoconjugate (for 2 different values of the power law coefficient *b*), using the Extended Fujita method of Harding, Schuck, et al. (2011). *B is reprinted from Harding, Schuck, et al. (2011) with permission from Elsevier.*

render nonideality effects negligible. Nonideality, besides lowering the apparent weight average molecular weight of a distribution, can also lead to skewing of the distribution due to hypersharpening of the sedimenting boundary, namely faster moving components slowed down by sedimenting through a solution of the slower moving components—as considered in Section 5.1.1—another reason for working at as low a concentration as is possible (see Section 5.2).

Taking a value for b for alginate ~0.33 (Harding, Schuck, et al., 2011; Harding, Smith, et al., 2011), and taking $M_w = 280{,}000$ (from SEC-MALS) and $s = s_{20,w}$ (at 0.03 mg/ml) = 4.3S, this yields a value for $\kappa_s = 0.0685$. The distribution $f(M)$ versus M obtained in Fig. 6A corresponds to the $g(s)$ versus s distribution of Fig. 2B, and estimates for M_z/M_w and M_w/M_n are also given. The method has a large dynamic range, covering molecular weights $>10^8$ g/mol: Fig. 6B gives the distribution for a large glycoconjugate of *Streptococcus pneumoniae* capsular polysaccharide with tetanus toxoid protein. If there is uncertainty in b, then this can be adjusted to monitor effects on the distribution, as shown in Fig. 6B. If b is adjusted, it is important that κ_s is adjusted too in Eq. (5) so that the known weighted average values for M and s used to calibrate the plot still correspond.

5.4 Conformation and Flexibility Analysis

There are a number of options open for the analysis of the overall conformation of a carbohydrate-based macromolecule (in terms of general models such as sphere, rod, and coil) and in terms of the flexibility of the macromolecule (Harding, Vårum, et al., 1991). The inherent polydispersity of these substances appears to make measurements more difficult, although it can actually be beneficial as measurements of the sedimentation coefficient (and other parameters from other hydrodynamic measurements, such as the intrinsic viscosity $[\eta]$ or radius of gyration R_g) as a function of molecular weight within a homologous series can provide conformation and flexibility information. Complications can exist, such as through solvent potentially draining through a macromolecule, although as Tanford (1961) and others have long ago pointed out, the strong hydrodynamic interactions between elements within the particle and interstitial solvent generally dominate over these effects. Conformation and flexibility analysis is however always of low resolution.

5.4.1 Scaling or Power Law

Simple scaling law analysis uses the power law equation (cf. Eq. 5)

$$s = K_s M^b \tag{7}$$

with the limiting values of b for different conformation types given above. The variation in the exponent for different conformation types is however not as great compared with the variation of $[\eta]$ with M or R_g with M. Equation (7) can however be used in conjunction with other hydrodynamic scaling relations with M, for example, with $[\eta]$ and M or R_g and M (see Harding, 1995b; Harding, Vårum, et al., 1991). The relatively low dependence of the sedimentation coefficient on conformation is on the other hand advantageous in terms of the *Extended Fujita* method for molecular weight distribution analysis from sedimentation velocity (Section 5.3).

5.4.2 Wales–van Holde Relation

Another useful conformation parameter is the Wales–van Holde (Wales & van Holde, 1954) ratio R of the limiting (at $c=0$) concentration dependence coefficient k_s to the intrinsic viscosity $[\eta]$

$$R = \frac{k_s}{[\eta]} \tag{8}$$

This has approximate limits R ~0.2 for a rod and ~1.6 for a random coil/compact sphere (Creeth & Knight, 1965; Rowe, 1977), although for charged polysaccharides, the ionic strength of the supporting electrolyte needs to be sufficient such that polyelectrolyte effects are fully suppressed. An excellent recent example has been the application to chitosans of differing degrees of acetylation by Schütz, Käuper and Wandrey (2012), in agreement with earlier estimations on different chitosans by Morris, Castile, Smith, Adams, and Harding (2009a, 2009b).

5.4.3 Persistence Length Estimation

A more quantitative representation of particle flexibility comes from measurement of the *persistence length* L_p which has theoretical limits of 0 for a random coil and ∞ for a stiff rod (see García de la Torre & Harding, 2013). Practically, the limits are ~1–2 nm for a random coil and ~200–300 nm for a very stiff rod-shaped macromolecule. Persistence lengths, L_p, can in principle be estimated several different ways using sedimentation coefficient, intrinsic viscosity, or radius of gyration-based approaches. For example, the Yamakawa–Fuji relation (Yamakawa & Fujii, 1973):

$$s^0 = \frac{M_L(1-\bar{v}\rho_0)}{3\pi\eta_0 N_A}\left[1.843\left(\frac{M_w}{2M_L L_p}\right)^{1/2} + A_2 + A_3\left(\frac{M_w}{2M_L L_p}\right)^{-1/2} + \cdots\right] \tag{9}$$

where M_L is the mass per unit length, $\bar{v}$ the partial specific volume, ρ_o the solvent density, η_o the solvent viscosity, N_A Avogadro's number, and A_2 and A_3 are coefficients. The corresponding equation for viscosity is the Bushin–Bohdanecky equation (Bohdanecky, 1983; Bushin, Tsvetkov, Lysenko, & Emel'yanov, 1981):

$$\left(\frac{M_w^2}{[\eta]}\right)^{1/3} = A_0 M_L \Phi^{-1/3} + B_0 \Phi^{-1/3}\left(\frac{2L_p}{M_L}\right)^{-1/2} M_w^{1/2} \tag{10}$$

where Φ is the Flory–Fox coefficient (2.86×10^{23}/mol) and A_0 and B_0 are tabulated coefficients.

Yamakawa and Fujii (1973) showed that A_2 in Eq. (9) can be considered as $-\ln(d/2L_p)$ and $A_3 = 0.1382$ if the persistence length L_p is much higher than the chain diameter, d. Difficulties arise if the mass per unit length is not known, although both relations (Eqs. 9 and 10) have been built into an algorithm Multi_HYDFIT (Ortega & García de la Torre, 2007) which estimates the best range of values of L_p and M_L based on minimization of a target function Δ. An estimate for the chain diameter d is also required but extensive simulations have shown that the results returned for L_p are relatively insensitive to the value chosen for d (see García de la Torre & Harding, 2013 & references therein). An example for a *Streptococcal* capsular polysaccharide is given in Fig. 7 yielding a value for L_p of (6.2 ± 0.6) nm—a semiflexible structure.

5.4.4 Sedimentation Conformation Zoning

A check for consistency of the above results can be obtained from a *sedimentation conformation zone* plot of $k_s M_L$ versus $[s]/M_L$ introduced by Pavlov, Rowe, and Harding (Pavlov, Harding, & Rowe, 1999; Pavlov, Rowe, & Harding, 1997) where

$$[s] = \frac{s^0 \eta_o}{(1-\bar{v}\rho_o)} \tag{11}$$

and k_s is the concentration dependence or Gralen coefficient. The "zones" A: rigid rod, B: rod with limited flexibility, C: semiflexible coil, D: random coil, E: compact/globular are based on a series of polysaccharides of known

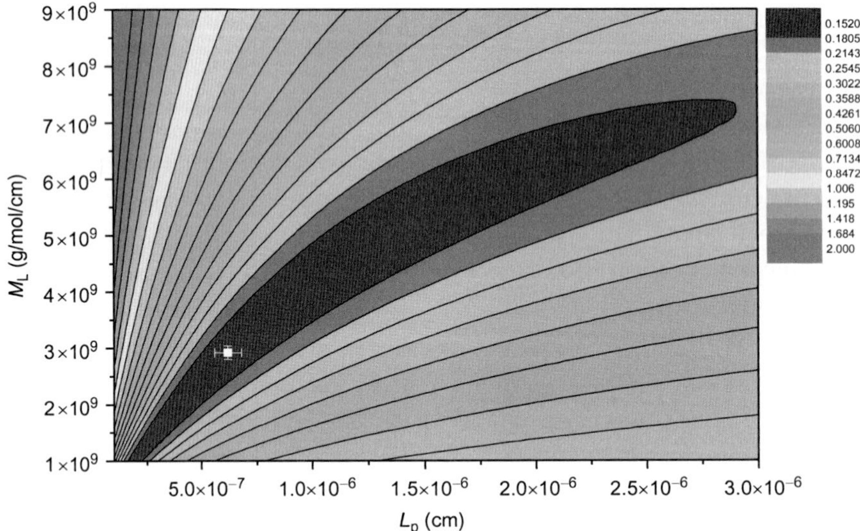

Figure 7 Global or "HYDFIT" estimation of the persistence length L_p versus mass per unit length M_L for a capsular *Streptococcus pneumoniae* polysaccharide SP(4). The white cross represents the minimum of a target function based on fitting the Yamakawa–Fujii and Bushin–Bohdanecky equations. The plot yields $L_p \sim 6.2$ (nm) and $M_L \sim 2.92 \times 10^9$ (g/mol/cm) at the minimum target (error) function value of 0.15. *Reprinted from Harding et al. (2012), with permission from Elsevier.* (See the color plate.)

$[s]$, k_s and M_L, and known conformation type. Figure 8 shows the results for capsular polysaccharides from *S. pneumoniae*—all those studied are clearly semiflexible chains. Similar results have been found for example for chitosans (Morris et al., 2009a, 2009b).

5.4.5 Quasi-Rigid Particle Modeling

The intrinsic viscosity and sedimentation coefficient can also give valuable shape information on the protein components of glycoconjugates, for example, those used in the production of glycoconjugate vaccines. As recently reviewed by García de la Torre and Harding (2013), a whole rafter of modeling strategies are available ranging from simple whole body or ellipsoid representations of the structures to more complex bead modeling (the latter usually requiring additional information such as from solution X-ray scattering). The software for this simple modeling—the "ELLIPS" suite of programs (Harding & Cölfen, 1995; Harding, Horton, & Cölfen, 1997) is now very simple to use (Harding, 2013b; Harding, Cölfen, & Aziz, 2005), and Fig. 9 shows such a representation for the tetanus toxoid protein in terms of a prolate ellipsoid of revolution (an ellipsoid with one major and

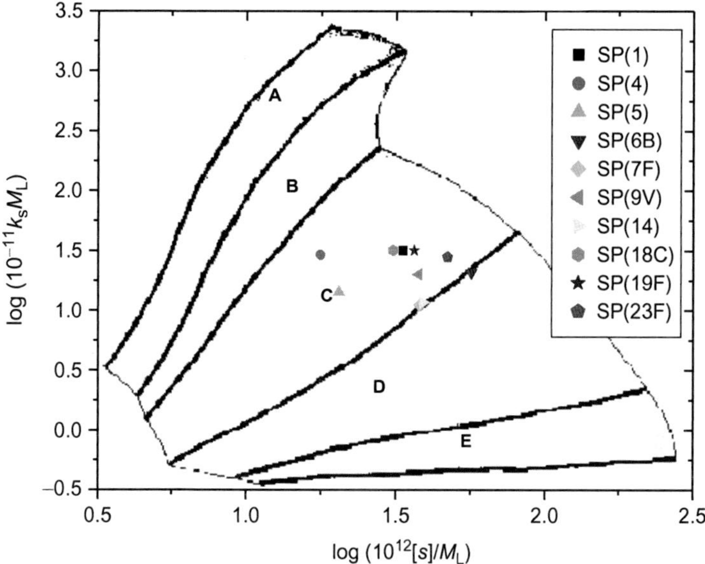

Figure 8 Conformation zoning plot. k_s is the concentration dependence sedimentation coefficient (ml/g), M_L is the mass per unit length and [*s*] is the intrinsic sedimentation coefficient. All 10 *Streptococcus pneumoniae* capsular polysaccharides have conformations in Zone C (semiflexible coil) or near the boundary with Zone D (highly flexible). *Reprinted from Harding et al. (2012), with permission from Elsevier.*

two equal minor axes) of axial ratio ~3:1. Coincidentally, the representation looked remarkably like the "guessed" cartoon representation given earlier by Astronomo and Burton (2010)!

Ellipsoidal representations also appear appropriate in describing the conformations of lignins from wood and nonwood sources. Favis and Goring (1984) had suggested that they may have plate-like structures and if so, Alzahrani et al. (2015) have recently found that a flat oblate model (two major axes and one minor axis) of axial ratio of ~30:1 seems appropriate, again on the basis of viscometry and sedimentation measurements.

6. EXAMPLES OF APPLICATIONS TO SPECIFIC CARBOHYDRATE SYSTEMS

Table 1 gives a comprehensive (but not exhaustive) summary of applications of AUC to carbohydrate systems from the work of Svedberg and Gralen on the molecular weights of cellulose in cuprammonium sulfate to modern applications to the molecular weight distribution and conformation of glycoconjugate vaccines. It should be stressed that AUC applies

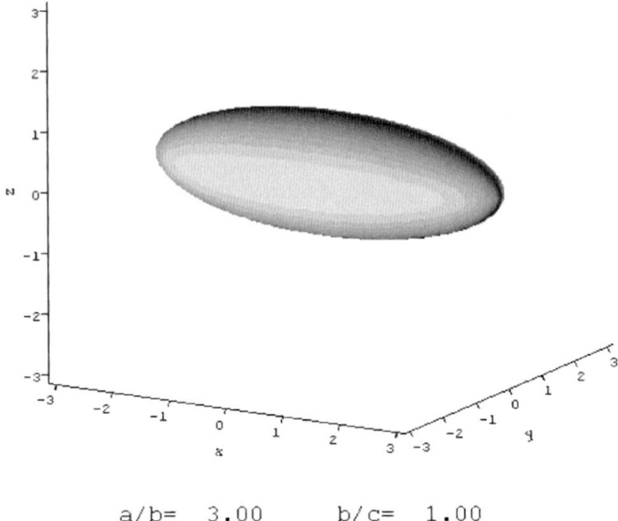

Figure 9 Low-resolution structure of tetanus toxoid conjugation protein using the simple ellipsoid modeling routine ELLIPS1 showing an extended form of (prolate) axial ratio ~3:1. *Reprinted from Abdelhameed et al. (2012), with permission from Elsevier.*

principally to systems of macromolecules or macromolecular assemblies *in solution*, although particles of sedimentation coefficient >3000S (such as mucin–chitosan complexes) can be registered (Deacon et al., 1998). However for such very large particles, "disc centrifugation" procedures (essentially preparative sedimentation in a sucrose or other density gradient, with a turbidity detector to register the sedimentation rate) are more appropriate, as, for example, used in the study of the stability of chitosan–tripolyphosphate particulate systems considered for use in drug delivery (Morris et al., 2011).

Besides characterization of molecular weight, conformation, and flexibility, the most important application of AUC in carbohydrate-based systems has been in the study of macromolecular interaction phenomena, and for this final section of this chapter we will now consider some aspects of this.

7. ANALYTICAL ULTRACENTRIFUGE ANALYSIS OF INTERACTIONS INVOLVING CARBOHYDRATE POLYMERS

7.1 Carbohydrate Self-Association

Certain classes of carbohydrate polymer are known to form double or triple helices (for example, xanthan and schizophyllan, respectively) but these tend

Table 1 Analytical Ultracentrifuge Studies on Carbohydrate Polymers

Macromolecular System	Type of Study	References
Acanthus ebracteatus polysaccharides	Molecular weight, conformation	Hokputsa,Harding, Inngjerdingen, et al. (2004)
Alginate	Novel combination of sedimentation equilibrium with size exclusion chromatography for molecular weight distribution	Ball, Harding, and Mitchell (1988)
	Thermodynamic nonideality study	Horton, Harding, Mitchell, & Morton-Holmes (1991)
	Novel use of XL-A ultracentrifuge for Schlieren study	Dhami, Cölfen, et al. (1995)
Aminocelluloses	Protein-like self-association	Nikolajski et al. (2014)
	Completely reversible tetramerization of AEA-1	Heinze et al. (2011a, 2011b)
Amylovoran	Molecular weight and shape	Jumel, Geider, and Harding (1997)
Arabinoxylan	Effect of uv-tagging on properties	Morris, Ebringerova, Harding, and Hromadkova (1999)
	AUC study of a weak self-association	Patel et al. (2007)
Blue dextran	Use of absorption optics on the XL-A analytical ultracentrifuge	Errington, Harding, and Rowe (1992)
Capsular polysaccharides from *Streptococcus pneumoniae*	AUC/hydrodynamic analysis of mol wt., heterogeneity, and conformation of a series of serotypes	Harding et al. (2012)
Carbohydrate-coated dendrimers	Solution conformation	Pavlov, Korneeva, Jumel, et al. (1999)

Continued

Table 1 Analytical Ultracentrifuge Studies on Carbohydrate Polymers—cont'd

Macromolecular System	Type of Study	References
Carbohydrate-protein recognition	Ligand-mediated dimerization	Flint et al. (2004)
Carboxymethyl chitin	Molecular weight and conformation	Korneeva, Vichoreva, Harding, and Pavlov (1996)
	Solution conformation at high ionic strength	Pavlov, Korneeva, Harding, and Vichoreva (1998); Pavlov, Korneeva, Vikhoreva, and Harding (1998)
Carboxymethyl cellulose	Molecular flexibility study on dendronized derivatives	Pohl, Morris, Harding, and Heinze (2009)
Charged polysaccharides	Extent of charge screening	Winzor et al. (2004)
Cellulose	Sedimentation velocity and diffusion measurements	Gralen (1944)
	Nitrocellulose	Jullander (1945, 1987)
Chitosan	Interactions with DNA and xanthan	Almutairi et al. (2015)
	Effect of fluorophore 9-anthraldehyde on hydrodynamic properties	Cölfen, Harding, and Vårum (1996)
	Interactions with lysozyme	Cölfen, Harding, Vårum, and Winzor (1996)
	Interactions with mucin—sedimentation and atomic force microscopy study	Deacon et al. (2000)
	Specific interactions with mucins from different regions of stomach	Deacon et al. (1999)
	Correlation of AUC with SEC-MALS data	Fee et al. (2003)
	Interactions with mucin	Fiebrig, Harding, Stokke, et al. (1994); Fiebrig, Harding, and Davis (1994); Fiebrig, Davis, and Harding (1995)

Table 1 Analytical Ultracentrifuge Studies on Carbohydrate Polymers—cont'd

Macromolecular System	Type of Study	References
	Depolymerization with temperature	Morris et al. (2009a)
	Global hydrodynamic analysis of flexibility	Morris et al. (2009b)
	Sedimentation behavior in relation to other properties/ nanogels	Schütz, Käuper, and Wandrey (2012)
	Analysis by sedimentation velocity of mucin complexation	Harding (1997a)
	Comparison of different degrees of acetylation	Errington, Harding, Vårum, and Illum (1993)
	Storage stability study of chitosan-tripolyphosphate nanoparticles	Morris, Castile, Adams, and Harding (2011)
Chondroitin sulfate	Structural integrity after laser irradiation	Jumel et al. (2002); Sobol et al. (2000)
Citrus pectins	AUC and light scattering evidence for extended conformation	Harding, Berth, Ball, and Mitchell (1990)
	Molecular weight distribution and conformation	Harding, Berth, Ball, Mitchell, and Garcia de la Torre (1991)
	Global hydrodynamic analysis of flexibility	Morris et al. (2008)
Cyanobacterial exopolysaccharides	Solution properties in relation to use	Li, Harding, and Liu (2001)
	Comparison of hydrodynamic properties of exopolysaccharide from *Aphanothece halophtica* GR02 with those of xanthan	Morris et al. (2001)
Cystic fibrosis mucin	Self-association, polydispersity, and nonideality study	Harding and Creeth (1982)

Continued

Table 1 Analytical Ultracentrifuge Studies on Carbohydrate Polymers—cont'd

Macromolecular System	Type of Study	References
Desert algae polysaccharides	Molecular weight, conformation	Hokputsa, Hu, Paulsen, and Harding (2003)
Dextran	Protein diffusion through incompatible PEG/dextran phase systems	Harding and Tombs (1989, 2002)
Dextran T-500	Novel combination of sedimentation equilibrium with size exclusion chromatography for molecular weight distribution	Ball, Harding, and Simpkin (1990)
Diethylaminoethyl-dextran	Interactions with mucin	Anderson, Harding, and Davis (1989)
Di-iodotyrosine dextran	Characterization using the XL-A analytical ultracentrifuge	Errington, Harding, Illum, and Schact (1992)
Dietary fiber polysaccharides—various	AUC for probing possible interactions with gliadin	Kök, Harding, Adams, Tatham, and Morris (2010); Kök et al. (2012); Adams et al. (2012)
Durian rinds polysaccharides	Molecular weight, conformation	Hokputsa, Hu, Paulsen, et al. (2004)
Food polysaccharides	Analysis by AUC	Harding (1994a, 1995c)
Galactomannans	Global hydrodynamic analysis of flexibility	Morris et al. (2008)
	Effect of pressure assisted solubilization on AUC	Patel, Picout, Harding, and Ross-Murphy (2006)
	Interactions with gliadin studied by sedimentation velocity	Seifert, Heinevetter, Cölfen, and Harding (1995)
Glycoconjugate vaccine: tetanus toxoid protein	Dimerization and conformation study	Abdelhameed et al. (2012)
Glycogen	Molecular weight and conformation	Morris et al. (2008)

Table 1 Analytical Ultracentrifuge Studies on Carbohydrate Polymers—cont'd

Macromolecular System	Type of Study	References
Guar	Thermal degradation effects on sedimentation coefficient and molecular weight	Bradley, Ball, Harding, and Mitchell (1989)
	Effect of gamma irradiation on macromolecular integrity	Jumel, Harding, Mitchell (1996)
Heteroxylan	Molecular weight, conformation	Dhami, Harding, Elizabeth, et al. (1995)
Hyaluronic acid	Comparison of AUC with SEC-MALS	Hokputsa, Jumel, Alexander, and Harding (2003a)
	Molecular weight, conformation of chemically degraded material	Hokputsa, Jumel, Alexander, and Harding (2003b)
Hydroxypropylmethyl cellulose	Molecular weight and shape	Jumel, Harding, Mitchell, et al. (1996)
Inulin	Molecular weight and shape in DMSO	Azis, Chin, Deacon, Harding, and Pavlov (1999)
Kappa-carrageenan	Size, shape, and hydration	Harding, Day, Dhami, and Lowe (1997)
Konjac glucomannan	Molecular weight and flexibility using global analysis	Kök, Abdelhameed, Ang, Morris, and Harding (2009)
Lambda-carrageenan	Conformation in solution	Almutairi et al. (2013)
Lignin	AUC and viscometry study of molecular weight distribution and conformation of soda, Alcell, kraft, and soda lignins	Alzahrani et al. (2015)
Locust bean gum	Comparison of hot and cold water-soluble fractions	Gaisford, Harding, Mitchell, and Bradley (1986)
Methyl galactan (synthetic)	Molecular weight	Koschella et al. (2008)

Continued

Table 1 Analytical Ultracentrifuge Studies on Carbohydrate Polymers—cont'd

Macromolecular System	Type of Study	References
Methylcelluloses	Global hydrodynamic analysis of flexibility	Patel, Morris, Garcia de la Torre, et al. (2008)
Microbial polysaccharides	AUC in relation to structure and other properties	Morris and Harding (2009)
Mucins	New approach to molecular weight distribution analysis	Gillis, Adams, Wolf, et al. (2013)
	Interactions with mussel glue protein	Deacon, Davis, White, Waite, and Harding (1997), Deacon, Davis, Waite, and Harding (1998).
	Mucoadhesive delivery systems	Fiebrig et al. (1995); Fiebrig, Harding, and Davis (1994)
	Solution structure of pig colonic mucins	Fogg, Allen, Harding, and Pearson (1994); Fogg et al. (1996); Jumel, Fogg, et al. (1997)
	Use of AUC in mucoadhesion analysis	Harding (2003, 2006)
	Linear swollen coil model from AUC and electron microscopy	Harding, Creeth, and Rowe (1983); Harding, Rowe, and Creeth (1983)
	Sedimentation equilibrium analysis	Harding (1994a)
	Diffusion coefficients from AUC and correlation with gel hydration and dissolution	Dodd, Place, Hall, and Harding (1998)
	Study of polyelectrolyte behavior	Harding and Creeth (1983)
	Heterogeneity analysis	Harding (1984)
	Macrostructure from AUC and other methods	Harding (1989)
	Analysis by AUC of mucoadhesion	Harding (1998); Harding, Davis, Deacon, et al. (1999)
	Analysis by AUC of mucins and polysaccharides	Harding (1994b)

Table 1 Analytical Ultracentrifuge Studies on Carbohydrate Polymers—cont'd

Macromolecular System	Type of Study	References
Pectic polysaccharides from *Biophytum petersianum* Klotzsch	Solution properties in relation to immunology	Inngjerdingen et al. (2008)
Pectic polysaccharides from *Glinus oppositifolius*	Solution properties in relation to immunology	Inngjerdingen et al. (2005, 2007)
Pectin	Elevated temperature AUC and depolymerization	Morris, Butler, Foster, Jumel, and Harding (1999); Morris, Foster, and Harding (2002)
	Effect of degree of esterification on AUC properties	Morris, Foster, and Harding (2000)
	Effect of different storage temperatures on solutions and comparison with gels	Morris, Castile et al. (2010)
	Effect of uv-tagging on properties	Morris, Hromadkova, et al. (2002)
	Solution properties in relation to immunology of pectin and pectic arabinogalactan from *Vernonia kotschyana*	Negard et al. (2005)
	Stability	Morris, Adams, Harding, Castile, and Smith (2012)
Polysaccharide encapsulants	Review of properties including AUC	Wandrey, Bartkowiak, and Harding (2009)
Polysaccharide stability against bioprocessing	Use of AUC with other hydrodynamic tools	Harding (2010a, 2012a)
Polysaccharide–protein complexes	AUC and other analyses	Harding et al. (1993)
	Solution studies of interaction products after dry-heating of mixtures	Jumel, Harding, Mitchell, and Dickinson (1993); Kelly, Gudo, Mitchell, and Harding (1994)

Continued

Table 1 Analytical Ultracentrifuge Studies on Carbohydrate Polymers—cont'd

Macromolecular System	Type of Study	References
Pullulans	Characterization of standard samples	Kawahara, Ohta, Miyamoto, and Nakamura (1984)
	Solution properties of pullulans recovered from agricultural waste	Israilides, Scanlon, Smith, Harding, and Jumel (1994)
Schizophyllan	Hydrodynamic study in water	Yanaki, Norisuye, and Fujita (1980)
Starch	Heterogeneity of starches from somatic embryo- and seed-derived plants	Azhakanandam et al. (2000)
	Preparative/analytical ultracentrifuge study on density gradient fractionation	Majzoobi, Rowe, Connock, Hill, and Harding (2003)
	Molecular weights of amyloses and amylopectins	Ong, Jumel, Tokarczuk, Blanshard, and Harding (1994)
	Sedimentation study on the effect of damage on starch solutions	Tester, Patel, and Harding (2006)
	Heterogeneity of starch from modified waxy maize	Desse et al. (2009)
	Sedimentation velocity characterization of heterogeneity—comparison with SEC-MALS for rice starch	Tongdang, Bligh, Jumel, and Harding (1999)
	Sedimentation velocity characterization of starch-modified magnetite nanoparticles	Soshnikova et al. (2013)
Stewartan	Molecular weight and shape	Jumel, Geider, Harding (1997)
Sugar beet pectins	Solution characterization	Morris, Ralet, Bonnin, Thibault, and Harding (2010)

Table 1 Analytical Ultracentrifuge Studies on Carbohydrate Polymers—cont'd

Macromolecular System	Type of Study	References
Ternary mixtures of konjac glucomannan, xanthan, and alginate	AUC study of ionic strength-dependent interactions	Harding, Smith, et al. (2011)
	Interaction—and effect of ionic strength	Abdelhameed et al. (2010)
Tomato pectin	Effect of genetic modification on molecular weights	Seymour and Harding (1987); Seymour, Harding, Taylor, Hobson, and Tucker (1987); Smith et al. (1990)
Xanthan	Molecular weight, heterogeneity	Dhami, Harding, Jones, et al. (1995)
	Comparative heterogeneity with different pyruvate and acetate content	Erten et al. (2014)
Xylinan (acetan)	AUC and DLS study	Harding et al. (1996)
Xyloglucans (irradiated)	Global hydrodynamic analysis of flexibility of irradiated material	Patel, Morris, Ebringerova, et al. (2008)
	Solution properties of pressure solubilized material from tamarind seed and detarium gum	Picout, Ross-Murphy, Errington, and Harding (2003)

to be nonreversible: i.e., lowering the concentration does not cause a reversible dissociation. Two systems have however been shown to give reversible behavior—one a very weak dimerization (heteroxylan) and the other a stronger oligomerization (aminocelluloses).

A very weak reversible dimerization was observed for heteroxylan—these are β(1–4)-linked xylose polymers with arabinose (sometimes esterified with phenolic acids) side chains (Patel et al., 2007). Evidence for dimerization comes not from the appearance of an extra component in the sedimentation coefficient distribution plots but from the concentration dependence of the sedimentation coefficient after allowing for nonideality. For example, one heteroxylan (PO2) gave a dissociation constant value $K_d \sim (340 \pm 50)\ \mu M$ at

20 °C. Intriguingly decreasing the temperature to 5 °C greatly suppressed the interaction ($K_d > 3000\ \mu M$), whereas raising the temperature to 30 °C increased the dimerization strength ($K_d \sim 140\ \mu M$): this decrease in K_d (increase in dimerization) with increase in temperature is systematic of a reversible hydrophobic interaction.

Even more unusual *protein-like* association has been observed in solutions of the aqueous soluble carbohydrates known as the 6-deoxy-6-(ω-aminoalkyl) aminocelluloses which had previously been reported to produce controllable self-assembling films for enzyme immobilization and other applications. $c(s)$ versus s distributions show multiple components (up to pentamers) all linked according to $s \sim M^{2/3}$ (Fig. 10). Not only had such oligomerization—of the sort seen, for example, in antibody aggregation processes—never been seen in polysaccharides before, but also the power law coefficient of 2/3 is more appropriate for globular proteins rather than carbohydrates (Heinze et al., 2011a, 2011b). The oligomerization was at least partially reversible (the proportion of the lower molecular weight components decreasing as the total concentration is increased). One particular aminocellulose exhibited a completely reversible tetramerization step in the associative process (Nikolajski et al., 2014) as deduced on the basis of (i) point average molecular weights (local estimates of molecular weight at individual radial positions, r), $M_n(r)$, $M_w(r)$, and $M_z(r)$ all converging to the same (monomer value) $M_1 = 3250$ g/mol at zero concentration (and converging to the tetramer value

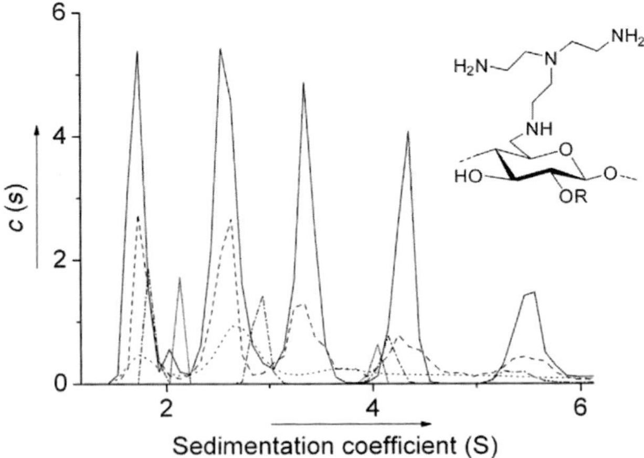

Figure 10 Sedimentation coefficient distributions of a 6-deoxy-6-amino cellulose (BAEA cellulose) at various concentrations: solid (—) 2.0 mg/ml; dash (– –) 1.0 mg/ml; dot (······) 0.5 mg/ml; dash dot (– · – ·) 0.25 mg/ml; short dot (······) 0.125 mg/ml. *Reproduced from Heinze et al. (2011a) with permission from Wiley.*

at higher concentration) and (ii) overlap of plots of $M_z(r)$ versus $c(r)$ for different loading concentrations. This completely reversible tetramerization (and subsequent assembly into larger structures) resembled more like the behavior of a protein self-assembly process—such as sickle cell deoxyhemoglobin formation— than the behavior of any carbohydrate, possibly revising traditional views of what is "protein-like" and what is "carbohydrate-like" behavior (Nikolajski et al., 2014).

7.2 Carbohydrate-Protein Interactions: Co-Sedimentation in the Analytical Ultracentrifuge

The principle of co-sedimentation for the evaluation of the extent of "binding" or association of macromolecules to other macromolecules or smaller ligands—using different optical systems or the UV-absorption optical system tuned at different wavelengths to pick up different chromophores—has been described by Harding and Winzor (2001) and applied to the quantification of protein–cofactor interactions (Marsh & Harding, 1993; Marsh, Harding, & Leadley, 1989) for the elucidation of protein–cofactor interactions. In addition, this has also been applied to encapsulation processes (Morgan, Harding, & Petrak, 1990a, 1990b; Wandrey et al., 2009), mucoadhesive systems (Harding, 2003, 2006), and the study of protein–polysaccharide interactions.

Recent work on carbohydrate–protein interactions has been focusing on the possible use of fiber polysaccharides as a macromolecular block trying to stop gliadins reaching the mucosal epithelia and causing an immune response in people with gluten intolerance or coeliac disease. The work is currently ongoing, but an assay procedure based on sedimentation velocity in the analytical ultracentrifuge monitoring for co-sedimentation has been developed (Kök et al., 2012). The assay procedure takes advantage of the fact that gliadins absorb ultraviolet light at a wavelength ~280 nm, whereas polysaccharides generally do not. Figure 11 shows an example for an interaction between iota-carrageenan and gliadin. A comprehensive survey of fiber polysaccharides as potential macromolecular barriers to gliadin will be published in the near future.

7.3 Carbohydrate—Mucin "Mucoadhesive" Interactions

Co-sedimentation has also been used extensively to the study of mucin–mucoadhesive interactions. Mucoadhesion has been considered as a route to making the delivery of drugs taken orally more efficient: by using a

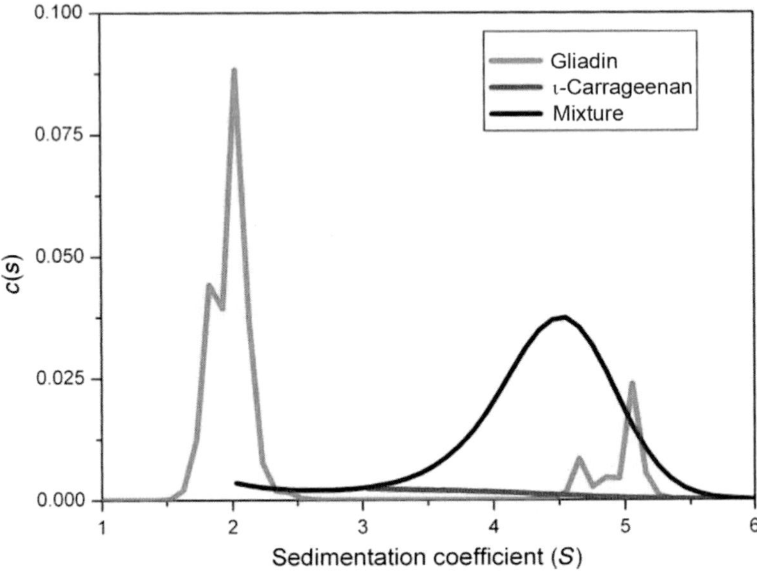

Figure 11 Sedimentation coefficient distribution diagrams for gliadins and iota-carrageenan in aqueous phosphate-chloride buffer. UV-absorption optics at 280 nm were used showing only the gliadins—and whatever they may have interacted with. Red line (light gray in the print version): gliadin only control at 5.0 mg/ml loading concentration showing material sedimenting at 2S and a small amount of aggregated material at ~5S. Blue line (dark gray in the print version): iota-carrageenan control at 1.0 mg/ml. The sedimenting material is almost transparent at 280 nm. Black line (same concentrations): mixture showing a substantial amount of material sedimenting at ~4.5S: this may indicate an interaction with gliadin. *Reproduced with permission from Kök et al. (2012).*

suitable mucoadhesive polymer carrier, the transit time through the maximum sites of absorption (usually the small intestine) can be increased through interactions of the carrier with the mucosal layer lining the gastrointestinal tract. Similar strategies are considered for application through the mucosal layer lining the nose. Table 1 includes a comprehensive list of the studies that have used co-sedimentation to probe the efficiency of candidate mucoadhesive polysaccharides—the most effective polysaccharide so far has been shown to be polycationic chitosan and the highly cationic mussel glue protein (see Harding, 2003, 2006 & references therein). The effects of changing the solvent environment (pH, ionic strength, bile salt concentration, etc.) can be easily probed, also the efficiency of binding of mucoadhesives with mucin glycoproteins from different regions of the alimentary tract (Deacon et al., 1999).

7.4 Carbohydrate—Nucleic Acid Interactions

There is increasing interest in the use of polycationic polysaccharides as histone analogues for condensing nucleic acids in DNA/RNA therapies against disease. Chitosans have been the focus of particular attention; these are soluble derivatives of chitin (poly *N*-acetyl glucosamine). Reducing the degree of acetylation (DA) yields the soluble polycationic form of chitin known as chitosan. In a recent study (Almutairi et al., 2015), the principle of co-sedimentation was used to explore the effectiveness of chitosans for binding to DNA, in a similar way to earlier studies applied to the study of chitosan "mucoadhesive" types of interaction with mucins. In the Almutairi et al. (2015) study, solutions of two chitosan samples of different degrees of acetylation, known as "CHIT5" (DA=25%) and "CHIT6" (15%), and different weight average molecular weights M_w (95,000 and 170,000 g/mol, respectively) from sedimentation equilibrium were characterized and then studied in mixtures with the DNA of M_w, estimated from the *Extended Fujita* method of Harding, Schuck, et al. (2011) and Harding, Smith, et al. (2011) to be approximately 300,000 g/mol.

Sedimentation velocity of separate 1.0 mg/ml solutions of chitosans CHIT5, CHIT6, and DNA at a temperature of 20 °C all gave unimodal distributions with respective weight average sedimentation coefficients $s_{20,w}$ of 1.9S, 2.3S, and 6.8S, respectively. Each chitosan preparation was then mixed in a 1:1 w/w ratio with the DNA to a total concentration of 1.0 mg/ml. A clear shift to a high s value is observed in both cases, with nothing sedimenting at the rate of uncomplexed chitosan suggesting that the interaction had gone to completion. From the shoulder of the complex, some unreacted DNA appeared remaining. Multi-Gaussian analysis of the $g(s)$ versus s distribution for the complex suggests that ~72% DNA had interacted with CHIT5 and ~83% for CHIT6. The s distribution of the complex is very broad, even on a logarithmic scale, with material in excess of 100S, with the larger molecular weight and highest positively charged chitosan CHIT6 showing the greatest degree of complexation (Fig. 12).

7.5 Carbohydrate–Carbohydrate Interactions

7.5.1 Chitosan–Xanthan Interactions

The advantageous properties of mixtures of chitosan and xanthan as hydrogels have been considered for somewhile but only very recently has the biophysical basis behind this interaction been studied by AUC, again using co-sedimentation (Almutairi et al., 2015). Xanthan, like DNA, is a double

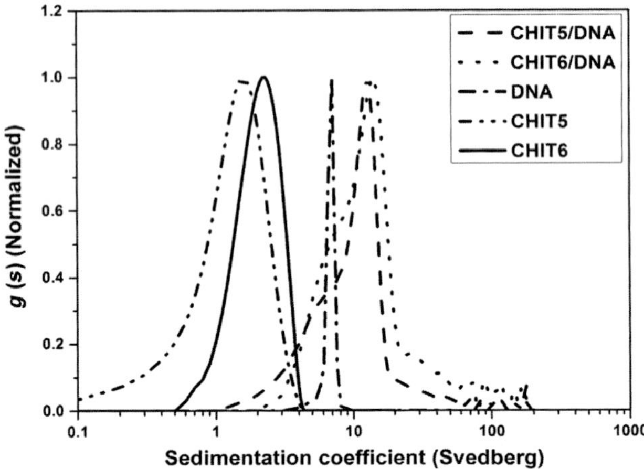

Figure 12 Normalized sedimentation coefficient distribution profiles obtained from sedimentation velocity experiments for unmixed controls and mixtures of chitosan with DNA. The shift of the complex to a higher sedimentation coefficient is seen. Multi-Gaussian analysis yields an estimate of the proportion of material complexed. *Reprinted from Almutairi et al. (2015), with permission from Elsevier.*

helical highly polyanionic polymer obtained from the microorganism *Xanthomonas campestris* and is widely used in the Food Industry as a thickener. The molecular weight of the particular xanthan used in the study of Almutairi et al. (2015)—known as "xanthan-STD"—had been found previously to be $\sim 3.2 \times 10^6$ g/mol using sedimentation equilibrium in the analytical ultracentrifuge (Erten et al., 2014).

As with the chitosan–DNA study, the sedimentation velocity profile of xanthan was first characterized at 1.0 mg/ml yielding a unimodal $g(s)$ versus s plot with weighted average $s_{20,w}$ of 4.8S. Then 1:1 by weight mixtures of chitosan with xanthan at a total concentration of 1.0 mg/ml were considered. As with chitosan–DNA, all the chitosan was complexed, but unlike chitosan–DNA, for xanthan–DNA there was no residual xanthan left behind for both CHIT5 and CHIT6 complexes (Almutairi et al., 2015). The weighted average $s_{20,w}$ for the complexes of 10.3S and 10.1S for CHIT5–xanthan and CHIT6–xanthan, respectively, was obtained.

7.5.2 PGX®

Sedimentation coefficient distribution profiling has also been used to good effect for detecting interactions in mixed polysaccharide systems used for health applications. An example is the proprietary commercial product

PGX® (PolyGlycopleX®) which is manufactured from konjac glucomannan, xanthan, and sodium alginate using a proprietary manufacturing process (EnviroSimplex®). This product has been successfully used as a dietary/satiety product for help toward obesity control. Detailed analysis by co-sedimentation (Abdelhameed et al., 2010; Harding, Smith, et al., 2011) showed clear evidence for a ternary interaction between the components at low ionic strength reinforcing experiments on the rheological properties of this system.

8. CONCLUSIONS

We have seen how important the analytical ultracentrifuge is for the characterization of large carbohydrate-based polymers. These materials still represent a considerable challenge, because of their polydispersity, great variety of conformation types, and large nonidealities, but there are nonetheless good strategies for dealing with these problems. One of the greatest recent impacts has been the development of software for the handling of both sedimentation velocity and sedimentation equilibrium data, greatly enhancing the resolving power of the instrument. In terms of future developments, one important area that is being explored is the reintroduction of Schlieren (refractive index gradient) optical system to complement the existing optical systems available. This optical system was once a feature of earlier commercial analytical ultracentrifuges such as the Beckman Model E, the MOM (Hungarian Optical Works) ultracentrifuge, and the MSE Centriscan ultracentrifuges (see Harding, Rowe, & Horton, 1992), instruments that are no longer available. For protein and nucleic acid work, the diversity of the optical systems present on current commercially available instruments—the Rayleigh interference, uv/visible absorption, and fluorescence optical systems—are particularly useful for protein-based systems. For polysaccharides, however, which lack appropriate chromophores or fluorophores (unless specifically labeled), only the Rayleigh system has been available, unless interactions with proteins or other materials with useable chromophores are being investigated—for example, in the interaction studies between polysaccharides with proteins, DNA, and mucins considered above. Incorporation of a Schlieren system—which like the Rayleigh system does not need chromophores—would greatly improve the power of the instrumentation for polysaccharide and synthetic polymer applications. Shortly after the introduction of the Beckman XL-A ultracentrifuge in 1990, it was shown that under certain situations, Schlieren profiles could be generated (Dhami,

Cölfen, et al., 1995; Dhami, Harding, Elizabeth, et al., 1995; Dhami, Harding, Jones, et al., 1995), and how this could now be extended into the production of a fully integrated system. Ultracentrifuge methods are also proving invaluable in increasing our understanding about the enzyme systems involved in both synthesis and degradation. These methods have already proven useful in early attempts at gene silencing of enzymes considered responsible for the softening of tomatoes on ripening (Seymour & Harding, 1987; Seymour et al., 1987; Smith et al., 1990) and now, taking advantage of the new developments in software it appears that they have helped with the identification of a specific pectate lyase (Uluisik et al., 2015), yet another example of the importance of the analytical ultracentrifuge.

REFERENCES

Abdelhameed, A. S., Ang, S., Morris, G. A., Smith, I., Lawson, C., Gahler, R., et al. (2010). An analytical ultracentrifuge study on ternary mixtures of konjac glucomannan supplemented with sodium alginate and xanthan gum. *Carbohydrate Polymers*, *81*, 145–148.

Abdelhameed, A. S., Morris, G. A., Adams, G. G., Rowe, A. J., Laloux, O., Cerny, L., et al. (2012). An asymmetric and slightly dimerized structure for the tetanus toxoid protein used in glycoconjugate vaccines. *Carbohydrate Polymers*, *90*, 1831–1835.

Adams, G. G., Kök, M. S., Imran, S., Harding, S. E., Ilyas, M., & Tatham, A. S. (2012). The interaction of dietary fibres with disulphide bonds (S-S) and a potential strategy to reduce the toxicity of the gluten proteins in coeliac disease. *Biotechnology & Genetic Engineering Reviews*, *28*, 115–130.

Almutairi, F. M., Adams, G. G., Kök, M. S., Lawson, C. J., Gahler, R., Wood, S., et al. (2013). An analytical ultracentrifugation based study on the conformation of lambda carrageenan in aqueous solution. *Carbohydrate Polymers*, *97*, 203–209.

Almutairi, F. M., Erten, T., Adams, G. G., Hayes, M., Mcloughlin, P., Kök, M. S., et al. (2015). Hydrodynamic characterisation of chitosan and its interaction with two polyanions: DNA and xanthan. *Carbohydrate Polymers*, *122*, 355–366.

Alzahrani, Q. E., Adams, G. G., Gillis, R. B., Besong, T. M. D., Kök, M. S., Fong, E., et al. (2015). Matrix-free hydrodynamic study on the size distribution and conformation of lignins from wood and non-wood. *Holtzforschung*. in press.

Anderson, M. T., Harding, S. E., & Davis, S. S. (1989). On the interaction in solution of a candidate mucoadhesive polymer, diethylaminoethyl-dextran, with pig gastric mucus glycoprotein. *Biochemical Society Transactions*, *17*, 1101–1102.

Astronomo, R. D., & Burton, D. R. (2010). Carbohydrate vaccines: Developing sweet solutions to sticky situations? *Nature Reviews. Drug Discovery*, *9*, 308–324.

Azhakanandam, K., Power, J. B., Lowe, K. C., Cocking, E. C., Tongdang, T., Jumel, K., et al. (2000). Qualitative assessment of aromatic Indica rice (Oryza sativa L.): Proteins, lipids and starch in grain from somatic embryo- and seed-derived plants. *Journal of Plant Physiology*, *156*, 783–789.

Azis, B. H., Chin, B., Deacon, M. P., Harding, S. E., & Pavlov, G. M. (1999). Size and shape of inulin in dimethyl sulphoxide solution. *Carbohydrate Polymers*, *38*, 231–234.

Ball, A., Harding, S. E., & Mitchell, J. R. (1988). Combined low-speed sedimentation equilibrium/gel permeation chromatography approach to molecular weight distribution analysis. *International Journal of Biological Macromolecules*, *10*, 259–264.

Ball, A., Harding, S. E., & Simpkin, N. J. (1990). On the molecular weight distribution of dextran T-500. *Gums and Stabilisers for the Food Industry*, *5*, 447–450.

Bohdanecky, M. (1983). New method for estimating the parameters of the wormlike chain model from the intrinsic viscosity of stiff-chain polymers. *Macromolecules*, *16*, 1483–1493.

Bradley, T. D., Ball, A., Harding, S. E., & Mitchell, J. R. (1989). The thermal degradation of guar gum. *Carbohydrate Polymers*, *10*, 205–214.

Brunow, G. (2001). Methods to reveal the structure of lignin. In M. Hofrichter & A. Steinbüchel (Eds.), *Biopolymers. Volume 1. Lignin, humic substances and coal*, Vol. 1 (p. 106); Weinheim, Germany: Wiley-VCH Verlag GmbH.

Bushin, S. V., Tsvetkov, V. N., Lysenko, Y. B., & Emel'yanov, V. N. (1981). Conformational properties and rigidity of molecules of ladder polyphenylsiloxane in solutions according to the data of sedimentation-diffusion analysis and viscometry. *Vysokomolekulyarnye Soedineniya*, *A23*, 2494–2503.

Clay, O., Carels, N., Douady, C. J., & Bernadi, G. (2005). Using analytical ultracentrifugation of DNA in CsCl gradients to explore large-scale properties of genomes. In D. J. Scott, S. E. Harding, & A. J. Rowe (Eds.), *Analytical ultracentrifugation. Techniques and methods* (pp. 231–252). Cambridge, UK: Royal Society of Chemistry.

Cölfen, H., & Harding, S. E. (1997). MSTARA and MSTARI: Interactive PC algorithms for simple, model independent evaluation of sedimentation equilibrium data. *European Biophysics Journal*, *25*, 333–346.

Cölfen, H., Harding, S. E., & Vårum, K. M. (1996). Investigation, using analytical ultracentrifugation, of the effect of the incorporation of the fluorophore 9-anthraldehyde on two chitosans of differing degrees of acetylation. *Carbohydrate Polymers*, *30*, 55–60.

Cölfen, H., Harding, S. E., Vårum, K. M., & Winzor, D. J. (1996). A study by analytical ultracentrifugation on the interaction between lysozyme and extensively deacetylated chitin (chitosan). *Carbohydrate Polymers*, *30*, 45–53.

Creeth, J. M. (1947). *Some physico-chemical studies on nucleic acids and related substances*. PhD Dissertation, UK: University College Nottingham.

Creeth, J. M. (1964). An approximate 'steady state' condition in the Ultracentrifuge. *Proceedings of the Royal Society of London*, *A282*, 403–421.

Creeth, J. M., Gulland, J. M., & Jordan, D. O. (1947). Deoxypentose nucleic acids. Part III. Viscosity and streaming birefringence of solutions of the sodium salt of the deoxypentose nucleic acid of calf thymus. *Journal of the Chemical Society*, *1947*, 1141–1145.

Creeth, J. M., & Harding, S. E. (1982a). A simple test for macromolecular heterogeneity in the analytical ultracentrifuge. *Biochemical Journal*, *205*, 639–641.

Creeth, J. M., & Harding, S. E. (1982b). Some observations on a new type of point average molecular weight. *Journal of Biochemical and Biophysical Methods*, *7*, 25–34.

Creeth, J. M., & Knight, C. G. (1965). On the estimation of the shape of macromolecules from sedimentation and viscosity measurements. *Biochimica et Biophysica Acta*, *102*, 549–558.

Daly, P., Maluk, M., Zwirek, M., & Halpin, C. (2012). Lignin biosynthesis and lignin manipulation. In S. E. Harding (Ed.), *Stability of complex carbohydrate structures: Biofuels, foods, vaccines and shipwrecks* (pp. 153–159). Cambridge, UK: Royal Society of Chemistry.

Dam, J., & Schuck, P. (2004). Calculating sedimentation coefficient distributions by direct modeling of sedimentation velocity concentration profiles. *Methods in Enzymology*, *384*, 185–212.

Deacon, M. P., Davis, S. S., Waite, J. H., & Harding, S. E. (1998). Structure and mucoadhesion of mussel glue protein in dilute solution. *Biochemistry*, *37*, 14108–14112.

Deacon, M. P., Davis, S. S., White, R. J., Nordman, H., Carlstedt, I., Errington, N., et al. (1999). Are chitosan-mucin interactions specific to different regions of the stomach? Velocity ultracentrifugation offers a clue. *Carbohydrate Polymers*, *38*, 235–238.

Deacon, M. P., Davis, S. S., White, R., Waite, J. H., & Harding, S. E. (1997). Monomeric behaviour of Mytilus edulin (mussel) glue protein in dilute solution. *Biochemical Society Transactions, 25*, 422S.

Deacon, M. P., McGurk, S., Roberts, C. J., Williams, P. M., Tendler, S. J. B., Davies, M. C., et al. (2000). Atomic force microscopy of gastric mucin and chitosan mucoadhesive systems. *Biochemical Journal, 348*, 557–563.

Desse, M., Ang, S., Morris, G. A., Abu-Hardan, M., Wolf, B., Hill, S. E., et al. (2009). Analysis of the continuous phase of the modified waxy maize suspension. *Carbohydrate Polymers*, 77, 320–325.

Dhami, R., Cölfen, H., & Harding, S. E. (1995). A comparitive "Schlieren" study of the sedimentation behaviour of three polysaccharides using the Beckman Optima XL-A and Model E analytical centrifuges. *Progress in Colloid and Polymer Science, 99*, 188–193.

Dhami, R., Harding, S. E., Elizabeth, N. J., & Ebringerova, A. (1995). Hydrodynamic characterisation of the molar mass and gross conformation of corn cob heteroxylan AGX. *Carbohydrate Polymers, 28*, 113–119.

Dhami, R., Harding, S. E., Jones, T., & Hughes, T. (1995). Physico-chemical studies on a commercial food-grade xanthan—I. Characterisation by sedimentation velocity, sedimentation equilibrium and viscometry. *Carbohydrate Polymers, 27*, 93–99.

Dodd, S., Place, G. A., Hall, R. L., & Harding, S. E. (1998). Hydrodynamic properties of mucins secreted by primary cultures of Guinea-pig tracheal epithelial cells: Determination of diffusion coefficients by analytical ultracentrifugation and kinetic analysis of mucus gel hydration and dissolution. *European Biophysics Journal, 28*, 38–47.

Errington, N., Harding, S. E., Illum, L., & Schact, E. H. (1992). Physico-chemical studies on di-iodotyrosine dextran. *Carbohydrate Polymers, 18*, 611–613.

Errington, N., Harding, S. E., & Rowe, A. J. (1992). Use of scanning absorption optics for sedimentation equilibrium analysis of labelled polysaccharides: Molecular weight of blue dextran. *Carbohydrate Polymers, 17*, 151–154.

Errington, N., Harding, S. E., Vårum, K. M., & Illum, L. (1993). Hydrodynamic characterization of chitosans varying in degree of acetylation. *International Journal of Biological Macromolecules, 15*, 113–117.

Erten, T., Adams, G., Foster, T. J., & Harding, S. E. (2014). Comparative heterogeneity, molecular weights and viscosities of xanthans of different pyruvate and acetate content. *Food Hydrocolloids, 42*, 335–341.

Favis, B. D., & Goring, D. A. I. (1984). A model for the leaching of lignin macromolecules from pulp fibres. *Journal of Pulp and Paper Science, 10*, J139–J143.

Fee, M., Errington, N., Jumel, K., Illum, L., Smith, A., & Harding, S. E. (2003). Correlation of SEC/MALLS with ultracentrifuge and viscometric data for chitosans. *European Biophysics Journal, 32*, 457–464.

Fiebrig, I., Davis, S. S., & Harding, S. E. (1995). Methods used to develop mucoadhesive drug delivery systems: Bioadhesion in the gastrointestinal tract (Chapter 18). In S. E. Harding, S. E. Hill, & J. R. Mitchell (Eds.), *Biopolymer mixtures*. Nottingham: Nottingham University Press.

Fiebrig, I., Harding, S. E., & Davis, S. S. (1994). Sedimentation analysis of potential interactions between mucins and a putative bioadhesive polymer. *Progress in Colloid and Polymer Science, 94*, 66–73.

Fiebrig, I., Harding, S. E., Stokke, B. T., Vårum, K. M., Jordan, D., & Davis, S. S. (1994). The potential of chitosan as a mucoadhesive drug carrier: Studies on its interaction with pig gastric mucin on a molecular level. *European Journal of Pharmaceutical Science, 2*, 185.

Flint, J., Nurizzo, D., Harding, S. E., Longman, E., Davies, G. J., Gilbert, H. J., et al. (2004). Ligand-mediated dimerization of a carbohydrate-binding module reveals a novel mechanism for protein-carbohydrate recognition. *Journal of Molecular Biology, 337*, 417–426.

Fogg, F. J., Allen, A., Harding, S. E., & Pearson, J. P. (1994). The structure of secreted mucins isolated from the adherent mucus gel: Comparison with the gene products. *Biochemical Society Transactions*, *22*, 229S.

Fogg, F. J. J., Hutton, D. A., Jumel, K., Pearson, J. P., Harding, S. E., & Allen, A. (1996). Characterization of pig colonic mucins. *Biochemical Journal*, *316*, 937–942.

Gaisford, S. E., Harding, S. E., Mitchell, J. R., & Bradley, T. D. (1986). A comparison between the hot and cold water soluble fractions of two locust bean gum samples. *Carbohydrate Polymers*, *6*, 423–442.

García de la Torre, J., & Harding, S. E. (2013). Hydrodynamic modelling of protein conformation in solution: ELLIPS and HYDRO. *Biophysical Reviews*, *5*, 195–206.

Gillis, R. B., Adams, G. G., Heinze, T., Nikolajski, M., Harding, S. E., & Rowe, A. J. (2013). MultiSig: A new high-precision approach to the analysis of complex biomolecular systems. *European Biophysics Journal*, *42*, 777–786.

Gillis, R. B., Adams, G. G., Wolf, B., Berry, M., Besong, T. M. D., Corfield, A., et al. (2013). Molecular weight distribution analysis by ultracentrifugation: Adaptation of a new approach for mucins. *Carbohydrate Polymers*, *93*, 178–183.

Gralen, N. (1944). *Sedimentation and diffusion measurements on cellulose and cellulose derivatives*. PhD Dissertation. Sweden: University of Uppsala.

Harding, S. E. (1984). An analysis of the heterogeneity of mucins. *Biochemical Journal*, *219*, 1061–1064.

Harding, S. E. (1988). Polysaccharide molecular weight determination: Which technique? *Gums and Stabilisers for the Food Industry*, *4*, 15–23.

Harding, S. E. (1989). The macrostructure of mucus glycoproteins in solution. *Advances in Carbohydrate Chemistry and Biochemistry*, *47*, 345–381.

Harding, S. E. (1992). Sedimentation analysis of polysaccharides. In S. E. Harding, A. J. Rowe, & J. C. Horton (Eds.), *Analytical ultracentrifugation in biochemistry and polymer science* (pp. 495–516). Cambridge: Royal Society of Chemistry.

Harding, S. E. (1993). Examining Cinderella (the 7th Harden discussion meeting on physical methods for glycopolymer characterisation, Eynsham Hall, 11–13 September 1992). *The Biochemist*. Jun/July, 30–31.

Harding, S. E. (1994a). Sedimentation equilibrium analysis of glycopolymers. *Progress in Colloid and Polymer Science*, *94*, 54–65.

Harding, S. E. (1994b). The analytical ultracentrifuge spins again. *Trends in Analytical Chemistry*, *13*, 439–446.

Harding, S. E. (1994c). The sedimentation equilibrium analysis of polysaccharides and mucins: A guided tour of problem solving for difficult heterogeneous systems. In T. M. Schuster & T. M. Laue (Eds.), *Modern analytical ultracentrifugation* (pp. 315–341). Boston USA: Birkhäuser.

Harding, S. E. (1994d). Shapes and sizes of food polysaccharides by sedimentation analysis—Recent developments. *Gums and Stabilisers for the Food Industry*, *7*, 55–65.

Harding, S. E. (1994e). Determination of macromolecular heterogeneity, shape and interactions using sedimentation velocity analytical ultracentrifugation. *Methods in Molecular Biology*, *22*, 61–73.

Harding, S. E. (1994f). Determination of absolute molecular weights using sedimentation equilibrium analytical ultracentrifugation. *Methods in Molecular Biology*, *22*, 75–84.

Harding, S. E. (1995a). Some recent developments in the size and shape analysis of industrial polysaccharides in solution using sedimentation analysis in the analytical ultracentrifuge. *Carbohydrate Polymers*, *28*, 227–237.

Harding, S. E. (1995b). On the hydrodynamic analysis of macromolecular conformation. *Biophysical Chemistry*, *55*, 69–93.

Harding, S. E. (1995c). Ultracentrifugation of food biopolymers. In E. Dickinson (Ed.), *New physico-techniques for the characterisation of complex food systems* (pp. 117–146). London, UK: Blackie Academic & Professional.

Harding, S. E. (1997a). Characterisation of chitosan-mucin complexes by sedimentation velocity analytical ultracentrifugation. In R. A. A. Muzzarelli & M. G. Peter (Eds.), *Chitin handbook* (pp. 457–466). Lyon, France: European Chitin Society.

Harding, S. E. (1997b). The intrinsic viscosity of biological macromolecules. Progress in measurement, interpretation and application to structure in dilute solution. *Progress in Biophysics and Molecular Biology, 68*, 207–262.

Harding, S. E. (1998). Mucin-biopolymer interactions for mucoadhesion (Abstract 53). *Glycobiology, 8*(11).

Harding, S. E. (2000). Polysaccharides/centrifugation. In C. Poole & M. Cooke (Eds.), *Encyclopaedia of separation science III* (pp. 3921–3929). Orlando, FL: Academic press.

Harding, S. E. (2003). Mucoadhesive interactions. *Biochemical Society Transactions, 31*, 1036–1041.

Harding, S. E. (2005a). Analysis of polysaccharide size, shape and interactions. In D. J. Scott, S. E. Harding, & A. J. Rowe (Eds.), *Analytical ultracentrifugation. Techniques and methods* (pp. 231–252). Cambridge, UK: Royal Society of Chemistry.

Harding, S. E. (2005b). Analysis of polysaccharides by ultracentrifugation: Size, conformation and interactions in solution. *Advances in Polymer Science, 186*, 211–254.

Harding, S. E. (2005c). Challenges for the modern analytical ultracentrifuge analysis of polysaccharides. *Carbohydrate Research, 340*, 811–826.

Harding, S. E. (2006). Trends in mucoadhesive analysis. *Trends in Food Science and Technology, 17*, 255–262.

Harding, S. E. (2010a). Some observations on the effects of bioprocessing on biopolymer stability. *Journal of Drug Targeting, 18*, 732–740.

Harding, S. E. (2010b). *Dr. Michael Creeth. Scientist who helped pave the way for Watson and Crick*. UK: The Independent p. 39.

Harding, S. E. (Ed.), (2012a). *Stability of complex carbohydrate structures: Biofuels, foods, vaccines and shipwrecks*. Cambridge, UK: Royal Society of Chemistry.

Harding, S. E. (2012b). Viscometry, analytical ultracentrifugation and light scattering probes for carbohydrate stability. In S. E. Harding (Ed.), *Stability of complex carbohydrate structures: Biofuels, foods, vaccines and shipwrecks* (pp. 80–98). Cambridge, UK: Royal Society of Chemistry.

Harding, S. E. (2013a). ELLIPS and COVOL. In G. G. K. Roberts & S. E. Harding (Eds.), *Encyclopedia of biophysics* (pp. 654–662). Berlin: Springer Verlag (Hydrodynamics section).

Harding, S. E. (2013b). Polysaccharides: Biophysical properties. In G. G. K. Roberts & S. E. Harding (Eds.), *Encyclopedia of biophysics* (pp. 1907–1912). Berlin: Springer Verlag (Hydrodynamics section).

Harding, S. E., Abdelhameed, A. S., & Morris, G. A. (2010). Molecular weight distribution evaluation of polysaccharides and glycoconjugates using analytical ultracentrifugation. *Macromolecular Bioscience, 10*, 714–720.

Harding, S. E., Abdelhameed, A. S., Morris, G. A., Adams, G. G., Laloux, O., Cerny, L., et al. (2012). Solution properties of capsular polysaccharides from Streptococcus pneumoniae. *Carbohydrate Polymers, 90*, 237–242.

Harding, S. E., Berth, G., Ball, A., & Mitchell, J. R. (1990). Hydrodynamic evidence for an extended conformation for citrus pectins in dilute solution. *Gums and Stabilisers for the Food Industry, 5*, 257–271.

Harding, S. E., Berth, G., Ball, A., Mitchell, J. R., & Garcia de la Torre, J. (1991). The molecular weight distribution and conformation of citrus pectins in solution studied by hydrodynamics. *Carbohydrate Polymers, 16*, 1–15.

Harding, S. E., Berth, G., Hartmann, J., Jumel, K., Cölfen, H., & Christensen, B. E. (1996). Physicochemical studies on Xylinan (Acetan). III. Hydrodynamic characterization by analytical ultracentrifugation and dynamic light scattering. *Biopolymers, 39*, 729–736.

Harding, S. E., & Cölfen, H. (1995). Inversion formulae for ellipsoid of revolution macromolecular shape functions. *Analytical Biochemistry*, *228*, 131–142.

Harding, S. E., Cölfen, H., & Aziz, Z. (2005). The ELLIPS suite of whole-body protein conformation algorithms for Microsoft Windows. In D. J. Scott, S. E. Harding, & A. J. Rowe (Eds.), *Analytical ultracentrifugation. Techniques and methods* (pp. 460–483). Cambridge, UK: Royal Society of Chemistry.

Harding, S. E., & Creeth, J. M. (1982). Self association, polydispersity and non-ideality in a cystic fibrotic glycoprotein. *IRCS Medical Science*, *10*, 474–475.

Harding, S. E., & Creeth, J. M. (1983). Polyelectrolyte behaviour in mucus glycoproteins. *Biochimica et Biophysica Acta*, *746*, 114–119.

Harding, S. E., Creeth, J. M., & Rowe, A. J. (1983). Modelling the conformation of mucus glycoproteins in solution. In A. Chester, D. Heinegard, A. Lundblad, & S. Svensson (Eds.), *Proceedings 7th International Glycoconjugates Conference* (pp. 558–559), Sweden: Olsson Reklambyra.

Harding, S. E., Davis, S. S., Deacon, M. P., & Fiebrig, I. (1999). Biopolymer mucoadhesives. *Biotechnology and Genetic Engineering Reviews*, *16*, 41–86.

Harding, S. E., Day, K., Dhami, R., & Lowe, P. M. (1997). Further observations on the size, shape and hydration of kappa-carrageenan in dilute solution. *Carbohydrate Polymers*, *32*, 81–87.

Harding, S. E., Horton, J. C., & Cölfen, H. (1997). The ELLIPS suite of macromolecular conformation algorithms. *European Biophysics Journal*, *25*, 347–360.

Harding, S. E., Horton, J. C., & Morgan, P. J. (1992). MSTAR: A FORTRAN algorithm for the model independent molecular weight analysis of macromolecules using low speed or high speed sedimentation equilibrium. In S. E. Harding, A. J. Rowe, & J. C. Horton (Eds.), *Analytical ultracentrifugation in biochemistry and polymer science* (pp. 275–294). Cambridge: Royal Society of Chemistry.

Harding, S. E., Jumel, K., Kelly, R., Gudo, E., Horton, J. C., & Mitchell, J. R. (1993). The structure and nature of protein-polysaccharide complexes. *Food Proteins: Structure and Functionality*, *8*, 216–226.

Harding, S. E., Rowe, A. J., & Creeth, J. M. (1983). Further evidence for a flexible and highly expanded spheroidal model for mucus glycoproteins in solution. *Biochemical Journal*, *209*, 893–896.

Harding, S. E., Rowe, A. J., & Horton, J. C. (Eds.), (1992). *Analytical ultracentrifugation in biochemistry and polymer science*. Cambridge, UK: Royal Society of Chemistry.

Harding, S. E., Schuck, P., Abdelhammed, A. S., Adams, G., Kök, M. S., & Morris, G. A. (2011). Extended Fujita approach to the molecular weight distribution of polysaccharides and other polymer systems. *Methods*, *54*, 136–144.

Harding, S. E., Smith, I. H., Lawson, C. J., Gahler, R. J., & Wood, S. (2011). Studies on macromolecular interactions in ternary mixtures of konjac glucomannan, xanthan gum and sodium alginate. *Carbohydrate Polymers*, *83*, 329–338.

Harding, S. E., & Tombs, M. P. (1989). Protein diffusion through interfaces. In K. D. Schwenke & B. Raab (Eds.), *Interactions in protein systems* (pp. 65–72). Berlin: Abhandlungen der Akademie der Wissenschaften der DDR, Akademie-Verlag.

Harding, S. E., & Tombs, M. P. (2002). The analytical ultracentrifuge as a probe for interface transport phenomena. *Biotechnology & Genetic Engineering Reviews*, *19*, 55–69.

Harding, S. E., Vårum, K. M., Stokke, B. T., & Smidsrød, O. (1991). Molecular weight determination of polysaccharides. *Advances in Carbohydrate Analysis*, *1*, 63–144.

Harding, S. E., & Winzor, D. J. (2001). Sedimentation velocity analytical ultracentrifugation. In S. E. Harding & B. Z. Chowdhry (Eds.), *Protein-ligand interactions: Hydrodynamics and calorimetry* (pp. 75–103). Oxford, UK: Oxford University Press.

Harding, S. E., & Winzor, D. J. (2010). James Michael Creeth, 1924–2010. *Macromolecular Bioscience*, *10*, 696–699.

Heinze, T., Nikolajski, M., Daus, S., Besong, T. M. D., Michaelis, N., Berlin, P., et al. (2011a). Proteinähnliche oligomerisierung von Kohlenhydraten. *Angewandte Chemie*, *87*, 8761–8763.

Heinze, T., Nikolajski, M., Daus, S., Besong, T. M. D., Michaelis, N., Berlin, P., et al. (2011b). Protein-like oligomerisation of carbohydrates. *Angewandte Chemie, International Edition*, *50*, 8602–8604.

Hokputsa, S., Gerddit, W., Pongsamart, S., Inngjerdingen, K., Heinze, T., Koschella, A., et al. (2004). Water-soluble polysaccharides with pharmaceutical importance from Durian rinds (Durio zibethinus Murr.): Isolation, fractionation, characterisation and bioactivity. *Carbohydrate Polymers*, *56*, 471–481.

Hokputsa, S., Harding, S. E., Inngjerdingen, S., Jumel, K., Michaelsen, T. E., Heinze, T., et al. (2004). Bioactive polysaccharides from the stems of the Thai medicinal plant Acanthus ebracteatus: Their chemical and physical features. *Carbohydrate Research*, *339*, 753–762.

Hokputsa, S., Hu, C., Paulsen, B. S., & Harding, S. E. (2003). A physico-chemical comparative study on extracellular carbohydrate polymers from five desert algae. *Carbohydrate Polymers*, *54*, 27–32.

Hokputsa, S., Jumel, K., Alexander, K., & Harding, S. E. (2003a). A comparison of molecular mass determination of hyaluronic acid using SEC/MALLS and sedimentation equilibrium. *European Biophysics Journal*, *32*, 450–456.

Hokputsa, S., Jumel, K., Alexander, C., & Harding, S. E. (2003b). Hydrodynamic characterisation of chemically degraded hyaluronic acid. *Carbohydrate Polymers*, *52*, 111–117.

Horton, J. C., Harding, S. E., & Mitchell, J. R. (1991). Gel permeation chromatography—Multi angle laser light scattering characterization of the molecular mass distribution of "Pronova" sodium alginate. *Biochemical Society Transactions*, *19*, 510–511.

Horton, J. C., Harding, S. E., Mitchell, J. R., & Morton-Holmes, D. F. (1991). Thermodynamic non-ideality of dilute solutions of sodium alginate studied by sedimentation equilibrium ultracentrifugation. *Food Hydrocolloids*, *5*, 125–127.

Inngjerdingen, K. T., Debes, S., Inngjerdingen, M., Hokputsa, S., Harding, S. E., Rolstad, B., et al. (2005). Bioactive pectic polysaccharides from Glinus oppositifolius (L.) Aug. DC., a Malian medicinal plant, isolation and partial characterization. *Journal of Ethnopharmacology*, *101*, 204–214.

Inngjerdingen, M., Inngjerdingen, K. T., Patel, T. R., Allen, S., Chen, X., Rolstad, B., et al. (2008). Pectic polysaccharides from Biophytum petersianum Klotzsch, and their activation of macrophages and dendritic cells. *Glycobiology*, *18*, 1074–1184.

Inngjerdingen, K. T., Patel, T., Chen, X., Kenne, L., Allen, S., Morris, G. A., et al. (2007). Immunological and structural properties of a pectic polymer from Glinus oppositifolius. *Glycobiology*, *17*, 1299–1310.

Israilides, C., Scanlon, B., Smith, A., Harding, S. E., & Jumel, K. (1994). Characterization of pullulans produced from agro-industrial waste. *Carbohydrate Polymers*, *25*, 203–209.

Jullander, I. (1945). Studies on nitrocellulose, including the construction of an osmotic balance. *Arkiv för Kemi, Mineralogi och Geologi*, *21A*, 132.

Jullander, I. (1987). Svedberg and the polysaccharides. In B. Ranby (Ed.), *Physical chemistry of colloids and macromolecules* (pp. 29–36). Oxford: Blackwell Scientific.

Jumel, K., Fiebrig, I., & Harding, S. E. (1996). Rapid size distribution and purity analysis of gastric mucus glycoproteins by size exclusion chromotography/multi angle laser light scattering. *International Journal of Biological Macromolecules*, *18*, 133–139.

Jumel, K., Fogg, F. J. J., Hutton, D. A., Pearson, J. P., Allen, A., & Harding, S. E. (1997). A polydisperse linear random coil model for the quaternary structure of pig colonic mucin. *European Biophysics Journal*, *25*, 477–480.

Jumel, K., Geider, K., & Harding, S. E. (1997). The solution molecular weight and shape of the bacterial exopolysaccharides amylovoran and stewartan. *International Journal of Biological Macromolecules, 20,* 251–258.

Jumel, K., Harding, S. E., & Mitchell, J. R. (1996). Effect of gamma irradiation on the macromolecular integrity of guar gum. *Carbohydrate Research, 282,* 223–236.

Jumel, K., Harding, S. E., Mitchell, J. R., & Dickinson, E. (1993). Evidence for protein-polysaccharide complex formation as a result of dry-heating of mixtures. In E. Dickinson & P. Walstra (Eds.), *Food colloids and polymers: Stability and mechanical properties* (pp. 156–160). Cambridge: Royal Society of Chemistry.

Jumel, K., Harding, S. E., Mitchell, J. R., To, K.-M., Hayter, I., O'mullane, J. E., et al. (1996). Molar mass and viscometric characterisation of hydroxypropylmethyl cellulose. *Carbohydrate Polymers, 29,* 105–109.

Jumel, K., Harding, S. E., Sobol, E., Omel'chenko, A., Sviridov, A., & Jones, N. (2002). Aspects of the structural integrity of chondroitin sulphate after laser irradiation. *Carbohydrate Polymers, 48,* 241–245.

Kawahara, K., Ohta, K., Miyamoto, H., & Nakamura, S. (1984). Preparation and solution properties of pullulan fractions as standard samples for water-soluble polymers. *Carbohydrate Polymers, 4,* 335–356.

Kelly, R., Gudo, E. S., Mitchell, J. R., & Harding, S. E. (1994). Some observations on the nature of heated mixtures of bovine serum albumin with an alginate and a pectin. *Carbohydrate Polymers, 23,* 115–120.

Kök, M. S., Abdelhameed, A. S., Ang, S., Morris, G. A., & Harding, S. E. (2009). A novel global hydrodynamic analysis of the molecular flexibility of the dietary fibre polysaccharide konjac glucomannan. *Food Hydrocolloids, 23,* 1910–1917.

Kök, M. S., Gillis, R., Ang, S., Lafond, D., Tatham, A. S., Adams, G., et al. (2012). Can dietary fibre help provide safer food products for sufferers of gluten intolerance? A well-established biophysical probe may help towards providing an answer. *BMC Biophysics, 5,* 10.

Kök, M. S., Harding, S. E., Adams, G. G., Tatham, A. S., & Morris, G. (2010). Exploring the potential of dietary fibres: Can protein-polysaccharide interactions be the key to providing a new solution to dietary disorders? *Journal of Biotechnology, 150,* 101.

Korneeva, E. V., Vichoreva, G. A., Harding, S. E., & Pavlov, G. M. (1996). Hydrodynamic study of carboxymethyl chitin. *Abstracts of Papers of the American Chemical Society, 212*(Part 1) 75-cell.

Koschella, A., Inngjerdingen, K., Paulsen, B. S., Morris, G. A., Harding, S. E., & Heinze, T. (2008). Unconventional methyl galactan synthesized via the thexyldimethylsilyl intermediate: Preparation, characterization, and properties. *Macromolecular Bioscience, 8,* 96–105.

Li, P., Harding, S. E., & Liu, Z. (2001). Cyanobacterial exopolysaccharides: Their nature and potential biotechnological applications. *Biotechnology & Genetic Engineering Reviews, 18,* 375–404.

Majzoobi, M., Rowe, A. J., Connock, M., Hill, S. E., & Harding, S. E. (2003). Partial fractionation of wheat starch amylose and amylopectin using zonal ultracentrifugation. *Carbohydrate Polymers, 52,* 269–274.

Marsh, E. N., & Harding, S. E. (1993). Methylmalonyl-CoA mutase from Propionibacterium shermanii: Characterisation of the cobalamin-inhibited form and subunit-cofactor interactions studied by analytical ultracentrifugation. *Biochemical Journal, 290,* 551–555.

Marsh, E. N., Harding, S. E., & Leadley, P. F. (1989). Subunit interactions in Propionibacterium shermanii methylmalonyl-CoA mutase studied by analytical ultracentrifugation. *Biochemical Journal, 260,* 353–358.

Meselson, M., & Stahl, F. W. (1958). The replication of DNA in Escherichia coli. *Proceedings of the National Academy of Sciences of the United States of America, 44,* 671–682.

Mitra, S. (2014). Detecting RNA tertiary folding by sedimentation velocity analytical ultracentrifugation. *Methods in Molecular Biology*, *1086*, 265–288.
Morgan, P. J., Harding, S. E., & Petrak, K. (1990a). Hydrodynamic properties of a polyisoprene/polyoxyethylene block polymer. *Macromolecules*, *23*, 4461–4464.
Morgan, P. J., Harding, S. E., & Petrak, K. (1990b). Interactions of a model block copolymer drug delivery system with two serum proteins and myoglobin. *Biochemical Society Transactions*, *18*, 1021–1022.
Morris, G. A., Adams, G. G., Harding, S. E., Castile, J. D., & Smith, A. (2012). Stability of pectin-based drug delivery systems. In S. E. Harding (Ed.), *Stability of complex carbohydrate structures: Biofuels, foods, vaccines and shipwrecks* (pp. 99–109). Cambridge, UK: Royal Society of Chemistry.
Morris, G. A., Ang, S., Hill, S. E., Lewis, S., Shafer, B., Nobbmann, U., et al. (2008). Molar mass and solution conformation of branched α(1 → 4), α(1 → 6) glucans. Part I: Glycogens in water. *Carbohydrate Polymers*, *71*, 101–108.
Morris, G. A., Butler, S. N. G., Foster, T. J., Jumel, K., & Harding, S. E. (1999). Elevated-temperature analytical ultracentrifugation of a low-methoxy polyuronide. *Progress in Colloid and Polymer Science*, *113*, 205–208.
Morris, G. A., Castile, J., Adams, G. G., & Harding, S. E. (2011). The effect of prolonged storage at different temperatures on the particle size distribution of tripolyphosphate (TPP)–chitosan nanoparticles. *Carbohydrate Polymers*, *84*, 1430–1434.
Morris, G. A., Castile, H., Smith, A., Adams, G. G., & Harding, S. E. (2009a). The kinetics of chitosan depolymerisation at different temperatures. *Polymer Degradation and Stability*, *94*, 1334–1348.
Morris, G. A., Castile, J., Smith, A., Adams, G. G., & Harding, S. E. (2009b). Macromolecular conformation of chitosan in dilute solution: A new global hydrodynamic approach. *Carbohydrate Polymers*, *76*, 616–621.
Morris, G. A., Castile, J., Smith, A., Adams, G. G., & Harding, S. E. (2010). The effect of different storage temperatures on the physical properties of pectin solutions and gels. *Polymer Degradation and Stability*, *95*, 2670–2673.
Morris, G. A., Ebringerova, A., Harding, S. E., & Hromadkova, Z. (1999). UV tagging leaves the structure integrity of an arabino-(4-O-methylglucurono) xylan polysaccharide unaffected. *Progress in Colloid and Polymer Science*, *113*, 201–204.
Morris, G. A., Foster, T. J., & Harding, S. E. (2000). The effect of the degree of esterification on the hydrodynamic properties of citrus pectin. *Food Hydrocolloids*, *14*, 227–236.
Morris, G. A., Foster, T. J., & Harding, S. E. (2002). A hydrodynamic study of the depolymerisation of a high methoxy pectin at elevated temperatures. *Carbohydrate Polymers*, *48*, 361–367.
Morris, G. A., García de la Torre, J., Ortega, A., Castile, J., Smith, A., & Harding, S. E. (2008). Molecular flexibility of citrus pectins by combined sedimentation and viscosity analysis. *Food Hydrocolloids*, *22*, 1435–1442.
Morris, G. A., & Harding, S. E. (2009). Polysaccharides, microbial. In M. Schaechter (Ed.), *Encyclopedia of microbiology* (3rd ed., pp. 482–494). Amsterdam: Elsevier.
Morris, G. A., & Harding, S. E. (2013). Hydrodynamic modelling of carbohydrate polymers. In G. C. K. Roberts (Ed.), *Encyclopedia of biophysics* (pp. 1006–1014). Berlin: Springer Verlag.
Morris, G. A., Hromadkova, Z., Ebringerova, A., Malovikova, A., Alfoldi, A., & Harding, S. E. (2002). Modification of pectin with UV-absorbing substituents and its effect on the structural and hydrodynamic properties of the water-soluble derivatives. *Carbohydrate Polymers*, *48*, 351–359.
Morris, G. A., Li, P., Puaud, M., Liu, Z., Mitchell, J. R., & Harding, S. E. (2001). Hydrodynamic characterisation of the exopolysaccharide from the halophilic cyanobacterium

Aphanothece halophytica GR02: A comparison with xanthan. *Carbohydrate Polymers*, *44*, 261–268.

Morris, G. A., Patel, T. R., Picout, D. R., Ross-Murphy, S. B., Ortega, A., Garcia de la Torre, J., et al. (2008). Global hydrodynamic analysis of the molecular flexibility of galactomannans. *Carbohydrate Polymers*, *72*, 356–360.

Morris, G. A., Ralet, R. C., Bonnin, E., Thibault, J. F., & Harding, S. E. (2010). Physical characterisation of the rhamnogalcturonan and homogalacturonan fractions of sugar beet (Beta vulgaris) pectin. *Carbohydrate Polymers*, *82*, 1161–1167.

Negard, C. S., Matsumoto, T., Inngjerdingen, M., Inngjerdingen, K., Hokputsa, S., Harding, S. E., et al. (2005). Structural and immunological studies of a pectin and a pectic arabinogalactan from Vernonia kotschyana Sch. Bip. ex Walp. (Asteraceae). *Carbohydrate Research*, *340*, 115–130.

Nikolajski, M., Adams, G. G., Gillis, R. B., Besong, D. T., Rowe, A. J., Heinze, T., et al. (2014). Protein-like fully reversible tetramerisation and super-association of an aminocellulose. *Nature: Scientific Reports*, *4*, 3861.

Ong, M. H., Jumel, K., Tokarczuk, P. F., Blanshard, J. M., & Harding, S. E. (1994). Simultaneous determinations of the molecular weight distributions of amyloses and the fine structures of amylopectins of native starches. *Carbohydrate Research*, *260*, 99–117.

Ortega, A., & García de la Torre, J. (2007). Equivalent radii and ratios of radii from solution properties as indicators of macromolecular conformation, shape, and flexibility. *Biomacromolecules*, *8*, 2464–2475.

Patel, T. R., Harding, S. E., Ebringerova, A., Deszczynski, M., Hromadkova, Z., Togola, A., et al. (2007). Weak self-association in a carbohydrate system. *Biophysical Journal*, *93*, 741–749.

Patel, T. R., Morris, G. A., Ebringerová, A., Vodenicarová, M., Velebny, V., Ortega, A., et al. (2008). Global conformation analysis of irradiated xyloglucans. *Carbohydrate Polymers*, *74*, 845–851.

Patel, T. R., Morris, G. A., Garcia de la Torre, J., Ortega, A., Mischnick, P., & Harding, S. E. (2008). Molecular flexibility of methylcelluloses of differing degree of substitution by combined sedimentation and viscosity analysis. *Macromolecular Bioscience*, *8*, 1108–1115.

Patel, T. R., Picout, D. R., Harding, S. E., & Ross-Murphy, S. B. (2006). Pressure cell assisted solution characterization of galactomannans. 3. Application of analytical ultracentrifugation techniques. *Biomacromolecules*, 7, 3513–3520.

Pavlov, G. M., Harding, S. E., & Rowe, A. J. (1999). Normalized scaling relations as a natural classification of linear macromolecules according to size. *Progress in Colloid and Polymer Science*, *113*, 76–80.

Pavlov, G. M., Korneeva, E. V., Harding, S. E., & Vichoreva, G. A. (1998). Dilute solution properties of carboxymethylchitins in high ionic-strength solvent. *Polymer*, *39*, 6951–6961.

Pavlov, G. M., Korneeva, E. V., Jumel, K., Harding, S. E., Meijer, E. W., Peerlings, H. W. I., et al. (1999). Hydrodynamic properties of carbohydrate-coated dendrimers. *Carbohydrate Polymers*, *38*, 195–202.

Pavlov, G. M., Korneeva, E. V., Vikhoreva, G. A., & Harding, S. E. (1998). Hydrodynamic and molecular characteristics of carboxymethylchitin in solution. *Polymer Science, Series A*, *40*, 1275–1281.

Pavlov, G. M., Rowe, A. J., & Harding, S. E. (1997). Conformation zoning of large molecules using the analytical ultracentrifuge. *Trends in Analytical Chemistry*, *16*, 401–404.

Picout, D. R., Ross-Murphy, S. B., Errington, N., & Harding, S. E. (2003). Pressure cell assisted solubilization of xyloglucans: Tamarind seed polysaccharide and detarium gum. *Biomacromolecules*, *4*, 799–807.

Pohl, M., Morris, G. A., Harding, S. E., & Heinze, T. (2009). Studies on the molecular flexibility of novel dendronized carboxymethyl cellulose derivatives. *European Polymer Journal, 45*, 1098–1110.

Rowe, A. J. (1977). The concentration dependence of transport processes: A general description applicable to the sedimentation, translational diffusion and viscosity coefficients of macromolecular solutes. *Biopolymers, 16*, 2595–2611.

Schuck, P., Gillis, R. B., Besong, D., Almutairi, F., Adams, G. G., Rowe, A. J., et al. (2014). SEDFIT-MSTAR: Molecular weight and molecular weight distribution analysis of polymers by sedimentation equilibrium in the ultracentrifuge. *Analyst, 139*, 79–92.

Schütz, C., Käuper, P., & Wandrey, C. (2012). Stability of polysaccharide complexes: The effect of media and temperature on the physical characteristics and stability of chitosan-tripolyphosphate/alginate nanogels. In S. E. Harding (Ed.), *Stability of complex carbohydrate structures: Biofuels, foods, vaccines and shipwrecks* (pp. 110–124). Cambridge UK: Royal Society of Chemistry.

Scott, D. J., Harding, S. E., & Winzor, D. J. (2015). Evaluation of diffusion coefficients by means of an approximate steady-state condition in sedimentation velocity distributions. *Analytical Biochemistry* (in press).

Seifert, A., Heinevetter, L., Cölfen, H., & Harding, S. E. (1995). Characterisation of gliadin-galactomannan incubation mixtures by analytical ultracentrifugation. Part l: Sedimentation velocity. *Carbohydrate Polymers, 28*, 325–332.

Seymour, G. B., & Harding, S. E. (1987). Analysis of the molecular size of tomato (Lycopersicon esculentum Mill) fruit polyuronides by gel filtration and low-speed sedimentation equilibrium. *Biochemical Journal, 245*, 463–466.

Seymour, G. B., Harding, S. E., Taylor, A. J., Hobson, G. E., & Tucker, G. A. (1987). Polyuronide solubilization during ripening of normal and mutant tomato fruit. *Phytochemistry, 26*, 1871–1875.

Smith, C. J., Watson, C. F., Morris, P. C., Bird, C. R., Seymour, G. B., Harding, S. E., et al. (1990). Inheritance and effect on ripening of antisense polygalcturonase genes in transgenic tomatoes. *Plant Molecular Biology, 14*, 369–379.

Sobol, E. B., Omel'chenko, A. I., Sviridov, A. P., Harding, S. E., Jumel, K., & Jones, N. (2000). Hydrodynamic study of the behaviour of chondroitin sulphate under nondestructive laser irradiation of cartilage. *Proceedings of SPIE, 3914*, 88–93.

Soshnikova, Y. M., Roman, S. G., Chebotareva, N. G., Baum, O. I., Obrezkova, M. V., Gillis, R. B., et al. (2013). Starch-modified magnetite nanoparticles for impregnation into cartilage. *Journal of Nanoparticle Research, 15*, 2092.

Stafford, W. F., III, & Correia, J. (2015). Sedimentation velocity: Fundamental principles. *Methods in Enzymology, 562*, 49–80.

Tanford, C. (1961). *Physical chemistry of macromolecules*. New York: J. Wiley & Sons, p343.

Tester, R. F., Patel, T., & Harding, S. E. (2006). Damaged starch characterisation by ultracentrifugation. *Carbohydrate Research, 341*, 130–137.

Tombs, M. P., & Harding, S. E. (1998). *An introduction to polysaccharide biotechnology*. London: Taylor & Francis.

Tongdang, T., Bligh, H. F. J., Jumel, K., & Harding, S. E. (1999). Combining sedimentation velocity with SEC-MALLS to probe the molecular structure of heterogeneous macromolecular systems that cannot be preparatively separated: Application to three rice starches. *Progress in Colloid and Polymer Science, 113*, 185–191.

Uluisik, S., Chapman, N., Smith, R., Poole, M., Adams, G. G., Gillis, R. B., et al. (2015). *Tomato fruit softening is inhibited by silencing a gene encoding a pectate lyase*. (mss submitted).

Wales, M., & van Holde, K. E. (1954). The concentration dependence of the sedimentation constants of flexible macromolecules. *Journal of Polymer Science, 14*, 81–86.

Wandrey, C., Bartkowiak, A., & Harding, S. E. (2009). Materials for encapsulation. In N. J. Zuidam & V. Nedovic (Eds.), *Encapsulation technologies for active food ingredients and food processing* (pp. 31–100). New York: Springer.

Watson, J. D. (2012). *Double Helix*. (Annotatedand Illustrated Edition, A. Gann & J. Witkowski, Eds.), New York: Simon Schuster, pp. 197–335.

Winzor, D. J., Carrington, L. E., Deszczynski, M., & Harding, S. E. (2004). Extent of charge screening in aqueous polysaccharide solutions. *Biomacromolecules*, *5*, 2456–2460.

Wyatt, P. J. (1992). Combined differential light scattering with various liquid chromatography separation techniques. In S. E. Harding, D. B. Sattelle, & V. A. Bloomfield (Eds.), *Laser light scattering in biochemistry*.Cambridge UK: Royal Society of Chemistry (chapter 3).

Wyatt, P. J. (2013). Multiangle light scattering from separated samples (MALS with SEC or FFF). In G. G. K. Roberts & S. E. Harding (Eds.), *Encyclopedia of biophysics* (pp. 1618–1637). Berlin: Springer Verlag (Hydrodynamics section).

Yamakawa, H., & Fujii, M. (1973). Translational friction coefficient of wormlike chains. *Macromolecules*, *6*, 407–415.

Yanaki, T., Norisuye, T., & Fujita, H. (1980). Triple helix of Schizophyllum commune polysaccharide in dilute solution 3. Hydrodynamic properties in water. *Macromolecules*, *13*, 1462–1466.

CHAPTER NINETEEN

Analytical Ultracentrifugation and Its Role in Development and Research of Therapeutical Proteins

Jun Liu, Sandeep Yadav, James Andya, Barthélemy Demeule, Steven J. Shire[1]
Late Stage Pharamaceutical and Processing Development, Genentech
[1]Corresponding author: e-mail address: shire.steve@gene.com

Contents

Abstract

The historical contributions of analytical ultracentrifugation (AUC) to modern biology and biotechnology drug development and research are discussed. AUC developed by Svedberg was used to show that proteins are actually large defined molecular entities and also provided the first experimental verification for the semiconservative replication model for DNA initially proposed by Watson and Crick. This chapter reviews the use of AUC to investigate molecular weight of recombinant-DNA-produced proteins, complex formation of antibodies, intermolecular interactions in dilute and high concentration protein solution, and their impact on physical properties such as solution viscosity. Recent studies using a "competitive binding" analysis by AUC have been useful in critically evaluating the design and interpretation of surface plasmon resonance

Methods in Enzymology, Volume 562
ISSN 0076-6879
http://dx.doi.org/10.1016/bs.mie.2015.04.008

measurements and are discussed. The future of this technology is also discussed including prospects for a new higher precision analytical ultracentrifuge.

1. THE DEVELOPMENT OF ANALYTICAL ULTRACENTRIFUGATION AND ITS ROLE IN BIOLOGY AND BIOTECHNOLOGY

Initially, Svedberg developed the analytical ultracentrifuge to study the size distributions in colloidal gold sols (van Holde & Hansen, 1998). In 1924, Svedberg in collaboration with Robin Fåhraeus used analytical ultracentrifugation (AUC) to investigate the size distribution of proteins, and in particular hemoglobin (Svedberg & Fåhraeus, 1926). At the time of these experiments, most researchers believed that proteins were inhomogeneous with a wide distribution of molecular weights, i.e., proteins were not considered as large molecules with a defined molecular weight. Analysis of the concentration gradients at equilibrium resulted in a molecular weight of 66,800 Da. Thus, this was the first time that it was clearly shown that proteins were large macromolecules with a narrow size distribution. The next major contribution to our current understanding of molecular biology was the seminal experiments by Meselson and Stahl on the mechanism of replication of DNA (Meselson & Stahl, 1958). In these classic experiments, the replication model proposed by Watson and Crick was verified using ^{14}N and ^{15}N isotopic labels.

Since these milestone experiments, the analytical ultracentrifuge has been used to analyze macromolecule size distributions and remains one of the best first principle methods for the quantitative study of interactions between macromolecules (Laue & Stafford, 1999; Lebowitz, Lewis, & Schuck, 2002; Schachman, 1989). The intense investigation of macromolecule structure during the 1950s and 1960s added greatly to our understanding of large molecule interactions, particularly of proteins. There are many excellent review articles and case studies using AUC (Alavattam et al., 2013; Harding, Rowe, & Horton, 1992; Lewis & Weiss, 1975; Schachman, 1959; Schuster & Laue, 1994; Teller, 1973; van Holde & Hansen, 1998; Varley, Brown, Dawkes, & Burns, 1997; Yadav et al., 2013). However, with the advent of recombinant DNA (rDNA) technology to produce proteins in bacterial and mammalian host cells that led to a large increase in the availability of highly purified protein molecules, many of the studies of these proteins were more qualitative in nature and the use of AUC decreased in favor of more high throughput and simple methods such as sodium dodecyl sulfate (SDS) gel electrophoresis and high-performance size-exclusion chromatography (hpSEC).

Much of this history of AUC is nicely summarized in a book chapter by Howard Schachman entitled "Is there a future for the analytical ultracentrifuge?" (Schachman, 1992). In particular, although AUC continued to play a critical role in characterizing biological systems, the complexity of maintaining the AUC instrumentation, known as the Beckman Model E, as well as the analysis of data, caused it to fall into disfavor in the late 1980s. In particular, the AUC hardware did not undergo any significant advancements and the basic model E continued to be the main AUC in use. Some researchers did modify the instrument for on-line data acquisition (Crepeau, Conrad, & Edelstein, 1976; Williams, 1976), but these developments were not readily accessible and thus remained in the hands of the originators. Fortunately, researchers such as David Yphantis and Howard Schachman and Stuart Edelstein continued to use the AUC in their work and developed new hardware and software analysis for the AUC that were available for the general user of AUC. The most important improvements included development of a commercially available split-beam photoelectric scanner to obtain UV absorbance output as a function of radial position during the centrifugation (Schachman & Edelstein, 1973) and nonlinear regression analysis software to investigate protein interactions (Johnson, Correia, Yphantis, & Halvorson, 1981).

Eventually, more quantitative results were needed and some biotechnology labs were set up to use the old Model E. However, for many start-up companies, this was a huge challenge since usually a full time scientist was needed to run the experiments and analyze the data. In addition, the Model E required people to maintain and keep the instrument in running condition. Despite some of the improvements in AUC technology, the method was not high throughput and usually not automated. Although researchers modified the instrument to interface with a computer, this was not widely done and the use of the Model E analytical ultracentrifuge continued to decrease. Many researchers seeing the slow death of the Model E came to the realization that a more modern AUC was needed that could be completely interfaced with a computer for more automated data collection and analysis. All of this led to a very important meeting at UC Berkeley where plans were discussed for a new AUC based on Beckman's Optima preparative series that used UV absorbance optics for detection (Giebler, 1992). The development of the new AUC, the XL-A, with user-friendly computer interface led to its increased use in addressing the challenges of developing protein drugs. The addition of a fluorescent detection system (FDS) increased the sensitivity of detection and also allowed for detection of low concentration of proteins resulting in the ability to determine high-affinity binding (Kroe & Laue, 2009). The impact of other macromolecules on kinetics and thermodynamics of the protein due to excluded volume considerations has been discussed (Minton, 2005) and evaluation of protein interactions in serum has been done using the FDS (Demeule et al., 2009). Addition of a refractive optical system (a refractive optical system was also available for the Model E) expanded the use of the new centrifuge to macromolecules with low or no UV absorbance. Unfortunately, obtaining parts for the upkeep of the XLA/I are becoming more of a challenge, and once again a new AUC is needed using modern analytical technology. Currently, this is being addressed by Professor Tom Laue at the University of New Hampshire. A new AUC with "open architecture" that allows for addition of different detection systems is being developed and promises to have much greater signal-to-noise detection (Colfen et al., 2010; Spin Analytical, 2014).

In this chapter, we will focus on examples of using AUC to aid in development of protein drugs from our lab and will not review the extensive work performed by other research and development groups.

2. ANALYSIS OF AGGREGATES BY AUC

SDS PAGE provided rapid and sensitive detection under conditions where the protein was unfolded and thus was used mainly to ascertain purity and molecular weights of rDNA-produced proteins. This method also provided a rapid way to investigate disulfide cross-linking by using SDS PAGE under reducing and nonreducing conditions. Although SDS PAGE is a rapid screening tool, it only provided limited information about protein self-association since the technique involved binding of a fixed amount of SDS/g resulting in unfolded protein preparations. Development of size-exclusion chromatography (SEC) as a high-performance and high-throughput technique allowed for determination of size distribution under conditions where the protein was properly folded. However, the molecular weight was indirectly determined by using a set of reference standards since the SEC resulted in the separation of molecules by hydrodynamic volume (Andrews, 1970; Whitaker, 1963). Thus, the molecular weights of several proteins were erroneously determined because of the large hydrodynamic volumes that were not adequately determined by the use of compact simple globular protein standards. Development of in-line static light scattering (SLS) detectors circumvented this problem since the weight average molecular weight of a protein in the separate peaks generated and detected by SEC could be determined without the use of globular protein standards (Wen, Arakawa, & Philo, 1996). Despite these improvements, interactions with the column matrix and often requirements to use solution conditions that were quite different than final formulated solutions of proteins still led to erroneous results, e.g., underestimation of aggregate content or incorrect determination of molecular weight (Carpenter et al., 2010; Philo, 2009). In addition, the self-association of proteins can be very concentration dependent resulting in only detection of monomer for weak interaction by SEC because of the large dilution that often occurs during sample preparation as well as during the chromatography. Another concern with the use of SEC is the possible filtration of very large aggregated species.

In general, protein aggregation is undesirable because it can lead to a reduction in activity and altered pharmacokinetics (Klotz & Thomas, 1993; Maggio, 2010; Shao, Li, Krishnamoorthy, Chermak, & Mitra, 1993). Aggregates have also been implicated in a heightened immune response (Moore & Leppert, 1980; Ratner, Phillips, & Steiner, 1990; Ring, Stephan, & Brendel, 1979; Underwood, Voina, & Van Wyk, 1974), and adverse events due to protein aggregates have also been reported (Demeule, Gurny, & Arvinte, 2006; Ryan, Webster, & Statler, 1996; Seidl et al., 2012). For these reasons, regulatory agencies are concerned that SEC results may not be sufficient to assess the actual molecular size distribution in protein solutions. Often the regulatory agencies are requesting alternative biophysical measurements such as SLS, field flow fractionation, dynamic light scattering or AUC to demonstrate that the SEC method in the control system is adequate to monitor protein aggregates. The limits of precision and accuracy in AUC measurements have been explored using three mAbs of unspecified origin (Pekar & Sukumar, 2007). This study showed a precision of ±0.3% over a measured range of 0.6–67% aggregate. Good accuracy, as ascertained by aggregate spiking experiments, could be achieved down to aggregate levels as low as 1.5%. Gabrielson and Arthur have also discussed additional studies using AUC that explore precision, accuracy, and factors that affect the measurements at low aggregate concentrations (Gabrielson & Arthur, 2011), and much of this work is reviewed and updated in another chapter in this current release of Methods in Enzymology.

Examples of the use of AUC to confirm SEC results for proteins include studies on two humanized mAbs stored at −70 and 40 °C (Liu, Andya, & Shire, 2006) and a study of aggregates of a humanized mAb (Andya, Liu, & Shire, 2010), and analysis of aggregates in recombinant glycoproteins (Hughes et al., 2009). In the study by Andya et al., SEC and AUC were used to determine percentage of aggregates in the same mAb sample. Although the SEC showed a single monomer peak at 99.6% with a small amount of aggregated species, the AUC analysis showed a significant aggregate peak (Fig. 1A). This amount of aggregate was concentration dependent as shown by sedimentation velocity analysis at different

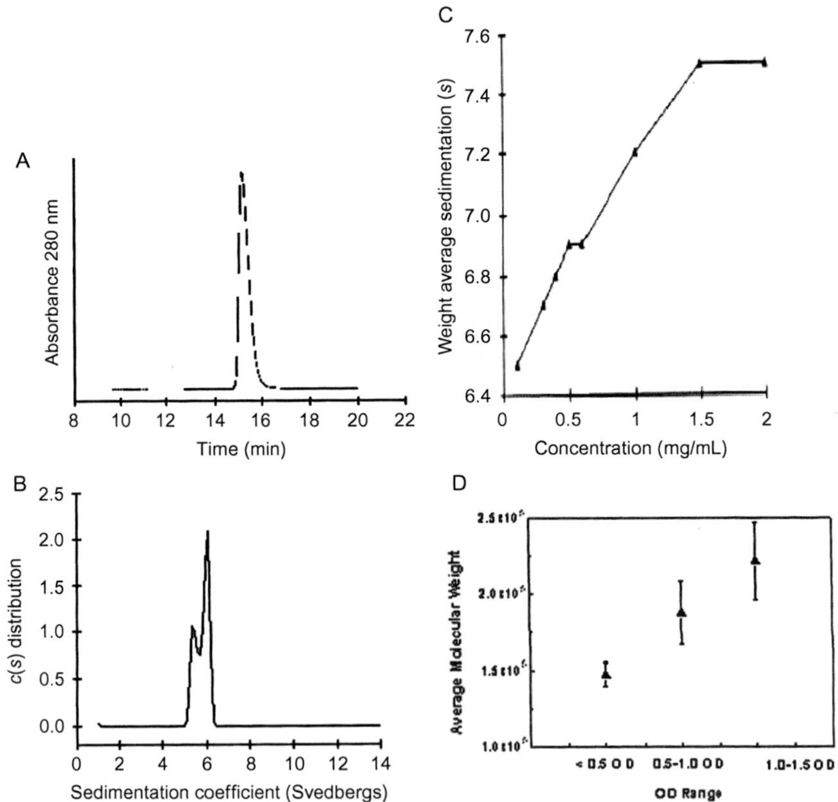

Figure 1 Size distribution of a full-length mAb determined by (A) SEC and (B) AUC sedimentation velocity. (C) Weight average sedimentation coefficient versus protein concentration observed during AUC sedimentation velocity analysis of the mAb in PBS. (D) Weight average molecular weight determined by AUC sedimentation equilibrium of the mAb in PBS. *Reproduced from Andya et al. (2010).*

loading concentrations (Fig. 1B) as well as a determination of weight average molecular weight by sedimentation equilibrium (Fig. 1C). This again illustrates that significant aggregation due to reversible protein self-association can be missed by SEC analysis due to dissociation in the high-ionic strength environment of the SEC mobile phase or dilution during the chromatography. Both of these case studies show that for biopharmaceutical applications, AUC provides a distinct advantage of carrying out a protein size distribution analysis in buffer conditions that are more relevant to product formulation and delivery.

3. EARLY STUDIES OF THE USE OF AUC IN DETERMINING MOLECULAR WEIGHT AND SIZE DISTRIBUTION OF PROTEIN DRUGS

3.1 "Simple" Globular Proteins

In the early years of biotechnology, many companies were trying to develop antiviral proteins such as the interferons. In particular, Human Leukocyte Interferon 1A (rhLIFα) produced by recombinant techniques in *Escherichia coli* had a molecular weight of ~19,400 Da as determined by SDS PAGE which was consistent with the DNA sequence (Wetzel et al., 1981). As already mentioned, SDS PAGE can be used to investigate purity and determine if the molecular weight is consistent with that expected from the DNA sequence. However, the technique provides limited information on the size distribution of the protein in solution. Sedimentation velocity AUC was used to assess size distribution in solution of rhLIFα (Shire, 1983). Interferon solutions at different solution pH (2.5–10) and loading concentrations (1–6.0 mg/mL) were centrifuged in an AnD rotor in a model E analytical ultracentrifuge equipped with schlieren and photoelectric scanner optics. A photograph using schlieren optics after ~30 min at 60,000 rpm shows a highly skewed asymmetric boundary and undoubtedly represents a "reaction boundary" (Gilbert & Gilbert, 1973) rather than a unique structural aggregate (Fig. 2A). Further quantitation of the sedimentation was accomplished by determining the apparent sedimentation coefficients by plotting the natural logarithm (ln) of the radial position of the boundary mid-point measured with the photoelectric scanner versus the time of sedimentation as based on the following equation:

$$s = \frac{1}{\omega^2}\frac{d\ln\bar{x}}{dt} \tag{1}$$

A

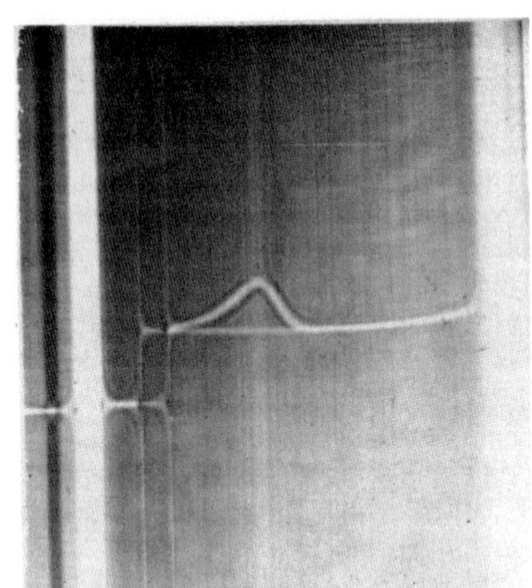

B

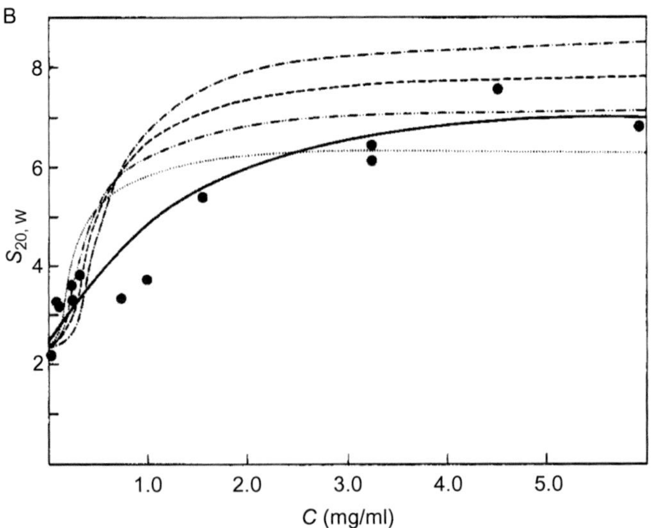

Figure 2 (A) Photograph using schlieren optics of rhLIFα at 3.2 mg/mL, 23 °C, pH 7.0 (0.01 m*M* $K(H)PO_4 + 0.09$ m*M* KCl) after ~30 min at 60,000 rpm in a Model E AnD rotor. (B) Corrected sedimentation coefficients (solid circles) for rhLIFα at pH 6.9–7.0 (0.01 *M* $K(H)PO_4 + 0.09$ *M* KCl), 23 °C, as a function of the loading protein concentration. Curves are computer-simulated weight-average sedimentation coefficients. Monomer molecular weight = 19,400, $\bar{V}_p = 0.73$, and ρ and η are for water at 20 °C. The intrinsic sedimentation coefficient of monomer = 2.4 S and $s_{n\text{-mer}} = s_{\text{monomer}}\, n^{2/3}$. For monomer to *n*-mer tight association, *K*(*n*) for *nM* (– – – –) *M*, was set equal to 1000.0 $M^{-(n-1)}$, and all *K*(*I*), *I* = 2, *n*−1 were set equal to 0.00001 $M^{-(I-1)}$: (– –) *n* = 8; (– – –) *n* = 7; (···) *n* = 6; (–) *n* = 5. The solid line through the data points was computed for a monomer to octamer association. $K(8) = 2000.0\ M^{-7}$, $K(7) = 500.0\ M^{-6}$, $K(6) = 400.0\ M^{-5}$, and $K(5) = K(4) = K(3) = K(2) = 4\ M^{-(n-1)}$. *From Shire (1983).*

where $\omega = 2\pi\nu$, ν is the frequency of rotation, $\bar{x}$ is the radial position of the mid-point of the boundary, and t is the time of sedimentation. The determined sedimentation coefficient s is dependent on the temperature, viscosity η, and density ρ of the solution and needs to be corrected to standard conditions of water at 20 °C to obtain $s_{20,w}$ using (Schachman, 1959)

$$s_{20,\mathrm{w}} = s\frac{\eta}{\eta_{20,\mathrm{w}}}\frac{(1-\bar{V}\rho)_{20,\mathrm{w}}}{(1-\bar{V}\rho)} \tag{2}$$

where η is the viscosity of the solvent at the temperature of the experiment and $\eta_{20,w}$ is the viscosity of water at 20 °C. ρ is the density of the solution at the temperature of the experiment and $\rho_{20,w}$ is the density of water at 20 °C. $\bar{V}$ is the partial specific volume of the protein and can be determined by high precision density measurements as a function of protein concentration. Physicochemical methods such as pycnometry require an inordinately large amount of material. It is possible to obtain $\bar{V}$ values using high precision density meters, which use a mechanical oscillator technique (Kratky, Leopold, & Stabinger, 1973), but this technique also requires a fair amount of protein, very tight temperature control and can be very time consuming. Alteration of solvent density using $D_2^{16}O$ or $D_2^{18}O$ in sedimentation velocity or sedimentation equilibrium experiments also can be used to determine $\bar{V}$ (Edelstein & Schachman, 1967). An alternate method is to compute the partial specific volume from a weight average of the partial specific volumes of the amino acid composition. Calculated and experimentally determined partial specific volumes of many proteins are in good agreement (Charlwood, 1957; McMeekin & Marshall, 1952). Thus, the partial specific volume was estimated from the amino acid composition and a new listing of partial specific volumes as suggested by Perkins (1986). The sedimentation coefficient at infinite dilution was compared to values computed, using the equation (Caspar, 1963; Schachman, 1959):

$$s = \frac{(1-\bar{V}_{\mathrm{p}}\rho)M_{\mathrm{p}}(b/a)^{2/3}}{6\pi N_0\eta R_0}\ln\left[\left[1+\left(1-b^2/a^2\right)^{1/2}\right]/(b/a)\right]/\left(1-b^2/a^2\right)^{1/2} \tag{3}$$

where N_0 is Avogadro's number, M_p is the protein molecular weight, and $R_0 = (ab^2)^{1/3}$ where a and b are the length of semimajor and semiminor axes, respectively, of a prolate ellipsoid. The resulting calculated sedimentation coefficient at infinite dilution in water (η and ρ of water at 20 °C) varies from 1.8 to 2.4 S, depending on the axial ratio, and is in good agreement with the experimental $s_{20,w}$ value of 2.0 S at 0.25 mg/mL and pH of 2.5. A plot of the experimental $s_{20,w}$ as a function of concentration along with computer-simulated weight average sedimentation coefficients as described by Cox (1969) is shown in Fig. 2B. Different association schemes were simulated and the curve with the best fit to the data suggests a monomer to octamer association, i.e., a distribution of molecular sizes. In this case, AUC unequivocally established the multimeric nature of rhLIFα in solution.

Human relaxin, a pregnancy hormone, and human interleukin-8 (rhIL-8), a chemokine, were two other proteins where protein self-association was investigated by AUC (Lowman et al., 1997; Shire, Holladay, & Rinderknecht, 1991). Human relaxin is produced *in vivo* the same way as insulin, i.e., as a single polypeptide chain that is converted to two disulfide-linked chains by enzymatic removal of a polypeptide strand termed the connecting peptide. The cysteine residues in the relaxin molecule are in exact alignment with the same residues in the insulin molecule, and the molecular weight of relaxin is similar to insulin (~6450 Da) as determined by fast atom bombardment mass spectroscopy (FAB-MS). Relaxin initially was produced by chemical synthesis of the two chains, which were then recombined to form a two-chain relaxin molecule (Canovadavis, Baldonado, & Teshima, 1990). Analysis using SEC at a loading concentration of 0.4 mg/mL results in one peak with an elution time of a monomer at 6500 Da (Fig. 3A). At a greater loading concentration at 2 mg/mL, the relaxin began eluting from the column as a dimer (Canovadavis et al., 1990). The result was corroborated using sedimentation equilibrium AUC (SE-AUC), where the resulting concentration gradient at equilibrium was best fitted with a monomer–dimer self-association model as described in Eq. (5) by Shire et al. (1991).The results are plotted as the ratio of weight average molecular determined by SE-AUC over monomer molecular weight versus protein loading concentration (Fig. 3B). The determined K_a for the reversible association was used to compute the amount of dimer at 0.5 mg/mL and 10 μg/mL relaxin, in the far-UV and at 0.5 and 20 μg/mL in the near-UV (Fig. 4). Upon dilution, the far UV CD (Fig. 4A) showed little change, suggesting that the secondary structure for monomer versus the protein in the aggregate was similar. The near-UV CD shows large changes upon dilution (Fig. 4B) and there is a clear decrease in the CD signal at 277 and 284 nm, whereas there is little change in the broad band at 295 nm, which is due to tryptophan residues. Thus, it appears that upon dissociation of the dimer, there is a change in the chiral environment of the lone tyrosine residue resulting in a greater mobility of the aromatic amino acid residues

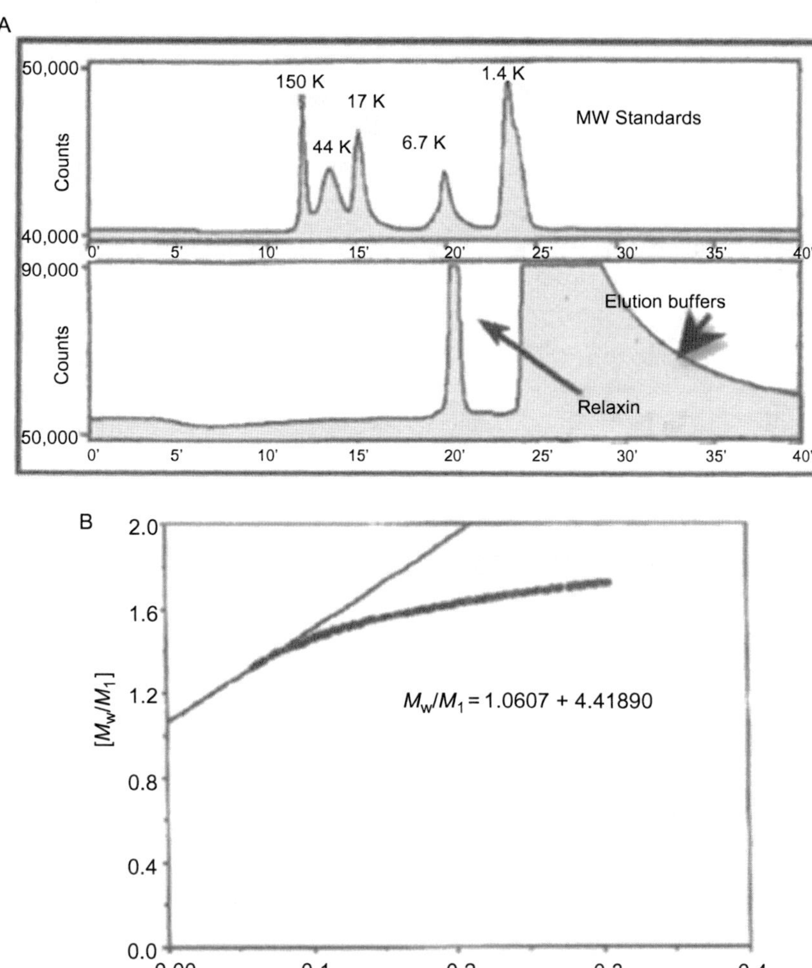

Figure 3 (A) Sizing chromatography of human relaxin. 50 μL at <0.4 mg/mL loaded onto a TSK G2000 SWXL column (300 mm × 7.5 mm I.D.) equilibrated with 10 m*M* sodium citrate, pH 5.0 in 0.25 *M* sodium chloride. Flow rate at 0.5 mL/min with detection at 214 nm. (B) Weight average molecular weight (*M*w) determined by sedimentation equilibrium analytical ultracentrifugation divided by the monomer molecular weight (M_1) determined from amino acid composition of human relaxin. Human relaxin at concentrations ranging from 0.05 to 0.2 mg/mL in 10 m*M* sodium citrate, 0.15 *M* sodium chloride at pH 5.0 was centrifuged at 22 or 32 k rpm at 19–21 °C for 18 h and concentration gradients detected at 280 nm.

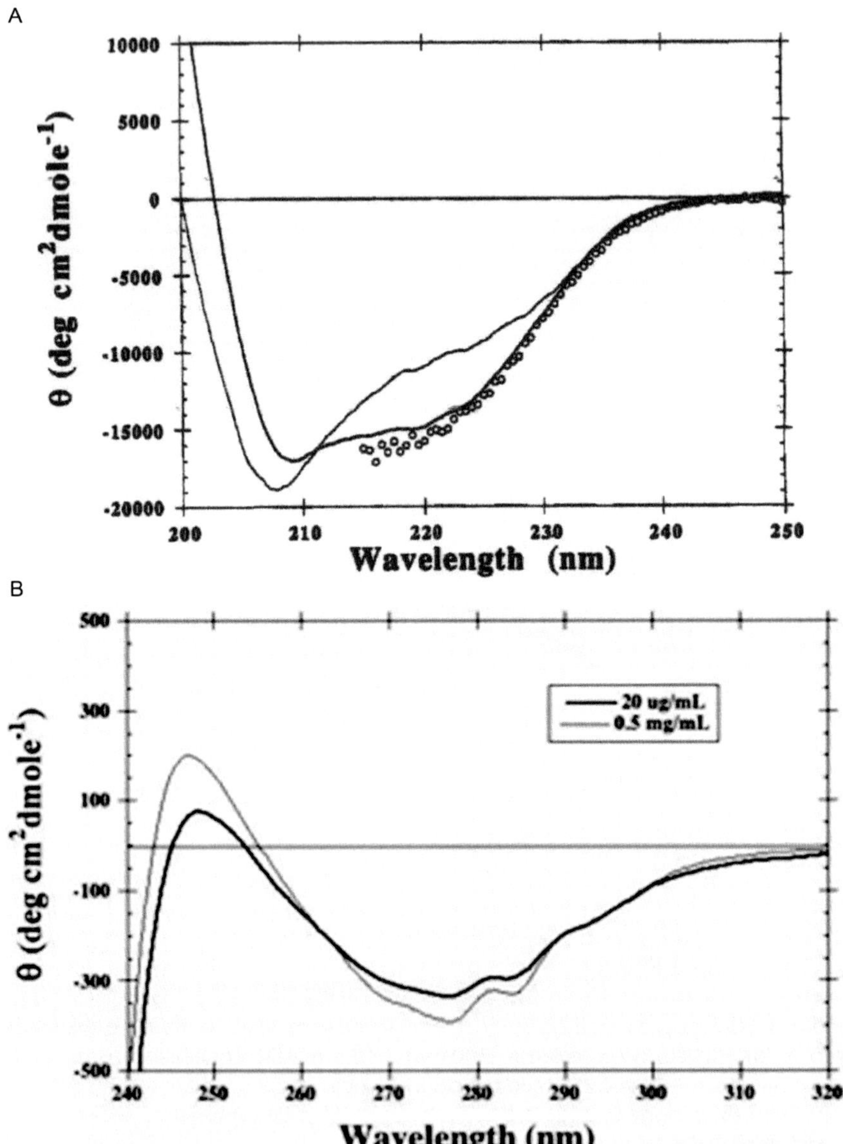

Figure 4 (A) Far-UV circular dichroism of human relaxin at 0.5 mg/mL (solid line) and 10 μg/mL (open circles). The 0.5-mg/mL data were collected in a thermostated 0.01 cm cylindrical cell at 20 °C as described in this figure. The CD data for relaxin at 10 μg/mL were obtained in a 1-cm thermostated cylindrical cuvette and are the result of an average of 10 scans using an average time for each single data point collection of 10 s. The weight fraction of monomer estimated from the determined association constant of 100 $(g/L)^{-1}$ is 0.13 at 0.5 mg/mL and 0.62 at 10 μg/mL. (B) Near-UV circular

and loss of signal. This conclusion was confirmed by the X-ray crystal structure of human relaxin, which crystallized as a dimer with the lone tyrosine from each monomer at the dimer interface (Eigenbrot et al., 1991). Thus, the solution studies were in good agreement with the crystal studies, suggesting that the determined crystal structure is very similar to the structure of the protein in solution.

rhIL-8, an 8 kDa single-chain protein that functions as a chemokine, was produced by secretion into the periplasmic space of *E. coli* and purified as described previously (Hebert et al., 1990). This protein has been observed to have dimeric structures by NMR and X-ray crystallography. A series of genetic variants were designed with mutations at the dimer interface. The folding of these mutants appears to be essentially the same as the wild-type protein as detected by NMR. Sedimentation equilibrium experiments showed that the protein does self-associate into dimers with dissociation constant values ranging from micromolar to millimolar. The wild-type rhIL-8 had a K_d value of ~10 μ*M* under physiological conditions (37 °C and ~150 m*M* ionic strength). Affinities of the different monomeric variants to the erythrocyte chemokine receptor DARC and the neutrophil IL-8 receptors CXCR1 and CXCR2 were similar to that obtained for wild type showing that the monomeric IL-8 is functionally equivalent to the wild type. Studies by other laboratories have reported a large range for K_d, and the AUC studies which were also done as a function of temperature and ionic strength show that the K_d value is highly dependent on solution conditions and can explain the larger range of K_d values reported.

3.2 More Complex Glycoproteins

As already mentioned, prior to the use of an on-line light scattering detector apparent molecular weight for SEC was estimated using a set of globular

Figure 4—Cont'd dichroism of human relaxin at 0.5 mg/mL (thin solid line) and 20 μg/mL (bold solid line). Relaxin at 0.5 mg/mL was thermostated at 20 °C in a 1-cm cell, whereas relaxin at 20 μg/mL was in an unthermostated 10-cm cylindrical cuvette. The temperature in the sample compartment was ~27 °C during the data collection process. The CD data were collected at 0.25-nm intervals at a spectral bandwidth of 0.5 nm and are the result of an average of three scans using an average time for each single data point collection of 5 s for the 0.5 mg/mL samples and the result of an average of 10 scans using an average time for each single data point collection of 10 s for the 20 μg/mL sample. The weight fraction of human relaxin monomer estimated from the determined association constant of 100 $(g/L)^{-1}$ is 0.13 at 0.5 mg/mL and 0.50 at 20 μg/mL.

protein standards. The addition of polysaccharide chains to proteins occurs mainly at Asn residues in particular sequences. Many of these proteins termed "glycosylated proteins" were produced in Chinese Hamster Ovary cells since *E. coli* is not capable of such posttranslational modifications. Although the level of glycosylation can be small such as in antibodies (~2% of the mass), there are cases where the glycosylation contributes a significant portion of the molecular weight. For rhgp120, the coat glycoprotein of the human immunodeficiency virus, glycans make up about 50% by weight of the molecule. More importantly, the molecule has a very large hydrodynamic volume due to the added polysaccharides, and since SEC actually separates by hydrodynamic volume, this analysis can lead to erroneous conclusions. Early work using SEC suggested that rhgp120 self-associates into a dimeric structure and upon deglycosylation using glycosidases result in monomeric molecules, suggesting that the attached polysaccharides promote dimerization (personal communication, Dr. Tim Gregory). Sedimentation equilibrium was used to determine molecular weight by analyzing the absorbance gradient as a single ideal sedimenting species. The absorbance, A, at any radial position, r, is related to the molecular weight, M_w by

$$A = A_0 \exp\left[\left(\frac{\omega^2}{2RT}\right) M_w (1 - \bar{v}\rho)\left(r^2 - r_0^2\right)\right] \tag{4}$$

where A_0 is the absorbance at a radial reference distance r_0, ω is the angular velocity, R is the gas constant, T is the Kelvin temperature, $\bar{v}$ is the partial specific volume, and ρ is the density of solution. A nonlinear regression analysis using Eq. (4) results in determination of the buoyant molecular weight, $M_w(1 - \bar{v}\rho)$. In order to convert these values to absolute molecular weights, it is necessary to determine the partial specific volume for the glycoprotein. As discussed previously, the partial specific volume of a protein can be estimated using a weight average of the partial specific volumes of the component amino acid residues. This estimation can be extended to glycoproteins with known average carbohydrate composition using the partial specific volumes for the different sugars that make up the linked polysaccharide chains, and it has been shown that there is good agreement between experimental and calculated values for glycoproteins (Gibbons, 1972). The molecular weight for rgp120 determined by sedimentation equilibrium using computed values for $\bar{v}$ is ~104 kDa, which is in very good agreement with the expected molecular weight of 102 kDa based on amino acid and average carbohydrate composition. This agreement is quite good considering that the partial specific volume was estimated from amino acid and average carbohydrate composition. In situations where the carbohydrate composition is not known, an estimate of $\bar{\nu}$ can still be made. The types of oligosaccharide structures usually found in N-linked glycosylated proteins are known and the $\bar{\nu}$ for each structure range from 0.62 to 0.64 cm^3/g, and thus an estimate for an average contribution of 0.63 cm^3/g is not unreasonable. An example of using this approach is that of the soluble form of the tumor necrosis factor receptor I (sTNF-R1), where the carbohydrate composition was unknown. Calculated molecular weights assuming 10%, 20%, and 30% carbohydrate with an average partial specific volume of 0.63 cm^3/g show that the final values are relatively insensitive to the carbohydrate content and show that this glycoprotein is a monomer rather than a dimer as suggested by SEC experiments (Pennica et al., 1992).

4. AUC AND FORMULATION DEVELOPMENT

4.1 Using AUC to Investigate Surfactant–Protein Interactions

Proteins in aqueous solutions containing surfactants have been characterized using sedimentation equilibrium and sedimentation velocity AUC (Tanford, Nozaki, Reynolds, & Makino, 1974). Nonionic surfactants are important components of protein formulation, and often added to mitigate problems of proteins aggregating at air–water interfaces as a result of solution agitation. The surfactant molecules migrate to the air–water interface inhibiting adsorption of proteins to the interface. At concentrations greater than the critical micelle concentration, the surfactants form organized structures, termed micelles. These micelles in turn may interact with specific regions of a proteins resulting in "mixed micelle" structures. In an extreme case, where a protein is normally incorporated into the cell membrane, there is a hydrophobic region of the protein that interacts with the surfactant micelle. An example of this is the blood coagulation protein tissue factor, which has a well-defined transmembrane sequence that is buried in the cell membrane (Nemerson, 1988). Sedimentation velocity with UV absorption detection at ~280 nm and a labeled electron paramagnetic resonance (EPR) technique (Jones, Cipola, Liu, Shire, & Randolph, 1999) was used to characterize surfactant formulations for tissue factor, and shows the versatility of AUC in studying such interactions.

5. USING AUC TO CHARACTERIZE IMMUNE COMPLEXES

Engineered antibodies have become a large focus for the biotechnology industry. These antibodies are often designed to interact with a variety of targets responsible directly or indirectly for different cancers as well as immunologically based disorders such as multiple sclerosis, arthritis, and asthma. Although many of the targets are on cell surfaces, antibodies have been developed to bind and neutralize circulating molecules in serum, which result in diseases such as asthma. An example of this is an anti-IgE antibody that helps to treat IgE-mediated allergic reactions. IgE makes up $<2\%$ of the total human immunoglobulin population and has a short circulating half-life (6 h to ~2 days) (Achatz et al., 2010; Stone, Prussin, & Metcalfe, 2010) compared to IgG1 which can have a half-life as long as 3 weeks (Mould & Sweeney, 2007). IgE generated upon exposure to an allergen binds to high-affinity Fc receptors on basophils and mast cells (Fig. 5A). Upon reexposure to the allergen, these receptors are cross-linked via Fab-allergen binding which leads to a rupture of the cells releasing histamines and leukotrienes that trigger asthmatic and respiratory symptoms. Standard therapies, such as antihistamines, antileukotrienes, cromoglycates,

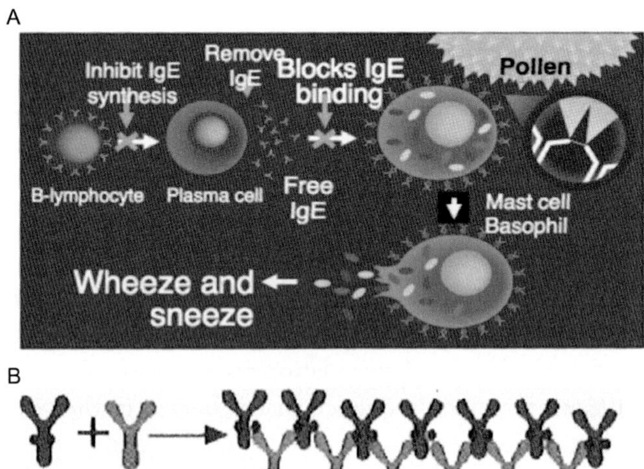

Figure 5 (A) Mechanism of action for an anti-IgE molecule to treat IgE allergic-mediated asthma. (B) The theoretical interaction of IgE (dark gray) with an anti-IgE monoclonal antibody (light gray) via binding of the two high-affinity Fc receptor sites on IgE. (See the color plate.)

and bronchodilators, which in severe cases are not sufficient to mitigate the disease symptoms (Scottish Intercollegiate Guidelines Network, 2014). Thus, an anti-IgE mAb was developed to treat IgE-mediated immune responses. The anti-IgE antibody is an IgG1 molecule that specifically binds to the Fc region of the IgE that is responsible for binding to the high-affinity receptors. An important property of this molecule is that the anti-IgE antibody also prevents the IgE from binding to the high-affinity receptors (Fig. 5A). Since the anti-IgE mAb has two antigen-binding sites each of which could combine with one of two sites on the target IgE molecule (located on each Fc heavy chain), the complexes could become very large (Fig. 5B). The size distribution of the IgE:anti-IgE complexes was investigated using sedimentation velocity and sedimentation equilibrium (Liu, Lester, Builder, & Shire, 1995). The sedimentation velocity analysis was done with the d*c*/d*t* method (Stafford, 1992). This analysis does not correct for diffusion, which results in a spread of the sedimenting boundaries. However, since the complexes are very large, the resulting sedimentation is sufficient to allow for resolution of the boundaries. The sedimentation velocity experiments were done at different molar ratios of IgE:anti-IgE and the results are summarized in Fig. 6. Sedimentation equilibrium data were also generated as a function of IgE to anti-IgE molar ratio (Fig. 7). Essentially, these results showed the determined molecular weight for the largest complex, which occurred at a 1:1 molar ratio of IgE, and anti-IgE was a hexamer

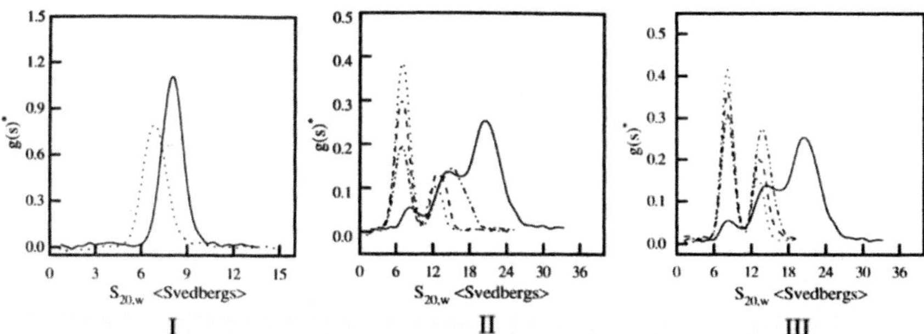

Figure 6 The apparent sedimentation distribution function ($g(s)^*$) obtained from d*c*/d*t* analysis of sedimentation velocity data. IgE (solid line) and anti-IgE (dotted line) monomers at 0.64 mg/mL (I); IgE and anti-IgE complexes at various molar ratios (II and III) in PBS at 10 °C. The molar ratios of IgE:anti-IgE were as follows: (II) 1:1 (solid line), 1:3 (dash-dotted line), 1:6 (dashed line), and 1:10 (dotted line) and (III) 1:1 (solid line), 3:l (dash-dotted line), 6:1 (dashed line), and 10:1 (dotted line). The sedimentation coefficients have been corrected to the standard condition of water at 20 °C. No faster moving species was observed in early scanning. *Previously published in Liu et al. (1995).*

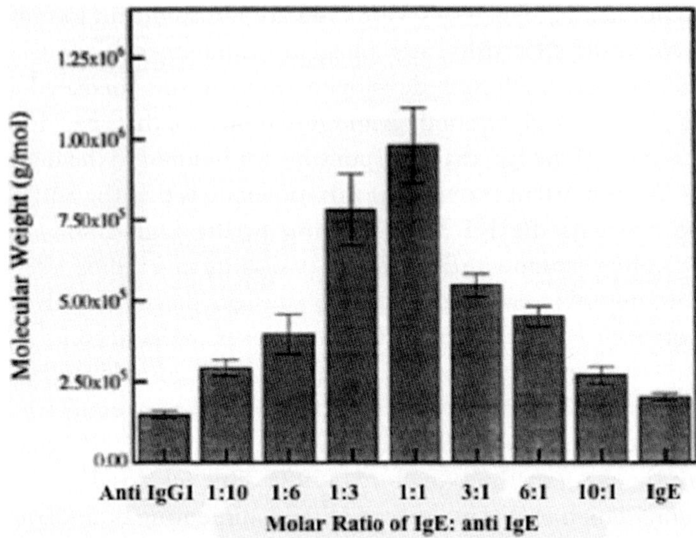

Figure 7 Sedimentation equilibrium analysis of IgE and anti-IgE complex formation in PBS at 10 °C. The weight average molecular weights of complexes at different molar ratios were obtained by analyzing the data from three different rotor speeds (5000, 7000, and 10,000 rpm) as a single ideal species simultaneously. The error bars correspond to a 95% confidence interval. *Reproduced from Liu et al. (1995).*

consisting of three IgE and three anti-IgE antibodies. The determined molecular weights of the intermediate species were consistent with that of a trimeric species which either has one IgE with two anti-IgE when there are in large excesses of anti-IgE, or two IgE with one anti-IgE, when there are in large excesses of IgE. A comparison of the experimental results to molecular weight computations for the largest observed species is shown in Fig. 8. The complexes formed with IgE and soluble high-affinity Fc receptor were also characterized. Hydrodynamic computations using bead models combined with the sedimentation equilibrium and sedimentation velocity analysis of the different complexes yields a model for complex formation as drawn in Fig. 9A. IgE is a slightly bigger molecule than an IgG1 (molecular weight of ~190 kDa compared to 150 kDa), and this is reflected in a value of $S_{20,w}$ for IgE at 8.2 S versus 7.1 S for anti-IgE. Thus, two types of trimers are formed where at an excess of IgE, two IgE molecules bind via their high-affinity receptor-binding site to one anti-IgE molecule (13.8 S) and at an excess of anti-IgE, two anti-IgE molecules bind via their Fab regions to the high-affinity-binding site on one IgE molecule (13.4 S) (Fig. 9A). At a 1:1 molar ratio, the largest sized complex whose

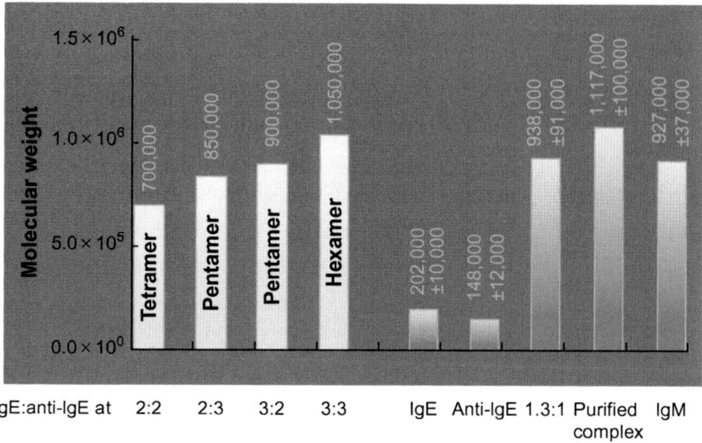

Figure 8 Comparison of computed of molecular weight assuming tetramers, pentamers, and hexamers with that of experimentally determined molecular weight of an IgM pentamer and largest complex at 1:1 molar ratio of IgE to anti-IgE after chromatographic purification. *Adapted from Liu et al. (1995).*

sedimentation coefficient is consistent with a computed cyclic hexamer is formed. Recently, the presence of such a structure was confirmed using transmission electron microscopy (Fig. 9B). In this case, AUC combined with modeling data enabled the determination of antibody–antigen complexes shape and stoichiometry.

6. ASSESSMENT OF DOSING OF ANTI-IgE BY AUC COMPETITIVE BINDING EXPERIMENTS

The formation of the different complexes, which are dependent on the molar ratio of IgE to anti-IgE dictates the pharmacokinetics of the drug therapy since IgE has a half-life of 6 h to 2 days, whereas an IgG1, due to the binding to Fc neonatal receptor (FcRn), has a typical half-life of ~3 weeks in serum. However, IgE takes on an extended circulating half-life when it is bound to IgG1 as a complex. Thus, the concentration of free IgE circulating in plasma and hence the dose of anti-IgE required to bind to free IgE should be related to the clearance rate of anti-IgE:IgE complexes, free IgE, unbound anti-IgE and the relative binding affinities of high-affinity receptor for IgE and that of IgE with anti-IgE. The anti-IgE monoclonal antibody was designed to have similar binding affinity to the IgE as the high-affinity Fc receptor, and thus the early formulation was at a relatively low

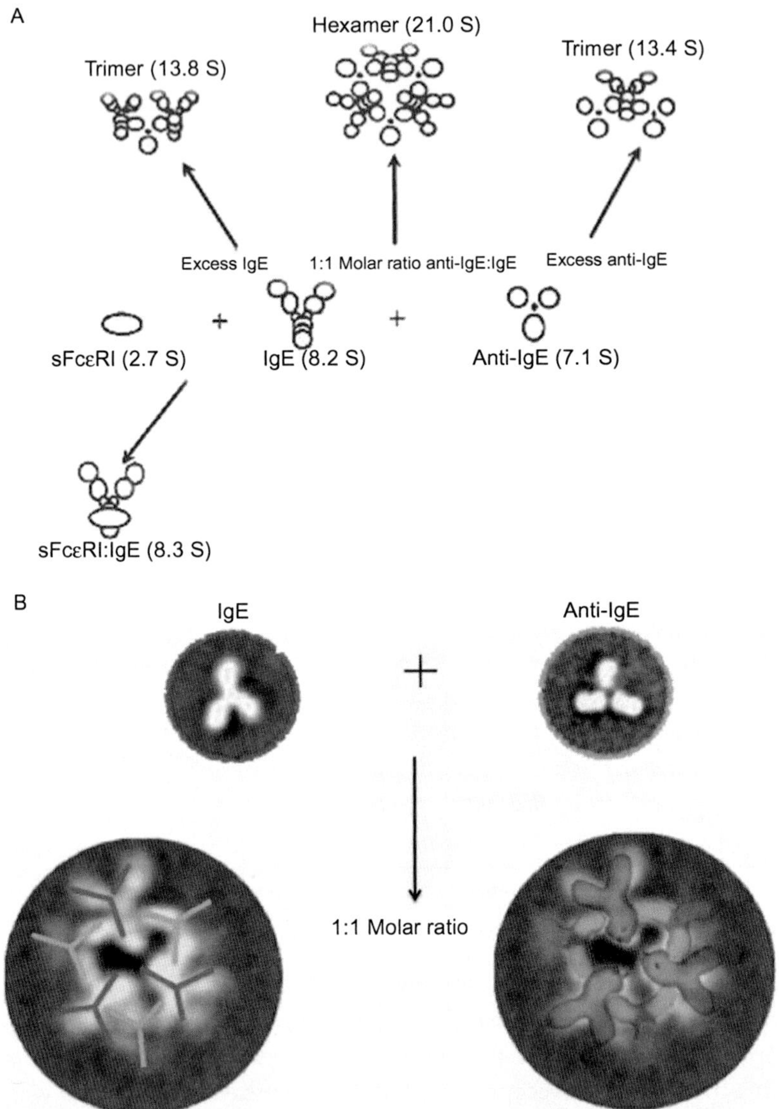

Figure 9 (A) Schematic diagram of complex formation by IgE and anti-IgE and IgE and soluble high-affinity Fc receptor, sFcεRIα (Liu, Ruppel, & Shire, 1997). (B) Transmission electron microscopy (TEM) of IgE–anti-IgE complexes formed at 1:1 molar ratio. A 3-μL solution of the sample was adsorbed onto an EM grid. After the sample solution was gently wicked away leaving a thin layer of solution, the grid was either plunged into liquid ethane or treated with a heavy metal stain. The preserved sample is then imaged in the TEM. Automated analysis of hundreds of images provides sufficient data for adequate statistical characterization. (TEM was kindly provided by NanoImaging Corporation.)

concentration of anti-IgE, ~5 mg/mL. Initial clinical results suggested that the dose needed to be higher and this raised several questions as to the accuracy of the affinity measurements using surface plasmon resonance (SPR) technology. Although SPR has been one of the best methods to determine high-affinity-binding constants, the experimental design needs to be thoroughly explored because of many artifacts that can occur, mainly as the results of immobilizing of one of the reactants to the carboxymethyldextran on the gold surface used in SPR (O'Shannessy, 1994; Trilling, Harmsen, Ruigrok, Zuilhof, & Beekwilder, 2013). In order to address this issue, a competitive binding AUC method was developed. In this method, preformed IgE:anti-IgE complexes were analyzed by sedimentation velocity AUC before and after addition of a soluble form of the Fc high-affinity receptor (sFcεRIα) (Yadav et al., 2013). The complexes were formed at a molar ratio of 6:1, anti-IgE:IgE where there is a predominance of trimeric species consisting of two anti-IgE molecules bound to one IgE molecule that sediments at ~13.3 S (Fig. 10A). After addition of soluble receptor at a ratio of 0.1:1, soluble receptor to IgE:anti-IgE complex, there is a reduction of this 13.3 S peak as well as a slight increase in the baseline between 8 and 9 S along with an increase of the unbound anti-IgE peak at ~7 S (Fig. 10A). With addition of increasing amounts of soluble receptor, there are additional increases in the anti-IgE peak, and the appearance of a peak at ~9 S which is likely the dimer peak consisting of one IgE and one soluble receptor (Liu et al., 1997). These data were analyzed using the differential sedimentation method of Stafford (Stafford, 1992), which does not take into account diffusion resulting in a broadening of the peak, and is the likely reason that at lower concentrations it is difficult to detect a single peak representative of the receptor:IgE dimer. This competitive binding analysis clearly showed that the receptor has a greater affinity for binding at the high-affinity Fc site on IgE than that of anti-IgE, thus requiring potentially greater dosing of the anti-IgE drug. The utility of the competitive binding AUC format was further shown by analyzing another anti-IgE mAb genetically engineered for higher binding affinity to IgE, referred to as anti-IgE2. The AUC analysis shows that the anti-IgE2 does indeed have largely improved affinity since even after increasing the molar ratio of soluble receptor to IgE–anti-IgE complex from 0.1:1 to 1:1, there is very little disruption of the IgE and anti-IgE complex (Fig. 10B). The competitive binding analysis performed using AUC was a crucial element in determining an appropriate dosage for the anti-IgE drug candidate.

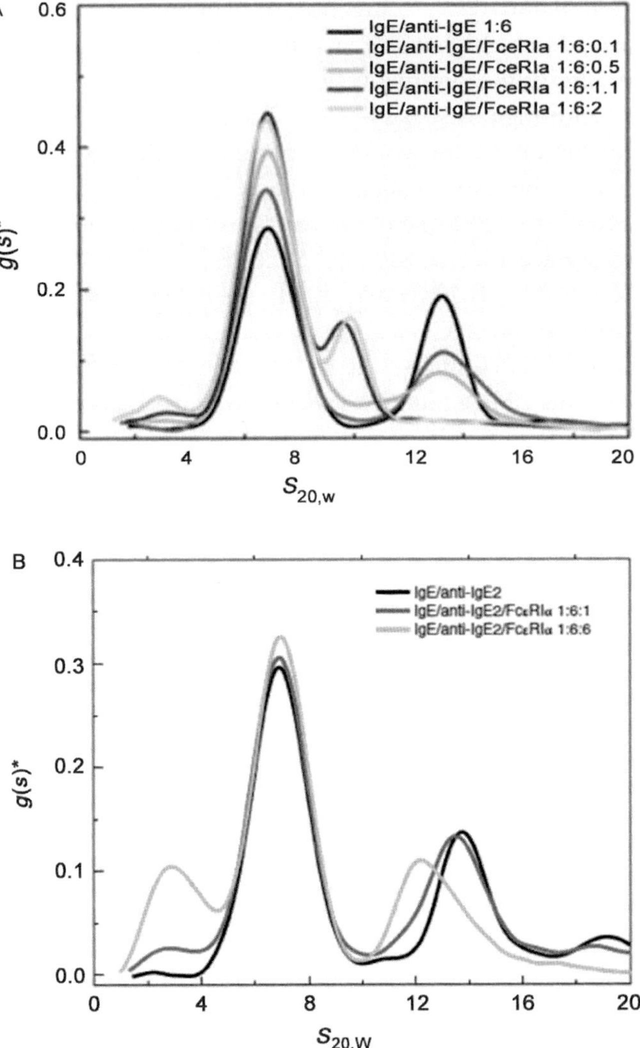

Figure 10 Differential sedimentation coefficient distribution of anti-IgE, FcεRIα, IgE:anti-IgE, and IgE:FcεRIα complexes (A) and anti-IgE2, FcεRIα, IgE:anti-IgE2, and IgE:FcεRIα complexes (B). Assessment of competition of binding of soluble high-affinity receptor, FcεRIα, with either preformed IgE:anti-IgE or IgE:anti-IgE2 complexes using AUC sedimentation velocity (Yadav et al., 2013). (See the color plate.)

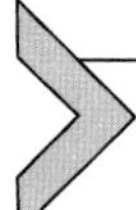

7. USING THE COMPETITIVE BINDING AUC METHOD TO CRITICALLY EVALUATE SPR MEASUREMENTS

As already mentioned, although SPR technology has been used successfully to evaluate high-affinity protein–protein interactions, there are many concerns associated with the binding of one of the reactants to the carboxymethyldextran on the gold surface used in SPR detection. While the determination of binding constants by AUC can be challenging, the competitive binding AUC method can be used to investigate the relative binding affinities in solution phase to determine if they are consistent with values reported using SPR. For example, SPR was used to evaluate the binding of three different VEGF inhibitors (VEGF Trap (also known as aflibercept), ranibizumab, and bevacizumab). The SPR results were purported to show that aflibercept compared to ranibizumab and bevacizumab had "markedly higher affinity for VEGF" (Papadopoulos et al., 2012). Recently, a thorough study using different SPR format designs showed a high dependency on outcome based on format used (Yang et al., 2014). It was also demonstrated that ranibizumab appeared to be a very tight VEGF binder in all formats tested. Overall, these SPR data suggested that the relative binding affinities of the different inhibitors still needed to be determined and raised the issue as to whether aflibercept does indeed have greater binding affinity than ranibizumab for VEGF. The competitive binding AUC method was therefore used to confirm or reject the previously reported SPR results. The data were obtained using an Optima XLI analytical ultracentrifuge equipped with UV absorbance optics, interference optics (Beckman Coulter, Fullerton, CA), and fluorescence detection system (Aviv Biomedical). All samples were centrifuged in centrifuge cells with 12-mm graphite-filled Epon centerpieces (Spin Analytical, Durham, NH) at 20 °C and rotor speed of 40,000 rpm using an An-50 Ti rotor (Beckman Coulter, Fullerton, CA). The concentration gradients as a function of time were analyzed using the program Sedfit (Schuck, Perugini, Gonzales, Howlett, & Schubert, 2002; Schuck & Rossmanith, 2000), which fits the data to the Lamm equation. This analysis takes into account the diffusion that occurs during the sedimentation experiment, resulting in narrower peaks and higher resolution than what was obtained previously for the anti-IgE studies using dc/dt analysis. The results are summarized in Fig. 11. The sedimentation coefficients in PBS corrected for standard conditions of water at 20 °C for VEGF, ranibizumab, and aflibercept show that the sedimentation coefficients in

PBS were 3.0, 3.7, and 5.2, respectively (Fig. 11A). After addition of VEGF to ranibizumab at 2:1 molar ratio and to aflibercept at 1:1 molar ratio, the sedimentation coefficient distribution showed sharp peaks at 6.0 and 6.5 S, respectively (Fig. 11A), with no significant amount of free VEGF, ranibizumab, or aflibercept. These results strongly support the binding of one VEGF molecules to two ranibizumab and only one VEGF to aflibercept

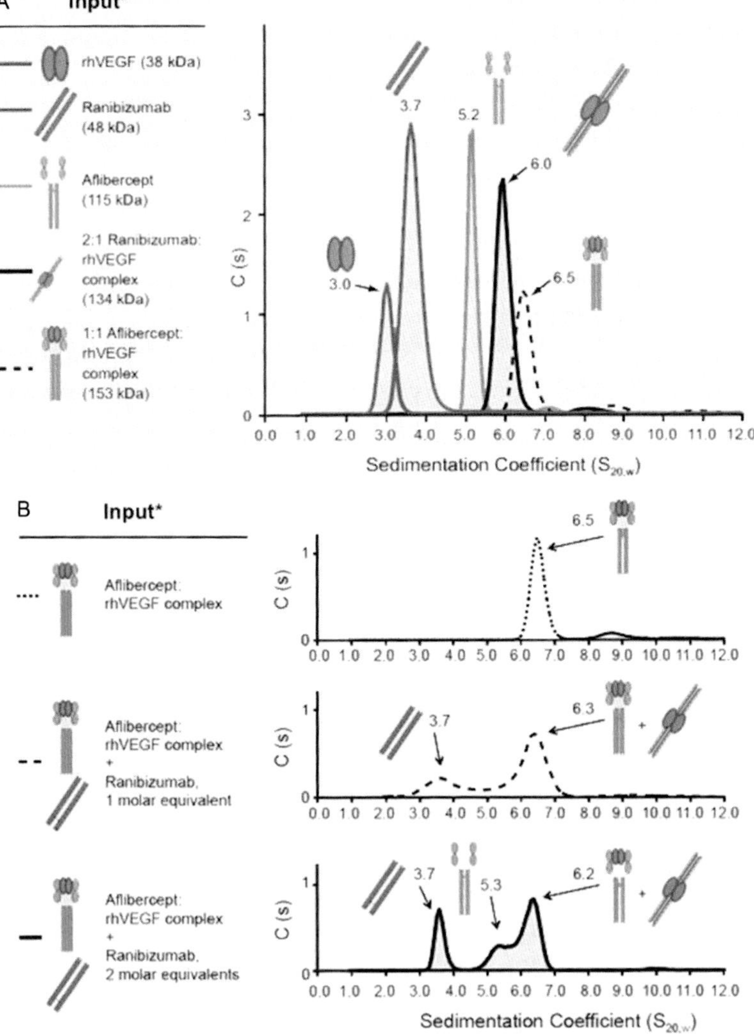

Figure 11 (A) Sedimentation coefficient determination of VEGF inhibitors alone and in complex with VEG-F. (B) Sedimentation coefficient determination of aflibercept/VEGF complex titrated with either 1 or 2 *M* equivalents of ranibizumab.

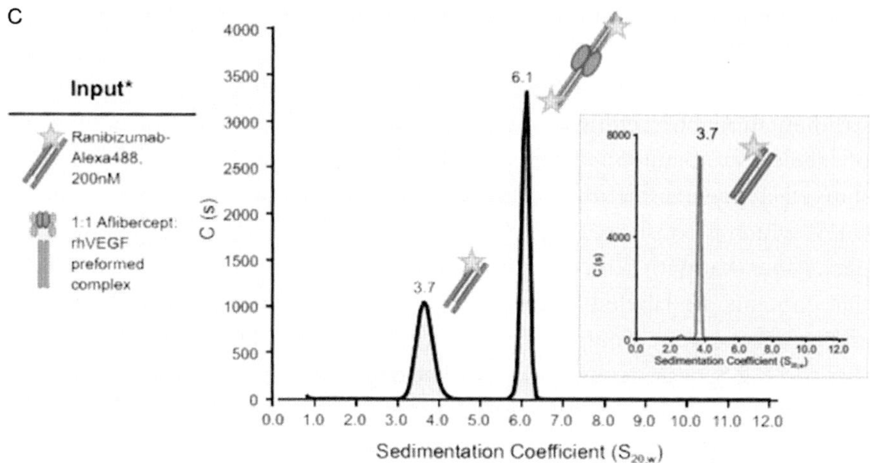

Figure 11—Cont'd (C) Sedimentation coefficient determination of aflibercept/VEGF complex after addition of fluorescently labeled ranibizumab. *Panel (A) is adapted from Yang et al. (2014).*

as expected. Most importantly, each of the free inhibitors and the different complexes with VEGF are clearly resolved. In an AUC competition assay, a preformed 1:1 complex of aflibercept/VEGF was titrated with 1 mol equivalent of free ranibizumab (Fig. 11B: dotted line, middle panel). The result was a major peak at 6.2 S, a peak at 3.6 S and a raised baseline between 4 and 6 S. The peak at 3.6 S is consistent with that seen for ranibizumab alone (Fig. 11A). The main peak at 6.3 S likely represents a mixture of 2:1 complex of ranibizumab/VEGF (~6 S) and a 1:1 complex of aflibercept/VEGF. Free aflibercept may also contribute to the observed raised baseline. In a separate experiment, ranibizumab was added at a 2 *M* ratio to the preformed 1:1 aflibercept–VEGF complex resulting in a peak at 6.2 S with a distinct shoulder at 5.3 S and a peak at 3.7 S (Fig. 11B: solid line bottom panel). The shoulder peak probably consists of free aflibercept (5.2 S in Fig. 11A) and the peak at 3.7 S with that of free VEGF (Fig. 11A). These competition experiments by SV-AUC show the distribution of species present after several hours of coincubation. Although it has not been established whether the system is at equilibrium, the interpretation of these results is not dependent on having attained equilibrium. In particular, it is evident that there is sufficient time for ranibizumab to displace aflibercept from a preformed complex with VEGF. Thus, these results show that ranibizumab was able to displace aflibercept from the preformed aflibercept/VEGF complex at similar concentration. Further confirmation of this conclusion was obtained by using a fluorescently labeled VEGF and running the competition experiment with the FDS (Fig. 11C). The sedimentation coefficient for the labeled VEGF ($S_{20,w}$ = 3.7 S, inset) confirmed that the labeling procedure did not alter the sedimentation characteristics of ranibizumab. Addition of 2 *M* equivalents of labeled ranibizumab to the preformed 1:1 aflibercept/VEGF complex showed a free ranibizumab peak at 3.7 S and a peak with $S_{20,w}$ value at 6.1 S, which is consistent with the presence of the ranibizumab/VEGF complex. Note that with fluorescence optics, only the fluorescently labeled species that contains the labeled ranibizumab are visible, and any free aflibercept or aflibercept-associated complex will not be detected. This definitively shows that the 6.2 S peak observed in the last panel of Fig. 11B does contain the ranibizumab/VEGF complex again showing that ranibizumab can displace aflibercept from preformed complexes.

8. INVESTIGATION OF PROTEIN–PROTEIN INTERACTIONS AND IMPACT ON VISCOSITY

As already discussed, previous studies on anti-IgE monoclonal antibodies using competitive binding by AUC have shown that higher dosing of the anti-IgE molecule was needed. This coupled with the need to develop a more convenient form of administration by the subcutaneous (SC) route led to development of high concentration formulations (>100 mg/mL). Early studies of anti-IgE mAb at high concentration showed that the formulated mAb had an unusually high concentration dependency on viscosity, which made for more difficult SC administration as well as the ability to manufacture the formulation (Shire, Shahrokh, & Liu, 2004). The anti-IgE mAb termed mAb1 was compared to another mAb called mAb2. The viscosity–concentration profile showed that mAb2 could be fitted with the Mooney equation that accounts only for excluded volume effects, whereas mAb1 data could not be fitted by that equation, suggesting that other interactions were involved that contributed to the high viscosity. These interactions are highly concentration dependent and reversible such that at high dilution, the mAb is essentially monomeric, and this is reflected in the concentration dependency of the viscosity. Many of the assays for size distribution such as SEC require measurements at low concentration. The analytical ultracentrifuge optics cannot accommodate the high concentration of the formulated mAb1, and thus an analysis was done using a preparative centrifuge sedimentation equilibrium method (Minton, 1989). In this method, the samples are centrifuged in a preparative centrifuge rotor until equilibrium is established (a few centrifuge runs at different times are used to determine if equilibrium has been attained). The contents of the centrifuge tubes are then fractionated using a microfractionator that results in a protein concentration measurement as a function of radial position (Fig. 12A). An apparent molecular weight was then obtained from these data using Eq. (4).

Assuming that mAb2 exists as monomeric molecule in solution, the non-ideality correction for the charge and excluded volume effects was obtained

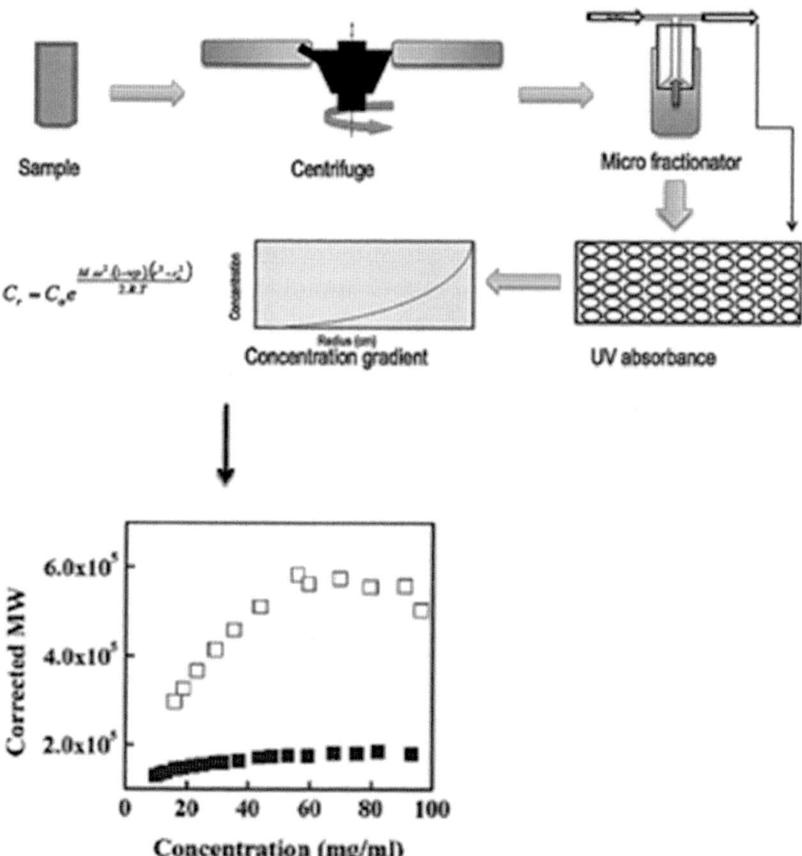

Figure 12 (A) Overview of a preparative AUC experiment. Centrifugation was carried out in a Beckman XL/A without the installed optics using a swinging bucket rotor. After attainment of sedimentation, equilibrium samples were fractionated using a Brandell micro fractionator, collected into a 96-well quartz plate and diluted to generate an absorbance versus radial position gradient that was used to generate a concentration for analysis to determine weight average molecular weight. (B) Corrected MW with no added salt (open squares) and 150 m*M* NaCl (solid squares) versus concentration for mAb1. *Overview drawing of preparative AUC provided by Sandeep Yadav, corrected MW plot from Liu, Nguyen, Andya, and Shire (2005).*

in the absence of added NaCl for an IgG1 monomer with 150 kDa molecular weight. These corrections were obtained from the relationship between apparent molecular weight, M_a at weight/volume concentration c, and actual molecular weight, M (Chatelier & Minton, 1987):

$$M = M_a[\mathbf{1} + c(\mathrm{d}\ln g/\mathrm{d}c)] \quad (5)$$

where g is the activity coefficient of the monoclonal antibody. Thus, this results in a multiplicative correction factor when M is set equal to 150 kDa. The resulting corrected M_w versus concentration suggests that the mAb1 undergoes self-association, which can be detected at high concentrations (Fig. 13B; Liu et al., 2005).

mAb1 and mAb2 were constructed using the same human IgG1 Fc framework and thus the differences in the antibodies are within the framework residues adjacent to as well as the complementarity determining regions of the mAbs (CDRs). Investigation of these sequences showed that there was an additional insertion of four amino acid residues in mAb1 and several charged residues in the CDR's, which are not present in mAb2. Mutant mAbs were then designed to investigate the impact on viscosity from the different charged amino acid residues. In addition, a mutant, designated as M-1, was created without the four amino acid insertion, and an aglycosylated version of mAb1, termed M-11, was generated to determine if differences in CDR length and glycosylation of the mAbs would alter

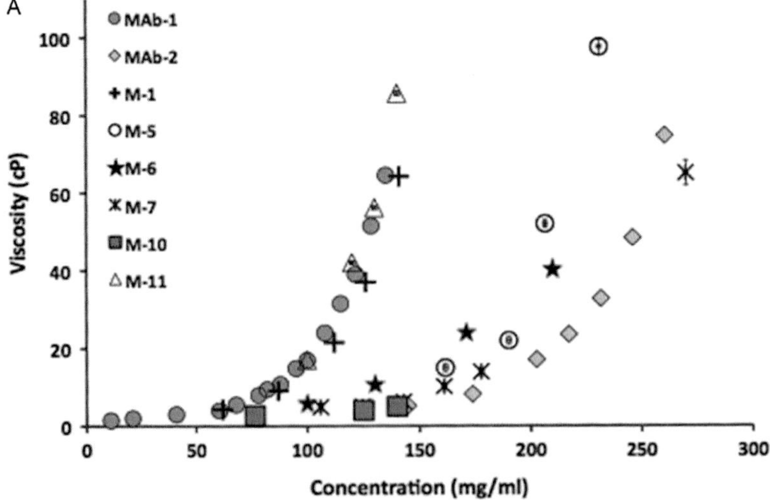

Figure 13 See legend on opposite page.

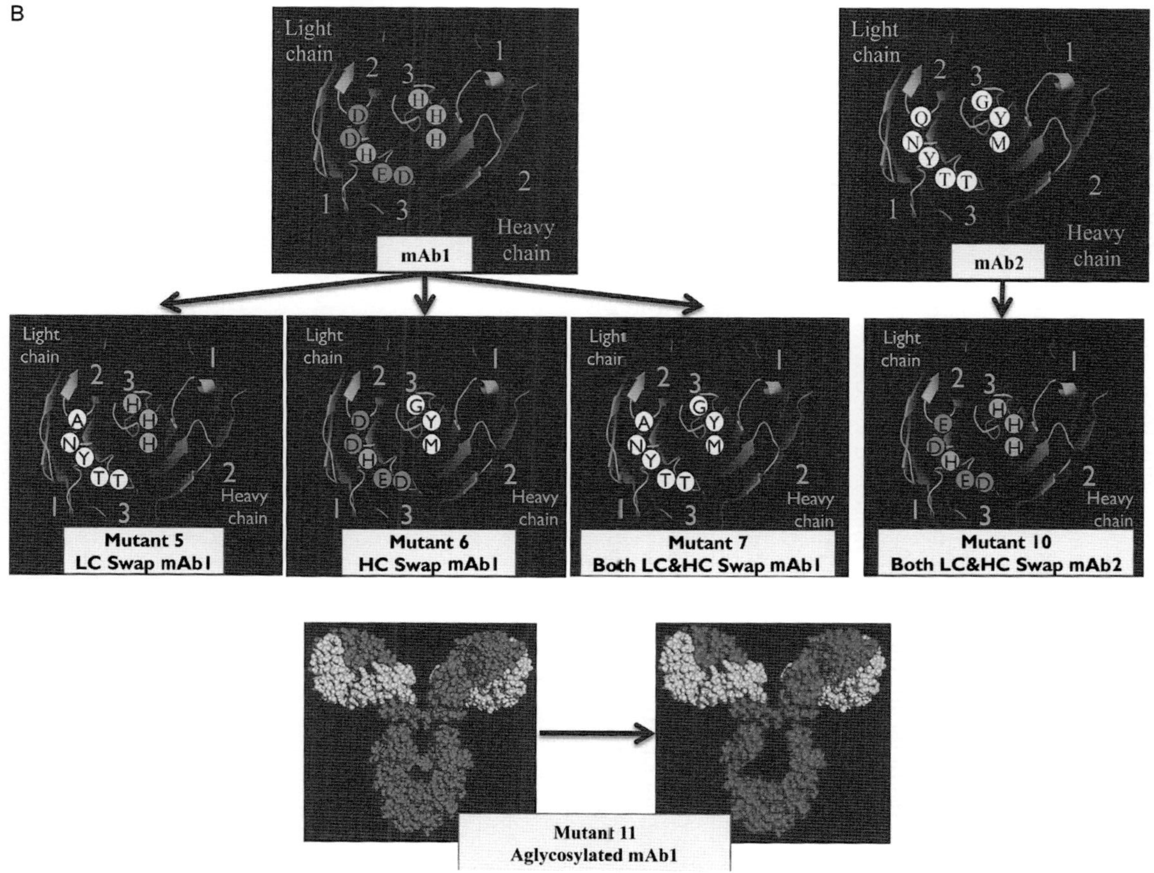

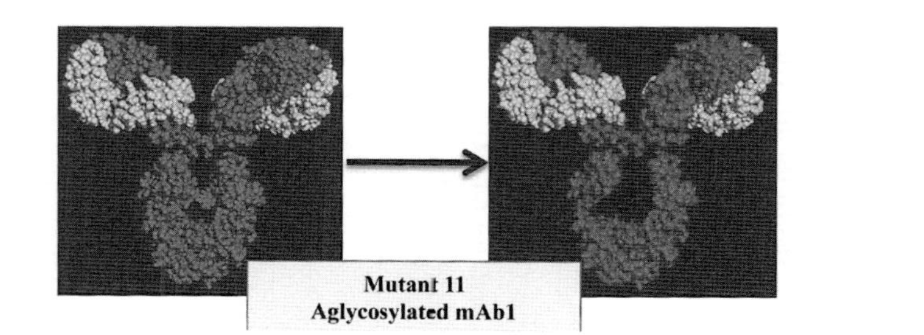

Figure 13 (A) Viscosity as a function of mAb concentration of mAb1 and mAb2 mutants with (B) schematics showing amino acid substitutions. All mabs and mutants are at pH 6 and 15 m*M* ionic strength. Viscosity determined using a cone-plate rheometer at a shear rate of 1000/s at 25 °C. *Adapted from Yadav et al. (2011).* (See the color plate.)

viscosity. The results are summarized in Fig. 13A and B and are provided with more details in Yadav et al. (2011). The deletion of the four amino acid sequence in the CDR of mAb1 (M-1) and the aglycosylated mutant (M-11) has no impact on the viscosity and thus these specific differences do not account for the different viscosity–concentration profile compared to mAb2. The viscosity for mutants with changes in either the VL chain (M-5) or in the VH chain (M-6) decreased although not to the values observed for mAb2. This shows the linkage of reversible self-association to viscosity and the fact that the attractive interactions are predominantly electrostatic in nature. The M-10 mutant created by replacing the non-charged residues in the mAb2 CDR with the charged residues from mAb1 was expected to have an increased viscosity similar to mAb1. However, the viscosity did not increase and was similar to that observed for mAb2. Previously, a linkage between reversible self-association at high concentration with increased viscosity was established using the preparative AUC technique (Liu et al., 2005). The corrected weight average molecular weights (corrected as described previously using mAb2 data) for mAb1, the M-7 and the M-10 mutants were determined by the preparative AUC technique at low (15 m*M*) and high-ionic strength (150 m*M*) (Fig. 14; Yadav

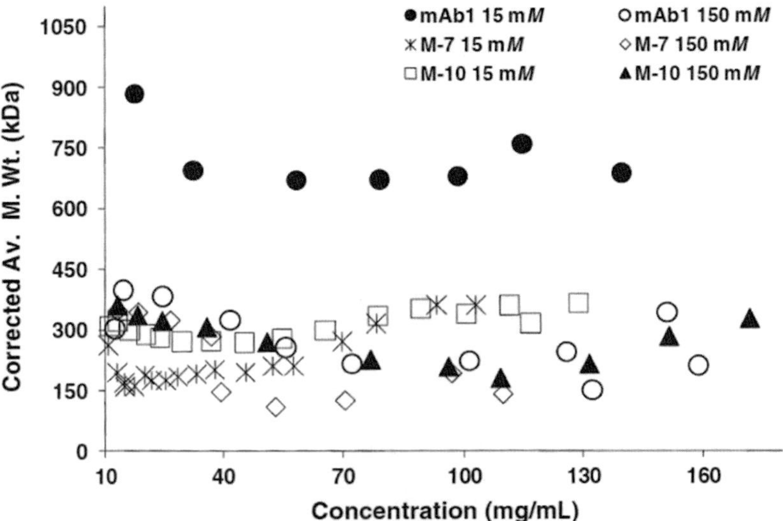

Figure 14 The corrected weight average molecular weights for mAb1 and mutants, using the nonideality corrections for mAb2 (Eq. 5). The measurement was conducted at pH 6.0, histidine hydrochloride buffer in 15 and 150 m*M* ionic strength (made up by addition of NaCl) at 12,000 rpm at 20 °C. *From Yadav et al. (2011).*

et al., 2011). The corrected molecular weight for mAb1 at 15 mM ionic strength is greater than M-7 and M-10 showing that mAb1 self-associates to a greater extent than M-7 and M-10 under these conditions. Replacement of the mAb1-charged residues in both the VL and VH chain resulted in a lower molecular weight and was similar to the M-10 mutant. Thus, the self-association behavior of mutant M-10 where residues were substituted into the mAb2 structure was consistent with the observed low viscosity–concentration profile. The question, however, still remained as to why the charged residue substitutions into mAb2 did not result in increased self-association and increased viscosity.

The self-association that occurs at high concentration reflects the radial distribution of a mAb whereby the movement of one mAb results in a concerted movement with the surrounding mAbs. This association has been termed a "network" or clustering of the mAbs (Yadav et al., 2013). Viscosity is influenced by the extent of clustering and by the nature of the protein–protein interactions (e.g., long range vs. short range). An *ab initio* computation using coarse-grained modeling showed that the mutant M-10 resulted in clustering but the distribution of clusters at 120 mg/mL for mAb1 resulted in a more extensive self-association network than mAb2 (Chaudhri et al., 2012, 2013).

9. SUMMARY AND CONCLUSIONS

In this chapter, we have discussed the different ways AUC has been used at Genentech, and early studies included solution molecular weight determinations for simple small proteins and more complex glycoproteins produced using rDNA technology. Later studies included interaction with formulation components such as surfactants, evaluation of antibody complexes, investigation of proteins interactions and impact on physical properties such as viscosity at high concentrations, and further development of a competitive binding technique using AUC that can be used to critically analyze high-affinity determinations by SPR technology.

The analytical ultracentrifuge technique still remains one of the best analytical methods to assess protein–protein interactions as well as interaction with formulation excipients. Regulatory agencies now routinely request that SEC methods be verified by using orthogonal biophysical techniques. Light scattering and AUC are the most common methods used and the AUC will continue to be often used to address the SEC issues. As mentioned in the introduction, the current XLA/I is now out of date and requires development of a new centrifuge, which will ensure continued use of AUC in the biotechnology sector. The AUC is a labor-intensive machine with a high price tag, list price as much as $350,000, and requires a dedicated person(s) to run the instrument and analyze the data. Currently, many small biotech companies use contract labs that are set up with the appropriate personnel to analyze their proteins. Development of a new AUC with good interactive software and improvement of precision in the optical systems will ensure that there is a huge future role for AUC in development of protein drugs.

REFERENCES

Achatz, G., Achatz-Straussberger, G., Feichtner, S., Koenigsberger, S., Lenz, S., Peckl-Schmid, D., et al. (2010). The biology of IgE: Molecular mechanism restraining potentially dangerous high serum IgE titres in vivo. In M. L. Penuchet & E. Jensen-Jorolim (Eds.), *Cancer and IgE* (pp. 13–36). New York: Springer.

Alavattam, S., Demeule, B., Liu, J., Yadav, S., Cromwell, M., & Shire, S. J. (2013). Biophysical analysis in support of development of protein pharmaceuticals. In L. O. Nahri (Ed.), *Biophysics for therapeutic protein development* (pp. 173–204). New York: Springer.

Andrews, P. (1970). Estimation of molecular size and molecular weights of biological compounds by gel filtration. *Methods of Biochemical Analysis*, *18*, 1–53.

Andya, J. D., Liu, J., & Shire, S. J. (2010). Analysis of irreversible aggregation, reversible self-association, and fragmentation of monoclonal antibodies by analytical ultracentrifugation. In S. J. Shire, W. G. Gombotz, K. Bechtold-Peters, & J. Andya (Eds.), *Current trends in monoclonal antibody development and manufacturing: Vol. XI.* (pp. 207–227). New York: AAPS Press/Springer.

Canovadavis, E., Baldonado, I. P., & Teshima, G. M. (1990). Characterization of chemically synthesized human relaxin by high-performance liquid-chromatography. *Journal of Chromatography*, *508*(1), 81–96.

Carpenter, J. F., Randolph, T. W., Jiskoot, W., Crommelin, D. J. A., Middaugh, C. R., & Winter, G. (2010). Potential inaccurate quantitation and sizing of protein aggregates by size exclusion chromatography: Essential need to use orthogonal methods to assure the quality of therapeutic protein products. *Journal of Pharmaceutical Sciences*, *99*(5), 2200–2208.

Caspar, D. L. (1963). Assembly and stability of the tobacco mosaic virus particle. *Advances in Protein Chemistry*, *18*, 37–121.

Charlwood, P. A. (1957). Partial specific volume of proteins in relation to composition and environment. *Journal of the American Chemical Society*, *79*, 776–781.

Chatelier, R. C., & Minton, A. P. (1987). Sedimentation equilibrium in macromolecular solutions of arbitrary concentration. I. Self-associating proteins. *Biopolymers*, *26*(4), 507–524.

Chaudhri, A., Zarraga, I. E., Kamerzell, T. J., Brandt, J. P., Patapoff, T. W., Shire, S. J., et al. (2012). Coarse-grained modeling of the self-association of therapeutic monoclonal antibodies. *Journal of Physical Chemistry B*, *116*(28), 8045–8057.

Chaudhri, A., Zarraga, I. E., Yadav, S., Patapoff, T. W., Shire, S. J., & Voth, G. A. (2013). The role of amino acid sequence in the self-association of therapeutic monoclonal antibodies: Insights from coarse-grained modeling. *Journal of Physical Chemistry B*, *117*(5), 1269–1279.

Colfen, H., Laue, T. M., Wohlleben, W., Schilling, K., Karabudak, E., Langhorst, B. W., et al. (2010). The open AUC project. *European Biophysics Journal with Biophysics Letters*, *39*(3), 347–359.

Cox, D. J. (1969). Computer simulation of sedimentation in the ultracentrifuge. IV. Velocity sedimentation of self-associating solutes. *Archives of Biochemistry and Biophysics*, *129*(1), 106–123.

Crepeau, R. H., Conrad, R. H., & Edelstein, S. J. (1976). UV laser scanning and fluorescence monitoring of analytical ultra centrifugation with an on-line computer system. In M. S. Lewis & G. H. Weiss (Eds.), *Proceedings of the conference "50 years of the ultracentrifuge"* (pp. 27–40). Amsterdam: North-Holland Publishing Company.

Demeule, B., Gurny, R., & Arvinte, T. (2006). Where disease pathogenesis meets protein formulation: Renal deposition of immunoglobulin aggregates. *European Journal of Pharmaceutics and Biopharmaceutics*, *62*(2), 121–130.

Demeule, B., Shire, S. J., & Liu, J. (2009). A therapeutic antibody and its antigen form different complexes in serum than in phosphate-buffered saline: A study by analytical ultracentrifugation. *Analytical Biochemistry*, *388*(2), 279–287.

Edelstein, S. J., & Schachman, H. K. (1967). The simultaneous determination of partial specific volumes and molecular weights with microgram quantities. *The Journal of Biological Chemistry*, *242*, 306–311.

Eigenbrot, C., Randal, M., Quan, C., Burnier, J., Oconnell, L., Rinderknecht, E., et al. (1991). X-ray structure of human relaxin at 1.5-a—Comparison to insulin and implications for receptor-binding determinants. *Journal of Molecular Biology*, *221*(1), 15–21.

Gabrielson, J. P., & Arthur, K. K. (2011). Measuring low levels of protein aggregation by sedimentation velocity. *Methods*, *54*(1), 83–91.

Gibbons, R. A. (1972). Physico-chemical methods for the determination of the purity, molecular size and shape of glycoproteins. In A. Gottschalk (Ed.), *Glycoproteins, part A* (2nd ed., pp. 12–140). Amsterdam: Elsevier.

Giebler, R. (1992). The optima XL-a: A new analytical ultracentrifuge with a novel precision absorption optical system. In S. E. Harding, A. J. Rowe, & J. C. Horton (Eds.), *Analytical ultracentrifugation in biochemistry and polymer science* (pp. 16–25). Melksham, Wiltshire, England: The Royal Society of Chemistry.

Gilbert, L. M., & Gilbert, G. A. (1973). Sedimentation velocity measurement of protein association. *Methods in Enzymology*, *27*, 273–296.

Harding, S. E., Rowe, A. J., & Horton, J. C. (Eds.), (1992). *Analytical ultracentrifugation in biochemistry and polymer science*. Cambridge, England: The Royal Society of Chemistry.

Hebert, C. A., Luscinskas, F. W., Kiely, J. M., Luis, E. A., Darbonne, W. C., Bennett, G. L., et al. (1990). Endothelial and leukocyte forms of Il-8—Conversion by thrombin and interactions with neutrophils. *Journal of Immunology*, *145*(9), 3033–3040.

Hughes, H., Morgan, C., Brunyak, E., Barranco, K., Cohen, E., Edmunds, T., et al. (2009). A multi-tiered analytical approach for the analysis and quantitation of high-molecular-weight aggregates in a recombinant therapeutic glycoprotein. *AAPS Journal*, *11*(2), 335–341.

Johnson, M. L., Correia, J. J., Yphantis, D. A., & Halvorson, H. R. (1981). Analysis of data from the analytical ultra-centrifuge by non-linear least-squares techniques. *Biophysical Journal*, *36*(3), 575–588.

Jones, L. S., Cipola, D., Liu, J., Shire, S. J., & Randolph, T. W. (1999). Investigation of protein-surfactant interactions by analytical ultracentrifugation and electron paramagnetic resonance: The use of recombinant human tissue factor as an example. *Pharmaceutical Research*, *16*(6), 808–812.

Klotz, A. V., & Thomas, B. A. (1993). N5-methylasparagine and asparagine as nucleophiles in peptides—Main-chain vs side-chain amide cleavage. *Journal of Organic Chemistry*, *58*(25), 6985–6989.

Kratky, O., Leopold, H., & Stabinger, H. (1973). *The determination of the partial specific volume of proteins by the mechanical oscillator technique*. New York: Academic Press.

Kroe, R. R., & Laue, T. M. (2009). NUTS and BOLTS: Applications of fluorescence-detected sedimentation. *Analytical Biochemistry*, *390*(1), 1–13.

Laue, T. M., & Stafford, W. F. (1999). Modern applications of analytical ultracentrifugation. *Annual Review of Biophysics and Biomolecular Structure*, *28*, 75–100.

Lebowitz, J., Lewis, M. S., & Schuck, P. (2002). Modern analytical ultracentrifugation in protein science: A tutorial review. *Protein Science*, *11*(9), 2067–2079.

Lewis, M. S., & Weiss, G. H. (1975). *Fifty years of the ultracentrifuge*. Amsterdam: North-Holland Publishing Company.

Liu, J., Andya, J. D., & Shire, S. J. (2006). A critical review of analytical ultracentrifugation and field flow fractionation methods for measuring protein aggregation. *AAPS Journal*, *8*(3), E580–E589.

Liu, J., Lester, P., Builder, S., & Shire, S. J. (1995). Characterization of complex formation by humanized anti-IgE monoclonal antibody and monoclonal human IgE. *Biochemistry*, *34*(33), 10474–10482.

Liu, J., Nguyen, M. D., Andya, J. D., & Shire, S. J. (2005). Reversible self-association increases the viscosity of a concentrated monoclonal antibody in aqueous solution. *Journal of Pharmaceutical Sciences*, *94*(9), 1928–1940.

Liu, J., Ruppel, J., & Shire, S. J. (1997). Interaction of human IgE with soluble forms of IgE high affinity receptors. *Pharmaceutical Research*, *14*(10), 1388–1393.

Lowman, H. B., Fairbrother, W. J., Slagle, P. H., Kabakoff, R., Liu, J., Shire, S., et al. (1997). Monomeric variants of IL-8: Effects of side chain substitutions and solution conditions upon dimer formation. *Protein Science*, *6*(3), 598–608.

Maggio, E. T. (2010). Use of excipients to control aggregation in peptide and protein formulations. *Journal of Excipients and Food Chemistry*, *1*(2), 40–49.

McMeekin, T. L., & Marshall, K. (1952). Specific volumes of proteins and their relationship to their amino acid contents. *Science*, *116*, 142–143.

Meselson, M., & Stahl, F. W. (1958). The replication of DNA in Escherichia Coli. *Proceedings of the National Academy of Sciences*, *44*, 671–682.

Minton, A. P. (1989). Analytical centrifugation with preparative ultracentrifuges. *Analytical Biochemistry*, *176*(2), 209–216.

Minton, A. P. (2005). Influence of macromolecular crowding upon the stability and state of association of proteins: Predictions and observations. *Journal of Pharmaceutical Sciences*, *94*(8), 1668–1675.

Moore, W. V., & Leppert, P. (1980). Role of aggregated human growth hormone (hGH) in development of antibodies to hGH. *The Journal of Clinical Endocrinology and Metabolism*, *51*(4), 691–697.

Mould, D. R., & Sweeney, K. R. D. (2007). The pharmacokinetics and pharmacodynamics of monoclonal antibodies—Mechanistic modeling applied to drug development. *Current Opinion in Drug Discovery & Development*, *10*(1), 84–96.

Nemerson, Y. (1988). Tissue factor and hemostasis. *Blood*, *71*, 1–8.

O'Shannessy, D. J. (1994). Determination of kinetic rate and equilibrium binding constants for macromolecular interactions: A critique of the surface plasmon resonance literature. *Current Opinion in Biotechnology*, *5*(1), 65–71.

Papadopoulos, N., Martin, J., Ruan, Q., Rafique, A., Rosconi, M. P., Shi, E. G., et al. (2012). Binding and neutralization of vascular endothelial growth factor (VEGF) and related ligands by VEGF trap, ranibizumab and bevacizumab. *Angiogenesis*, *15*(2), 171–185.

Pekar, A., & Sukumar, M. (2007). Quantitation of aggregates in therapeutic proteins using sedimentation velocity analytical ultracentrifugation: Practical considerations that affect precision and accuracy. *Analytical Biochemistry*, *367*(2), 225–237.

Pennica, D., Kohr, W. J., Fendly, B. M., Shire, S. J., Raab, H. E., Borchardt, P. E., et al. (1992). Characterization of a recombinant extracellular domain of the type

I tumor necrosis factor receptor: Evidence for tumor necrosis factor-a induced receptor aggregation. *Biochemistry, 31*, 1134–1141.

Perkins, S. J. (1986). Protein volumes and hydration effects—The calculations of partial specific volumes, neutron-scattering match points and 280-nm absorption-coefficients for proteins and glycoproteins from amino-acid-sequences. *European Journal of Biochemistry, 157*(1), 169–180.

Philo, J. S. (2009). A critical review of methods for size characterization of non-particulate protein aggregates. *Current Pharmaceutical Biotechnology, 10*(4), 359–372.

Ratner, R. E., Phillips, T. M., & Steiner, M. (1990). Persistent cutaneous insulin allergy resulting from high-molecular-weight insulin aggregates. *Diabetes, 39*(6), 728–733.

Ring, J., Stephan, W., & Brendel, W. (1979). Anaphylactoid reactions to infusions of plasma-protein and human-serum albumin—Role of aggregated proteins and of stabilizers added during production. *Clinical Allergy, 9*(1), 89–97.

Ryan, M. E., Webster, M. L., & Statler, J. D. (1996). Adverse effects of intravenous immunoglobulin therapy. *Clinical Pediatrics, 35*(1), 23–31.

Schachman, H. K. (1959). *Ultracentrifugation in biochemistry*. New York: Academic Press.

Schachman, H. K. (1989). Analytical ultracentrifugation reborn. *Nature, 341*(6239), 259–260.

Schachman, H. K. (1992). Is there a future for the ultracentrifuge. In S. E. Harding, A. J. Rowe, & J. C. Horton (Eds.), *Analytical ultracentrifugation in biochemistry and polymer science* (pp. 3–15). Melksham, Wiltshire, England: The Royal Society of Chemistry.

Schachman, H. K., & Edelstein, S. J. (1973). Ultracentrifugal studies with absorption optics and a split-beam photoelectric scanner. *Methods in Enzymology, 27*, 3–59.

Schuck, P., Perugini, M. A., Gonzales, N. R., Howlett, G. J., & Schubert, D. (2002). Size-distribution analysis of proteins by analytical ultracentrifugation: Strategies and application to model systems. *Biophysical Journal, 82*(2), 1096–1111.

Schuck, P., & Rossmanith, P. (2000). Determination of the sedimentation coefficient distribution by least-squares boundary modeling. *Biopolymers, 54*(5), 328–341.

Schuster, T. M., & Laue, T. M. (Eds.), (1994). *Emerging biochemical and biophysical techniques. Modern analytical ultracentrifugation*. Boston: Birkhauser.

Scottish Intercollegiate Guidelines Network. (2014). *British guideline on the management of asthma: A national clinical guideline*. London: British Thoracic Society. http://www.brit.-thoracic.org.uk.

Seidl, A., Hainzl, O., Richter, M., Fischer, R., Bohm, S., Deutel, B., et al. (2012). Tungsten-induced denaturation and aggregation of epoetin alfa during primary packaging as a cause of immunogenicity. *Pharmaceutical Research, 29*(6), 1454–1467.

Shao, Z. Z., Li, Y. P., Krishnamoorthy, R., Chermak, T., & Mitra, A. K. (1993). Differential-effects of anionic, cationic, nonionic, and physiological surfactants on the dissociation, alpha-chymotryptic degradation, and enteral absorption of insulin hexamers. *Pharmaceutical Research, 10*(2), 243–251.

Shire, S. J. (1983). pH-dependent polymerization of a human-leukocyte interferon produced by recombinant deoxyribonucleic-acid technology. *Biochemistry, 22*(11), 2664–2671.

Shire, S. J., Holladay, L. A., & Rinderknecht, E. (1991). Self-association of human and porcine relaxin as assessed by analytical ultracentrifugation and circular-dichroism. *Biochemistry, 30*(31), 7703–7711.

Shire, S. J., Shahrokh, Z., & Liu, J. (2004). Challenges in the development of high protein concentration formulations. *Journal of Pharmaceutical Sciences, 93*(6), 1390–1402.

Spin Analytical (2014). Centrifugal fluid analyzer, www.spinanalytical.com/.

Stafford, W. F., 3rd. (1992). Boundary analysis in sedimentation transport experiments: A procedure for obtaining sedimentation coefficient distributions using the time derivative of the concentration profile. *Analytical Biochemistry, 203*(2), 295–301.

Stone, K. D., Prussin, C., & Metcalfe, D. D. (2010). IgE, mast cells, basophils, and eosinophils. *Journal of Allergy and Clinical Immunology, 125*(2), S73–S80.
Svedberg, T., & Fåhraeus, R. (1926). A new method for the determination of the molecular weight of the proteins. *The Journal of the American Chemical Society, 48*, 430–438.
Tanford, C., Nozaki, Y., Reynolds, J. A., & Makino, S. (1974). Molecular characterization of proteins in detergent solutions. *Biochemistry, 13*(11), 2369–2376.
Teller, D. C. (1973). Characterization of proteins by sedimentation equilibrium in the analytical ultracentrifuge. *Methods in Enzymology, 27*, 346–441.
Trilling, A. K., Harmsen, M. M., Ruigrok, V. J. B., Zuilhof, H., & Beekwilder, J. (2013). The effect of uniform capture molecule orientation on biosensor sensitivity: Dependence on analyte properties. *Biosensors & Bioelectronics, 40*(1), 219–226.
Underwood, L. E., Voina, S. J., & Van Wyk, J. J. (1974). Restoration of growth by human growth hormone (Roos) in hypopituitary dwarfs immunized by other human growth hormone preparations: Clinical and immunological studies. *The Journal of Clinical Endocrinology and Metabolism, 38*(2), 288–297.
van Holde, K. E., & Hansen, J. C. (1998). Analytical ultracentrifugation from 1924 to the present: A remarkable history. *Chemtracts. Biochemistry and Molecular Biology, 11*, 933–943.
Varley, P. G., Brown, A. J., Dawkes, H. C., & Burns, N. R. (1997). A case study and use of sedimentation equilibrium analytical ultracentrifugation as a tool for biopharmaceutical development. *European Biophysics Journal with Biophysics Letters, 25*(5–6), 437–443.
Wen, J., Arakawa, T., & Philo, J. S. (1996). Size-exclusion chromatography with on-line light-scattering, absorbance, and refractive index detectors for studying proteins and their interactions. *Analytical Biochemistry, 240*(2), 155–166.
Wetzel, R., Perry, L. J., Estell, D. A., Lin, N., Levine, H. L., Slinker, B., et al. (1981). Properties of a human alpha-interferon purified from Escherichia coli extracts. *Journal of Interferon Research, 1*(3), 381–390.
Whitaker, J. R. (1963). Determination of molecular-weights of proteins by gel-filtration on sephadex. *Analytical Chemistry, 12*, 1950–1953.
Williams, R. C. J. (1976). Improvements in precision of sedimentation-equilibrium experiments with an on-line absorption system. In M. S. Lewis & G. H. Weiss (Eds.), *Proceedings of the conference "50 years of the ultracentrifuge"* (pp. 19–26). Amsterdam: North-Holland Publishing Company.
Yadav, S., Liu, J., Scherer, T. M., Gokarn, Y., Demeule, B., Kanai, S., et al. (2013). Assessment and significance of protein–protein interactions during development of protein biopharmaceuticals. *Biophysical Reviews, 5*, 121–136.
Yadav, S., Sreedhara, A., Kanai, S., Liu, J., Lien, S., Lowman, H., et al. (2011). Establishing a link between amino acid sequences and self-associating and viscoelastic behavior of two closely related monoclonal antibodies. *Pharmaceutical Research, 28*(7), 1750–1764.
Yang, J. H., Wang, X. D., Fuh, G., Yu, L. L., Wakshull, E., Khosraviani, M., et al. (2014). Comparison of binding characteristics and in vitro activities of three inhibitors of vascular endothelial growth factor A. *Molecular Pharmaceutics, 11*(10), 3421–3430.

CHAPTER TWENTY

Guidance to Achieve Accurate Aggregate Quantitation in Biopharmaceuticals by SV-AUC

Kelly K. Arthur, Brent S. Kendrick, John P. Gabrielson[1]
Attribute Sciences, Amgen Inc., Longmont, Colorado, USA
[1]Corresponding author: e-mail address: jgabriel@elionlabs.com

Contents

Abstract

The levels and types of aggregates present in protein biopharmaceuticals must be assessed during all stages of product development, manufacturing, and storage of the finished product. Routine monitoring of aggregate levels in biopharmaceuticals is typically achieved by size exclusion chromatography (SEC) due to its high precision, speed, robustness, and simplicity to operate. However, SEC is error prone and requires careful method development to ensure accuracy of reported aggregate levels. Sedimentation velocity analytical ultracentrifugation (SV-AUC) is an orthogonal technique that can be used to measure protein aggregation without many of the potential inaccuracies of SEC. In this chapter, we discuss applications of SV-AUC during biopharmaceutical development and how characteristics of the technique make it better suited for some applications than others. We then discuss the elements of a comprehensive analytical control strategy for SV-AUC. Successful implementation of these analytical control elements ensures that SV-AUC provides continued value over the long time frames necessary to bring biopharmaceuticals to market.

Methods in Enzymology, Volume 562
ISSN 0076-6879
http://dx.doi.org/10.1016/bs.mie.2015.06.011

1. INTRODUCTION

Protein aggregation in biopharmaceuticals is a pervasive and usually deleterious degradation mechanism that can occur under a variety of manufacturing, storage, and sample handling conditions, and across a wide range of protein concentrations and formulations. Maintaining sufficiently low aggregate levels in biopharmaceuticals is critical to the quality of these products, as the presence of aggregates can alter product potency and elicit immune responses in patients. As such, the aggregate content, nature, and distribution must be fully characterized, quantified, and routinely monitored as part of a successful analytical control strategy for a biopharmaceutical product (Berkowitz, Engen, Mazzeo, & Jones, 2012). While analytical ultracentrifugation (AUC) was first developed in the early 1900s, in the past 15 years significant improvements have been made to data analysis capabilities (software algorithms as well as speed of computer processors) to enable the emergence of AUC as an important method for protein aggregate quantitation (Brown & Schuck, 2006; Dam, Schuck, Michael, & Ludwig, 2004; Laue & Stafford, 1999; Schuck, 2000; Schuck, Perugini, Gonzales, Howlett, & Schubert, 2002). Currently, sedimentation velocity AUC (SV-AUC) with SEDFIT *c*(*s*) distribution analysis is widely applied for quantitation of aggregate levels in biopharmaceuticals (Schuck, 2000).[1]

While SV-AUC has become an important tool for quantitation of aggregates in biopharmaceutical products, it is not well suited for all applications (Berkowitz et al., 2012). For example, chromatography-based separation methods are usually better suited for product release and stability testing in a quality control laboratory. In part, this is because the SV-AUC method has lower throughput and requires more sample, instrumentation, and data manipulations that can impact its precision compared to chromatographic separation methods. However, there are many other applications, e.g., lead candidate selection, formulation development, and product characterization, for which SV-AUC offers distinct advantages over size exclusion chromatography (SEC) due to its ability to analyze samples under various solution conditions, with minimal surface interactions, and with high

[1] We have found SEDFIT *c*(*s*) analysis to be particularly suitable for quantitation of low levels of protein aggregation, as is typically the case with biopharmaceutical protein products. Therefore, throughout this chapter, all references to data analysis parameters will refer to application of the SEDFIT *c*(*s*) distribution model to the raw data.

resolving power (Carpenter et al., 2010; Gabrielson, Brader, et al., 2007; Philo, 2006).

In this chapter, we discuss appropriate and recommended applications of SV-AUC in the context of aggregate quantitation for biopharmaceutical products. We then present critical operating parameters and necessary controls to optimize the method and ensure accurate results reporting. As previously mentioned, SV-AUC requires more analyst oversight and entails more complicated data analysis than SEC. Therefore, without a proper understanding of the potential sources of error, and suitable controls for each, SV-AUC results cannot be used confidently to make well-informed decisions. Finally, we propose a method for SV-AUC lifecycle management, specifically practices that can be implemented to aid in controlling SV-AUC performance over time. Developing a strategy for continuous method improvement through a combination of method design, qualification, and monitoring is important to enable accurate results throughout the lifecycle of a biopharmaceutical product.

2. APPLICATIONS OF SV-AUC

AUC is a critical analytical tool for analysis of biopharmaceutical products. While it may never achieve the speed and capacity that other size separation methods provide, AUC offers unique capabilities that enable verification of the accuracy of protein aggregate measurements by other methods under a variety of solution conditions. SEC is the most commonly applied analytical tool for quantitation of aggregates and has many advantages in terms of throughput, automation, capacity, precision, and sensitivity. However, SEC is susceptible to many artifacts, which if not characterized properly (using the AUC technique) may lead to gross errors in aggregate quantitation. Examples of SEC artifacts include (1) adsorption of protein size variants to the SEC stationary phase (often large molecular weight aggregates adsorb preferentially to the stationary phase, resulting in a significant overestimate of product purity); (2) significant dilutional effects when a sample is injected into the SEC mobile phase, which can result in dissociation of reversible protein aggregates; and (3) mobile phase-induced changes in the aggregation profile due to differences in pH, salt concentration, or lack of stabilizing excipients relative to the formulation buffer (Carpenter et al., 2010; Gabrielson, Brader, et al., 2007; Philo, 2006). On the other hand, AUC is not subject to the potential artifacts of the SEC technique, and when used appropriately as described in this chapter, can guide SEC method development to minimize these artifacts.

SV-AUC enables *in situ* analysis of soluble aggregates in final formulation buffer conditions under which the product is stored and delivered to patients. Analyzing the aggregation profile in formulation buffer and, if possible, without sample dilution avoids many potential issues encountered during SEC analysis, where dilutional effects and chromatography stationary phase interactions can disrupt the aggregate profile (Carpenter et al., 2010; Gabrielson, Brader, et al., 2007; Philo, 2006). SEC mobile phases typically require high salt concentrations and/or organic solvents to minimize nonspecific interactions with the column packing. The high ionic strength of the SEC mobile phase and/or the presence of organic components in the buffer have the potential either to disrupt weakly associated aggregates or generate aggregates on the column during analysis. Thus, the measurement itself has the potential to alter the aggregate profile during SEC analysis. The *in situ* analysis of aggregates in formulation conditions by SV-AUC removes these analysis-based artifacts and thereby allows more accurate quantitation of aggregate levels.

With SEC analysis, protein aggregates can differentially adsorb to the SEC stationary phase. This can occur when SEC mobile phase conditions have not been optimized for a given protein, and will result in underreporting of aggregate content (Carpenter et al., 2010; Gabrielson, Brader, et al., 2007; Philo, 2006). Conversely, SV-AUC is not appreciably impacted by nonspecific protein adsorption to surfaces during analysis. During early development of the SEC method, therefore, the product's aggregation profile should be verified by SV-AUC to establish the accuracy of the SEC method. Using this approach, significant redevelopment work and costly delays can be avoided later in product development.

SV-AUC also offers a significant improvement in the detection and resolution of large size variants, i.e., those aggregates in the size range above ca. 20 nm or 500 kDa (Philo, 2006). This improved resolution provides important information about both the size distribution and levels of large aggregates, which can provide insight into aggregation mechanisms, i.e., whether aggregate formation is due primarily to an increase in dimer, an increase in size variants larger than dimer, or a mixture of both. It can also help assess the capability of SEC to accurately report such aggregates, which often elute in either the void volume of the column or worse, are filtered out by the SEC column frit or stationary phase (Carpenter et al., 2010; Gabrielson, Brader, et al., 2007; Philo, 2006).

Some of the unique advantages of AUC enable it to be suitable for many purposes, including but not limited to: early-phase molecule selection, clone selection, assessment of molecule manufacturability, formulation development, process development, analytical development, product aggregation profile characterization, comparability/biosimilarity assessments, product characterization, and reference standard qualification. Early in clinical development of biopharmaceuticals (i.e., during lead compound selection, clone selection, production of toxicology, and first-in-human lots), there is typically little investment in analytical method development, and even less knowledge of the potential performance problems with analytical methods such as SEC prior to formal method development. These early development samples are often very diverse in terms of purity, number, and types of sequence variants and posttranslational modifications, all of which can impact the aggregation profile. Furthermore, extreme stress conditions such as high temperature and pH extremes may be used during candidate selection in order to select the most stable candidate under a range of stresses it might encounter during manufacturing and storage (Jiang, Li, Li, Gabrielson, & Wen, 2015). Extreme stresses are also routinely applied during formulation development to identify excipient conditions that stabilize against various degradation pathways, including aggregation. Utilization of SV-AUC enables an assessment of not only aggregation propensity under these conditions, but also the suitability of the SEC method for these diverse sample types. By applying SV-AUC in early development, the drug developer helps ensure a product, its formulation, and its manufacturing process are not developed with a potentially devastating undetected aggregation issue.

After early candidate selection, formulation, and process development, AUC should be utilized periodically to ensure that late-stage manufacturing process changes do not impact the product's aggregation profile. Even though SEC methods are typically well developed in later stages of process development, SV-AUC provides a critical orthogonal technique to verify the aggregation profile is acceptable with the commercial process. It also has other important uses and applications during late-stage product characterization, such as an orthogonal method for comparability and biosimilarity studies, reference standard qualification, elucidation of structure, degradation pathways, and impurities characterization. For elucidation of structure and degradation pathways characterization, SV-AUC is particularly useful because it can quantify a larger dynamic size range of aggregates and reversible self-association that may be undetected with SEC analysis alone (Carpenter et al., 2010; Gabrielson, Brader, et al., 2007; Philo, 2006). Much of the information from these studies is submitted as product characterization information in an applicant's biologics license application. A summary of common applications of SV-AUC during biopharmaceutical development is provided in Table 1.

As indicated in Table 1, SV-AUC offers an important array of capabilities that should be utilized throughout the product development lifecycle in order to characterize and establish a reliable control strategy for aggregates. There are a few limitations of the method, namely that the quantitation limit (QL) for small aggregates (below ca. 500 kDa) is roughly 3.0% (Gabrielson & Arthur, 2011; Gabrielson et al., 2010; Gabrielson, Randolph, Kendrick, &

Table 1 Common Applications of SV-AUC During Biopharmaceutical Development

<table>
<tr><th>Application</th><th>Advantages</th><th>Limitations</th></tr>
<tr><td>Lead molecular candidate selection and manufacturability assessments</td><td rowspan="4">Reliable quantitation of aggregate profile, especially with undeveloped or unqualified SEC methods; quantitation limit (QL) is lower (more sensitive) than SEC for large molecular weight aggregates (>500 kDa)</td><td rowspan="7">Requires highly specialized equipment and operators; less precise than SEC (leading to a higher QL for small aggregates, e.g., dimers–tetramers of antibody-sized products)</td></tr>
<tr><td>Clone selection</td></tr>
<tr><td>Early process development</td></tr>
<tr><td>Early formulation development</td></tr>
<tr><td>Analytical SEC method development and validation</td><td>Orthogonal method to quantify and elucidate aggregation profile to verify accuracy of SEC method for a variety of sample types</td></tr>
<tr><td>Comparability and biosimilarity studies</td><td>Orthogonal method to quantify aggregate levels; provides additional assurance of comparable product profile</td></tr>
<tr><td>Elucidation of structure, degradation pathways, and impurities</td><td>Provides a size and aggregate profile over a large dynamic range</td></tr>
</table>

Stoner, 2007; Pekar & Sukumar, 2007), although it is much more sensitive for larger aggregates. Also, the technique requires skilled scientists to execute the analysis due to the complex nature of deconvoluting the sedimentation profiles to derive the aggregation profile. Operators of the method as well as end-users of the data should be aware of these risks in order to appropriately control them (a detailed discussion of key operating parameters is provided in the following section). Overall, however, the ability to assess the product aggregation profile directly in formulation buffer without perturbing the aggregation state, the wide dynamic range of the technique, and the ability to run the product over a dilution series and in various buffers enable a thorough understanding of the distribution, levels, and nature of aggregates (e.g., reversibility, covalent/noncovalent linkages).

3. OPTIMIZING SV-AUC METHOD PARAMETERS AND APPLYING REAL-TIME CONTROLS

The previous section discussed applications of SV-AUC to enhance and enable a thorough understanding of aggregation in biopharmaceuticals. In order for the method to be meaningful in the context of aggregate quantitation, the data reported must be both accurate and precise, i.e., variability in the measurement must be reduced as much as possible. Achieving accurate and precise results is paramount for informed decision making. Unfortunately, insufficient or improper system controls can compromise results, leading to inaccurate conclusions. For example, some experimental parameters are known to falsely over-report aggregate levels, while others can mask the presence of true aggregates present in solution. However, with optimized method parameters and application of real-time control strategies (meaning system and result suitability assessments applied at the time of data acquisition and analysis), SV-AUC can be a powerful and useful tool for aggregate quantitation. This section provides guidance on identifying and managing experimental risk factors that could lead to inaccurate aggregate quantitation as well as recommended real-time controls for achieving high-quality SV-AUC data.

3.1 Avoiding False Aggregate Reporting

As previously stated, there are known SV-AUC operating and data analysis parameters that lead to both falsely over- and underreporting the true aggregate content. Here we discuss, and recommend control strategies for, three major error sources: cell alignment, centerpiece integrity, and the presence of cosedimenting solutes in solution.

In our experience, incorrect cell alignment is the primary experimental factor that can lead to false over-reporting of aggregates. In a previous study that examined potential sources of inter- and intrarun variability for the detection of protein aggregates by SV-AUC (Arthur, Gabrielson, Kendrick, & Stoner, 2009), external cell alignment was found to be the most significant source of measurement variability.[2] Other effects evaluated included deliberate addition of random and time-invariant noise, thermal convection, and meniscus placement during data analysis (real-time controls for these other factors are discussed later).

Misalignment of the cells, and therefore centerpieces, results in convective transport which is not modeled during data analysis. Cell misalignment manifests during data analysis in two ways: first, as elevated levels of aggregates natively present in solution, and second, as false larger aggregates that would not be observed if cells were correctly aligned. The increase in measured aggregation levels caused by both of these errors is directly proportional to the degree of misalignment. Importantly, the effect is not directionally dependent; clockwise and counter clockwise misalignments result in equal over-reporting of the true aggregate content (Gabrielson & Arthur, 2011). For example, in a monoclonal antibody monomer–dimer system containing 96.5% monomer and 3.5% dimer, for every degree of misalignment the dimer content was shown to increase by 1.4% and the presence of aggregates larger than dimer, not present under perfectly aligned conditions, increased by 1.7% for each degree misalignment (Arthur et al., 2009). Similar results obtained using bovine serum albumin support that the effect is inherent to the degree of misalignment and not the protein product (Gabrielson & Arthur, 2011).

It is important to note that, while the effect of misalignment on false aggregate reporting is significant, it is not currently possible to adequately diagnose misalignment during data analysis. The effect of cell misalignment on fit quality, e.g., high RMSD of fit and patterns in the residual bitmaps, is not perceptible until 4° misalignment (Arthur et al., 2009). Therefore, due to the dramatic effect external cell misalignment can have on the accuracy and variability of results (i.e., different degrees of misalignment across cells leading to different levels of aggregation measured by the technique), it is

[2] External cell alignment refers to the angular rotation of an assembled cell in the rotor hole and is measured with respect to the center of rotation of the rotor.

imperative to have a control strategy in place for this operating parameter. Per the instrument manufacturer's recommendation, alignment is performed by aligning scribe lines located on the rotor and cell housing. However, in our experience, visual inspection of the scribe lines does not adequately control and ensure proper cell alignment. The human eye has difficulty perceiving misalignments less than or equal to 0.5°, which is approximately equivalent to 0.1 mm displacement (Gabrielson et al., 2010). Recall, this degree of misalignment can result in up to approximately 1.5% over-reporting of total aggregate. In addition, it has been observed previously that the scribe lines are not always accurate (Gabrielson et al., 2010). Therefore, we highly recommend that a custom tool capable of controlling, and ideally measuring, external cell alignment be used. Alignment tools can be designed for custom use or purchased commercially.

While external cell alignment can be controlled using a tool, internal misalignment of the centerpiece channels is not easily controlled or monitored. Internal misalignment can occur as the result of slight play between the centerpiece keyway and ridge on the cell housing or manufacturing variability resulting in misaligned channels. Internal misalignment theoretically affects the accuracy of aggregate reporting similarly to external misalignment. Furthermore, defects in centerpiece integrity, such as septum deviations and channel inconsistencies, have been shown to lead to inaccurate aggregate quantitation, detection of false larger aggregates, and increased variability in the measurement (Gabrielson & Arthur, 2011; Gabrielson et al., 2010; Pekar & Sukumar, 2007). In some cases, lack of centerpiece integrity has resulted in over-reporting of aggregates by as much as 3% (Pekar & Sukumar, 2007). These defects have been shown to result from both age and manufacturing variability. Furthermore, there is evidence that the cell housing–centerpiece combination can also affect aggregate quantitation; changing only the cell housing can alter the levels of detected aggregate by as much as 1–2% (Pekar & Sukumar, 2007). This effect is likely the result of either changes in the amount of play between the centerpiece and keyway resulting in internal alignment changes, or possibly related to deformation of the centerpiece and cell housings when under high gravitational force. Consequently, without an awareness of the potential for internal misalignment and centerpiece/cell integrity issues, and a control strategy to address these issues, the analyst may inaccurately report the levels and distribution of aggregates. Unfortunately, centerpiece defects are often visually imperceptible and can only reliably be detected by specialized tools such as microscopic imaging. Therefore, we recommend the following controls to minimize the risk of false aggregate reporting due to internal misalignment and centerpiece/cell integrity: (1) keep all cell components (cell housing, windows, centerpieces) together for the lifetime of the cell, (2) use flow-through cleaning within an experiment to keep internal alignment consistent, (3) run multiple replicates of samples (at least three) to aid in identifying potential outliers, and (4) implement a result monitoring program (refer to Section 4) to aid in identifying changes and outliers.

Finally, while cell alignment and lack of centerpiece integrity lead to over-reporting of aggregates, the presence of cosedimenting solutes, such as sugars or sugar alcohols, in the protein formulation can mask aggregates and therefore lead to overestimation of product purity. Sugars are a common class of excipients used in biopharmaceutical formulations and are frequently present at high concentrations, 5% or greater. During ultracentrifugation, these excipients form dynamic density and viscosity gradients. If the cosedimentation is not accounted for, the true aggregate content present in solution will be underreported. At sorbitol concentrations of 10% (w/v), the aggregate levels have been shown to be underreported by as much as 77%, and at 5% (w/v) for the same protein systems, the aggregate was underreported by 17–52% (Gabrielson, Arthur, Kendrick, Randolph, & Stoner, 2009). Fortunately, there are two data analysis strategies which have been shown to account for dynamic gradients and improve aggregate reporting accuracy. The first analysis approach leverages the inhomogeneous solvent fitting option in SEDFIT (Schuck, 2004), which models the sedimentation of macromolecules in a dynamic concentration gradient of a cosolute. This model is the most rigorous and accurate approach to improve aggregate quantitation accuracy; however, additional experimentation is required to measure excipient sedimentation and diffusion coefficients as well as to determine the functional dependence of solvent density and viscosity on excipient concentration. The second approach is to exclude data near the bottom of the cell from the SEDFIT/*c*(*s*) analysis (Gabrielson et al., 2009). By moving the bottom fitting limit inward to approximately 6.7 cm (for a dual channel Beckman centerpiece with a standard channel base of ~7.0 cm), data from the region of the solution column where the model inadequacy is most pronounced are excluded, and therefore, aggregate quantitation accuracy is considerably improved. Finally, application of TI noise decomposition during data analysis using the SEDFIT/*c*(*s*) distribution can be a useful tool for diagnosing the presence of density gradients. Model inadequacies, resulting from dynamic gradients, are captured as time-invariant noise that produces a characteristic upward tailing of the TI noise profile near the bottom of the cell (Gabrielson et al., 2009).

3.2 Recommended Operational Parameters and Real-Time Controls

Due to the negative impact of cell alignment, cell component integrity, and cosedimenting solutes on the accuracy and precision of the SV-AUC method, it is important not only to be aware of these potential pitfalls but to have control strategies in place for each. However, there are additional operating parameters which can also impact SV-AUC data quality. These parameters include both physical prerun controls as well as postrun data analysis controls and goodness-of-fit inspections. If real-time system suitability controls are not in place for these parameters, both the accuracy and variability of the method can be compromised. Examples of prerun controls include checking for concentration-dependent nonideality, equipment integrity confirmations, reducing the potential for thermal fluctuations during the run, ensuring cell alignment, and selecting a replication strategy appropriate for the study (discussed later in the chapter). Examples of postrun controls include selecting the appropriate model parameters, inspections for goodness of fit, identifying and modeling the presence of cosedimenting solutes, and thorough data documentation. For each of these aforementioned parameters, example control strategies are provided in Table 2, along with the potential consequence if the parameter is not controlled. Utilizing real-time control strategies will aid in reducing the variability of the method as well as identifying inaccuracies in the results.

Importantly, even with real-time controls in place, the method may not be sufficiently precise for some applications. In fact, fully understanding the method precision is critical to reporting useful results. In our experience, when real-time controls are applied, the precision of the assay for aggregate quantitation is approximately 0.3–0.4% (Gabrielson & Arthur, 2011; Gabrielson et al., 2010). Further discussion of the analytical control strategy for SV-AUC is provided in Section 4.

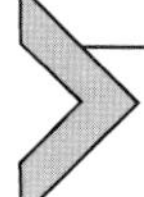

4. METHOD LIFECYCLE MAINTENANCE: MONITORING AND CONTROLLING SV-AUC METHOD PERFORMANCE OVER TIME

Analytical methods used to test biopharmaceuticals, like the products themselves, have lifecycles that must be actively maintained to ensure consistent performance. Successful lifecycle management involves careful method design, rigorous demonstration of the suitability of the method for its intended purpose, and continual monitoring of method performance. When implemented well, this process is cyclical in nature because active

Table 2 Example Real-Time Controls for Pre- and Postrun SV-AUC Operating and Analysis Parameters to Reduce Measurement Variability and Improve Result Accuracy

Parameter	Prerun Controls	Consequence If Not Controlled
Sample concentration	Screen samples at multiple concentrations the first time the protein-buffer system is evaluated. Look for evidence of concentration-dependent self-association If using absorbance optics, target OD $\sim$0.5–1.0	Concentration-dependent nonideality, inaccurate aggregate quantitation
Cell integrity	Visually inspect all cell components for cracks, scratches, chips, and overall cleanliness Use flow-through cleaning within an experimental protocol to maintain the same internal alignment across runs	Cell defects are typically associated with overestimation of true aggregate content Increased variability of replicate measurements
Thermal convection	Preequilibrate rotor to run temperature (at least 4 h), followed by a 30 min in-chamber equilibration after the cells have been loaded into the rotor Ensure actual temperature is at the set point temperature before starting the run	Convection associated with overestimation of true aggregate content
Cell alignment	Verify/set cell alignment using a tool. Visual inspection of alignment is insufficient and can result in inaccuracies of $\pm 0.5°$ misalignment	Overestimation of true aggregate content, generation of false aggregates, increased variability of replicate measurements
Replication	Analyze samples with replication so that outliers can be identified	Increased sample-to-sample variability and inaccuracy of reported results

Parameter	Postrun Controls	Consequence If Not Controlled
General $c(s)$ model recommendations	Enable TI noise deconvolution Allow the meniscus to vary during data analysis When cosedimenting solutes are present in solution, adjust fit limits or use in homogeneous solvent fitting option	Inaccurate meniscus position can lead to over- and underreporting of true aggregate content depending on direction of meniscus placement. Improper modeling of cosedimenting solutes masks true aggregate content
Inspections for goodness of fit[a]	Verify RMS noise is <0.01 OD Inspect RMS noise versus RMSD of fit. If RMSD of fit $\gg$ RMS noise, model inadequacies exist Look for patterns in residual bitmaps which can arise from nonideality Look for patterns in TI noise which can indicate back diffusion, cosedimenting solutes, or other model inadequacies Inspect parameters allowed to vary during data fitting. Did any move to an unreasonable position? Were polysorbate levels above CMC? If so, micelles can manifest as slow sedimenting species	Degraded signal-to-noise typically leads to underreporting of true aggregate content Higher variability and inaccuracies can result from improper modeling

[a]Documenting all model parameters before and after fit, and capturing TI noise profiles, residual bitmaps, and distribution profiles to help inspect goodness of fit are strongly recommended.

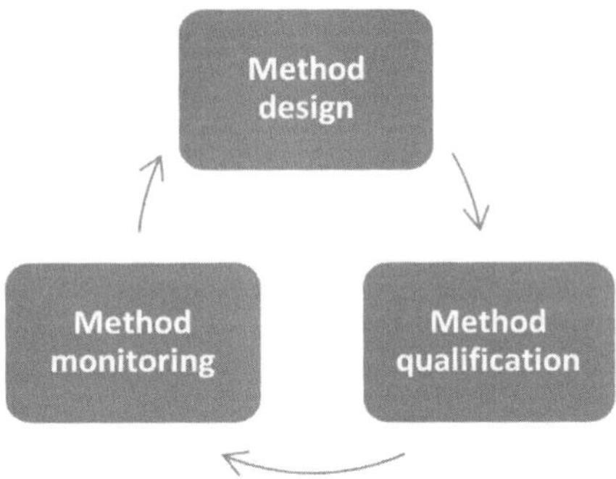

Figure 1 Illustration of the cyclical nature of analytical method lifecycle management. Active monitoring, in which the results of the monitoring program are acted upon at defined intervals, results in method redesign, leading to continuous method improvement.

method monitoring naturally results in method redesign, leading to continuous method improvement as illustrated in Fig. 1.

For SV-AUC, implementing a method lifecycle management program is critical. As discussed in Section 3, results of SV-AUC can change dramatically depending on how the method is run. Thus, applying real-time controls is a crucial step needed to increase the accuracy and precision of aggregate levels measured by SV-AUC. But demonstrating that the method remains in a state of control over time is equally important. In this section, we follow the cyclical process of method lifecycle management, starting with method design and progressing through qualification and monitoring; within this framework, we discuss how the performance of the SV-AUC method can be controlled over the long time periods necessary to bring biopharmaceuticals to market.

4.1 Designing an SV-AUC Method for Long-Term Success

There are two design objectives that must be satisfied to create a robust and well-controlled SV-AUC method for applications in the biopharmaceutical industry: (1) to minimize error[3] and (2) to maximize precision. The method needs to be designed such that it can be used over a time period of at least 10 years to enable direct data comparisons of a product from preclinical

[3] Accuracy and precision are the two most important characteristics of many analytical methods. Rather than use the term "maximize accuracy" we chose "minimize error" as the second design objective to reflect the fact that for SV-AUC, inaccuracy is often the result of undesirable yet identifiable (as opposed to random) error sources, which can be minimized or avoided. These errors can typically be assigned to human, instrument, and cell component sources (refer to Section 3).

development through registration, and ideally continuing through post-approval changes to the product.

A method procedure should be written to provide both step-by-step instruction and requirements for real-time controls applied by the analyst when running SV-AUC (as discussed at length in Section 3). Although necessary, a written method procedure is insufficient to ensure lack of "drift" in SV-AUC data over time. Data drift is characterized by a change in either the mean level or variance of aggregate measurements of a stable control sample, and it may be caused by a new analyst running the method, degradation of cell components, or even changes in instrument performance over time. While much of the potential drift can be controlled through implementation of a monitoring program (covered later in this chapter), some of the potential drift can be eliminated up front by designing an analyst training program and implementing system suitability requirements and result acceptance criteria. Thus, in addition to controls the analyst can apply real time to minimize the likelihood for error and to improve the precision of the results, supplemental controls can be applied to ensure the method maintains a state of control over much longer time periods.

The SV-AUC technique is intensive, requiring multiple steps to be performed properly and sequenced correctly to obtain the highest quality results.[4] Not surprisingly, analyst training is therefore an important element in the overall method control strategy. We consider this to be a method design component because, through careful training of new AUC users, the precision of the method can be improved and potential errors can be minimized or even eliminated. We recommend an interactive training program in which the trainee is encouraged to actively participate in a combination of classroom training and hands-on laboratory work. Employing a "watch, help, do, explain" format may be beneficial. In this model, the trainee first observes, and asks questions, as the instructor performs the method in its entirety. Next, the two individuals work together to perform one or more additional runs. Third, the trainee performs the method, while the instructor observes, intervening as necessary. Finally, the training is complete when the new analyst is able to explain the reasoning for each step of the method, including data analysis. In effect, this model is one that facilitates

[4] Use of the phrase "highest quality results" here and elsewhere in this chapter is synonymous with "most accurate and precise results," notwithstanding the inherent limitations of the technique. The "highest quality" results are achieved when the design objectives, minimizing errors and maximizing precision, are met.

a role reversal: by the end, the trainee "instructs" his/her trainer. This type of training is particularly conducive to one-on-one training sessions, but similar approaches can be used for small group training. Even after completing an exacting training program, such as the one described here, the newly trained analyst is still a relative novice and will inevitably learn much more over time.

With a detailed procedure written and a training program implemented, the design portion of the overall method control strategy is pending one final element: implementation of system suitability requirements and result acceptance criteria. Even when a trained analyst follows a written procedure, the results of SV-AUC analysis of biopharmaceuticals can be compromised, albeit infrequently, by poor data quality. The underlying causes of poor data quality may arise from various sources, including the instrument, cell components, or data analysis (lack of fit). The goal of implementing system suitability and result acceptance criteria is to objectively ascertain whether or not data quality meets a minimum standard for reporting. While this goal is straightforward, the selection of specific data parameters to measure, from which a data quality decision can be made, and the determination of the numeric criteria are far from straightforward. The criteria one chooses cannot be made to invalidate high-quality data simply because the result is unexpected. In other words, the result acceptance criteria must be based on *data quality* and not sample quality. We recommend as few criteria as necessary to demonstrate sufficient data quality. Some examples are listed in Table 3.

Prior to testing samples, it is typically beneficial to define a process by which an invalidated result can be investigated, and if warranted, a sample retest protocol can be executed. It is important to define this process before an invalid test result occurs to avoid the perception of reactive testing into compliance.

In summary, the overall method design consists of the following three features:

- A detailed, written procedure which guides and instructs the analyst in performing the method in a consistent fashion, including requirements for real-time controls;
- A thorough and rigorous training program that involves interactive guidance on both laboratory best practices and data analysis aspects; and
- A relatively small but sufficient set of system suitability and result acceptance requirements, which enable detection of poor data based on objective criteria.

Table 3 Example Long-Term Method Performance Control Strategies

Parameter	Action Required	Purpose of Action	Frequency
Preventative maintenance (PM)	Regularly PM the instrument, or preform comparable system suitability verifications. Keep records of PMs and additional service	Verify and maintain the system functionality	PM at least once/year
Baseline noise	Verify signal intensity is within manufacturer specification and root mean square (RMS) noise is <0.01 OD. Create a control chart to track changes over time	Allows identification of degradation in system performance and triggers need for service. Increased noise can lead to underreporting of aggregates	Each run
Centerpiece control	Keep all cell components together as a set (same windows, centerpiece, cell housing, rotor hole for the cell lifetime). Track date of first use, number of runs, and performance[a] for each centerpiece	Identification of "bad" centerpieces/ cells. Cell components distort together while under high gravitational force. Keeping the components as a set aids in control charting/integrity monitoring	Each run
Run a standard in one cell	Verify expected aggregate content and sedimentation coefficient obtained. Create a control chart of standard results	Deviations from expected results could indicate issues with instrument performance and/or centerpiece integrity issues	Once per protocol
Run a standard in all cells	Verify expected aggregate content and sedimentation coefficient obtained. Use results in combination with standard and centerpiece control charts	Aids in identifying cell/centerpiece integrity issues	Once per month to once per quarter depending on frequency of instrument use

Continued

Table 3 Example Long-Term Method Performance Control Strategies—cont'd

Parameter	Action Required	Purpose of Action	Frequency
Standard deviation	Document the standard deviation across all measurements for a sample/standard within a run. Trend the results using a control chart (may be broken down by product)	Aids in understanding typical method performance/ variability. Deviations from normal indicate possible instrument or centerpiece performance degradation	Each run
Training	Establish initial training program and have periodic review/discussions among analysts	Allows analysts to ask questions, demonstrate knowledge, and share best practices	Every 6 months to once per year
Method review	Periodically review and update the method to ensure it reflects current practices	Enables implementation of new best practices and ensures consistency across users	Once per year

[a]The centerpiece performance should be a metric established by the analyst to allow for comparisons across centerpieces and runs to identify degradation in centerpiece integrity. Typically, degradation in centerpiece integrity is evidenced by higher levels of aggregate and greater variability in measurements from run to run.

4.2 Method Performance Qualification

The second major element of a robust SV-AUC method control strategy is qualification of the method. For product purity methods like SV-AUC, a thorough qualification[5] study comprises evaluation of the following method performance characteristics: accuracy, precision, specificity, detection limit

[5] ICH Q2(R1) specifies method performance characteristics to evaluate during method validation. Many biopharmaceutical characterization methods, like SV-AUC, do not require formal validation. Nevertheless, method qualification is often beneficial, sometimes necessary, and requires evaluation of the same method performance characteristics. Although the distinction between method qualification and validation is subtle, we consider qualification to be an investigative study in which method characteristics are evaluated without predefined expectations; conversely, for validation, method performance is shown to achieve predefined performance requirements.

(DL), QL, linearity, and range (ICH Q2(R1), 2005). The goal of the study is to demonstrate that SV-AUC is fit for its primary purpose in testing biopharmaceuticals: to quantitatively measure aggregate levels in the finished product. Much has been published on the basic elements of a comprehensive and successful qualification of SV-AUC (Gabrielson & Arthur, 2011; Gabrielson et al., 2010; Gabrielson, Randolph, et al., 2007; Pekar & Sukumar, 2007). For the purposes in this chapter, we address only one aspect of qualification—method precision—and consider an array of useful outcomes that can be achieved after determining the precision of the method in a holistic fashion.

As stated previously, our two design objectives are to minimize error and to maximize precision. Through careful method design (the first method lifecycle stage, Fig. 1), we can accomplish the first design objective through a combination of written procedures, training, and sample result acceptance criteria. The second design objective, maximizing precision, is addressed during the second method lifecycle stage, qualification. During this stage, we cannot actually improve the precision, but we can define the method precision under best-case, yet realistic, conditions of use. Following method qualification, we can state that the method is (or is not) well suited to each of the applications described in Section 2.

For example purposes only, we will assume that a thorough set of method qualification experiments results in intermediate precision of 0.3% aggregate levels. Based on equations available in ICH Q2(R1), DL and QL can be calculated directly from the precision, in this case leading to DL = 1.0% and QL = 3.0%. Using these values in various ways, several decisions can now be made to guide how we apply SV-AUC for testing biopharmaceuticals, including selection of appropriate method applications, replication strategies, and result reporting requirements—all of which can be derived directly from our estimate of method precision.

Each study for which SV-AUC is employed has a consumer of the data. Often, the data consumer is not the analyst, yet he or she has a need for the data, presumably to make a technical decision. It is the job of the analyst to provide the data in a useful format, without complicated caveats about the method, in order to help the data consumer make an informed decision. This means the analyst, not the ultimate user of the data, needs to employ the right sample replication and result reporting strategies to ensure that the data are easily interpretable, even self-explanatory. The following example clarifies why data presentation is important and how replication and reporting can be leveraged by the analyst to the ultimate advantage of the end user.

One common use for SV-AUC is to evaluate biopharmaceutical product comparability following a manufacturing change. SV-AUC is almost always used as an orthogonal method to SEC, a far more precise method. Because poorly developed SEC methods have been shown to cause inaccurate quantitation of aggregate levels (Carpenter et al., 2010; Gabrielson, Brader, et al., 2007; Philo, 2006), SV-AUC is considered the standard for accuracy. For example purposes, assume a protein product has historically had very low aggregate levels (<1.0%). Following a manufacturing change, the SEC method measures a small increase in aggregate levels (to ~1.5%), but still below the specification limit (<2.0%). With a DL of 1.0% and QL of 3.0%, SV-AUC is ill suited to provide quantitative confirmation of the small increase in aggregate levels measured by SEC. But, importantly, we do not need to use it for that purpose. Rather, in this case, we apply SV-AUC to confirm that levels of aggregate in the finished product, after the manufacturing change, are in fact below the product specification limit. The study is therefore statistically designed to test a defined number of replicates of a defined number of product lots in order to report a very simple result: aggregate levels measured by SV-AUC are less than 2.0% (with a desired degree of statistical confidence). The matter of lack of comparability (by SEC) still must be investigated and explained, but the SV-AUC method provides important orthogonal assurance of product quality.

Three subtle points should be noted about this example. First, the study design is largely based on an estimate of method precision. Without a very thorough method qualification, the study described, although well designed, may have failed to deliver the required statistical power to meet its end point. Second, the analyst reports that the aggregate level in the product is less than a numerical value (2.0%), which itself is below the stated QL of the method. Moreover, based on the study design, the analyst could have reported a number (e.g., 1.7%) below the QL. The implication from this is that the QL of the method is a dynamic quantity and varies based on the number of replicates measured.[6] Third, the result is reported to the tenths place (<2.0%; refer to ASTM Standard E29-02, 2002). It would not be appropriate to report <2% or <2.00%. Thus, the apparently simple result (aggregate level <2.0%) betrays the complexity required to report it correctly. Yet the

[6] For random (nonsystematic) error, the variance, σ^2, is inversely proportional to the number of replicates, n. Thus, the precision of the method can be improved by $\frac{1}{\sqrt{n}}$ simply by taking more replicates. The QL scales proportionally to the precision (QL $= 10\sigma$) and is therefore a function of n (ICH Q2(R1)).

result itself, and not the steps required to determine it, is the only piece of information that the data consumer needs to make his or her decision.

In summary, the purpose of SV-AUC method qualification is to define, in quantitative terms, how the method is expected to perform under actual conditions of use. Establishing the precision of the method is the most critical component of the qualification exercise because future use of the method, including study design and result reporting, will depend directly on the calculated method precision.

4.3 Establishing a Method Monitoring Program

The final stage of comprehensive method lifecycle management is implementation of a monitoring program. As noted previously, an effective method monitoring program leads to method redesign, hence the cyclical process defined in Fig. 1. The term "monitoring" is very broad; we will use this term to describe the process by which the SV-AUC method can be shown, continually through time, to operate within a state of control. We recommend a three-part monitoring program:

- Monitoring results and data quality metrics over time,
- Monitoring the performance of the instrument and cell components over time, and
- Monitoring analyst training effectiveness (in a nonpunitive way).

Classically, analytical methods are monitored using control charts (i.e., trend charts). Nelson Rules (Nelson, 1984) can be applied to ensure that the method remains in a state of statistical control. In the case of SV-AUC, we recommend a modified use of trend charts along with additional monitoring. The use of trend charts is most conducive to the first part of the monitoring program (monitoring results), but it is not immediately clear what method response(s) should be plotted over time.

In order to monitor results and data quality metrics, it is imperative that one or more standards be run periodically. This may be a product reference standard, a nonproprietary standard (e.g., NIST IgG1 control), or a batch of protein degraded to create high levels of a stable aggregate. The frequency with which the standard is run by SV-AUC can vary, but it is beneficial to obtain at least 30 measurements of the standard over time to define control limits for the method. A subset of applicable Nelson Rules can then be applied to future analyses of the standard to ensure that results remain within statistical control. Depending on how the results will be used, different types of results can be monitored, including:

- Aggregate levels of a standard analyzed in every run, using the same centerpiece,
- Aggregate levels of a standard analyzed every 3 months,
- Aggregate levels measured for a specific centerpiece over time,
- Range or standard deviation of aggregate levels for all samples in a run (if those samples are replicates taken from the same parent sample), and/or
- Raw data baseline noise level for each sample in every run.

This list is not exhaustive, yet clearly, there is a wide range of possible results and data quality metrics that can be routinely monitored. Forethought is required to determine which parameter(s) should be monitored, and with what frequency, in order to derive as much value as possible from the monitoring program.

In addition to monitoring results and data quality metrics, the state of the instrument and cell components should be periodically inspected. A centerpiece monitoring program with the objective of detecting poor-performing centerpieces can easily be implemented using the same trending approach used for result monitoring. Centerpieces and other cells components should also be frequently inspected by the analyst, ideally prior to every run. Cell components with obvious signs of damage (e.g., scratches) should be removed from service. From an overall maintenance perspective, we recommend periodic preventive maintenance of the instrument, along with usable life policies for rotors, based on the number of revolutions, and centerpieces, based on the number of runs.

Analyst training verification is an important and often overlooked element of the monitoring program. Although a strict approach can be taken, we recommend a more growth-oriented training verification model. For example, two trained analysts can work together for a series of four runs, taking turns running the method and then observing it being run by the other analyst. The two can easily compare practices (tips and tricks) that they have learned over time. The objective with this model is to encourage a continuous learning culture, where results continue to improve over time.

Overall, a method monitoring program is a critical final element within a holistic method lifecycle management plan for SV-AUC. Various aspects of method monitoring, e.g., trending results, inspecting components, and verifying training, should be applied actively and consistently across runs over many months and years to demonstrate method control. It is expected that weak signals observed as part of the monitoring program will lead to changes in the design of the method. This is the essence of a cyclical method lifecycle based on continuous improvement.

5. CONCLUSIONS

SV-AUC has numerous advantages compared to SEC, including the ability to analyze biopharmaceuticals directly in their formulation, as well as to separate size variants with improved resolution and without a stationary phase. It therefore serves as an important tool to confirm the accuracy of SEC methods during development and validation. When an SV-AUC method is designed well and used appropriately, it adds significant value to the analytical strategy for control of biopharmaceutical products.

During early stages of product development, SV-AUC can be used effectively to screen molecular candidates, quantify aggregation in various formulations, and guide SEC method development. In later stages of product development, SV-AUC is most useful when applied as a complementary analytical tool to confirm the accuracy of SEC during product characterization and after manufacturing process changes. However, in order for the SV-AUC method to be most useful for these applications, it must be carefully designed with operational controls, sample result acceptance criteria, and an active method monitoring program. Only when the limitations of the SV-AUC method are well understood and mitigated with real-time and long-term controls can the full value of the method be realized for biopharmaceutical development.

REFERENCES

Arthur, K. K., Gabrielson, J. P., Kendrick, B. S., & Stoner, M. R. (2009). Detection of protein aggregates by sedimentation velocity analytical ultracentrifugation (SV-AUC): Sources of variability and their relative importance. *Journal of Pharmaceutical Sciences*, *98*(10), 3522–3539.

ASTM Standard E29-02. (2002). *Standard practice for using significant digits in test data to determine conformance with specifications*.

Berkowitz, S. A., Engen, J. R., Mazzeo, J. R., & Jones, G. B. (2012). Analytical tools for characterizing biopharmaceuticals and the implications for biosimilars. *Nature Reviews. Drug Discovery*, *11*(7), 527–540.

Brown, P. H., & Schuck, P. (2006). Macromolecular size-and-shape distributions by sedimentation velocity analytical ultracentrifugation. *Biophysical Journal*, *90*(12), 4651–4661.

Carpenter, J. F., Randolph, T. W., Jiskoot, W., Crommelin, D. J., Middaugh, C. R., & Winter, G. (2010). Potential inaccurate quantitation and sizing of protein aggregates by size exclusion chromatography: Essential need to use orthogonal methods to assure the quality of therapeutic protein products. *Journal of Pharmaceutical Sciences*, *99*(5), 2200–2208.

Dam, J., Schuck, P., Michael, L. J., & Ludwig, B. (2004). Calculating sedimentation coefficient distributions by direct modeling of sedimentation velocity concentration profiles. *Methods in Enzymology*, *384*, 185–212, Academic Press.

Gabrielson, J. P., & Arthur, K. K. (2011). Measuring low levels of protein aggregation by sedimentation velocity. *Methods*, *54*(1), 83–91.

Gabrielson, J. P., Arthur, K. K., Kendrick, B. S., Randolph, T. W., & Stoner, M. R. (2009). Common excipients impair detection of protein aggregates during sedimentation velocity analytical ultracentrifugation. *Journal of Pharmaceutical Sciences*, *98*(1), 50–62.

Gabrielson, J. P., Arthur, K. K., Stoner, M. R., Winn, B. C., Kendrick, B. S., Razinkov, V., et al. (2010). Precision of protein aggregation measurements by sedimentation velocity analytical ultracentrifugation in biopharmaceutical applications. *Analytical Biochemistry*, *396*(2), 231–241.

Gabrielson, J. P., Brader, M. L., Pekar, A. H., Mathis, K. B., Winter, G., Carpenter, J. F., et al. (2007). Quantitation of aggregate levels in a recombinant humanized monoclonal antibody formulation by size-exclusion chromatography, asymmetrical flow field flow fractionation, and sedimentation velocity. *Journal of Pharmaceutical Sciences*, *96*(2), 268–279.

Gabrielson, J. P., Randolph, T. W., Kendrick, B. S., & Stoner, M. R. (2007). Sedimentation velocity analytical ultracentrifugation and SEDFIT/c(s): Limits of quantitation for a monoclonal antibody system. *Analytical Biochemistry*, *361*, 24–30.

ICH Q2(R1). (2005). Validation of analytical procedures: Text and methodology Q2(R1). ICH harmonised tripartite guideline. In *International conference on harmonisation of technical requirements for registration of pharmaceuticals for human use*.

Jiang, Y., Li, C., Li, J., Gabrielson, J. P., & Wen, J. (2015). Technical decision making with higher order structure data: Higher order structure characterization during protein therapeutic candidate screening. *Journal of Pharmaceutical Sciences*, *104*, 1533–1538.

Laue, T. M., & Stafford, W. F. (1999). Modern applications of analytical ultracentrifugation. *Annual Review of Biophysics and Biomolecular Structure*, *28*, 75–100.

Nelson, L. S. (1984). Technical aids: The Shewhart control chart—Tests for special causes. *Journal of Quality Technology*, *16*(4), 237–239.

Pekar, A., & Sukumar, M. (2007). Quantitation of aggregates in therapeutic proteins using sedimentation velocity analytical ultracentrifugation: Practical considerations that affect precision and accuracy. *Analytical Biochemistry*, *367*, 225–237.

Philo, J. S. (2006). Is any measurement method optimal for all aggregate sizes and types? *The AAPS Journal*, *8*(3), E564–E571.

Schuck, P. (2000). Size-distribution analysis of macromolecules by sedimentation velocity ultracentrifugation and Lamm equation modeling. *Biophysical Journal*, *78*(3), 1606–1619.

Schuck, P. (2004). A model for sedimentation in inhomogeneous media. I. Dynamic density gradients from sedimenting co-solutes. *Biophysical Chemistry*, *108*(1–3), 187–200.

Schuck, P., Perugini, M. A., Gonzales, N. R., Howlett, G. J., & Schubert, D. (2002). Size-distribution analysis of proteins by analytical ultracentrifugation: Strategies and application to model systems. *Biophysical Journal*, *82*(2), 1096–1111.

CHAPTER TWENTY-ONE

Protein Assembly in Serum and the Differences from Assembly in Buffer

John J. Hill*,[1,2], Thomas M. Laue†
*Amgen, Seattle, Washington, USA
†Center to Advance Molecular Interaction Science, University of New Hampshire, Durham, New Hampshire, USA
[1]Corresponding author: e-mail address: hilljb@u.washington.edu

Contents

Abstract

This chapter illustrates how analytical ultracentrifugation methods, coupled with the fluorescence detection system, are an excellent approach to characterizing and comparing protein-binding interactions in dilute solution and concentrated, crowded solutions like serum. We show that in serum, the binding and assembly states for a pair of endogenous protein ligands and an antibody inhibitor are dramatically different than those observed in dilute, simple buffers. This type of analysis approach may be helpful in research efforts intent at discerning the underpinnings to a therapeutic's activity and pharmacokinetic properties *in vivo*.

[2] Current address: Department of Bioengineering, University of Washington, Seattle, WA 98195.

Methods in Enzymology, Volume 562
ISSN 0076-6879
http://dx.doi.org/10.1016/bs.mie.2015.06.012

ABBREVIATIONS

Ab antibody
AUC analytical ultracentrifugation
Da Daltons
FDS fluorescence detection system
HSA human serum albumin
IgG immunoglobulin G
JO Johnston–Ogston
Label Alexa Fluor® 488 fluorescent dye
OPG osteoprotegerin
OPG_D osteoprotegerin dimer
OPG_M osteoprotegerin monomer
PBS phosphate-buffered saline
PKPD pharmacokinetic and pharmacodynamic
RANK receptor activator of NF-κB
RANKL receptor activator of NF-κB ligand
S Svedberg
S sedimentation coefficient
SV sedimentation velocity

1. INTRODUCTION

With the development of automated microfluidic approaches coupled with the advancements in biophysical and analytical technologies and methods, binding-analysis measurements have become much simpler to implement in biochemistry laboratories. Consequently, the process of discovering and developing therapeutic molecules to modulate biological systems and contain disease states, now routinely involves numerous *in vitro* experiments focused on analyzing kinetic, affinity, assembly, and activity relationships in dilute simple buffers. The binding characteristics often are applied toward sorting and choosing potential drug candidates based on potency, selectivity for similar targets, and for species cross-reactivity. These binding studies act as assessments that aid in preclinical pharmacodynamic, pharmacokinetic, and toxicity studies. Coupled with the relatively high throughput of modern protein design and engineering efforts, large numbers of protein variants often are analyzed, generating immense data sets of binding- and activity-related parameters. A reductionist, structure–function-based approach, which uses highly purified proteins and simple solution conditions for these analyses, clearly has worked well for discovering and developing many therapeutics. Yet, the question remains whether this type of apparent *in vitro* binding information truly is sufficient for suitable descriptions of the mechanistic underpinnings to the activity and disposition properties occurring *in vivo*.

In vivo, macromolecules do not exist as highly purified and isolated sets of physical objects that randomly diffuse unobstructed through a simple, dilute environment. Rather, they exist in complex mixtures with many other types of molecules creating a concentrated molecular fluid, which may be defined as one in which the distances between molecules are on the same order of magnitude as the size of the molecules. For example, in serum, where the total protein concentration is ~70 mg/ml, with the majority of it being serum albumin at ~35–50 mg/ml (Garbett, Miller, Jenson, & Chaires, 2008; Howard, Hill, Galluppi, & McLean, 2010), the average distance between molecules is only a few nanometers, for the van der Waals surface dimensions. They are even closer when accounting for the electrostatic "edges" that extend out ~1 nm further, in physiological salt conditions. Here, any given type of macromolecule that is present at a relatively low concentration actually exists in a highly crowded and confined space in proximity to many other "background" molecules of varying electrostatic properties (Laue, 2012). The impact of crowding effects on the energetics and kinetics of macromolecular diffusion, assembly, and stability have been studied through theoretical models and simulations for a variety of concentrations and crowding conditions (Elcock, 2010). For example, molecular crowding theory provides a framework of models based around the entropic penalties that come from steric exclusion or "excluded volume" effects. Some models incorporate the impact of repulsive interactions from charge–charge and other electrostatic contributions or "effective excluded volume" effects in the framework (Zhou, Rivas, & Minton, 2008). Although these types of entropic models appear to describe simple solutions like concentrated protein formulations, well, they are much less able to describe the biological fluids like serum (Laue, 2012; Minton, 2013). The discontinuity appears because biological fluids exist as complex mixtures of macromolecules with different signs and magnitudes of charges, strengths and orientations of dipole moments, and other electrostatic features that produce attractive and repulsive interactions; such that it is challenging to properly describe all of the different electrostatic effects by excluded volume functions alone. Proximity energy theory, on the other hand, may be a more applicable approach to describing crowding effects in complex solutions (Laue & Demeler, 2011; Laue 2012). This framework goes beyond exclusion effects by incorporating the different pairwise, distance-dependent enthalpies of interaction (proximity energies) that occur between like and unlike molecules with different electrostatic features, and it appears applicable to both complex biological solutions and the simpler systems like protein formulations (Laue, 2012). However, no matter the theoretical model or simulation approach used to describe a system, there still is a need for experimental approaches that provide a practical characterization of molecular interactions in crowded conditions. This is especially critical for complex biological fluids where the solution composition can vary significantly between individuals and disease states, which makes robust simulations even more challenging or impractical to conduct (Garbett et al., 2008).

In this chapter, by way of an example system of proteins, we show through a novel experimental approach how binding and assembly properties can be impacted by crowded serum conditions. In particular, the trimeric molecule, receptor activator of NF-κB ligand (RANKL), its dimeric cognate inhibitor osteoprotegerin (OPG) (Walsh & Choi, 2014), and a discovery research human immunoglobulin G1 (IgG1) inhibitory antibody (Ab) that binds RANKL, have been successfully analyzed in serum and in dilute buffer by analytical ultracentrifugation (AUC) methods. The results illustrate how analyses done in more concentrated, crowded solution environments like serum may be important in efforts to characterize complex biological therapeutic systems. Additionally, this AUC approach is applicable to many types of concentrated solutions of varying compositional complexities, including protein formulations.

2. MATERIALS AND METHODS

2.1 Analytical Ultracentrifugation Approach

The use of AUC approaches to characterize protein binding and assembly phenomena has had a long and successful history (Brown, Balbo, & Schuck, 2009; Cole, Lary, Moody, & Laue, 2008; Laue & Stafford, 1999). The most hydrodynamically and thermodynamically detailed analyses are obtained from both sedimentation equilibrium (SE) and sedimentation velocity (SV) approaches with dilute solutions and simple buffers. In general terms, a solution is considered to be "dilute" if there is no significant hydrodynamic or thermodynamic nonideality occurring. In the case of many globular proteins with a moderate charge ($z \lesssim 15$) these criteria are met with total protein concentrations less than 2–3 mg/ml, and with buffers near physiological salt concentration (ionic strengths >100 mM), but also only if the resulting solvent concentration gradients are small enough to be neglected in the analysis. For most AUC studies with dilute solutions, the absorbance optical system and the Rayleigh interference optical system are useful. However, for typical macromolecules, optical sensitivity and deviations from the Beer–Lambert relationship limit their use to concentrations in the high n*M* to mid μ*M* range. Since the total macromolecular concentration of serum is on the order of 60–80 mg/ml, making it a highly concentrated medium that nearly is optically opaque at the wavelengths for typical protein absorptions, the absorbance and interference optical systems are not amenable for detection. However, we and others have found that the fluorescence detection system (FDS), when coupled with the SV approach, is particularly well suited for studies with concentrated solutions (Kroe & Laue, 2009; Lyons, Lary, Husain, Correia, & Cole, 2013; MacGregor, Anderson, & Laue, 2004; Zhao, Mayer, & Schuck, 2014). Moreover, the sensitivity and wide dynamic range of the FDS allow for a broader and more biologically relevant binding affinity space to be explored; notably, p*M*–μ*M*, and therefore, it can provide a more complete assessment of the biologically relevant distributions of assembled species that often occur over a broad concentration range, and in many different buffer compositions.

All AUC experiments discussed in this chapter were conducted at The Center to Advance Molecular Interactions Science (CAMIS), University of New Hampshire, Durham, NH, on a Beckman Coulter XL-A analytical ultracentrifuge outfitted with an FDS that was a prototype for the AU-FDS (Kroe & Laue, 2009), now available from Aviv Biomedical, Lakewood, NJ. SV experiments were run between 30k and 50k revolutions/min and 20 °C, using an 8-hole rotor, with SV cells assembled from 12 mm Spin60™ double-sector, charcoal epon centerpieces, and sapphire windows obtained from Spin Analytical, Inc. (Berwick, Maine). The raw data SV scans were analyzed using the *c*(*s*) method in the software program SEDFIT (v9.4 and v14.4fb) by Peter Schuck (Schuck, Perugini, Gonzales, Howlett, & Schubert, 2002). Scans, covering the full range of boundary movement, typically were analyzed with the data fitting limits set to minimize influences from the meniscus or material accumulation at the base of the cell. For dilute solutions and samples with single components, a conventional *c*(*s*) distribution was fit to a single frictional ratio (f/f_0) and meniscus position as well as the time-independent noise, with a fixed regularization confidence level of 0.68. A similar approach was taken for the samples composed of a mixture of observable sedimenting species except that the frictional ratio was then typically determined as a weight-average frictional ratio of all sedimenting material from the nonlinear optimization of this parameter. Fits were typically performed with a resolution of 100, but postfit "runs" often were performed at resolutions ranging between 200 and 300. Weight average sedimentation coefficients, S_w, were calculated by the method of second-moment integration option in the SEDFIT program. The $c(s)$ analysis for serum samples used a fixed meniscus and a value of 1.5 for f/f_0. In general, there were substantial systematic deviations in the errors of the fits in serum. Consequently, fitting for both f/f_0 and the meniscus position produced large f/f_0 values that did not have a physical meaning and meniscus values that were vastly different from where they physically appeared, which has been typical for the analysis of crowded solutions (MacGregor et al., 2004). In general, this modified $c(s)$ approach reproducibly recovered $c(s)$ distributions of comparable shape and number but with slightly different (±0.1 to 0.3 S) sedimentation coefficient values between runs; however, as has been noted, the results obtained in serum and detected with the FDS should not be over interpreted in terms of s value accuracy and the absolute quantitation of species (Kingsbury & Laue, 2011; Kroe & Laue, 2009; Lyons et al., 2013). All s values are reported as the uncorrected, raw values such that the contributions of solution density and viscosity were not normalized away.

2.2 Protein Labeling

Since this FDS uses the 488 nm line of an Ar^+ laser for an excitation source, and a pair of long-pass optical filters (>505 nm) for the emission optics, an extrinsic fluorescent probe was required for the fluorescence measurements. A number of suitable "green" dyes exist, but we found that the Alexa Fluor® 488 fluorescent dye by Molecular Probes® (label) worked very well given its large extinction coefficient (~71,000 M^{-1} cm^{-1}), relative insensitivity to pH and electrostatic changes, and that it undergoes nominal photobleaching. The tetrafluorophenyl ester moiety version of the label created a stable conjugation with protein amine groups, and the bond appeared to be stable during the course of the analyses with serum. In order to minimize potential label-related spurious binding results, we aimed for minimal labeling of the proteins. The conjugation reactions were conducted as recommended by the manufacturer but at a lower label:protein ratio of ~2.5:1 and at pH 8.3.

2.3 Protein and Solution Systems

Human RANKL (residues 143–317 of mRANKL) was produced in-house as a stable trimer of ~61 kDa, as previously reported (Ominsky et al., 2011). Human OPG, residues 22–401 (OPG_D), was produced in-house as a stable dimer of ~90 kDa MW, as previously reported (Lacey et al., 1998). A smaller, recombinant monomeric human OPG construct of residues 25–198 (OPG_M) with a MW of ~19.8 kDa was purchased from PeproTech® (Rocky Hill, NJ). Pooled human serum samples were obtained from BioreclamationIVT (Hicksville, NY). Human serum albumin (HSA) was obtained from Sigma-Aldrich as a lyophilized powder, essentially globulin free, ≥99%. Buffers were sparged with nitrogen gas in order to displace molecular oxygen and minimize potential quenching of fluorescent probes; albeit this approach did not render the samples completely anaerobic, it did appear to help in the analysis of fluorescent samples below ~1 nM in concentration. The buffer also contained ~0.1 mg/ml HSA as a protectant to manage nonspecific binding of the dilute labeled proteins to the container surface.

2.4 Special Considerations for Serum Studies

From both the safety and scientific perspectives, the handling procedures for biological materials are not as routine as those for buffers. Proper personal protective equipment and an observance of *Universal Precautions* should be implemented by trained staff when working with human or other primate materials. Although aspects to evaluating, implementing, and managing best safety practices may be specific to a particular institution and laboratory, some additional work practices and controls, unique to AUC, may be helpful to consider in designing an operating procedure, for example: (1) loading and unloading cells without needles; (2) disinfecting cells prior to cleaning/washing step; (3) and/or disinfecting and cleaning cells with an automated cell washer so as to minimize disassembly, assembly, and handling steps; (4) dedicating and using newer or best condition cell components that are properly assembled; (5) after loading cells and sealing the fill holes, and prior to loading the rotor, controlling for any potential cell leakage through ascertaining it in a vacuum chamber that is separate from the AUC and capable of less than 10 μm pressure, may also be helpful.

Since settling and separation of some serum components can occur during its storage, it is important that the serum be thoroughly but gently mixed to uniformity. As with any concentrated and viscous liquid, careful pipetting technique is notably necessary in order to maximize mixing and quantitative transfers, while also minimizing air bubbles and potential oxidation and denaturation issues. Additionally, overhandling and improper storage of serum can lead to compositional changes that are well known to impact serum protein-binding interactions; this includes pH changes and also lipolysis reactions that increase the free fatty acid content of the serum (Howard et al., 2010). We recommend assessing the serum pH at the temperature of interest, and gently correcting for it back to a value of ~7.4 prior to the various mixing and loading steps. Increasing the pH can be accomplished by addition of a drop of 1 N or lower NaOH or extra exposure to humidified inert gas, so as to gas exchange some of the CO_2. Decreasing the pH can be accomplished with 1 N or lower HCl or exposure to CO_2 gas (Howard et al., 2010).

From a spectroscopic perspective, serum often contains endogenous "green" fluorescent molecules that typically are bound to proteins like albumin, and these molecules can contribute signal to the SV analysis and result in background peaks. One notable contributor is the oxidized bilirubin–albumin adduct that has fluorescence excitation and emission spectral properties that can be observed with the FDS (Lamola & Russo, 2014). The contribution of this bilirubin–albumin adduct to serum studies can vary since the amount of this poorly fluorescent adduct in serum is rather variable between sources and the shelf life age and oxidation state of the serum. Additionally, the mere observation and contribution of the endogenous contaminants, to the total fluorescence, depend on the magnitude of the FDS voltage and gain settings. For example, we have found it to be nearly undetectable in some experiments with murine, cynomolgus, or human serum sources. In general, when we have optimized the FDS optical settings to be most beneficial for assessing the added exogenous fluorescently labeled protein of interest at >200 p*M*, the background albumin-adduct signal albeit present, often has not been significant, which is discussed further in Section 3.2.1. This has not always been the case for other researchers in the field; for example, one study illustrated a substantial contribution of the endogenous HSA peak to the analysis of labeled antibody studied at 10 n*M* in serum (Demeule, Shire, & Liu, 2009). The observed differences may simply be due to the serum source or other factors, such as the overall fluorescent dye:Ab ratio and corresponding specific activity of the protein, the photophysical properties of the fluorescent dye molecule at that conjugation site, or to emission optical differences between the prototype-FDS used in our study and their AU-FDS. In general, using the freshest serum possible, overlaying sample dead-volumes with an aliquot of inert gas like argon or nitrogen (noting not to perturb the pH), and good exogenous fluorescent dye conjugations seem to help alleviate the impact of serum background signals in the analysis.

3. RESULTS

3.1 Binding and Assembly in Dilute Solutions

3.1.1 OPG_Monomer and RANKL Interactions in PBS (phosphate-buffered saline) Buffer, pH 7.4, 0.1 mg/ml HSA

SV analysis and detection with the absorbance optical system is the simplest approach for assessing binding phenomena in dilute solution, and also for assessing any impact of the fluorescent dye on protein integrity. Figure 1A shows the *c*(*s*) results from an SV analysis with detection at 275 nm for samples of (1) unlabeled OPG_M (6.0 μ*M*), (2) RANKL (1.0 μ*M*), and (3) a mixture of RANKL + OPG_M (1.0 μ*M* + 6.0 μ*M*, respectively). The results are consistent with the expected behavior of the protein samples. The peak at ~1.8 S for the OPG_M sample is consistent with a MW of ~20 kDa in this simple buffer composition. Likewise, the sedimentation coefficient value of ~4.2 S for the RANKL sample is as expected for this molecule as a trimeric specie with an MW of ~103 kDa. The peak with a value of ~5.6 S found in the *c*(*s*) spectrum for sample (3), which contains both RANKL and OPG_M, is consistent with the trimeric RANKL binding three monomeric OPG molecules at a corresponding MW of ~163 kDa. The additional peak, at ~1.8 S, in sample (3), corresponds to the excess/unbound OPG. Similar SV results have been reported elsewhere, while utilizing slightly different constructs (Schneeweis, Willard, & Milla, 2005). SV analysis results for labeled RANKL sample alone and with excess OPG_M, monitored with the absorbance optics at 490 n*M* (Fig. 1B) and with the FDS (Fig. 1C), so as to observe just the sedimentation of the labeled RANKL species and not unlabeled protein, illustrate the utility of the FDS and integrity of the labeled protein. The peaks at ~4.2 and ~5.6 S are consistent with the results obtained for the unlabeled proteins shown in Fig. 1A for RANKL alone and RANKL with three OPG_M, respectively.

3.1.2 OPG_Dimer and RANKL Interactions in PBS Buffer, pH 7.4, 0.1 mg/ml HSA

The SV-AUC results illustrated in Fig. 1 represent simple analyses of fairly straightforward molecular assemblies that are being assessed at relatively high protein concentrations, which also illustrates that it is possible to compare labeled and unlabeled proteins via different detections systems. Next, we illustrate a more complex interacting system composed of the bivalent OPG_D and the trivalent RANKL, which produces a diverse assembly pattern that

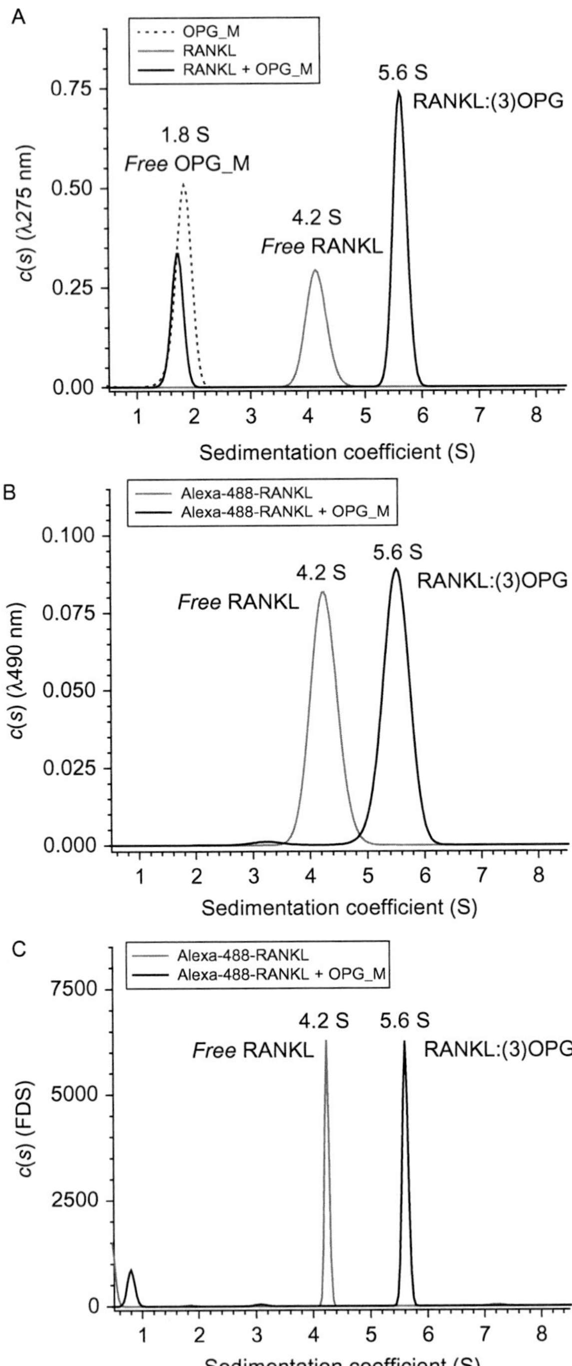

Figure 1 (A) SV analysis with the absorbance optical system at 275 nm for samples of 6.0 μ*M* unlabeled OPG_M (blue (dark gray in the print version) dotted line), 1.0 μ*M* RANKL (red (light gray in the print version)), and a mixture of RANKL+OPG_M (1.0 +6.0 μ*M*, respectively; black), as *c*(*s*), sedimentation coefficient distribution spectra; (B) 490 nm absorption detection; and (C) fluorescence detection.

is very concentration dependent. In this approach, labeled OPG_D at a biologically relevant, low concentration, and unlabeled RANKL at various concentrations, can be characterized in dilute solution by SV analysis with the FDS. Figure 2A provides a 3D graph showing the results from an analysis of 11 different concentrations of RANKL, 2 × serial diluted beginning at 492 n*M*, against a constant 233 p*M* concentration of labeled OPG_D, and the zero-RANKL control plotted as the "0.01 n*M* RANKL" value of that log space axis. All 12 samples were spun at the same time by utilizing both the A and B channels of six centrifugation cells. Scans were analyzed to 120 S in the *c*(*s*) analyses, in order to sufficiently capture the presence of larger sedimenting complexes. Prominent in many of the *c*(*s*) spectra, and most notable for the zero-RANKL control sample, is the ~4.8 S peak that corresponds to the uncomplexed (*free*) OPG_D. With increasing [RANKL], this peak decreases in area and new peaks begin to appear. The pattern of changes are consistent with *free* OPG_D binding RANKL, which initially is dominated by the 1:1 complex at the much lower [RANKL], as seen by the appearance of the ~8.6 S peak that is consistent with a 1:1 complex having a MW of ~193 kDa. Then, much larger sized species, representing higher order assembled complexes, become more apparent and prominent as the [RANKL] increases. The changes in the size of the sedimenting species are even more evident in the 2D graph shown in Fig. 2B, where the *c*(*s*) spectra at several concentrations of RANKL are directly compared. This assembly feature of the system is not inherently unique to the order of addition or the titration direction. Similar results have been obtained for the reverse binding experiment, wherein the OPG_D is titrated against ~200 p*M* labeled RANKL (data not shown). Although it is not revealed by the graphs in Fig. 2A and B, the *c*(*s*) analyses of the SV data sets with RANKL concentrations higher than ~1 n*M* require sedimentation coefficient assessments out to ~120 S in order to effectively fit the SV scans. As a consequence, the results reveal many higher order species, but at a lower apparent abundance than the species found below 35 S. Because of inherent uncertainties associated with the regularization routine in *c*(*s*) analysis and with the FDS (Lyons et al., 2013), the precision of the individual peak positions for the very large complexes (>30 S) is not as high as the more prominent and smaller species, so for clarity of the figures and to emphasize the point that complex size increase with concentration, we have minimized the size axis to the most accurate and abundant species. However, the appearance of the larger species is reproducible across replicate experiments, so they are unlikely to be simple fitting artifacts, *per se*.

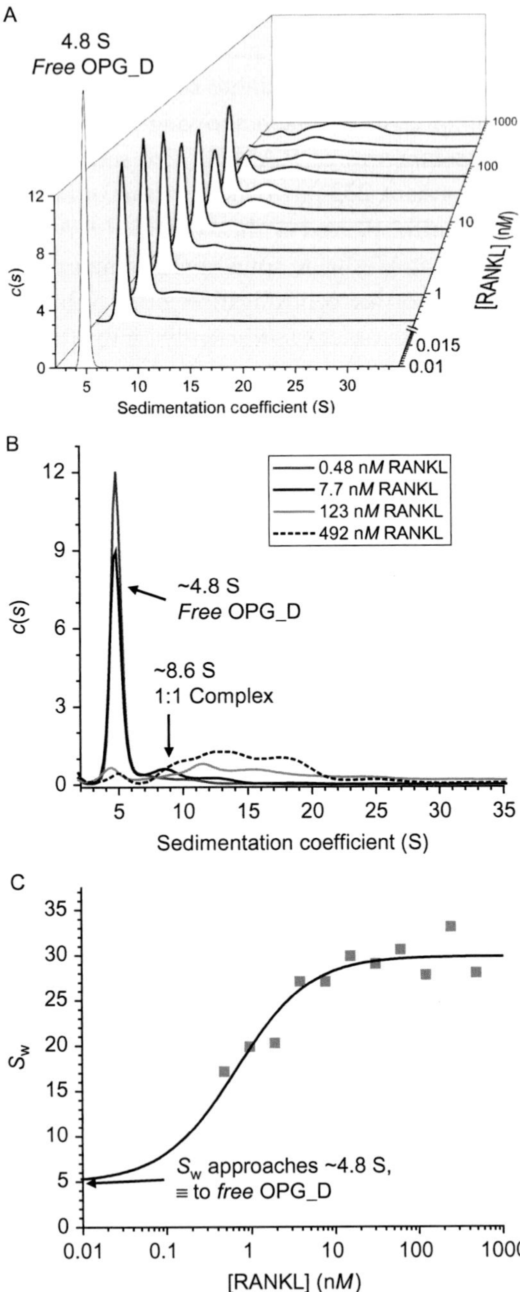

Figure 2 (A) A 3D graph showing the results from the *c*(*s*) analysis of 11 different concentrations of RANKL, 2 × serial diluted beginning at 492 n*M*, against a constant 233 p*M* Alexa-488 labeled OPG_D, and the zero RANKL control plotted as the "0.01 n*M* RANKL" value of that log space axis. (B) A 2D graph for a subset of the spectra in (A). (C) and *(Continued)*

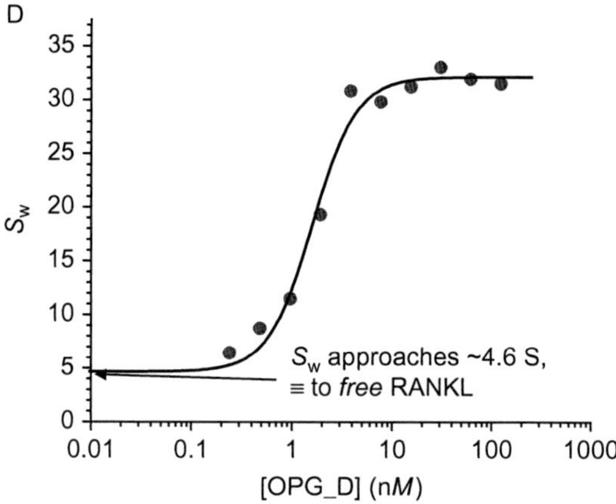

Figure 2—Cont'd (D) provide the S_w binding profiles for the assembly of Alexa-488 labeled OPG_D as a function of [RANKL], and the assembly of Alexa-488 labeled RANKL as a function of [OPG_D], respectively. The trend lines are not fits of the data, but rather are meant to guide the eye through the data profiles. (See the color plate.)

We recommend that when assessing and analyzing the assembly of multivalent proteins, it is worth optimizing the analysis procedure to effectively unveil the larger assembly complexes (e.g., through slower rotor speeds and expanded data analysis routines). Then, through calculation of the weight-average sedimentation coefficients, S_w, and their comparison across the various protein concentrations, one can better reveal trends to the assembly phenomena. For example, Fig. 2C and D provide the binding profiles for the assembly of labeled OPG_D as a function of [RANKL] and the assembly of labeled RANKL as a function of [OPG_D], respectively. Both profiles reveal a trend of increasing S_w with increasing titrant concentration to an S_w maximum around 30–32 S, which occurs around the ~100 n*M* titrant concentration position, as illustrated by the trend lines through the data points. The maximum S_w value indicates protein clusters with a MW near ~1.2 MDa, which might be formed by a hexamer of the 1:1 OPG_D:RANKL, for example. Clearly, a variety of different-sized clustered assemblies exist, but this apparent maximum in the size profile likely represents the most stable species in the distribution under this set of conditions. It is worth noting too that the RANKL-OPG_D binding and assembly process appears to either have relatively quickly equilibrated and reached an apparent equilibrium distribution of cluster sizes on the time scale of the sample preparations and rotor equilibration or that the initial instantaneous assembly of clusters does not significantly re-equilibrate in 24 h, since parallel titration samples held at 25 °C for 24 h prior to loading the rotor into the centrifuge (plus with a ~2 h rotor temperature equilibration prior to starting the SV run at 20 °C), also produced similar results (data not shown). It is important to note that not only for the protein system studied here, but in general terms, the midpoints of these types of binding and assembly profiles can vary significantly, depending on the concentration of the fixed protein species, the inherent binding affinity, and any geometric constraints imposed by the system. Consequently, the most abundant species that are observed in an SV analysis of a mixture of proteins is not necessarily the overall, thermodynamically most stable complex. The most stable complex per set of conditions is best revealed through assessing a distribution of protein concentrations and comparing the S_w values over various incubation times.

3.1.3 RANKL and Ab Interactions in PBS Buffer, pH 7.4, 0.1 mg/ml HSA

Multivalent therapeutic modalities, like antibodies, also exhibit complex assembly patterns. For example, Amgen's denosumab, a therapeutic monoclonal IgG2 type antibody that also binds to RANKL blocking its interaction with RANK, is well established at forming clustered assemblies with RANKL in dilute buffers (Arthur et al., 2012). Here, the most prominent specie apparently is a complex composed of two trimeric RANKL and three bivalent denosumab molecules, but other smaller and larger species can exist, where their abundance depends on the relative protein ratios used in the analysis. In this section, we expand on the multivalent binding and assembly analysis studies with dilute simple buffers by assessing the interactions of a human IgG1-type antibody, obtained from a discovery research effort at Amgen, which also binds RANKL. Figure 3A provides a 3D graph showing the results from an analysis of 11 different concentrations of Ab, 2× serial diluted beginning at 12.1 n*M*, against a constant 440 p*M* labeled RANKL, and the zero-Ab control plotted at the "0.01 n*M* Ab" value on that log space axis. Again, all 12 samples were spun at the same time by utilizing both the A and B channels of six centrifugation cells, and the run was started as soon as possible with respect to temperature equilibration (~2 h after the cell and rotor loading). Prominent in most of the $c(s)$ plots, and most notable for the zero-Ab control, is the ~4.6 S sedimenting specie that corresponds to the *free* RANKL. The 3D graph clearly shows that with increasing RANKL concentrations, the 4.6 S peak decreases while other sedimenting species, reflecting the various assembled complexes

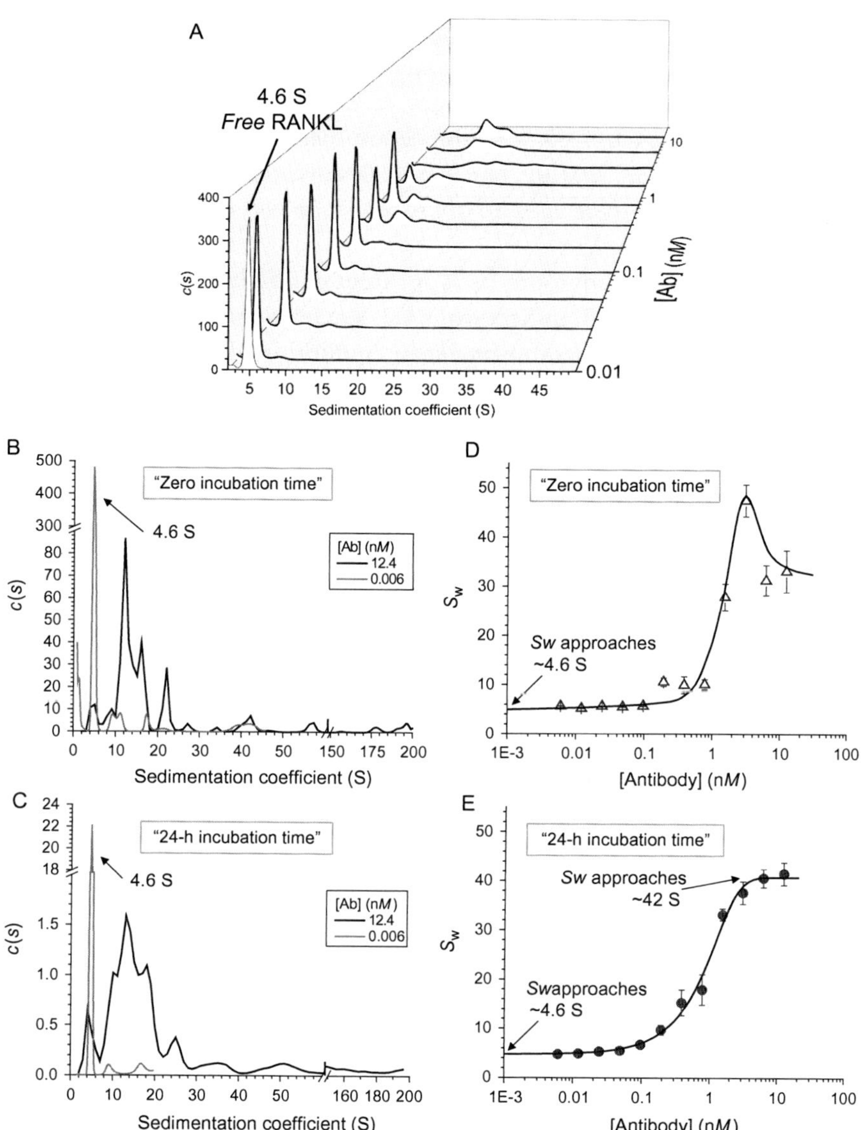

Figure 3 (A) A 3D graph showing the results from the *c*(*s*) analysis of 11 different concentrations of Ab, 2× serial diluted beginning at 12.1 n*M*, against a constant 440 p*M* Alexa-488 labeled RANKL, and the zero RANKL control plotted at the "0.01 n*M* Ab" value. (B) and (C) show a 2D graph containing the *c*(*s*) plots for several different Ab concentrations for the "0 incubation time" and "24-h incubation time" experiments. (D) and (E) provide the S_w binding profiles and their trend lines for the RANKL with Ab titrations for the "zero-incubation time" and "24-h incubation time" experiments, respectively. The trend lines are not fits of the data, but rather are meant to guide the eye through the data profiles and illustrate the "hook effect" in (D).

of Ab and RANKL, begin to appear to varying extents. Notably, species at 8.8, 12.0, and 16.0 S appear and becomes prominent during the course of the titration, which are similar values to the 8, 11.5, and 14 S species observed and characterized for the denosumab-RANKL ligand-binding studies (Arthur et al., 2012). The *c*(*s*) analyses for the Ab concentrations greater than 775 p*M* were assessed out to 200 S, in order to effectively reveal the higher order complexes.

Although it appeared not to be as apparent of a factor in the interactions of OPG_D and RANKL in Section 3.1.2, the relative rate at which multivalent binding systems assemble into their equilibrium distribution of assembled species often can be rather slow, and on the order many hours. This is due in part to both the starting concentrations of the proteins relative to their intrinsic bimolecular association rate, and also from so-called "avidity" effects; wherein the association, dissociation, and rearrangement events of the various individual binding and assembly processes are highly cooperatively linked and correspondingly slow. In order to assess a relative time dependence to the Ab- and RANKL-clustered assembly process, a "sister" set of SV cells were loaded with aliquots from the titration samples prepared for the results given in Fig. 3A. This second set of cells was incubated at ~25 °C for 24 h and then subjected to the same SV setup and rotor equilibration steps as the "zero-incubation-time" samples, prior to starting the run. Figure 3B and C provide 2D graph views of selected *c*(*s*) spectra for a more direct comparison of the "0 incubation time" and "24-h incubation time" experiments, respectively. The graphs reveal that significant changes occur in the abundance of a given species during the long incubation time. The differences between the two sets of incubated samples are even more pronounced when their S_w binding profiles are compared in Fig. 3D and E. Both S_w profiles appear to generate a maximum value at ~42 S, suggesting that this sized species, which likely is akin to a molecule with a MW of ca 1.5 + MDa, is the most stable specie to form under these conditions. A cluster of this size could be composed of three of the (2RANKL:3Ab) type clusters found for denosumab:RANKL (Arthur et al., 2012), a novel 6RANKL:6Ab, or some other combination of molecules of mass and shape to produce 42 S. Very conspicuously, the zero-incubation-time experiment shows a prominent "hook" at ~3 n*M* of Ab, where after which the S_w value decreases with higher [Ab], but the 24-h incubation-time experiment does not contain this "hook effect," as it is often referred too. In principle, the 24-h incubation data also will have a hook in the profile, but it will occur at higher [Ab]; this effect is very typical and well established as a diagnostic of the assembly of systems of multivalent molecules (Heidelberger, 1939). The hook effect mechanistically arises as follows: at lower Ab concentrations, the molecular species are predominantly populated by the "free" unbound RANKL and some simpler, smaller sized complexes such as 1:1 or 2:1 (RANKL:Ab); at the intermediate [Ab], higher order clusters dominate the populations; and then at very high [Ab], owing to the great excess of Ab, i.e., [Ab] $\gg$ [RANKL], the trimeric RANKL molecules will predominantly exist bound by a single Ab molecule, or if geometrically and sterically attainable, as the 1:2 or 1:3 RANKL:Ab complexes. Note, the midpoints of the profiles and their hook positions will be very dependent on the starting conditions, such as the protein concentrations and incubation times.

3.2 Binding and Assembly in Serum and Concentrated Albumin Solutions

3.2.1 Sedimentation Properties of Serum

SV analysis for concentrated solutions often is encumbered by Johnston–Ogston (JO) effects (Correia, Johnson, Weiss, & Yphantis, 1976). Briefly, the JO effect is observed for the sedimentation of mixtures of proteins at high concentrations, wherein fast-moving molecules must move through layers of slow-moving molecules as well as solvent, whereas the slowest molecules sediment in pure solvent. The slow-moving molecules also increase the apparent viscosity of the solvent, leading to a concentration-dependent decrease in the sedimentation coefficients by ~2–2.5 S for serum-like concentrations, from those in dilute solutions. Sometimes, the fast-moving molecules appear as two separate peaks, one in front of and the other with the slow-moving molecules. If the two molecules interact and bind, then it becomes more complicated and a detailed quantitative comparison of peak areas and the concentrations of sedimenting species is not possible, at least at the present time. Under these circumstances, the peak positions do reflect some hydrodynamic behavior of the molecular species; albeit, the actual sedimentation coefficients are inaccurate and calculations of fundamental hydrodynamic parameters are not always possible. In serum, JO effects mainly arise from albumin and IgG molecules, which make up ~50–70% and ~25–30%, respectively, of the total serum concentration (Garbett et al., 2008; Howard et al., 2010). Since HSA is the more abundant and slower moving of the two sized molecules, it is the most significant contributor to JO effects. As such, it is also a good one for monitoring and observing the general impact of serum effects and the increased apparent viscosity on sedimentation properties. Figure 4 shows the results from an

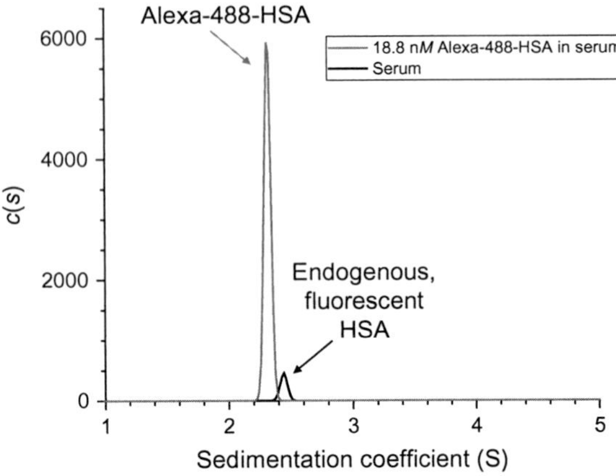

Figure 4 *c*(*s*) results from an SV-FDS analysis of an 18.8 n*M* aliquot of Alexa-488 labeled HSA monomer in human serum, and the neat human serum as a background control.

FDS-SV-AUC analysis of an 18.8 n*M* aliquot of labeled HSA monomer in human serum, which sediments at 2.3 S. HSA monomer typically sediments at ~4.5 S in dilute buffer, but consistent with the increased apparent viscosity of serum, its sedimentation decreases by 2.2 S. Also shown in Fig. 4 is the *c*(*s*) spectrum for the serum sample alone, without a labeled HSA aliquot; noteworthy, is that this sample and the labeled HSA sample were spun in the same SV-AUC run using the same FDS optical settings for detection. Here, the small peak at ~2.4 S arises from the portion of endogenous HSA that has bound bilirubin or another fluorescently active ligand or adduct, as discussed in Section 2.4. Additionally noteworthy is that Fig. 4 also successfully illustrates the approach of "doping" a low concentration of labeled molecule into a high concentration of the unlabeled form of that molecule, so as to act as a "reporter" molecule. As it appears in this example, when the dye label does not contribute spurious information to the analysis, then it can provide an ergodic assessment of the behavior of the unlabeled form of the molecules. This application of the FDS-SV-AUC approach, for example, may also have utility in efforts to assess, *in situ*, the formation of protein particulates that can often occur in highly concentrated therapeutic formulations.

3.2.2 RANKL Interactions With OPG_Monomer and OPG_Dimer in Concentrated Solutions

As discussed in Section 3.2.1, the high concentrations of HSA in serum can impact the sedimentation properties of a test molecule in a relatively nonspecific manner through crowding effects; however, it also can specifically bind to many exogenous molecules (Garbett et al., 2008; Howard et al.,

2010). With albumin concentrations approaching 700 μ*M* in serum, even very low affinity albumin binders could be mostly found in the bound form. Since there are many types of endogenous molecules in serum that also bind to albumin, significant competitive interactions can occur between the labeled test molecule, endogenous albumin binders, and albumin. The FDS-SV-AUC approach provides a unique opportunity to assess these complex HSA binding and assembly interactions, under different concentrated HSA conditions. For example, we have found a benefit to assessing and comparing the impact of different concentrated HSA solutions to different fold dilutions of serum, and in comparing between different concentrated HSA types such as lipid depleted, recombinant, and serum-purified forms, as well as in comparison to similar but unique albumins from different species such as ovalbumin. By way of an example and in order to compare the impact of specific albumin effects to those found in serum, we have assessed the sedimentation behaviors of the OPG molecules in different concentrated HSA solutions by the FDS-SV-AUC approach. Figure 5A shows the results from the SV analyses of 200 p*M* labeled OPG_M and labeled OPG_D in a very concentrated, 40 mg/ml, HSA solution. As a point of reference, HSA monomer sediments at 3.0 S in this same solution (data not shown); that this HSA *s* value is higher than that in serum, likely is due to the serum having a much higher overall concentration of molecules and a correspondingly higher apparent viscosity. Notably, both OPG constructs show two sedimentation peaks in this analysis. Since the OPG_D molecule and HSA have similar sedimentation coefficients, 4.8 S and 4.5 S, respectively, we expect the OPG_D to sediment near/with albumin (3.0 S). Indeed, the major peak at 3.7 S for OPG_D is consistent with this expectation; however, the peak at 5.5 S is not expected, and may reflect a direct interaction with albumin, indicating that this dimer construct of OPG can bind albumin at these higher albumin concentrations and crowded conditions. Similarly, the peak at 3.2 S for the OPG_M sample is also consistent with this construct binding to HSA. The peak at 1.8 S for the OPG_M is surprising, in that it is nearly the same *s* value as that recovered in dilute solution (Fig. 1A, Section 3.1.1). It is worth noting that Alexa-488 fluorescent dye alone does not appear to interact with HSA under these conditions and its apparent *s* value is recovered at <0.5 in this buffer (data not shown). Interestingly, this split peak pattern also was observed for sedimentation of the OPG constructs in 10 and 1.25 mg/ml solutions of HSA (data not shown), suggesting it is not entirely unique to the very high [HSA] concentrations or highly crowded environments. The peak areas for the "OPG-HSA" species appear to scale with [HSA], suggesting it may be a mass-action-driven process.

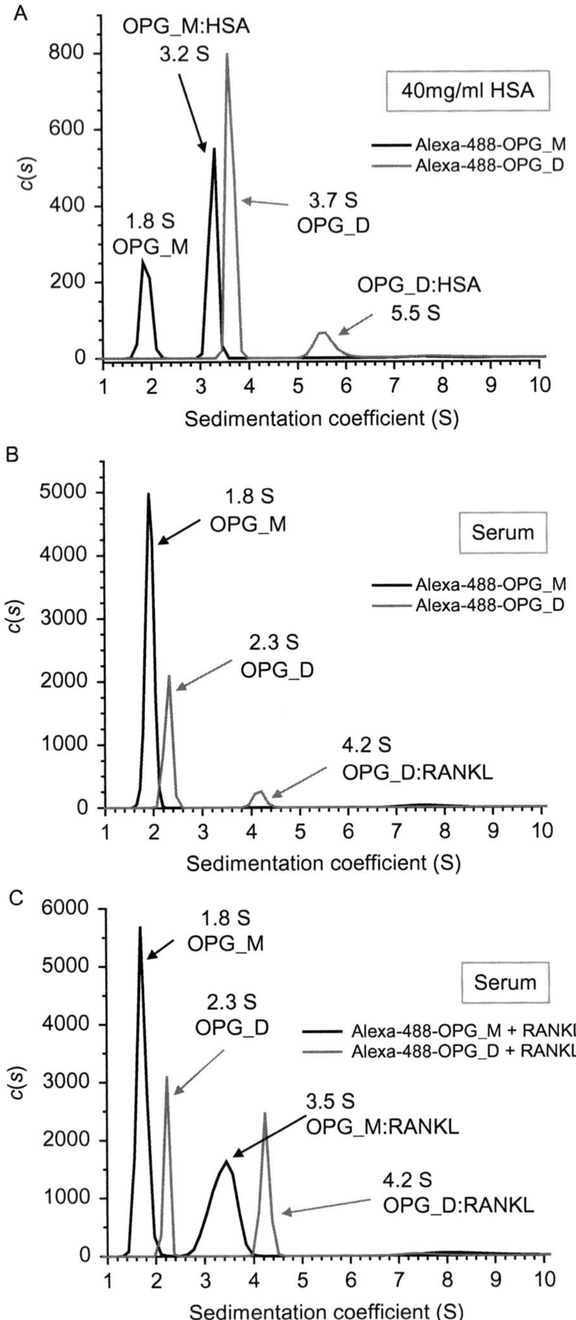

Figure 5 (A) *c*(*s*) results from the SV-FDS analyses of 200 p*M* Alexa-488 OPG_M and Alexa-488 OPG_D in a very concentrated, 40 mg/ml, HSA solution. (B) *c*(*s*) results from the SV-FDS analyses of 200 p*M* Alexa-488 OPG_M and 200 p*M* Alexa-488 OPG_D in human serum. (C) *c*(*s*) results from the SV-FDS analyses of 200 p*M* Alexa-488 OPG_M and Alexa-488 OPG_D in human serum with the addition of an aliquot recombinant HuRANKL at ∼85 n*M*.

The hydrodynamic behaviors of the OPG constructs in serum, at ~200 p*M*, are given in Fig. 5B. Both OPG constructs predominantly sediment below the HSA sedimentation value of 2.3 S (same serum sample as discussed in Section 3.2.1), but neither construct shows a peak that is consistent with binding to the endogenous HSA in the serum, as had been observed for the concentrated HSA solution, which may be due to endogenous compounds in serum binding to HSA and effectively competing with the OPG molecules (Garbett et al., 2008; Howard et al., 2010). The small peak around 4.2 S for the OPG_D sample likely is due to an interaction with the endogenous RANKL present in serum. This is corroborated by the results in Fig. 5C, where the samples are as in Fig. 5B, but contain an additional ~85 n*M* aliquot of unlabeled RANKL; wherein the resulting peaks at 3.5 and 4.3 S then are most consistent with the presence of an OPG_M: RANKL complex and an OPG_D:RANKL complex, respectively. The most notable and significant result of this figure is that in serum, OPG_D +RANKL do not appear to form the large clustered assemblies that had been so prominent in the dilute solutions (Fig. 2, Section 3.1.2). This loss of the large clustered assemblies has also been observed in the reverse titration; wherein unlabeled OPG_D is added to labeled RANKL in serum (data not shown), illustrating that the changes are not unique to the labeled molecule or their order of addition. Overall, the results suggest that factors associated with serum can abrogate the ability of OPG_D and RANKL to form stable, higher order complexes. For the sake of clarity and in consideration of the data analysis methods in general, it is worth noting that the apparent peaks around 8 S in the various spectra of Fig. 5 likely are fitting artifacts. Their presence, position, and magnitude are not reproducible between sample runs, as well as they vary between redundant fits of the same data sets (Lyons et al., 2013).

3.2.3 Antibody in Serum

In general, we and others (Demeule et al., 2009) have found that most fluorescently labeled IgG antibodies appear to sediment between 3.4 and 4.1 S (uncorrected sedimentation coefficients) in various sera (from human, cynomologous, or murine sources), where the *s* value variance appears to depend on the Ab composition and conjugates, and factors such as the total protein and lipid content of the sera. Consistent with this observation, the sedimentation coefficient for the antibody has a value of ~3.9–4.0 S in the human serum used in this study, as shown in Fig. 6 for two different concentrations of the antibody, at 1.9 n*M* and 19.0 n*M* respectively. The large

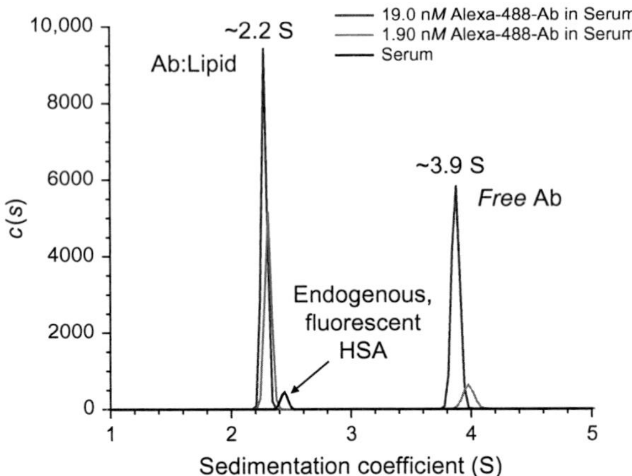

Figure 6 $c(s)$ analyses of 1.9 and 19.0 nM aliquots of Alexa-488 labeled antibody in human serum. (See the color plate.)

peaks at ~2.2 S are rather unexpected since they likely are not from bilirubin-bound HSA, which is shown again here by the very small peak in the sedimentation spectrum of neat serum, and as discussed in Section 3.2.1. In particular, the magnitudes of these 2.2 S peaks are much greater than the bilirubin–HSA contribution, and they scale with the 10-fold difference in Ab concentrations, illustrating they are due to the Ab. Albeit it is not possible from this analysis to identify the composition of the 2.2 S species, it may be due to the Ab interacting with lipids or lipoprotein particles, like exosomes for example, since the significantly lower density of lipids could "buoy" some of the Ab molecules and cause them to sediment slower. Alternatively, the peak could be due to JO effects splitting the *free* Ab sedimentation boundary; however, we and others have not routinely observed this behavior in other Ab-serum studies (Demeule et al., 2009) nor have we observed it with high [HSA] solutions (data not shown).

3.2.4 Antibody and RANKL Interactions in Serum

The abrogation of RANKL–OPG_D clusters in serum (Section 3.2.2) illustrates that crowded conditions can significantly influence the assembly state of endogenous multivalent binding partners. Similar assembly control has been observed in serum for the IgG1-type therapeutic antibody omalizumab and its multivalent target, human immunoglobin E (IgE), illustrating that the phenomenon is not unique to the RANKL–OPG system (Demeule et al., 2009). We also have explored the control of cluster assembly in serum for the anti-RANKL antibody and RANKL system.

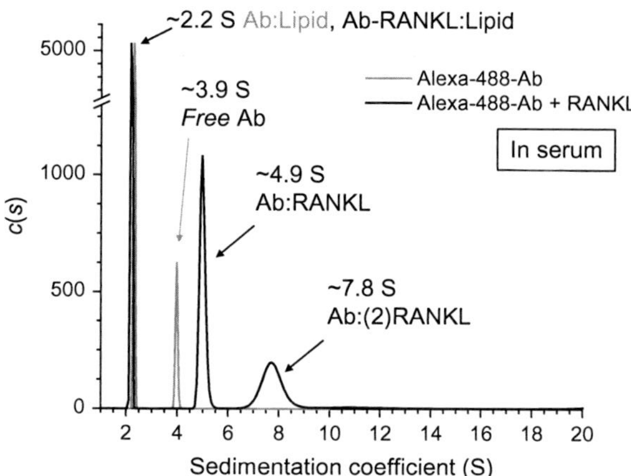

Figure 7 Results from adding 85 n*M* of unlabeled RANKL to serum containing 19 n*M* of labeled antibody, as in Fig. 6.

Figure 7 shows the results from adding 85 n*M* of unlabeled RANKL to serum containing 19 n*M* of the labeled Ab. The loss of the 3.9 S Ab peak and the appearance of peaks at 4.9 and 7.8 S represent the concomitant decrease of the unbound Ab species with the formation of the bound complexes as the 1:1 and 1:2 Ab:RANKL species. Notably, the clusters of Ab and RANKL that were very prominent in the dilute buffer titrations are also abrogated by these crowded conditions. Figure 7 also contains the apparent 2.2 S "Ab–lipid" peak, and from this analysis, it is not possible to assess whether the peak also contains an "Ab–RANKL–lipid" species.

4. CONCLUSIONS

From the FDS-SV-AUC studies discussed here and elsewhere (Demeule et al., 2009), it is apparent that crowded conditions, like serum, can significantly limit the size and distributions of assembled complexes, compared with the very large ones that can form in dilute buffers. The results with serum suggest several factors may be impacting the binding and assembly processes that occur in complex, concentrated solutions: (1) that the various electrostatic effects that become more pronounced in high-proximity conditions alter the apparent affinity and dynamics of protein assembly and modulate protein conformations, geometry, and assembly "architecture" (Laue & Demeler, 2011; Laue, 2012); (2) that nonspecific, excluded volume effects may influence assembly through molecules adopting different or more compact conformations that limit flexibility or the geometric space to participate in larger assemblies (Zhou et al., 2008); and (3) that specific interactions with serum components, such as albumin and lipids/lipid particles, can impact binding and assembly processes (Garbett et al., 2008). No matter the mechanistic interpretation of the results, the differences between the binding and assembly phenomena of dilute solution and serum can be significant. This is particularly notable since the complex, large clustered protein assemblies that occur in dilute solution inherently form as cooperative, high-affinity species with thermodynamics and dynamics that are very different than the simpler, smaller species that assemble in serum. Therefore, the binding characteristics observed in dilute *in vitro* assays are not necessarily, entirely interpretive of the interactions that actually occur in the crowded *in vivo* conditions. These differences indicate a necessity to characterize the assembly state and related binding interactions of therapeutic proteins in serum, which at this time appears to be uniquely accomplished by the FDS-SV-AUC analysis approaches.

Besides differences between dilute and serum conditions, there can be significant differences between serum sources that are worth assessing in regards to binding and assembly interactions by the FDS-SV-AUC approaches. With the exception of a few major components like albumin, the composition and concentrations of molecules in serum can vary significantly between individuals and their disease states (Garbett et al., 2008; Garbett, Mekmaysy, Helm, Jenson, & Chaires, 2009). As a result, some of the observed binding and assembly behaviors may be a consequence of the source of serum used in an analysis. For example, the behaviors of proteins that interact with lipid particles could be influenced by the concentrations of HDL, LDL, exosomes, or other lipid and lipoprotein particles that vary with diet and disease. In similar scenarios, the molecules that interact with albumin could be modulated by the presence or absence of endogenous albumin ligands that change with diet and disease (Garbett et al., 2008). The FDS-SV-AUC approaches may also help in sorting out the binding differences between multiple forms of a binding partner. For example, an antigen might exist as a soluble receptor, a clipped external domain, and also shed in full length with lipids as an exosome particle; the various compositions might be differentiated by their binding and unique sedimentation properties.

Serum-based binding studies may also have utility in pharmacokinetic and pharmacodynamic (PKPD) considerations, since changes in the concentrations of endogenous targets and the elimination of drug after administration can be intricately coupled to binding and assembly phenomena in serum

(Howard et al., 2010). Analytical procedures intent on quantitatively determining the amounts of free or bound proteins in biological fluids like serum, might recover different values when measurements are made with neat serum as compared to those made with serum diluted into buffers. Wherein, the decrease in crowding conditions, which accompanies the serum dilutions, likely imparts a change in the distribution of species from the free and simple bound species found in neat serum to the higher complexity, cooperatively driven, clustered species of dilute conditions. This redistribution process also likely has an inherently time-dependent hysteresis to it, as illustrated in Fig. 3, which might confound the analytical test procedure in a time-dependent manner. In this context then, a comparison of neat and dilute serum samples by FDS-SV-AUC analysis approaches might aid in developing robust bioanalytical analysis methods (Bashaw, DeSilva, Rose, Wang, & Shukla, 2014). The results of which may aid in refining clinical PKPD parameters, and preclinical measurements for so-called *in vitro–in vivo* PKPD correlations, as well as in allometric scaling efforts to relate PKPD parameters between different species. The FDS-SV-AUC approaches may also provide a significant contribution to the characterization scheme of protein engineering efforts intent on producing good "drug-like" properties for a therapeutic. This is notable since often the various designed protein molecules can have modulated charges and carbohydrate contents with unique proximity energy properties that potentially impact interactions with serum component (Garbett et al., 2009; Laue, 2012).

Overall, FDS-SV-AUC provides a rather unique and direct analysis approach for characterizing the binding and assembly properties of molecules in serum and other crowded solutions like highly concentrated protein formulations, as well as in dilute simple buffers; plus, it does so across a broad biochemically and biologically meaningful concentration space.

ACKNOWLEDGMENTS

The authors thank Hossein Salimi-Moosavi for the protein labeling work and helpful discussions, and also Jean Lee, Susan Pederson, Binodh Desilva, George Scott, and Larry Wienkers for helpful discussions and support of this work. We also are very grateful for the assistance from Sue Chase, who's dedicated and tireless scientific support and friendship made this project and long-distance collaboration successful.

REFERENCES

Arthur, K. K., Gabrielson, J. P., Hawkins, N., Anafi, D., Wypych, J., Nagi, A., et al. (2012). In vitro stoichiometry of complexes between the soluble RANK ligand and the monoclonal antibody denosumab. *Biochemistry*, *51*(3), 795–806.

Bashaw, E. D., DeSilva, B., Rose, M. J., Wang, Y. M., & Shukla, C. (2014). Bioanalytical method validation: Concepts, expectations and challenges in small molecule and macromolecule—A report of PITTCON 2013 symposium. *The AAPS Journal, 16*(3), 586–591.

Brown, P. H., Balbo, A., & Schuck, P. (2009). On the analysis of sedimentation velocity in the study of protein complexes. *European Biophysics Journal, 38*(8), 1079–1099.

Cole, J. L., Lary, J. W., Moody, T. P., & Laue, T. M. (2008). Analytical ultracentrifugation: Sedimentation velocity and sedimentation equilibrium. *Methods in Cell Biology, 84*, 143–179.

Correia, J. J., Johnson, M. L., Weiss, G. H., & Yphantis, D. A. (1976). Numerical study of the Johnston-Ogston effect in two-component systems. *Biophysical Chemistry, 5*(1–2), 255–264.

Demeule, B., Shire, S. J., & Liu, J. (2009). A therapeutic antibody and its antigen form different complexes in serum than in phosphate-buffered saline: A study by analytical ultracentrifugation. *Analytical Biochemistry, 388*(2), 279–287.

Elcock, A. H. (2010). Models of macromolecular crowding effects and the need for quantitative comparisons with experiment. *Current Opinion in Structural Biology, 20*(2), 196–206.

Garbett, N. C., Mekmaysy, C. S., Helm, C. W., Jenson, A. B., & Chaires, J. B. (2009). Differential scanning calorimetry of blood plasma for clinical diagnosis and monitoring. *Experimental and Molecular Pathology, 86*(3), 186–191.

Garbett, N. C., Miller, J. J., Jenson, A. B., & Chaires, J. B. (2008). Calorimetry outside the box: A new window into the plasma proteome. *Biophysical Journal, 94*(4), 1377–1383.

Heidelberger, M. (1939). Quantitative absolute methods in the study of antigen-antibody reactions. *Bacteriological Reviews, 3*(1), 49–95.

Howard, M. L., Hill, J. J., Galluppi, G. R., & McLean, M. A. (2010). Plasma protein binding in drug discovery and development. *Combinatorial Chemistry & High Throughput Screening, 13*(2), 170–187.

Kingsbury, J. S., & Laue, T. M. (2011). Fluorescence-detected sedimentation in dilute and highly concentrated solutions. *Methods in Enzymology, 492*, 283–304.

Kroe, R. R., & Laue, T. M. (2009). NUTS and BOLTS: Applications of fluorescence-detected sedimentation. *Analytical Biochemistry, 390*(1), 1–13.

Lacey, D. L., Timms, E., Tan, H. L., Kelley, M. J., Dunstan, C. R., Burgess, T., et al. (1998). Osteoprotegerin ligand is a cytokine that regulates osteoclast differentiation and activation. *Cell, 93*(2), 165–176.

Lamola, A. A., & Russo, M. (2014). Fluorescence excitation spectrum of bilirubin in blood: A model for the action spectrum for phototherapy of neonatal jaundice. *Photochemistry and Photobiology, 90*(2), 294–296.

Laue, T. (2012). Proximity energies: A framework for understanding concentrated solutions. *Journal of Molecular Recognition, 25*(3), 165–173.

Laue, T., & Demeler, B. (2011). A postreductionist framework for protein biochemistry. *Nature Chemical Biology, 7*(6), 331–334.

Laue, T. M., & Stafford, W. F., 3rd. (1999). Modern applications of analytical ultracentrifugation. *Annual Review of Biophysics and Biomolecular Structure, 28*, 75–100.

Lyons, D. F., Lary, J. W., Husain, B., Correia, J. J., & Cole, J. L. (2013). Are fluorescence-detected sedimentation velocity data reliable? *Analytical Biochemistry, 437*(2), 133–137.

MacGregor, I. K., Anderson, A. L., & Laue, T. M. (2004). Fluorescence detection for the XLI analytical ultracentrifuge. *Biophysical Chemistry, 108*(1–3), 165–185.

Minton, A. P. (2013). Quantitative assessment of the relative contributions of steric repulsion and chemical interactions to macromolecular crowding. *Biopolymers, 99*(4), 239–244.

Ominsky, M. S., Stouch, B., Schroeder, J., Pyrah, I., Stolina, M., Smith, S. Y., et al. (2011). Denosumab, a fully human RANKL antibody, reduced bone turnover markers and increased trabecular and cortical bone mass, density, and strength in ovariectomized cynomolgus monkeys. *Bone*, *49*(2), 162–173.

Schneeweis, L. A., Willard, D., & Milla, M. E. (2005). Functional dissection of osteoprotegerin and its interaction with receptor activator of NF-kappaB ligand. *The Journal of Biological Chemistry*, *280*(50), 41155–41164.

Schuck, P., Perugini, M. A., Gonzales, N. R., Howlett, G. J., & Schubert, D. (2002). Size-distribution analysis of proteins by analytical ultracentrifugation: Strategies and application to model systems. *Biophysical Journal*, *82*(2), 1096–1111.

Walsh, M. C., & Choi, Y. (2014). Biology of the RANKL–RANK–OPG system in immunity, bone, and beyond. *Frontiers in Immunology*, *5*, 511.

Zhao, H., Mayer, M. L., & Schuck, P. (2014). Analysis of protein interactions with picomolar binding affinity by fluorescence-detected sedimentation velocity. *Analytical Chemistry*, *86*(6), 3181–3187.

Zhou, H. X., Rivas, G., & Minton, A. P. (2008). Macromolecular crowding and confinement: Biochemical, biophysical, and potential physiological consequences. *Annual Review of Biophysics*, *37*, 375–397.

AUTHOR INDEX

Note: Page numbers followed by "*f*" indicate figures and "*t*" indicate tables.

C

E

F

G

H

I

J

K

L

M

N

O

P

Q

R

S

T

SUBJECT INDEX

Note: Page numbers followed by "*f*" indicate figures and "*t*" indicate tables.

B

C

D

E

K

L

M

N

Q

R

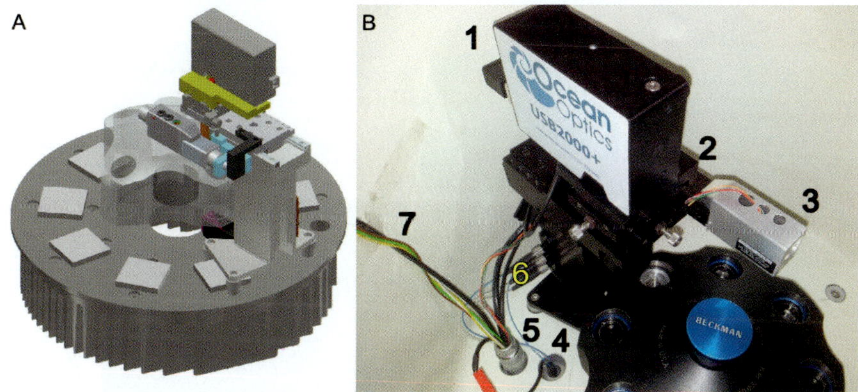

Joseph Z. Pearson *et al.*, Figure 1 (A) Schematic and (B) picture of Nanolytics detector hardware. (1) Ocean Optics USB2000+ spectrometer, (2) X–Y stage, (3) step motor, (4) optical feedthrough, (5) electrical feedthrough, (6) spare fibers, and (7) connections for optional interference optics.

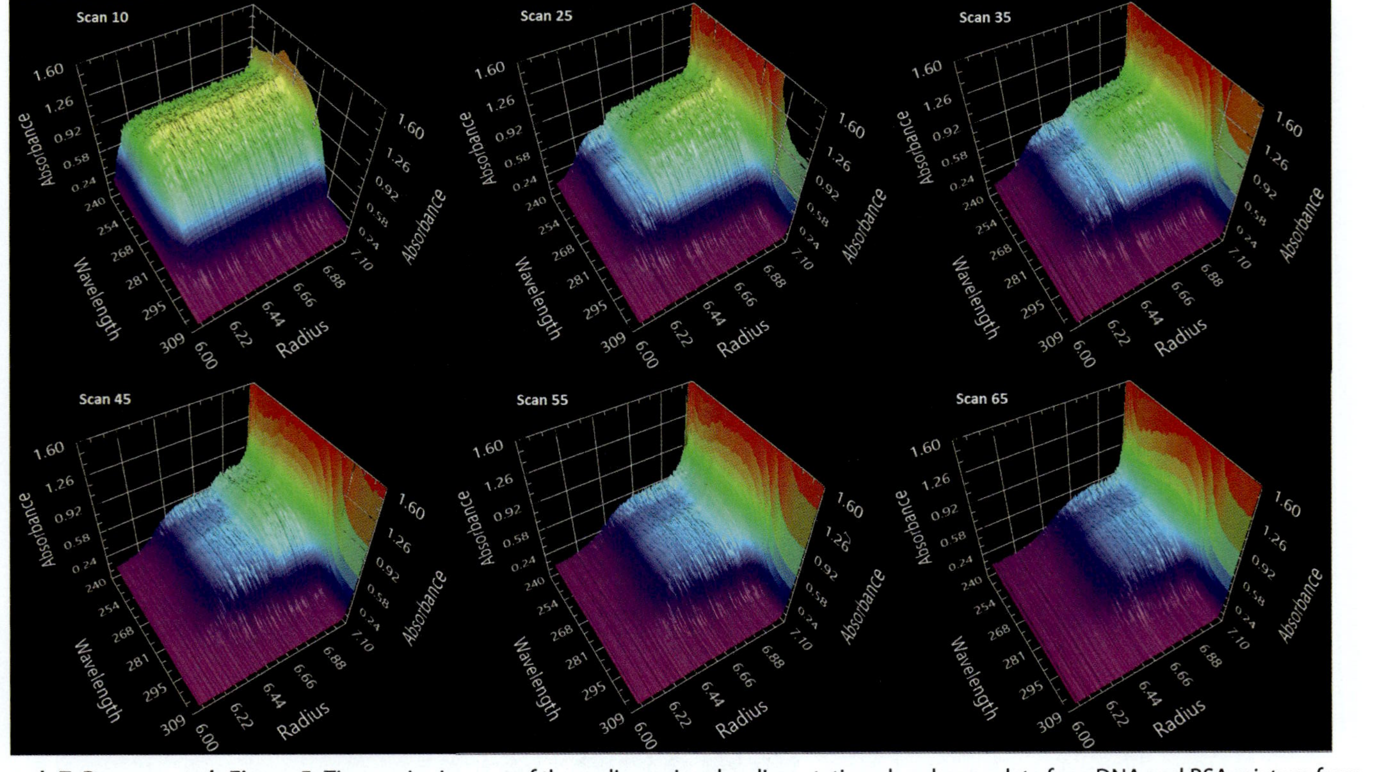

Joseph Z. Pearson *et al.*, Figure 5 Time series images of three-dimensional sedimentation absorbance data for a DNA and BSA mixture from open source MWL recorded with 50 μm radial step size. The intensity in OD units is dependent on the extinction coefficient at each wavelength. The absorbance spectra of the species in solution are evident across the wavelength range. The yellow coded species at ca. 270 nm across the spectra is indicative of a more rapidly sedimenting species, in this case a larger DNA fragment. *Images generated by LabView-based MWL data viewer written by Dirk Haffke.*

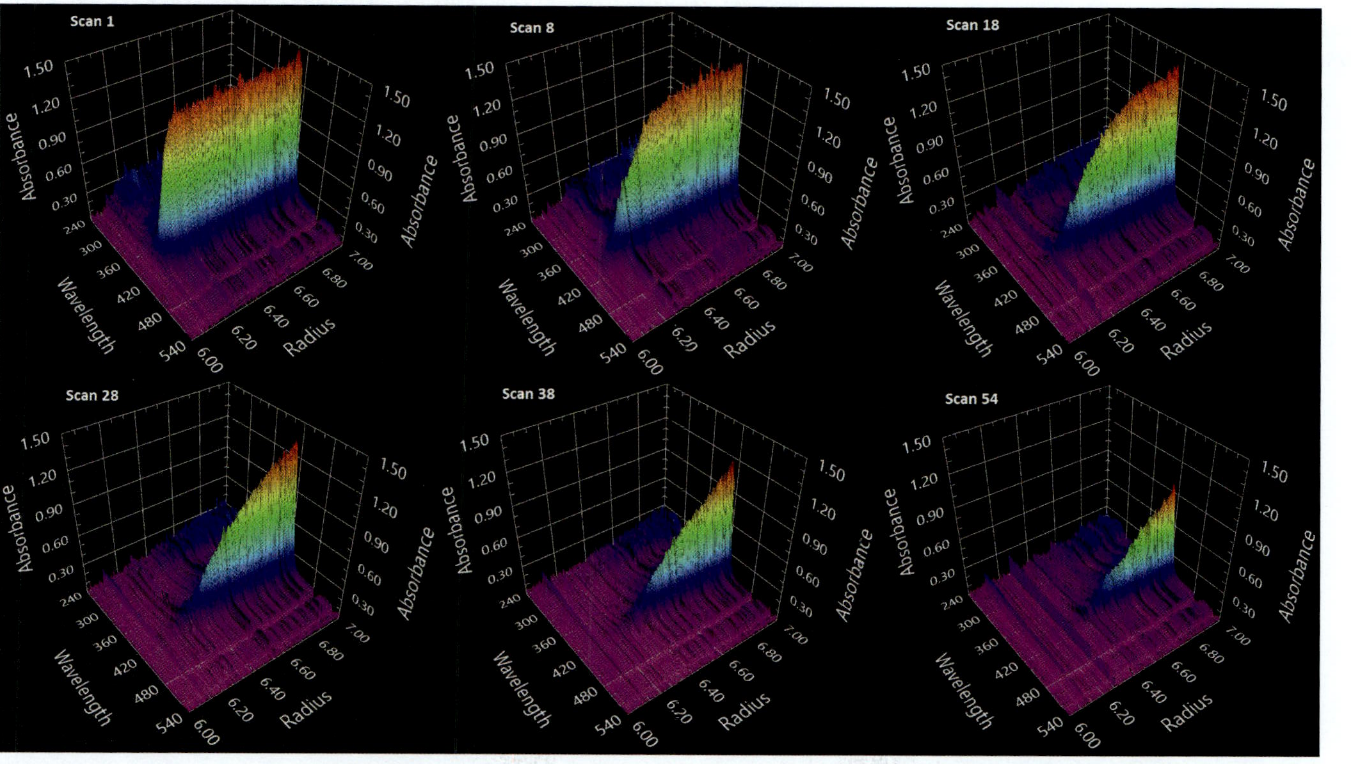

Joseph Z. Pearson *et al.*, Figure 6 Time series images of three-dimensional sedimentation absorbance data for a hemoglobin (Hb) and myoglobin (Mb) mixture. Data from Nanolytics second-generation MWL, using 16 bit USB2000+ Ocean Optics spectrometer with #1 grating and 10 μm radial step size. *Images generated by LabView-based MWL data viewer written by Dirk Haffke.*

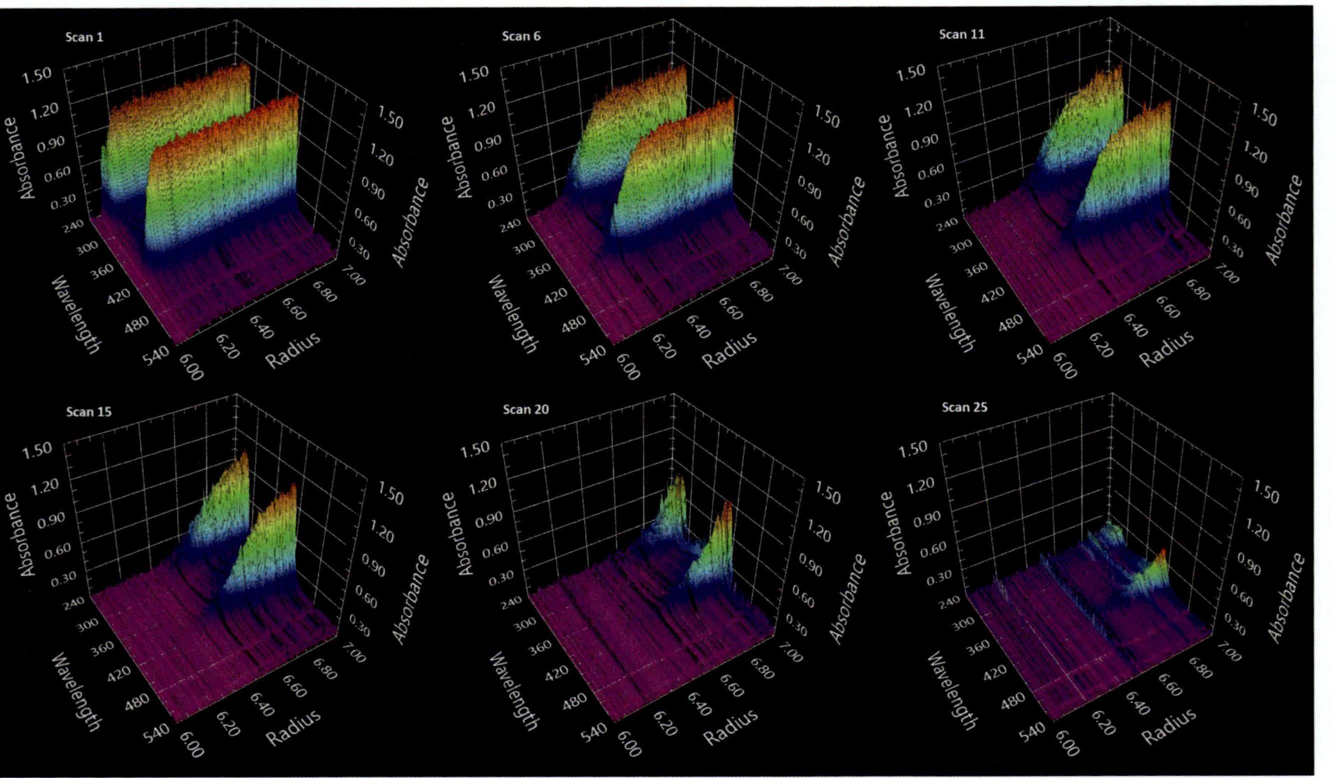

Joseph Z. Pearson *et al.*, Figure 7 Time series images of three-dimensional sedimentation absorbance data for a hemoglobin (Hb) and BSA mixture. Data from Nanolytics second-generation MWL, using 16 bit USB2000+ Ocean Optics spectrometer with #1 grating and 10 μm radial step size. *Images generated by LabView-based MWL data viewer written by Dirk Haffke.*

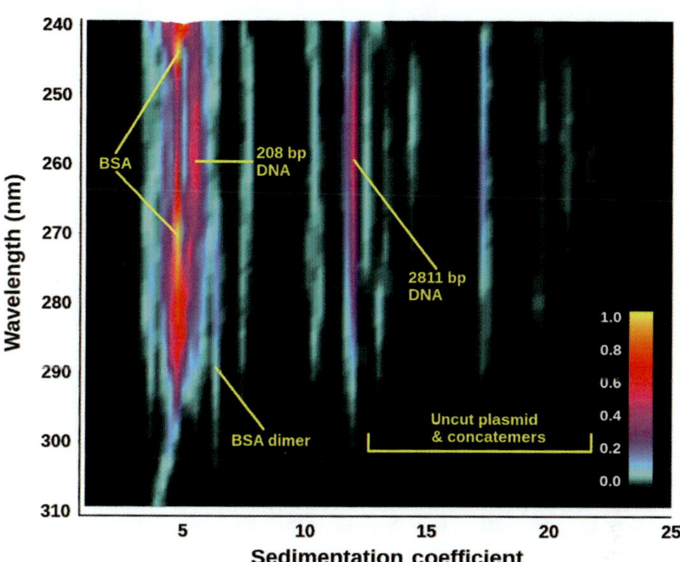

Gary E. Gorbet *et al.*, Figure 3 Projection view of the 2DSA-Monte Carlo sedimentation profile as a function of wavelength for the 50:50 DNA–BSA mixture. Remarkably, the protein absorbance spectrum at 4.3 S (two yellow peaks) can be clearly distinguished from the adjacent DNA peak with absorbance maximum around 258 nm, despite the proximity of the peaks (4.5 S vs. 5.2 S). Minor species can be identified based on their spectra. The straight lines attest to the high resolution and robustness of this approach to fit multiwavelength data (each wavelength is separately analyzed).

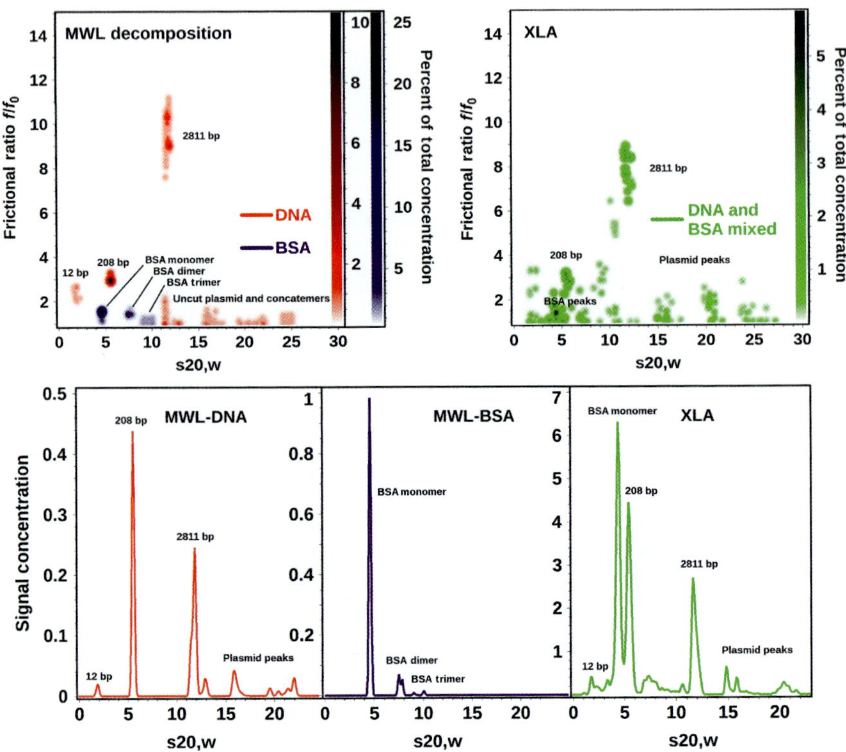

Gary E. Gorbet *et al.*, Figure 7 Global genetic algorithm Monte Carlo analysis of decomposition results obtained from six different DNA and BSA mixtures analyzed on the open AUC MWL instrument (top left) and the dual-wavelength results obtained from the Beckman-Coulter XL-A (right panel). The separate decomposition results for DNA (red) and BSA (blue) are combined in the left panel pseudo-3D plot to illustrate the exceptional separation achieved by spectral decomposition which even separates species with nearly identical sedimentation coefficients (the two major species sedimenting near 5 S). This approach demonstrates the superior resolution obtained from MWL analysis compared to the global two-wavelength analysis performed on the Beckman-Coulter XL-A (right panel, green). Lower panel: differential distributions from the same data shown above (red): decomposition for DNA, blue: decomposition for BSA, green: nonseparated XL-A data for DNA–BSA mixtures globally fitted to genetic algorithm–Monte Carlo analysis. A color representation of this figure is provided in the color plate section to allow the reader to distinguish the DNA and BSA contributions based on color.

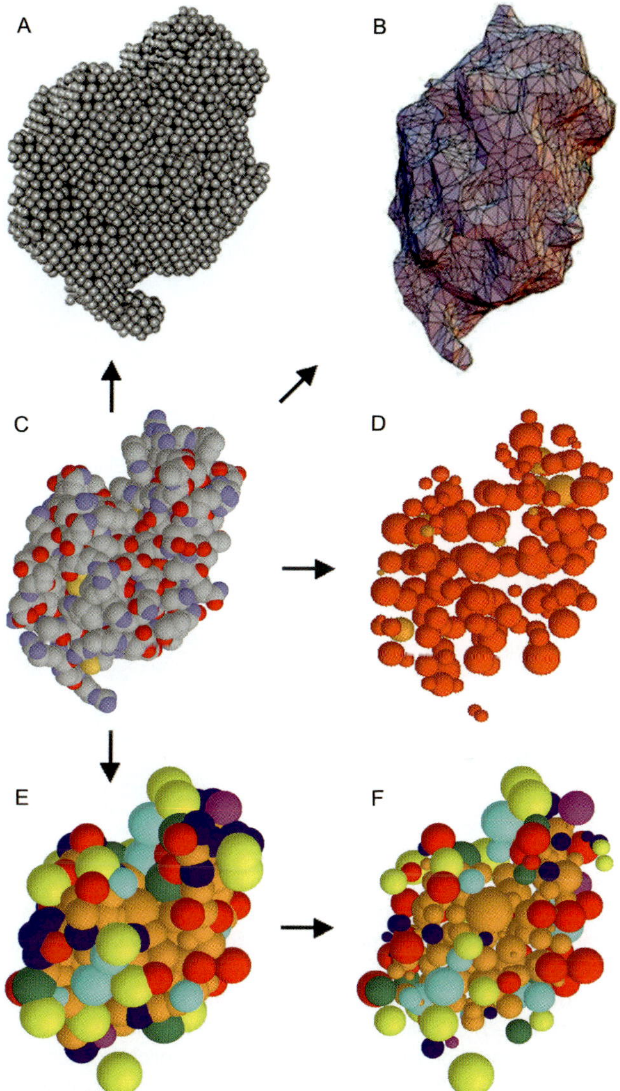

Mattia Rocco and Olwyn Byron, Figure 1 Hydrodynamic models generated by the available modeling programs starting from an atomic resolution structure ((C) lysozyme 6lyz.pdb). (A) A bead-shell model generated by HYDROPRO. (B) A tessellated model generated by BEST. (D) A 5 Å grid bead model generated by the AtoB method in US SOMO. (E) A direct correspondence bead model, with overlaps, generated by the SoMo method in US SOMO. (F) Same as in (E) after overlap removal. In (D), red and orange are the exposed and buried beads, respectively. In (E) and (F), the color coding is: blue, main-chain; cyan, hydrophobic; magenta, non-polar; red, polar; yellow, basic; green, acidic; white, fused beads; orange, buried beads.

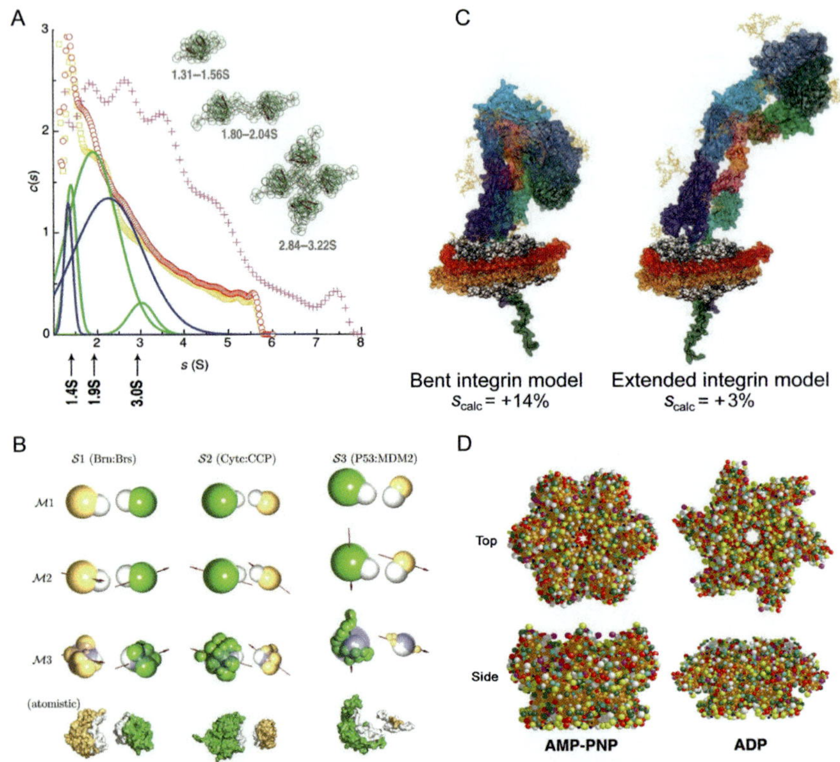

Mattia Rocco and Olwyn Byron, Figure 3 Examples of the use of AtoB (A, B) and SoMo (C, D) drawn from the literature. AtoB was used to (A) generate models for monomer, dimer, and tetramer for the SH3 domain of CRKL for which values of $s^0_{(20,w)}$ were computed (by SOLPRO) for a range of hydration levels and compared with $s^0_{(20,w)}$ determined by *c*(*s*) size distribution analysis of sedimentation velocity data and (B) provide one (M3) of three levels (M1–3) of coarse-graining of protein surfaces for use in a Langevin equation approach to modeling the formation of protein–protein encounter complexes. SoMo was used to generate (C) nanodisc-embedded integrin $\alpha_{IIb}\beta_3$ bent/closed (left) and extended/closed (right) models (their % deviation from experimental $s^0_{(20,w)}$ values is indicated), and (D) models of *N*-ethylmaleimide sensitive factor (NSF) in ADP- and AMP-PNP-bound states. *(A) Adapted from Harkiolaki et al. (2006) with permission. (B) Adapted from Schluttig, Alamanova, Helms, and Schwarz (2008) with permission. (C) Adapted from Rosano and Rocco (2010) with permission. (D) Adapted from Moeller et al. (2012) with permission.*

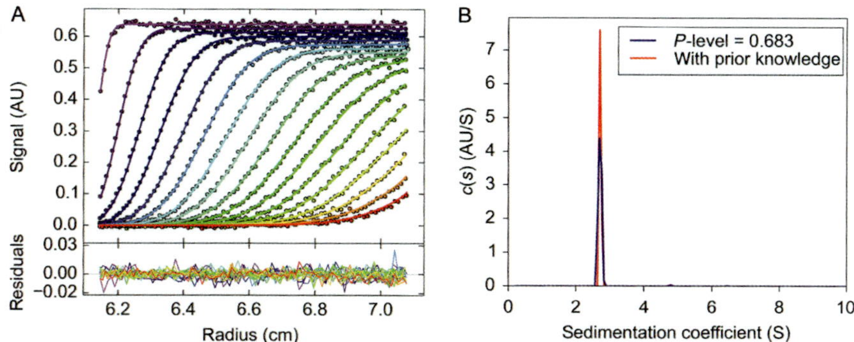

Chad A. Brautigam, Figure 2 The GUSSI svdfr and *c*(*s*) plots. (A) The default svdfr plot. In the upper panel, the individual data points are circles, and the fits to those data are shown as lines. The lower panel displays the residuals. Time-invariant noise features are subtracted from the data and fit lines. (B) A *c*(*s*) plot. The *c*(*s*) analysis for the data shown in (A) is displayed with the usual regularization (*P*-level = 0.683) along with the distribution after the application of prior knowledge (Brown, Balbo, & Schuck, 2007), i.e., that the species sedimenting at 2.7 S is a single, discrete species ("with prior knowledge").

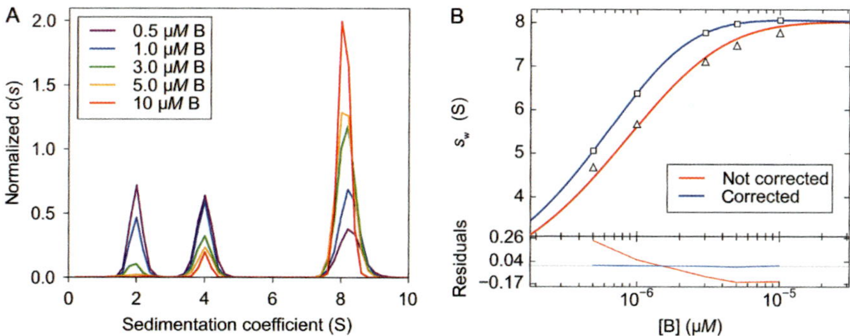

Chad A. Brautigam, Figure 4 Correction of s_w isotherms in GUSSI. The simulation assumes that species A - associates with species B to form a 1:1 AB complex with a dissociation constant of 1 μ*M*. Also, the sedimentation coefficient of A (s_A) is 2.0 S, s_B is 8.0 S, and s_{AB} is 8.65 S. The preparation of A is contaminated with a 4.0 S, inert species. The [A] was held constant at 1 μ*M* in the titration, while [B] was varied as described in the inset to part (A). (A) The *c*(*s*) distributions for the simulation. The simulated data were analyzed in SEDFIT. The distributions are normalized, and thus the contaminant at 4 S appears to have varying signal amplitudes. (B) Isotherm fitted to the data with (squares) or without (triangles) the correction applied as described in the text. Without the correction, K_D refined to 4.1 μ*M*, whereas the corrected data yielded the simulated value when evaluated, i.e., 1.0 μ*M*.

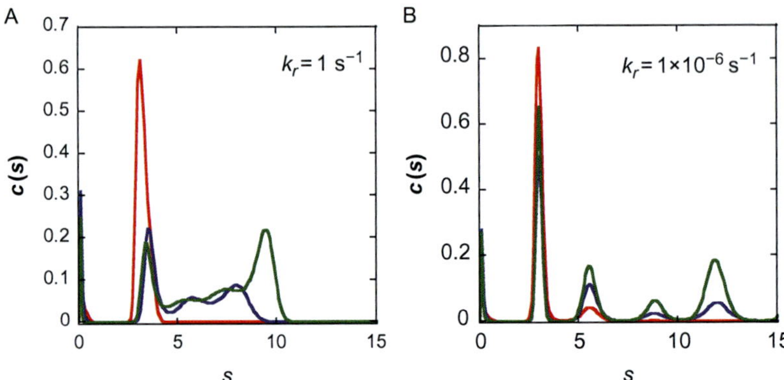

JiaBei Lin and Aaron L. Lucius, Figure 2 $c(s)$ distributions resulting from $c(s)$ analysis of data simulated from 1, 9, and 15 μ*M* protein. (A) Parameters are given in Table 1 and (B) parameters are given in Table 2. In both cases, the sedimentation coefficients used to simulate the data were $s = 3.0$, 5.6, 8.86, and 11.9 for monomer, dimer, tetramer, and hexamer, respectively. The extinction coefficient used in the simulation is 4.2 (mg/ml)$^{-1}$ cm^{-1} at 230 nm and 0.45 (mg/ml)$^{-1}$ cm^{-1} at 280 nm. Time between simulated scans is 4 min.

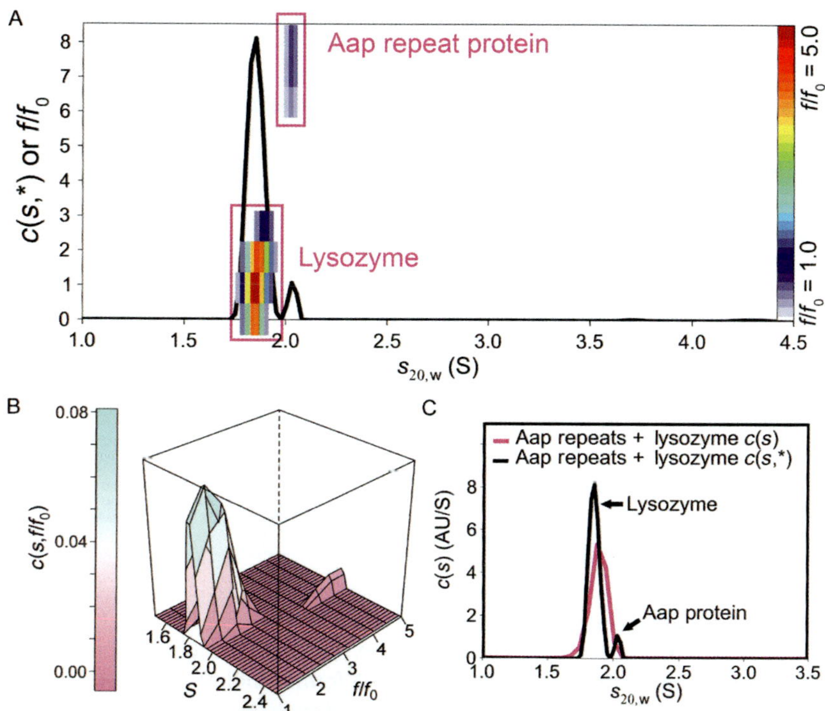

Catherine T. Chaton and Andrew B. Herr, Figure 2 Size-and-shape $c(s, f/f_0)$ analysis of test proteins. (A) The $c(s, f/f_0)$ distribution for the mixture of Aap and lysozyme is able to clearly resolve both species. Lysozyme appears at an $s_{20,w}$ of 1.84 and accounts for 90% of the loading signal. The highly elongated Aap protein sediments with an $s_{20,w}$ of 1.98 and accounts for 9% of the loading signal. (B) 3D surface representation of the $c(s, f/f_0)$ distribution visualized with the lattice package in R. (C) A comparison showing the original $c(s)$ distribution calculated for the mixture of Aap and lysozyme overlaid on the $c(s,*)$ distribution. The presence of the minor species, the elongated Aap protein, is only visible after decoupling the scaling relationship between s and M implicit in the standard $c(s)$ analysis.

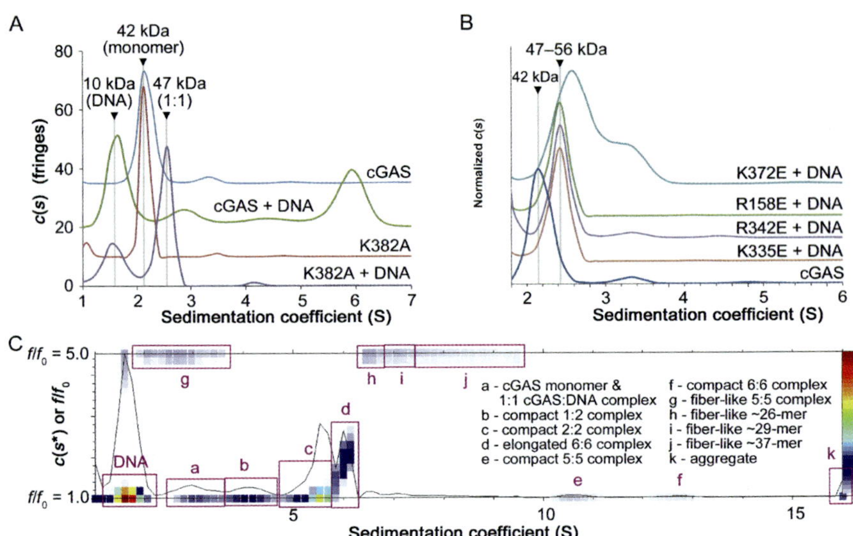

Catherine T. Chaton and Andrew B. Herr, Figure 4 Sedimentation analysis of cGAS interacting with dsDNA. (A) $c(s)$ distributions of WT and K382A mutants of mouse cGAS in the presence and absence of dsDNA. WT cGAS forms multiple oligomeric species in the presence of dsDNA. The K382A mutant, which disrupts H-bonds at the cGAS dimer interface, forms a 1:1 complex with the DNA but was unable to oligomerize. (B) Normalized $c(s)$ distributions of cGAS mutants altered in the DNA-binding sites, including A-site mutants (K372E and R158E) and B-site mutants (R34E and K335E), each sedimented in the presence of dsDNA. (C) Size-and-shape $c(s,f/f_0)$ analysis of WT cGAS: dsDNA demonstrates that multiple oligomeric forms of cGAS occur upon dsDNA binding. Both elongated and compact species are seen at a variety of possible stoichiometries, based on mass estimates from the $c(s,f/f_0)$ distribution. *This figure was reprinted from Li et al. (2013) with permission.*

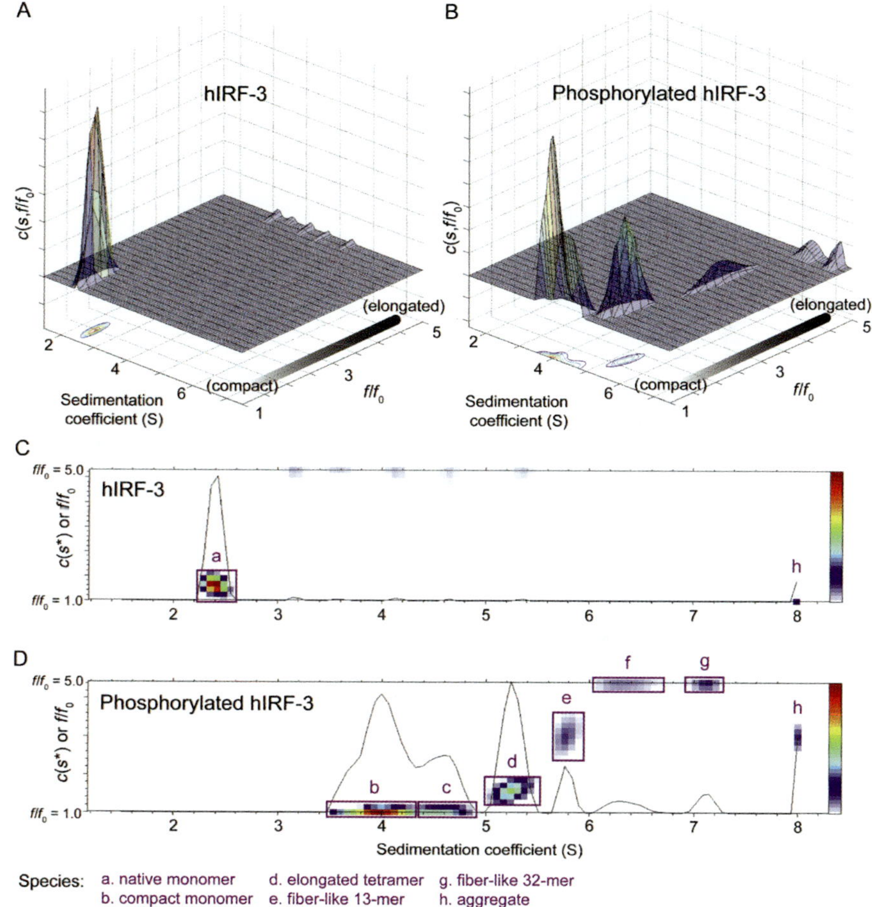

Catherine T. Chaton and Andrew B. Herr, Figure 5 Size-and-shape analysis of hIRF-3 phosphorylated by TBK-1. The $c(s, f/f_0)$ distribution was determined for native IRF-3 (A) and phosphorylated IRF-3 (B). A number of distinct species appear upon phosphorylation, including compact monomer and dimer forms as well as elongated species. Two-dimensional size-and-shape distributions in (C) and (D) show major peaks and the putative oligomeric states based on predicted mass from the $c(s, f/f_0)$ analysis. *This figure was reprinted from Shu et al. (2013) with permission.*

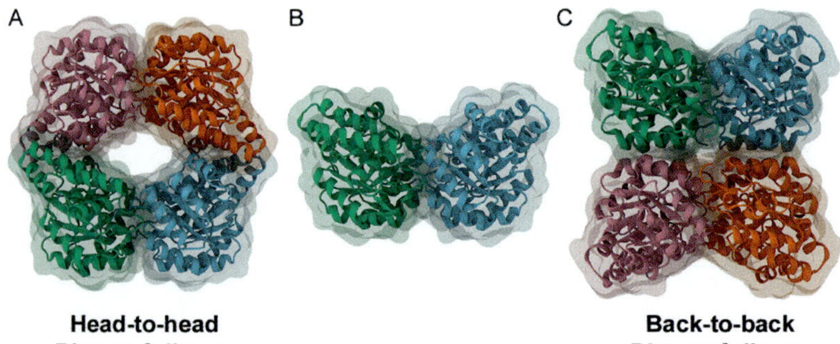

Tatiana P. Soares da Costa *et al.*, Figure 1 Quaternary structure of (A) *Bacillus anthracis* DHDPS (PDB ID: 3HIJ), (B) *S. aureus* DHDPS (PDB ID: 3DAQ), and (C) *Vitis vinifera* DHDPS (PDB ID: 3TUU).

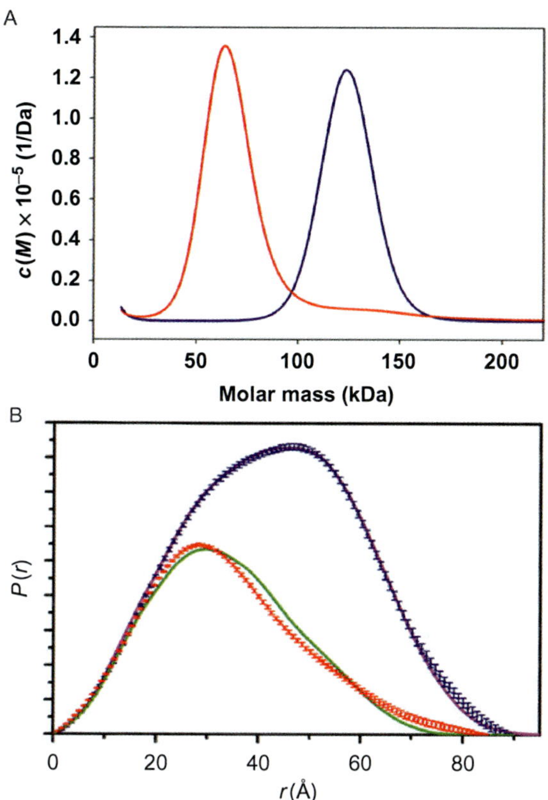

Tatiana P. Soares da Costa *et al.*, Figure 5 Structural characterization of wild-type *E. coli* DHDPS and L197Y. (A) Sedimentation velocity analyses showing the continuous mass [*c*(*M*)] distribution plotted as a function of molar mass (kDa) for wild-type *E. coli* DHDPS (blue line) and L197Y (red line). The L197Y mutant exists as a dimer (expected molecular mass of 62.5 kDa) compared to the tetrameric structure adopted by the wild-type enzyme (expected molar mass of 125 kDa). (B) Overlay of the interatomic distance distribution functions, *P*(*r*) *versus r*, derived from the scattering profiles for tetrameric wild-type *E. coli* DHDPS (blue symbols) and dimeric L197Y (red symbols) using GNOM (Svergun, 1992) alongside the profiles predicted from their corresponding crystal structures (wild-type, purple line; mutant, green line) using CRYSOL (Svergun, Barberato, & Koch, 1995). *Adapted from Griffin et al. (2008).*

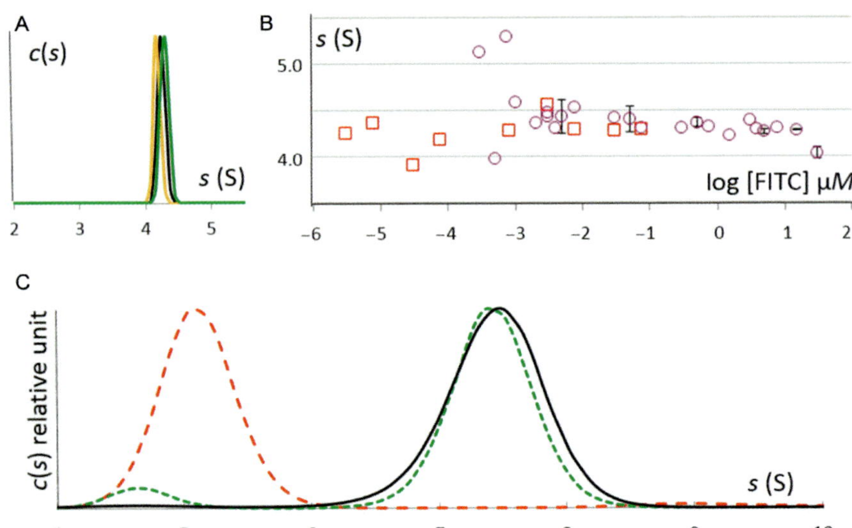

Aline Le Roy *et al.*, Figure 2 Sedimentation of labeled proteins using fluorescence optics. (A) *c*(*s*) superposition, from 280 nm absorbance, interference, and fluorescence data, for FITC-labeled BSA (12 and 14 μ*M*, respectively) in 30 m*M* phosphate pH 7.4, 100 m*M* NaCl. (B) *s* measured for BSA-FITC at different concentrations, with (red square) or without (purple) circle) 0.5 mg mL^{-1} (8 μ*M*) unlabeled BSA. (C) *c*(*s*) superposition for 2.5 μL GFP-pb5 crude extract, diluted at 0.5% for a final concentration of 8 n*M*, without (red dashed line) or with FhuA at a molar ratio FhuA/GFP-pb5 of 1 (green dotted line) and 5 (black, continuous line), in 50 m*M* Tris–HCl pH 8, 150 m*M* NaCl supplemented with 30 m*M* octyl glucoside. The AUC experiments were performed at 20 °C and 42,000 rpm using 12 mm path length cells.

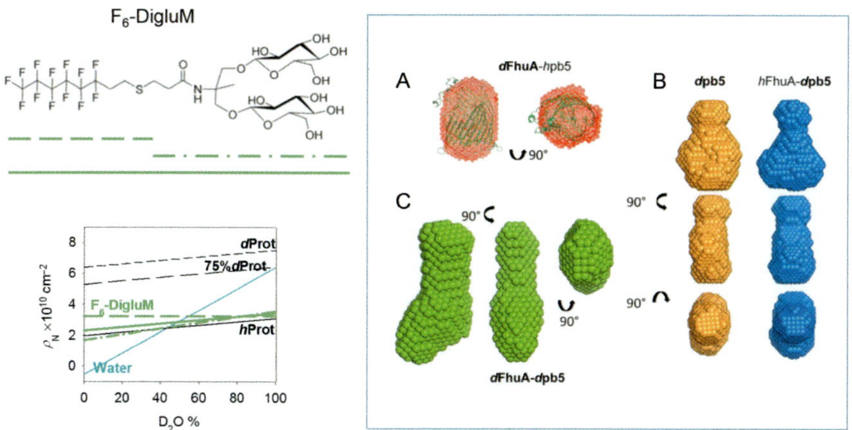

Aline Le Roy *et al.*, Figure 7 Low-resolution structures determined in SANS with the surfactant F_6-DigluM. Left top: chemical structure of F_6-DigluM. Left bottom: neutron scattering densities of water (mixture of H_2O and D_2O) (light blue); of protein (black lines) either hydrogenated (*h*Prot, continuous), 75% deuterated (75% *d*Prot, long dash), or deuterated (*d*Prot, small dash); and of F_6-DigluM (green lines, bold), with tail (green long dash) and head (green dash-dot) contributions, are plotted as a function of D_2O content. This figure was originally published in Breyton, Gabel, et al. (2013), with kind permission of *The European Physical Journal*. Right: *ab initio* envelopes from SANS at 46% D_2O, where F_6-DigluM and *h*proteins are masked, of *d*FhuA-*h*pb5 (red mesh), superimposed with the crystal structure of FhuA (2FCP, green) (A); *d*pb5 (orange) and *h*FhuA-*d*pb5 (blue) (B); and *d*FhuA-*d*pb5 (C). Index *d* and *h* are for hydrogenated and deuterated proteins. This research was originally published in Breyton, Flayhan, et al. (2013). © The American Society for Biochemistry and Molecular Biology.

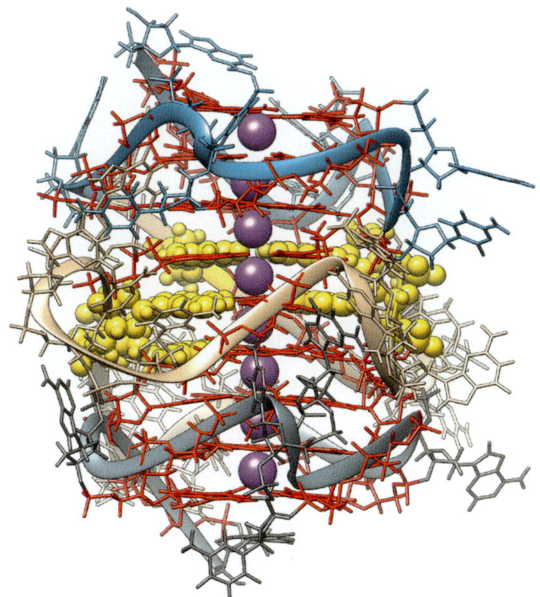

Jonathan B. Chaires *et al.*, Figure 5 Improved hTERT structure. Molecular model of a structure consistent with all biophysical data. Quadruplex guanine bases in red, potassium ions in purple, and guanine mutations in yellow ball and stick representation.

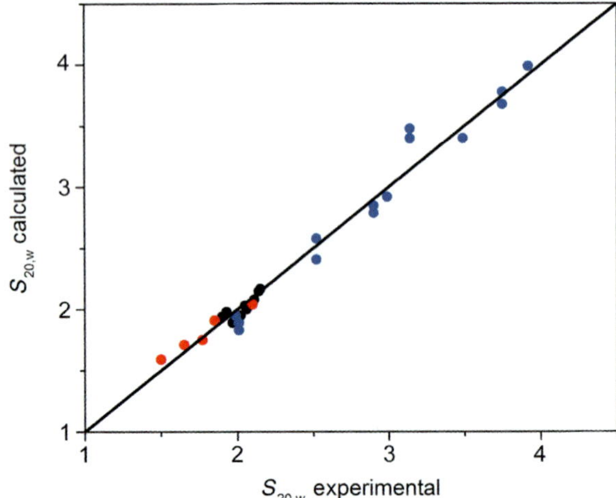

Jonathan B. Chaires *et al.*, Figure 8 Correlation of predicted and measured sedimentation coefficients. The points in red are for short DNA duplexes taken from Fernandes et al. (2002). The black points are for known human telomere structures taken from Le et al. (2014). The blue points are for a variety of higher order quadruplex structures determined in our laboratory.

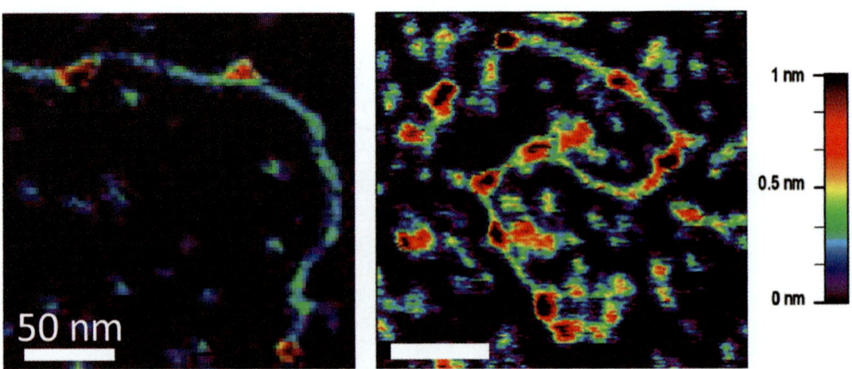

Ingrid Tessmer and Michael G. Fried, Figure 3 Short cooperative clusters formed by AGT on the 1000-bp DNA fragment. With increasing protein concentration (left 6 μ*M*, right 12 μ*M*), the frequency of DNA-bound clusters increases but their lengths appear to remain unchanged. Relative deflection of the AFM tip provides AFM topographical heights and is encoded in color with dark purple representing the smallest values and red-to-black the greatest, as shown in the color bar on the right. Scale bars in the images indicate 50 nm.

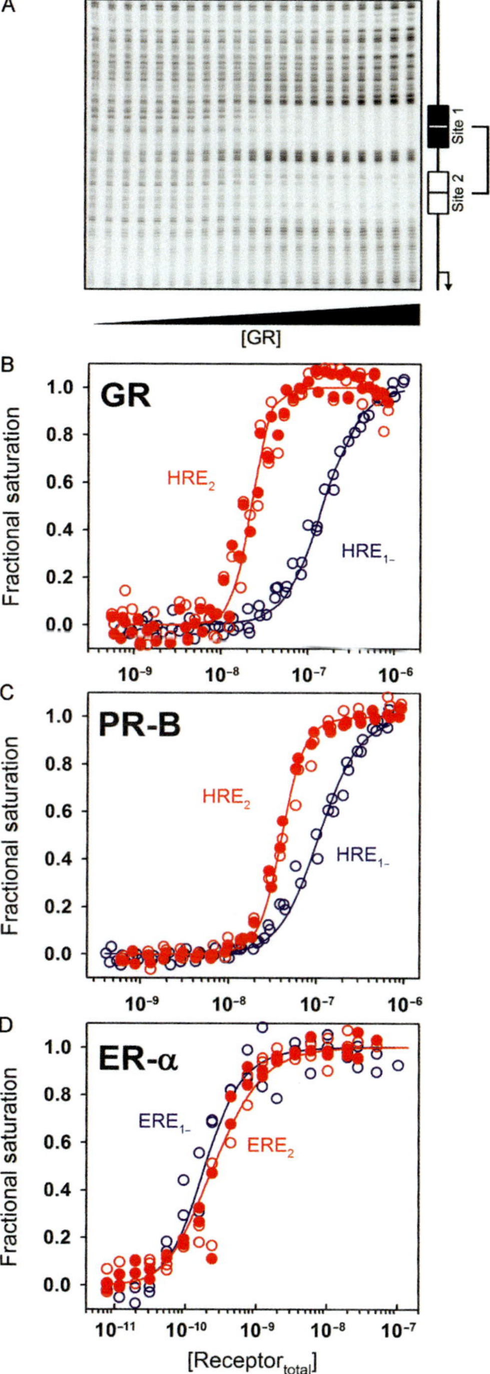

A
Site 1
Site 2
35 bp
[GR]
B
GR
Fractional saturation
HRE2
HRE1-
C
PR-B
HRE2
HRE1-
D
ER-α
ERE1-
ERE2
[Receptor total]

David L. Bain *et al.*, Figure 9 Quantitative DNase footprint titrations of GR, PR-B, and ER-α at the 35 bp center-to-center HRE_2 promoter. (A) Representative autoradiogram of GR binding at the 35 bp center-to-center HRE_2 promoter. GR concentration increases from left to right as depicted by the solid triangle. Positions of site 1 (filled rectangle) and site 2 (open rectangle) are depicted on the right. (B) Individual site-binding isotherms constructed from analysis of GR footprint titrations at the HRE_{1-} and HRE_2 promoters. Filled and open circles represent binding to sites 1 and 2 of the HRE_2 promoter, respectively. Open squares represent binding to site 2 of the HRE_{1-} promoter. Continuous lines represent best global fit to all isotherms using Eqs. (3) and (4). Fit lines for both HRE_2 sites overlay because the sequences of sites 1 and 2 are identical. (C) Individual site-binding isotherms constructed from analysis of PR-B footprint titrations at the HRE_{1-} and HRE_2 promoters under the same buffer conditions as GR. Symbols corresponding to binding site and promoter are identical to those described above. Data were fit to Eqs. (1) and (2). (D) Analysis of ER-α binding under identical conditions, except that the two response elements now correspond to AGATCAcagTGACCT rather than TGTACAggaTGTTCT. Data were fit using Eqs. (1) and (2). Symbols are the same as above.

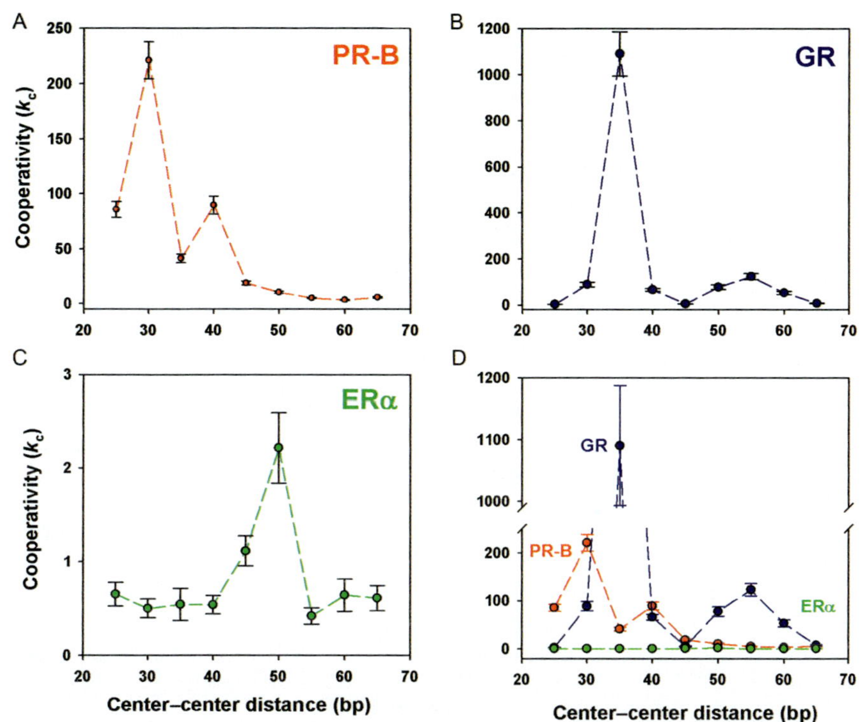

David L. Bain *et al.*, Figure 10 Intersite cooperativity as a function of promoter architecture for GR, PR-B, and ER-α. Resolved intersite cooperativity (k_c) as a function of center–center response element distance for (A) PR-B, (B) GR, and (C) ER-α. (D) Overlay plot of PR-B, GR, and ER-α. Error bars represent one standard deviation as determined by the program Scientist (Micromath).

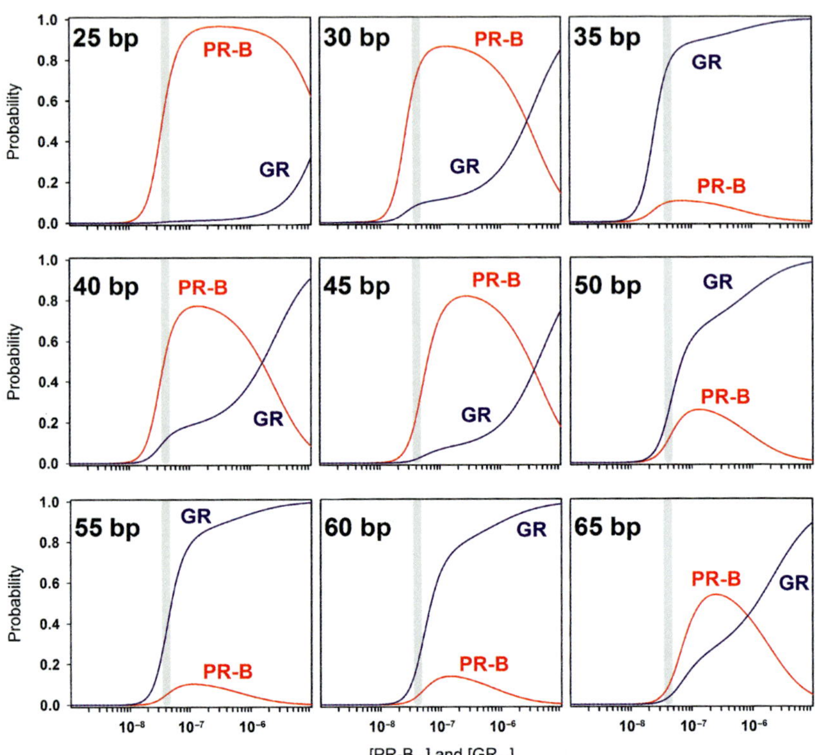

David L. Bain *et al.*, Figure 11 Simulated probabilities of GR and PR-B occupancy at the nine promoters under competitive binding conditions. Plots represent probability of the fully ligated state for GR (continuous line) and PR-B (broken line) assembly in the presence of equimolar concentrations of both receptors ranging from 1 n*M* to 10 μ*M*. Simulations were carried out with the energetics listed in Connaghan et al. (2014) and assuming no heterodimerization or heterocooperativity. Gray bar represents estimate of intracellular receptor concentration (40 n*M*) (Theofan & Notides, 1984).

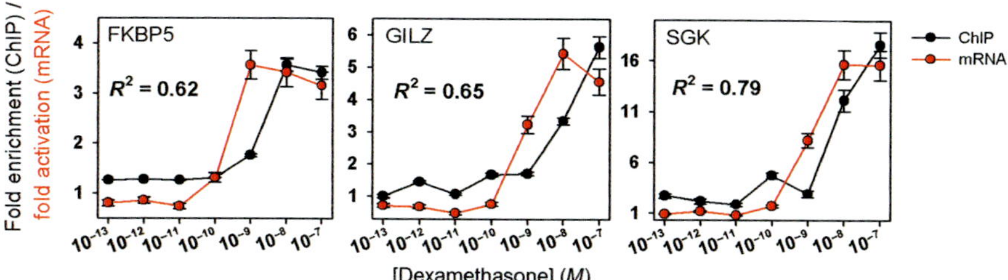

David L. Bain *et al.*, Figure 13 *In vivo* analyses of receptor–DNA occupancy and mRNA production at three endogenous promoters. Promoter occupancy and corresponding mRNA levels as a function of ligand concentration (dexamethasone) at three endogenous promoters (FKBP5, GILZ, and SGK). Promoter occupancy (filled circles) was determined by chromatin immunoprecipitation (ChIP). mRNA production (filled inverted triangles) was determined by quantitative reverse transcription PCR (RT-PCR). Error bars represent the standard error of the mean of three independent experiments. The coefficient of determination (R^2) represents the correlation between promoter occupancy (ChIP) and mRNA production (RT-PCR).

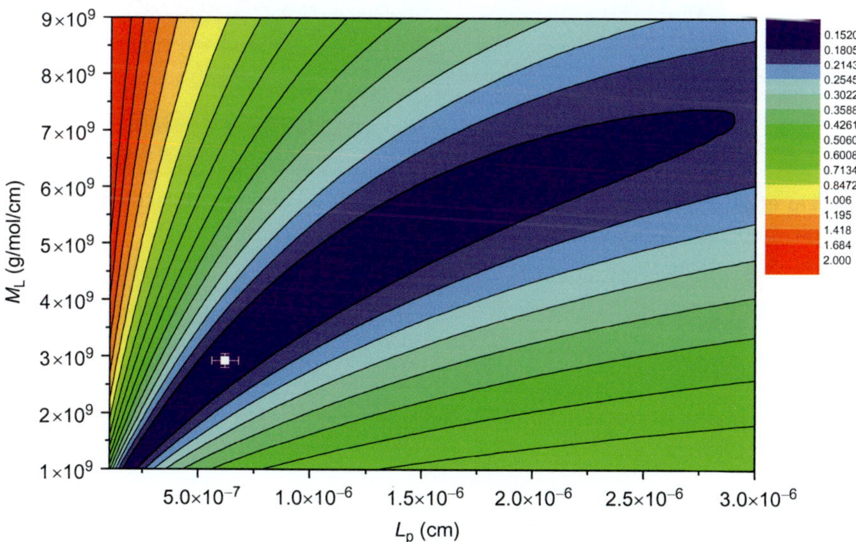

Stephen E. Harding *et al.*, Figure 7 Global or "HYDFIT" estimation of the persistence length L_p versus mass per unit length M_L for a capsular *Streptococcus pneumoniae* polysaccharide SP(4). The white cross represents the minimum of a target function based on fitting the Yamakawa–Fujii and Bushin–Bohdanecky equations. The plot yields $L_p \sim 6.2$ (nm) and $M_L \sim 2.92 \times 10^9$ (g/mol/cm) at the minimum target (error) function value of 0.15. *Reprinted from Harding et al. (2012), with permission from Elsevier.*

A

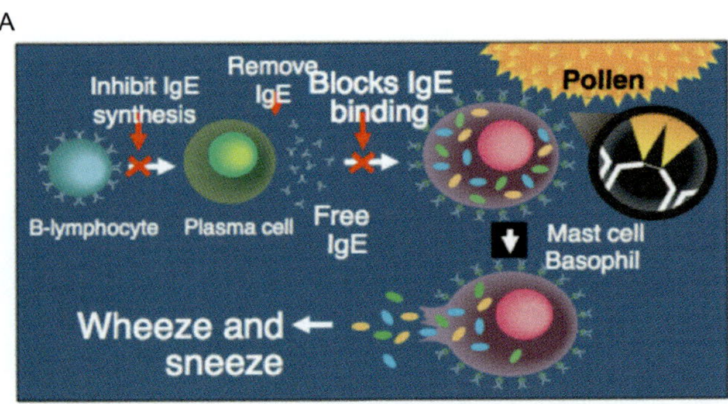

Jun Liu *et al.*, Figure 5 (A) Mechanism of action for an anti-IgE molecule to treat IgE allergic-mediated asthma.

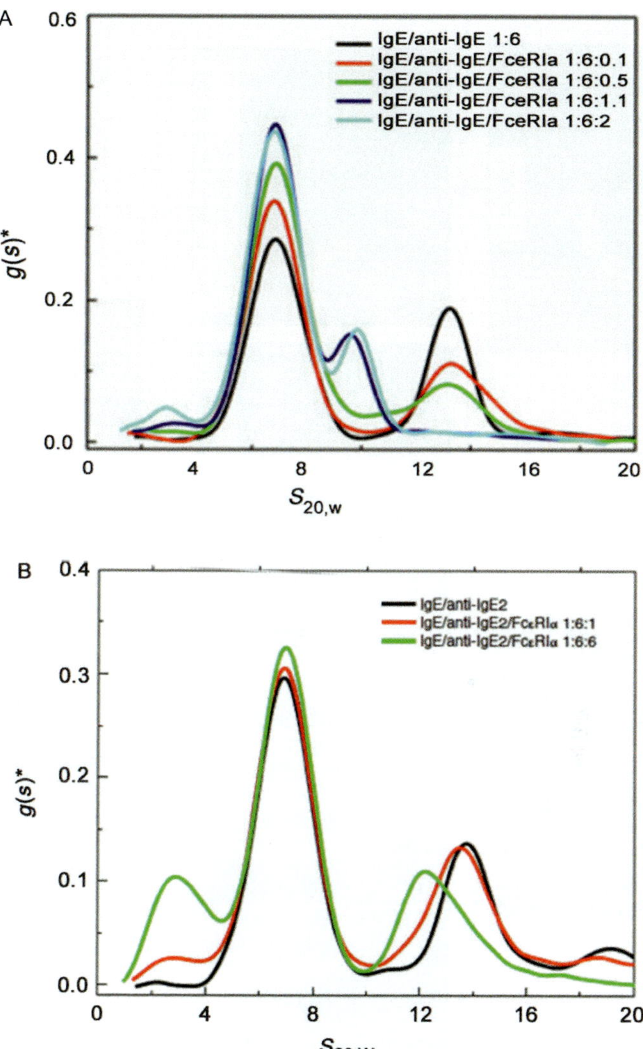

Jun Liu *et al.*, Figure 10 Differential sedimentation coefficient distribution of anti-IgE, FcεRIα, IgE:anti-IgE, and IgE:FcεRIα complexes (A) and anti-IgE2, FcεRIα, IgE:anti-IgE2, and IgE:FcεRIα complexes (B). Assessment of competition of binding of soluble high-affinity receptor, FcεRIα, with either preformed IgE:anti-IgE or IgE:anti-IgE2 complexes using AUC sedimentation velocity (Yadav et al., 2013).

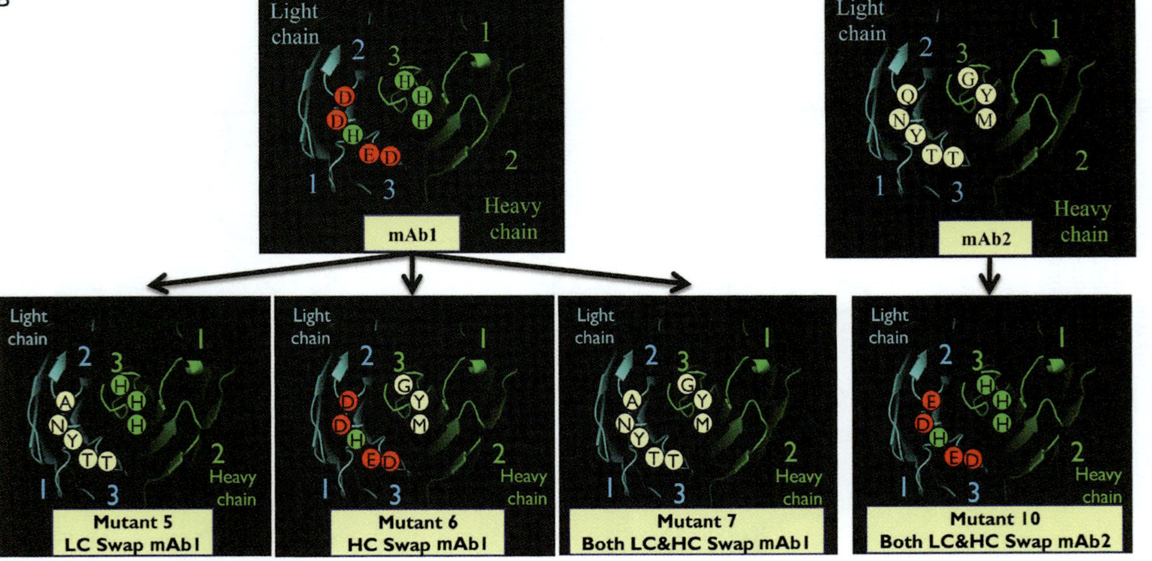

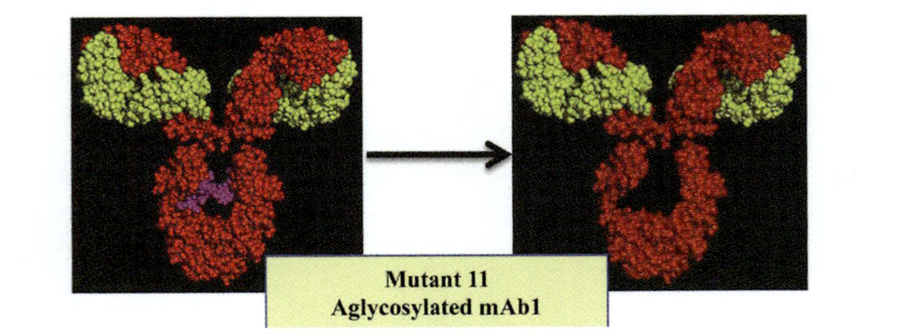

Jun Liu *et al.*, Figure 13 (B) schematics showing amino acid substitutions. All mabs and mutants are at pH 6 and 15 m*M* ionic strength. Viscosity determined using a cone-plate rheometer at a shear rate of 1000/s at 25 °C. *Adapted from Yadav et al. (2011).*

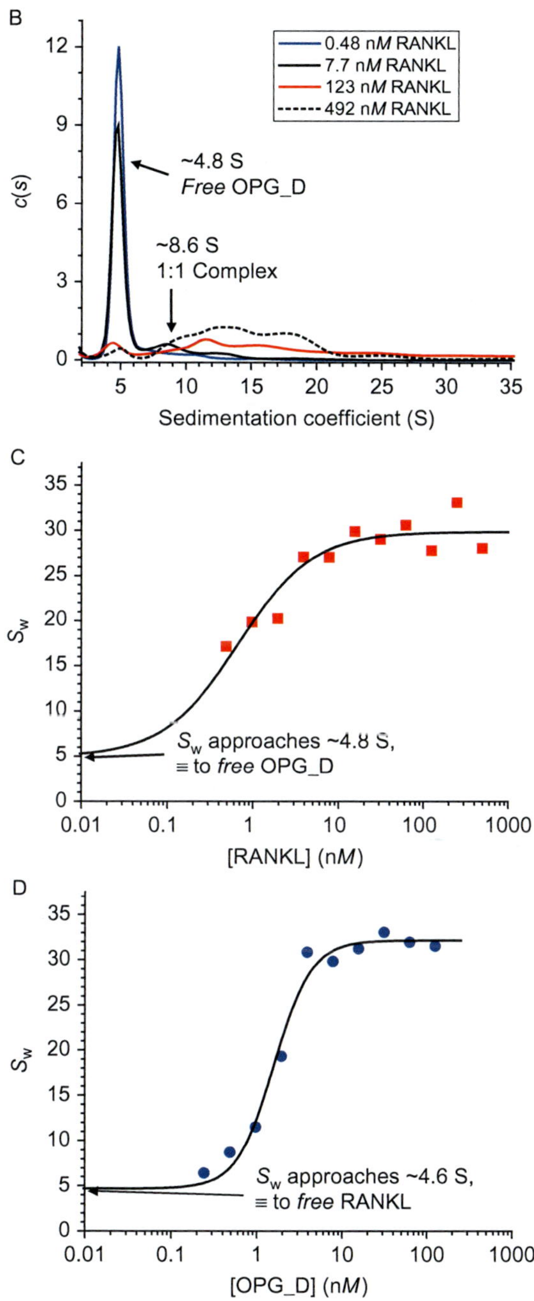

John J. Hill and Thomas M. Laue, Figure 2 (A) A 3D graph showing the results from the *c*(*s*) analysis of 11 different concentrations of RANKL, 2× serial diluted beginning at 492 n*M*, against a constant 233 p*M* Alexa-488 labeled OPG_D, and the zero RANKL control plotted as the "0.01 n*M* RANKL" value of that log space axis. (B) A 2D graph for a subset of the spectra in (A). (C) and (D) provide the S_w binding profiles for the assembly of Alexa-488 labeled OPG_D as a function of [RANKL], and the assembly of Alexa-488 labeled RANKL as a function of [OPG_D], respectively. The trend lines are not fits of the data, but rather are meant to guide the eye through the data profiles.

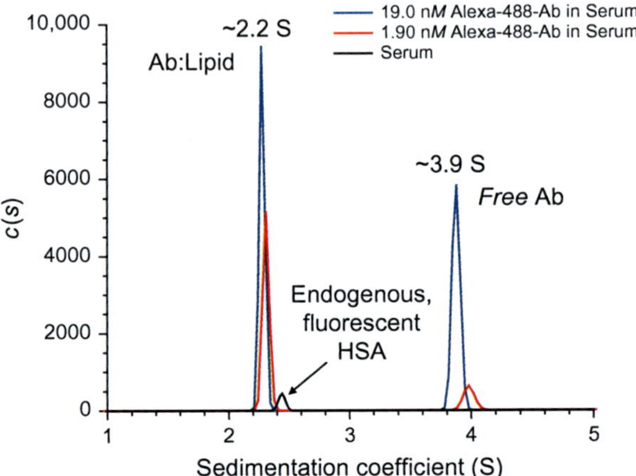

John J. Hill and Thomas M. Laue, Figure 6 *c*(*s*) analyses of 1.9 and 19.0 n*M* aliquots of Alexa-488 labeled antibody in human serum.

CPI Antony Rowe
Eastbourne, UK
September 27, 2015